Digital Communications

Introducing the fundamentals of digital communication with a robust bottom-up approach, this textbook is designed to equip senior undergraduate and graduate students in communications engineering with the core skills they need to assess, compare, and design state-of-the-art digital communication systems.

Key Features
- Provides a fast, concise grounding in key algorithms, concepts, mathematical principles, and mathematical tools for understanding state-of-the-art digital communications.
- Emphasizes readability and accessibility, to quickly familiarize students with the key topics of digital communications, and includes all relevant derivations.
- Presents over 70 carefully designed multi-part end-of-chapter problems with over 360 individual questions, to gauge student understanding and translate knowledge into real-world problem solving.
- Accompanied online by interactive visualizations of signals, downloadable MATLAB® code, and solutions for instructors.

Robert F. H. Fischer is a Full Professor and Director of the Institute of Communications Engineering at Universität Ulm. He has over 25 years of experience in teaching digital communications. He is a Senior Member of the IEEE.

Johannes B. Huber is a Full Professor at the Friedrich-Alexander-Universität Erlangen-Nürnberg, where he has more than 30 years of teaching experience. Additionally, he is regularly invited to the universities of Klagenfurt and Linz as a guest lecturer. He is a Fellow of the IEEE, an ITG Fellow, an Ordinary Member of the Bavarian Academy for Sciences and Humanities, and a Corresponding Fellow of the Royal Society of Edinburgh.

"Delve into the intricacies of digital communications with this comprehensive guide which caters to all levels. Offering a solid foundation, it meticulously covers essential aspects, ensuring readers have access to all the necessary background information. With a plethora of exercises and illustrative examples, it stands as an exceptional resource on digital communications."

Alexander Lampe, Mittweida University of Applied Sciences

"Accurate, concise, and very readable, this is a real gem for students and instructors alike. It focuses on the mathematical tools and algorithms for basic as well as advanced digital transmission schemes. The authors prove their claims in detail, and each chapter provides very helpful examples and problems. Filling a significant gap in the literature, it is highly recommended to everyone interested in really understanding digital communications."

Norbert Goertz, Vienna University of Technology

"Written with a wealth of experience in educating generations of students in electrical engineering, this is an ideal companion for undergraduates and graduates studying digital communications. This textbook gives an accessible and in-depth introduction to basic principles."

Gerhard Kramer, Technical University of Munich

Digital Communications
A Foundational Approach

Robert F. H. Fischer
Universität Ulm, Germany

Johannes B. Huber
Friedrich-Alexander-Universität Erlangen-Nürnberg, Germany

Shaftesbury Road, Cambridge CB2 8EA, United Kingdom

One Liberty Plaza, 20th Floor, New York, NY 10006, USA

477 Williamstown Road, Port Melbourne, VIC 3207, Australia

314–321, 3rd Floor, Plot 3, Splendor Forum, Jasola District Centre,
New Delhi – 110025, India

103 Penang Road, #05–06/07, Visioncrest Commercial, Singapore 238467

Cambridge University Press is part of Cambridge University Press & Assessment, a department of the University of Cambridge.

We share the University's mission to contribute to society through the pursuit of education, learning and research at the highest international levels of excellence.

www.cambridge.org
Information on this title: www.cambridge.org/highereducation/isbn/9781009429665

DOI: 10.1017/9781009429658

© Robert F. H. Fischer and Johannes B. Huber 2025

This publication is in copyright. Subject to statutory exception and to the provisions of relevant collective licensing agreements, no reproduction of any part may take place without the written permission of Cambridge University Press & Assessment.

When citing this work, please include a reference to the DOI 10.1017/9781009429658

First published 2025

A catalogue record for this publication is available from the British Library

A Cataloging-in-Publication data record for this book is available from the Library of Congress

ISBN 978-1-009-42966-5 Hardback

Additional resources for this publication at www.cambridge.org/fischer-huber

Cambridge University Press & Assessment has no responsibility for the persistence or accuracy of URLs for external or third-party internet websites referred to in this publication and does not guarantee that any content on such websites is, or will remain, accurate or appropriate.

Contents

Preface			*page* xiii
Acknowledgments			xix
1	**Introduction**		1
	1.1	Digital Communications	4
		1.1.1 Why "Digital": From Analog to Digital	4
		1.1.2 Why "Digital": Advantages of Digital Communications	5
	1.2	Examples of Digital Transmission Systems	7
		1.2.1 Digital Baseband Transmission over Metallic Wires	7
		1.2.2 Optical Transmission Using Light Waves	7
		1.2.3 Digital Carrier-Modulated (Wireless) Transmission	8
		1.2.4 Digital Information Storage	8
		1.2.5 Driving Forces in the Development	8
	1.3	Basics of Digital Communication Schemes	9
		1.3.1 Data Source	9
		1.3.2 Transmitter	10
		1.3.3 Channel	11
		1.3.4 Receiver	15
	1.4	Summary of Basic Definitions	16
		1.4.1 Parameters of a Digital Communication Scheme	18
		1.4.2 Optimality Criteria in Digital Communications	18
	1.5	Trade-Off between Power and Bandwidth Efficiency	19
	1.6	Problems	22
2	**Equivalent Complex Baseband Signals**		26
	2.1	Equivalent Complex Baseband Signals	27
		2.1.1 Basic Definitions	27
		2.1.2 ECB Transformation	30
		2.1.3 The Quadrature Components of ECB Signals	34
		2.1.4 Up/Downconversion with Frequency and Phase Errors	39
	2.2	LTI Systems and ECB Signals	40
		2.2.1 Equivalent Transfer Function	40
		2.2.2 ECB Systems and Quadrature Components	42
	2.3	Stochastic Processes in the Equivalent Complex Baseband	44
		2.3.1 ECB Model of White Noise	47
		2.3.2 Filtering of White Noise in the ECB Domain	48
	2.4	Problems	49

3 Digital Pulse-Amplitude Modulation — 54
- 3.1 Basic Principles — 54
  - 3.1.1 Modulation — 55
  - 3.1.2 Baseband and Carrier-Modulated Transmission — 56
- 3.2 Average Power Spectral Density of PAM Signals — 58
  - 3.2.1 Average Autocorrelation Function of $s(t)$ — 58
  - 3.2.2 Average Power Spectral Density of $s(t)$ — 60
  - 3.2.3 Uncoded Transmission — 61
- 3.3 Choice of the Signal Constellation — 63
  - 3.3.1 Amplitude-Shift Keying (ASK) — 64
  - 3.3.2 Phase-Shift Keying (PSK) — 64
  - 3.3.3 Digital Quadrature-Amplitude Modulation (QAM) — 65
- 3.4 Receiver Structure and Choice of the Basic Pulse Shape — 66
  - 3.4.1 Optimum Receive Filter for PAM — 66
  - 3.4.2 Intersymbol Interference-Free Transmission – the Nyquist Criterion — 73
  - 3.4.3 Eye Pattern — 82
  - 3.4.4 Crest Factor of PAM Transmit Signals — 84
  - 3.4.5 Remarks on the Roll-Off Factor and the Nyquist Bandwidth — 87
- 3.5 Detection of the PAM Signal Points — 88
  - 3.5.1 Maximum-a-Posteriori (MAP) Strategy — 88
  - 3.5.2 Maximum-Likelihood (ML) Strategy — 89
- 3.6 Bit and Symbol Error Ratio — 92
  - 3.6.1 Symbol Error Ratio of ASK — 94
  - 3.6.2 Symbol Error Ratio of General Constellations — 97
  - 3.6.3 Approximation of the Bit Error Ratio — 99
  - 3.6.4 Normalized Minimum Squared Euclidean Distance of PAM Constellations — 100
  - 3.6.5 Bit Error Ratio of Selected PAM Schemes — 101
- 3.7 Assessment of PAM Schemes in the Power/Rate and Power/Bandwidth Plane — 102
- 3.8 Problems — 106

4 Variants of PAM Transmission Schemes — 112
- 4.1 Schemes with Reduced Peak-to-Average Power Ratio — 112
  - 4.1.1 Offset QAM — 113
  - 4.1.2 Minimum-Shift Keying (MSK) — 119
  - 4.1.3 Gaussian Minimum-Shift Keying (GMSK) — 125
- 4.2 Schemes without Explicit I/Q Demodulation — 133
  - 4.2.1 "Carrierless" Amplitude and Phase Modulation — 133
  - 4.2.2 Coherent Reception by Energy Detection — 141
- 4.3 Problems — 149

5 Non-coherent Demodulation — 154
- 5.1 Phase and Frequency Errors in PAM Transmission — 154
  - 5.1.1 PAM Scheme with Phase Error — 155
  - 5.1.2 Phase Ambiguities — 156
- 5.2 Differential Phase-Shift Keying (DPSK) — 157
  - 5.2.1 Differential M-ary PSK — 157
  - 5.2.2 Differential Coherent Demodulation (DCPSK) — 158
  - 5.2.3 Differential Demodulation (DPSK) — 160
- 5.3 Non-coherent PAM Schemes — 169
  - 5.3.1 Non-coherent Demodulation of Unipolar ASK — 169
  - 5.3.2 Comparison with Coherent Demodulation — 173
- 5.4 Problems — 176

6 Signal Space Representation — 180
- 6.1 Coding and Modulation — 180
  - 6.1.1 Modulation — 181
  - 6.1.2 Average Power Spectral Density — 185
  - 6.1.3 Channel Coding — 188
  - 6.1.4 Equivalent Coding and Modulation — 191
- 6.2 Signal Representation via Orthonormal Basis Functions — 195
  - 6.2.1 Definition of a Signal Space — 195
  - 6.2.2 Energy and Signal Space — 202
  - 6.2.3 Remark on Normalization — 203
- 6.3 Gram–Schmidt Procedure — 204
- 6.4 Representation of the Receive Signal in Signal Space — 210
  - 6.4.1 Projections and Receiver Structure — 211
  - 6.4.2 End-to-End Model — 212
  - 6.4.3 Signal Space Representation with Reduced Dimensionality — 214
- 6.5 Maximum-Likelihood Detection and Decoding — 215
  - 6.5.1 Vector Demodulator — 215
  - 6.5.2 Correlation Receiver — 219
- 6.6 Combining Coding and Modulation — 223
  - 6.6.1 Coding for Error Correction – Power Efficiency — 223
  - 6.6.2 Coding for Bandwidth Efficiency — 225
  - 6.6.3 Line Coding — 228
- 6.7 Non-coherent Demodulation — 231
  - 6.7.1 Optimum Non-coherent Detection — 232
  - 6.7.2 Multiple-Symbol Differential Detection of DPSK — 237
- 6.8 Problems — 242

7 Digital Frequency and Phase Modulation — 247
- 7.1 General Concept of Frequency-Shift Keying — 247
  - 7.1.1 Signal Elements — 248
  - 7.1.2 Average Power Spectral Density of FSK — 252

	7.2	Binary Frequency-Shift Keying	253
		7.2.1 Discontinuous Binary FSK	253
		7.2.2 Binary Continuous-Phase FSK	259
		7.2.3 Assessment of Binary FSK in the Power/Bandwidth Plane	264
	7.3	M-ary Frequency-Shift Keying (M-ary FSK)	268
	7.4	Introduction to Continuous-Phase Modulation (CPM)	270
		7.4.1 CPM Signal and Its Interpretation	270
		7.4.2 Coherent Reception and Normalized Minimum Squared Euclidean Distance	278
		7.4.3 Average Power Spectral Densities of CPM	279
		7.4.4 Power/Bandwidth Plane for CPM	280
		7.4.5 Complexity Reduction of Coherent Receiver for CPM	281
	7.5	Problems	282
8	**Equalization of Dispersive Channels**		**287**
	8.1	Situation	287
		8.1.1 End-to-End Pulse	288
		8.1.2 Spectral Signal-to-Noise Ratio	290
		8.1.3 Receive Filter	291
		8.1.4 Discrete-Time End-to-End Model	291
	8.2	Optimum Receiver – Maximum-Likelihood Sequence Detection I	292
		8.2.1 Maximum-Likelihood Sequence Detection	292
		8.2.2 Power Efficiency of MLSD	297
	8.3	Linear and Nonlinear Equalization	301
		8.3.1 Matched-Filter Bound	303
		8.3.2 Discrete-Time Part of the Receive Filter	306
		8.3.3 Linear Equalization	306
		8.3.4 Whitened-Matched Filter and Decision-Feedback Equalization	310
		8.3.5 Noise Prediction	317
	8.4	Maximum-Likelihood Sequence Detection II	319
		8.4.1 End-to-End Model	320
		8.4.2 Maximum–Likelihood Approach	321
		8.4.3 Reduced-State Sequence Detection	323
	8.5	Transmitter-Side Channel Pre-equalization	324
		8.5.1 Transmitter-Side Preprocessing	324
		8.5.2 Linear Pre-equalization	326
		8.5.3 Tomlinson–Harashima Precoding (THP)	328
		8.5.4 Combined Transmitter- and Receiver-Side Linear Equalization	331
	8.6	Problems	338
9	**Orthogonal Frequency-Division Multiplexing**		**344**
	9.1	Basics of Multi-Carrier Transmission	344

	9.1.1	Single-Carrier Transmission	345
	9.1.2	Multi-Carrier Transmission – Frequency Multiplexing	345
	9.1.3	Denominations and Applications	345
9.2	Basic Principles of OFDM	347
	9.2.1	OFDM Transmitter	347
	9.2.2	Block Transmission	349
	9.2.3	Cyclic Repetition	351
	9.2.4	End-to-End Model of OFDM	357
	9.2.5	Discussion	358
	9.2.6	Average Power Spectral Density of OFDM Signals	359
	9.2.7	Signal-to-Noise Ratio	362
9.3	OFDM with Channel Knowledge at the Transmitter	363
	9.3.1	Situation	363
	9.3.2	Optimum Loading	364
	9.3.3	Signal-to-Noise Ratio	370
9.4	Comparison of Single- and Multi-Carrier Modulation	373
	9.4.1	No Channel State Information at the Transmitter	373
	9.4.2	Channel State Information at the Transmitter	374
	9.4.3	Summary	376
9.5	Peak-Power Reduction	377
	9.5.1	Gaussian Signal Model and Principal Coding Approach	378
	9.5.2	Peak-to-Average Power Reduction Schemes	381
9.6	Problems	383

Appendix A: Signals, Systems, and Processes — 386

A.1	Recapitulation of Signals and Systems		386
	A.1.1	Complex Numbers	386
	A.1.2	Continuous-Time Signals and Systems	387
	A.1.3	Discrete-Time Signals and Systems	391
A.2	Some Basic Correspondences and Properties		392
A.3	Causality, Minimum-Phase Systems, and Stability		393
	A.3.1	Causal Signals	393
	A.3.2	Minimum-Phase Systems	395
	A.3.3	Stable Systems	406
	A.3.4	Application of the Time–Frequency Duality	406
A.4	Stochastic Processes		406
	A.4.1	Parameters of Stochastic Processes	407
	A.4.2	Wide-Sense Stationarity	410
	A.4.3	Gaussian Processes	411
	A.4.4	LTI Systems and Stochastic Processes	413
	A.4.5	Sampling and Interpolation of Stochastic Processes	413
	A.4.6	Cyclo-stationary Stochastic Processes	414
A.5	Prediction		415

Appendix B: Short Introduction to Channel Coding 419
- B.1 Binary Block Codes 419
  - B.1.1 Basic Definitions 419
  - B.1.2 Linear Binary Block Codes 422
  - B.1.3 Hard-Decision Decoding of Linear Binary Block Codes 426
  - B.1.4 Soft-Decision Decoding of Simple Linear Binary Block Codes 427
  - B.1.5 Tanner Graph and Basics of Iterative Decoding 431
  - B.1.6 Log-Likelihood Ratio and Practical Considerations 434
  - B.1.7 Remarks on LDPC Codes and Belief Propagation 437
  - B.1.8 Polar Codes 437
- B.2 Trellis Codes 444
  - B.2.1 Basic Definitions 444
  - B.2.2 Convolutional Codes 446
  - B.2.3 The Viterbi Algorithm for Trellis Decoding 449
- B.3 General Version of Mapping 452

Appendix C: Supplements to OFDM 454
- C.1 Preliminaries 454
- C.2 Discrete-Time OFDM 455
- C.3 Continuous-Time Variant 457
- C.4 Comparison and Discussion 460
- C.5 Derivations 461
  - C.5.1 Continuous Time 461
  - C.5.2 Discrete Time 462
- C.6 Noise Variance in OFDM 463
- C.7 Remarks on Unique-Word OFDM 465

Appendix D: Time-Invariant and Time-Varying Channel Models 468
- D.1 Linear Time-Invariant Channel Models 468
- D.2 Models for Cable Transmission 469
  - D.2.1 Coaxial Cables 469
  - D.2.2 Symmetric Wire Pairs 470
- D.3 Radio Transmission 473
- D.4 Linear Time-Varying Systems 475
  - D.4.1 Time-Domain Characterization of LTV Systems 475
  - D.4.2 Frequency-Domain Characterization of LTV Systems 478
- D.5 Stochastic, Time-Varying Channel Models 480
  - D.5.1 Flat-Fading Channel 480
  - D.5.2 Frequency-Selective Channel 483

Appendix E: Implementation of PAM Transmission in Software 487
- E.1 Preliminaries and Simulation Environment 487
  - E.1.1 Transmission System 487

	E.1.2	MATLAB® Simulation Environment	487
E.2	PAM Transmitter		489
	E.2.1	Mapping	489
	E.2.2	Pulse Shaping	491
	E.2.3	Modulation to Radio Frequency	492
E.3	Channel		493
E.4	Receiver		495
	E.4.1	Demodulation to Baseband	495
	E.4.2	Matched Filter and Sampling	496
	E.4.3	Threshold Decision and Demapping	497
	E.4.4	BER Simulations	498
Bibliography			501
Index			512

Preface

This textbook presents a condensed introduction to digital communications, applicable to a first-semester Master's course in electrical or communication engineering. The material covers fundamental principles and addresses important topics that are built on them. However, since it is an introduction, some readers may find that certain items are not handled in sufficient detail. These readers are referred to the numerous books on the subject of digital communications (cf. the list of further reading at the end of the book). There are no basics of information theory included as these are assumed to be well known by the reader. The very important topic of channel coding is covered only very superficially and left to the further reading and the rich literature thereon. The main focus of our presentation of digital communications is high understandability combined with easy and fast perception, even for concepts which are rather sophisticated.

Overview

The contents of this textbook are the fundamentals of digital communications. We will cover the following topics, arranged in the listed chapters.

1. Introduction
   First, some introductory material is covered and examples of digital communication applications are given. The used notation is explained.
2. Equivalent Complex Baseband Signals
   When dealing with digital communications, most parts of the processing are done in the baseband domain; we define complex baseband signals, systems, and noise in this domain.
3. Digital Pulse-Amplitude Modulation
   The simplest but also most frequently used types of digital communication schemes belong to the class of digital pulse-amplitude modulation (PAM). A detailed description of the transmitter is given and the related optimum receiver is derived. This chapter is intended as an extension of the knowledge already gained during Bachelor's studies.
4. Variants of PAM Transmission Schemes
   Some important modulation formats used in practice can be seen as a generalization of conventional PAM schemes. We address offset schemes, minimum-shift keying, so-called carrierless amplitude and phase modulation, and I/Q demodulation solely based on energy detection.

5. Non-coherent Demodulation
   In some situations, synchronization of the receiver-side downmixing is not possible. Consequently, we study non-coherent reception and differential schemes.
6. Signal Space Representation
   This chapter provides a more general view of digital communications. It is shown how signals can be represented as points in a signal space, how general (non-PAM) schemes can be designed, and, especially, how the optimum receiver for any kind of digital communication scheme can be derived systematically.
7. Digital Frequency and Phase Modulation
   As an application of the signal space representation, we consider frequency-shift keying modulation formats including continuous-phase modulation, and discuss their difference from PAM schemes.
8. Equalization of Dispersive Channels
   In typical applications, the channel will be dispersive (i.e., produce echoes). Synonymously, the channel transfer function will not be flat. We address the problem of equalization, derive the optimum strategy, and discuss sub-optimum but low-complexity variants.
9. Orthogonal Frequency-Division Multiplexing
   As an alternative to equalization methods for channels with non-flat transfer function, so-called multi-carrier schemes can be used. The most important variant, orthogonal frequency-division multiplexing (OFDM), is derived and explained.

Five appendices offer basics on signal and systems, basics on channel coding, some supplements to OFDM, a brief overview of time-invariant and time-variant channel models, and a MATLAB® project on simulating PAM schemes.

Please note that supplementary material can be found on the companion website (www.cambridge.org/fischer-huber).

Aims and Approach

The aims of this textbook are:

- the assessment, comparison, and design of state-of-the-art digital communication schemes;
- the ability to operate with the required tools, such as equivalent complex baseband signals and systems, stochastic processes, and the representation of signals in a signal space;
- the derivation of the optimum receiver concept for any type of digital transmission scheme;
- the design and evaluation of equalizers for dispersive channels;
- the comparison of single- and multi-carrier transmission methods.

The approach taken to meet these aims is directly based on how state-of-the-art digital communication schemes are designed today. In contrast to analog systems,

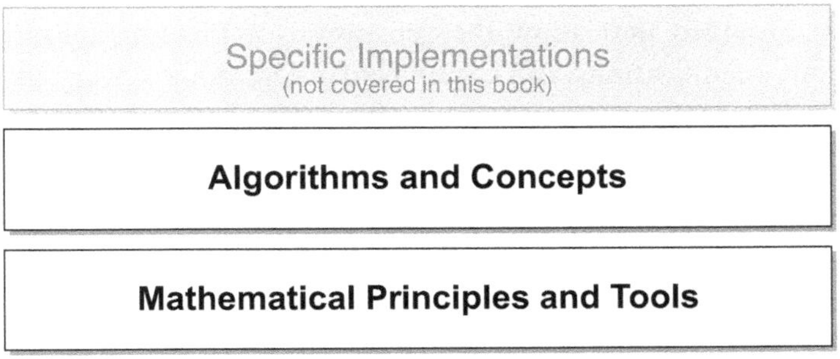

Figure P.1
Hierarchy of abstraction.

> **Specific Implementations**
> (not covered in this book)
>
> **Algorithms and Concepts**
>
> **Mathematical Principles and Tools**

which were purely implemented in hardware, the major parts of digital schemes are typically realized in software (running on digital processors), implementing mathematical algorithms. In summary:

- modern digital communication systems are directly based on specific *mathematical principles and concepts*;
- the components of the schemes are usually given as *algorithms*;
- implementing these algorithms in *software* immediately gives practical *realizations*.

This view of digital communication schemes is further visualized in Figure P.1. A hierarchy of abstraction is shown. The mathematical principles and tools establish the fundamentals. Thereon, the algorithms and concepts (e.g., for a receiver) are built. As we will see, these algorithms and procedures are direct consequences of the aims (performance criteria) we aspire to. The solution of the optimization directly results in guidelines how the processing steps have to be designed. In this textbook, we deal with these two layers of abstraction. The uppermost layer, comprising specific implementations (e.g., on specific hardware platforms such as FPGAs), is not part of the book.

This textbook adopts the following didactic concepts in presenting the material.

- Our pedagogical approach is strictly bottom-up (i.e., first a foundation is laid, then the material presented in earlier chapters is used as a basis for the subsequent exposition). The book is designed to outline the interactions and relationships between different principles and follows a common thread of developing the material out of the task to be solved.
- The contents are presented in text, equations, and a large number of supporting figures, with very high attention to readability and comprehensibility.
- The presented contents are accompanied by carefully designed problems, which provide a deeper understanding of the material as well as of the relation to applications in technical practice. Detailed solutions to these problems will be available for instructors.
- Interactive visualizations of signals that deepen the understanding will be available online at the accompanying website. In addition, MATLAB® scripts will

be provided that allow the generation of these visualizations and, in particular, modifications and experiments. The book covers the whole range from reading explanatory text, solving problems with paper and pencil, working with pre-designed visualizations to using software tools with own programming.

The didactic concept is also reflected in the selection of the content. The book goes well beyond a phenomenological description of the main concept of digital communications. The material is mathematically oriented; all relevant derivations are included and nothing remains without proof. All signals and systems are consistently presented in the equivalent complex baseband, which is neatly introduced at the beginning. To lay the foundation, pulse-amplitude modulation is studied in detail. This establishes the basis for the following generalizations, in particular for the signal space concept. Here, we clearly distinguish between modulation and coding. The separation between both aspects is unambiguously defined and the interaction between these building blocks is studied. Using this approach, the analysis of frequency modulation schemes (with continuous phase) is evident, and approaches for transmission over channels with intersymbol interference can readily be derived.

Prerequisites

In order to meet the above-mentioned aims, this textbook builds upon the student's knowledge from undergraduate introductory courses on signals and systems and an introduction to communications. In particular, we expect the reader to be familiar with

- basic calculus
  In particular, we heavily use complex numbers, need basic calculus (differentiation and integration, series expansion), and a knowledge of linear algebra (vectors and matrices, eigenvalues, bases).
- signals and systems
  We expect a sound knowledge of discrete-time and continuous-time signals, linear time-invariant systems, Fourier and z-transforms, and the sampling theorem with reconstruction.
- fundamentals of random variables and random processes
  In communications, the use of random variables (probability, pdf, cdf, expectation), maximum-likelihood and maximum a-posteriori detection principles, and stochastic processes (stationarity, acf, psd, filtering) is indispensable.
- fundamentals of communications
  We expect the reader to already have knowledge of the basics of analog and digital transmission, PAM (concepts, error ratio calculation), the notion of bandwidth and power efficiency, and some fundamentals of information theory and channel coding.

Please note that we do not cite references in the main text. Books and original papers that we consider as important in the field are listed in the Bibliography at the end of the book.

Notation

We use the notation that is standard in the field of communications. We follow ISO 80000-2:2009 for typesetting mathematical functions (e.g., sin, cos, etc.) and constants (in particular e, the base of the natural logarithm; j, the imaginary unit; d, the differential) in an upright font. Through notation, we clearly distinguish ordinary variables and signals from the concept of random variables and random processes. Moreover, we distinguish real-valued or complex-valued quantities from those drawn from a finite field, in particular the binary field $\mathbb{F}_2$.

In detail, we use the following conventions:

- real-valued or complex-valued variables or signals are typeset in italic font (e.g., R, $s(t)$, etc.)[1];
- vectors are typeset as lowercase bold letters (e.g., $\boldsymbol{v}$) and matrices are typeset as uppercase bold letters (e.g., $\boldsymbol{M}$);
- random variables and random processes are typeset in sans-serif font (e.g., r, x(t), etc.), while realizations of random variables and random processes are simple variables or signals and, thus, typeset in italic font;
- quantities from a finite field are typeset in Fraktur font, scalars in regular font, vectors and matrices as bold lower- and uppercase letters, respectively (e.g., $\mathfrak{c}_i$, $\mathbf{\mathfrak{c}}$, $\mathfrak{G}$, etc.).

In some places, important statements are summarized. In order to draw the reader's attention to these points, they are marked with a big exclamation mark in the left margin, as you can see here.

Hints for Instructors

The material in this textbook is suited for an early Master's course, or even a course towards the end of Bachelor studies. The scope of the book is larger than time permits for a typical one-semester class. The instructor may make a selection. To that end, Figure P.2 shows a dependency diagram of the chapters. The chapters shown above depend on (i.e., build upon) all those shown below them. For example, after going through Chapters 1, 2, and 3, a short course can study Chapters 4 and/or 5, which do not go so deep. A more comprehensive course can (after the first three basic chapters) cover Chapter 6 and a selection of Chapters 7, 8, or 9. The appendices are optional and, in a condensed version, provide some basis of topics relevant for selected chapters.

---

[1] The only exceptions are upper case Greek letters, which have been typeset in roman.

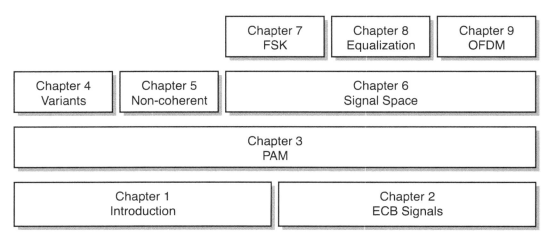

Figure P.2 Dependencies of the chapters.

Acknowledgments

The present book is the outcome of research and teaching over the last three decades. Over this time, a lot of people have influenced us, given comments and hints, and, thus, directly or indirectly had an impact on this book. We thank the many Master's and PhD students who attended the *Digital Communications* course in Erlangen and Ulm for their comments and suggestions for improvement.

A lot of people read drafts of parts of this book and pointed out typos or didactic problems to us. We cannot name them all, but want to acknowledge (in alphabetic order) Günther Haas, Alexander Lampe, Lutz Lampe, Stefan Müller-Weinfurtner, Johannes Pfeiffer, Andreas Schenk, Rebekka Schulz, Christian Siegl, Carmen Sippel, Elena Sterk, Sebastian Stern, Werner Teich, George Yammine and say a special thanks to Clemens Stierstorfer.

The authors would like to particularly mention the stimulating discussions with Gottfried Ungerböck, with whom we were both privileged to share professional time.

Many thanks to our wives Barbara and Monika, who had to give up many hours of companionship in favor of this textbook.

Finally, I, Johannes, devote many thanks to my former diploma, PhD, and habilitation student, and friend, Robert, for encouraging me to join this book project as a coauthor.

We also want to thank the publisher, in particular Elizabeth Horne and Naomi Chopra, for their useful hints and patience during the phase of finishing the manuscript.

1 Introduction

In this book, we deal with particular techniques for communication. What is considered as "communication"? In our context, we use the following definition:

> **Definition 1.1 *Communication***
>
> *Communication* is the activity of conveying information.

Thereby, messages can be communicated across

- *space* (from one location to another, i.e., *tele*communications)
- *time* (from one time instant to another, i.e., storage).

The basic block diagram of unidirectional (digital) communication is given in Figure 1.1, which in communications is considered as "Figure 1."

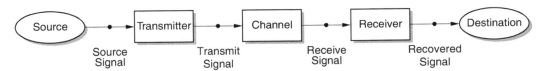

Figure 1.1 Block diagram of (digital) communication.

A *source* wants to convey its *source signal* to a *destination* or *sink*. Since the source and sink are not collocated (in time and space), this transfer has to be made via some transmission medium, abstracted as the *channel*. However, the source signal typically cannot be fed directly to the channel. Hence, a transmitter has to "translate" between the source and the channel. Similarly, the receiver has to generate a *recovered signal* from the obtained *receive signal* that is delivered to the sink.

In this book, all considerations are restricted to a single unidirectional communication link. Bidirectional and multilateral communication systems, as well as communication networks, are beyond the scope of this introductory text.

Nevertheless, such further issues are fully based on the methods for unidirectional communication derived and presented here.

Channel

The channel is a very import part of the communication system. What is a "channel"? J. L. Kelly gave a catchy definition:

> **Definition 1.2** *Channel*
>
> The *channel* is the part of the communication system that one is either unwilling or unable to change.

Typically, the channel acts destructively, in the sense that it

- attenuates the signal
  Only a small fraction of the transmitted energy is captured at the receiver.
- distorts (linearly or nonlinearly) the signal
  The transmitted waveform is distorted, for example, in case the signal is not only received directly but also via echos (linear distortions), nonlinear components will distort (e.g., clip) the signal too (nonlinear distortions).
- superimposes noise or interference
  The main source of noise is the receiver frontend (amplifiers); interference may be caused by other systems operating at the same time in the same frequency band.

Transmitter/Receiver

As already mentioned, in almost no cases can the source signal be communicated directly over the channel. As an example, in (digital) mobile telephony, the voice signal occupies (approximately) the spectrum from 0 to 8 kHz; the transmit signal should be concentrated around a carrier frequency of 1 or 2 GHz.

Hence, transmitter and receiver perform an *adaptation* of the source to the channel. This adaptation/translation has to be done such that the *resources*, which are typically (very) limited, are utilized most efficiently. The main resources we will treat (assess) in this book are

- *Signal Power*
  The transmit power is limited for regulatory reasons and, in mobile scenarios, because of limited battery power.[1]

---

[1] In cases where signals from other communication links are the main source of interference, nothing is gained by increasing the signal power.

- *Bandwidth*
  Typically, each radio-frequency transmission system is assigned a spectral band (with given bandwidth in hertz, measured one-sided on the positive side of the frequency axis), in which the transmit signal has to fit; bandwidth is a very scarce (and thus costly) resource.
- *Processing and Storage Capabilities*
  Eventually, a transmitter and receiver have to be implemented (in hardware or as software on a generic hardware platform); these must be realizable with the given processing power and the available memory.

Throughout this book we use the term *rate*. To make this specific, we give the following definition:

Definition 1.3 *Rate*

In communications engineering, the generic term *rate* is used to specify an average amount of information [bit] per something.

Examples of rates are the transmitted information per time, denoted as the data rate R_T [bit/s], the entropy of a discrete information source, also called the source rate $H(Q)$ [bit/symbol], the rate R [bit/symbol] per discrete time step of a digital communications scheme, the rate of the modulation R_m [bit/modulation step], or the code rate R_c [bit/code symbol].

Note that there is a common confusion with the expression *bandwidth* in electrical engineering and communications on the one hand, and computer science and often everyday language on the other hand.

Box 1.1 Comment on *Bandwidth* and *Data Rate*

In this book, we exclusively use the term *bandwidth* with its original meaning: the support of a Fourier spectrum, more specifically of a power or energy spectral density (psd) of a signal on the *positive side* of the frequency axis. Thus, the *bandwidth* B_{RF} of a signal is defined to be the width of the interval at positive frequencies where such a spectrum is (substantially) greater than zero. Bandwidth is measured single-sided for physical, real signals and expressed in hertz (Hz). As a consequence, this bandwidth B_{RF} shows up two-sided in the equivalent complex baseband representation of a signal in most cases (cf. Chapter 2).

For the speed of information flow R_T transmitted by a digital communication system we use the expression *data rate*, sometimes also *data speed* or *information flow*, but never *bandwidth*. Data rate is measured in bit/s.

Unfortunately, the terms *bandwidth* and *data rate* have commonly become confused and misunderstood.

> **Definition 1.4** *Spectral Efficiency*
>
> A communication scheme transporting information with data rate R_T by a signal whose spectrum occupies an interval B_{RF} on the positive frequency axis offers the *spectral efficiency*
>
> $$\Gamma = \frac{R_T}{B_{RF}} \quad \left[\frac{\text{bit/s}}{\text{Hz}}\right].$$

> **Definition 1.5** *Power Efficiency*
>
> The *power efficiency* of a digital communications scheme over a given channel is characterized by the minimum transmit power that is sufficient to guarantee a desired data reliability (error ratio). Preferably, it is expressed by the minimally required energy per transmitted bit.

1.1 Digital Communications

Since the 1980s, all newly designed information transmission systems are exclusively *digital*. By digital we mean that a *discrete-time binary sequence* (sequence of "binary digits") has to be communicated instead of directly using an analog signal.

1.1.1 Why "Digital": From Analog to Digital

Recall that each type of source signal can be converted into a binary data stream via *pulse-code modulation (PCM)*. To that end, the following steps are carried out.

- The signal is *sampled* (time discretization) using a sufficiently high sampling rate (obeying the sampling theorem).
- The samples are *quantized* to a given number of levels; each level is represented by a binary number.
- The stream of binary symbols obtained by concatenating the binary numbers of each sample is typically *compressed* (i.e., source coding is used to eliminate redundancy).

The compression step, which should always be used, leads to the possibility of ideally decoupling the source coding part (sampling, quantization, compression) from the subsequent channel coding and modulation part. At the interface between both blocks, without loss of generality, a *redundancy-free source sequence of binary symbols* can be assumed. This means that the symbols "0" and "1" are equally likely to occur and that within the sequence, subsequent symbols are statistically independent (identically and independently distributed, i.i.d.).

Hence, in what follows, the source in Figure 1.1 is assumed to include an ideal (binary) source coding scheme. The source delivers a binary i.i.d. sequence.

In contrast to colloquial language, a strict distinction is made between the terms *binary symbol* and *bit* throughout this book. The widespread malapropism of "bit" for binary symbol – fundamentally misunderstanding information theory and digital communications – is not continued here.

> **Box 1.2 Comment on *Binary Symbol* vs. *bit***
>
> A symbol or variable that can take two different values only is denoted as a *binary symbol*. In contrast, the expression *bit* is strictly used in its original sense as an auxiliary unit for Shannon's information measure when employing the binary logarithm. Thus, an abstract amount of information is expressed in [bit], usually not restricted to an integer number.

A binary symbol represents 1 bit of information if and only if both values have equal a-priori probabilities 1/2, otherwise less information is contained. In a sequence of binary symbols, each element represents 1 bit of information if and only if the a-priori probabilities are 1/2 and if there are no statistical interdependencies within the sequence, that is, the sequence consists of i.i.d. symbols and there is no memory within the sequence. Otherwise, the average information per binary symbol, the *entropy*, is less than 1 bit/symbol, as memory decreases entropy because the uncertainty on a next observation may be lower with knowledge of the previously observed symbol, exploiting a memory for prediction.

The "binary symbols" are abstract entities, usually denoted as "o" and "1" (equivalently they might be called "low" and "high" or "head" and "tail" or "yes" and "no" or "red" and "black" ...), which should not be confused with the natural numbers 0 and 1. Conversely, the transmit signal is a physically realizable, measurable signal. In order to emphasize that the source symbols are not natural numbers, but mathematically are drawn from the finite field $\mathbb{F}_2$ of cardinality 2, we typeset these elements and the variables containing them in Fraktur font, for example, q[l]. Real/complex numbers and elements of finite fields should not be mixed up, and no arithmetic operation is defined between such worlds.

1.1.2 Why "Digital": Advantages of Digital Communications

What are the main advantages of digital transmission schemes (performing the detour of transforming an analog source signal to a binary data stream and then again producing an analog transmit signal) over the direct route of analog modulation methods? There are three essential aspects.

- *Digital transmission schemes have a significantly higher power efficiency than analog transmission schemes*
  If we assume the same bandwidth for analog and digital schemes and require the same robustness of the sink signal in the presence of noise, digital schemes require much less transmission power (see example below).

- *Using digital transmission schemes, a continuous exchange between power efficiency and bandwidth efficiency is possible*
  For a given power efficiency (robustness), a significantly higher bandwidth efficiency (less required bandwidth) compared to analog transmission is achieved. Moreover, there is a very flexible trade-off between both quantities.
- *For digital transmission schemes, signal regeneration is possible*
  In contrast to analog schemes, arbitrarily long distances can be bridged without any loss in quality; digital copies are exactly the same as the original (e.g., think of copying an analog compact audio cassette compared to copying a digital compact disc (CD)).

Example 1.1 Audio Transmission

We consider the radio transmission of an audio signal (mono). The same quality at the receiver output is desired for both transmission methods. Here we assume a signal-to-noise ratio $\text{SNR} \cong 60$ dB, which is considered to be HiFi (high fidelity) quality. The channel is the same for analog and digital modulation (same signal attenuation, same signal bandwidth, same noise level).

The typically required transmit powers are

100 kW	analog FM transmission
≈ 30 W	PCM + digital transmission without source and channel coding
< 1 W	PCM + digital transmission with source and channel coding

In Figure 1.2 the effect of signal regeneration is depicted. The respective behavior can be described as follows:

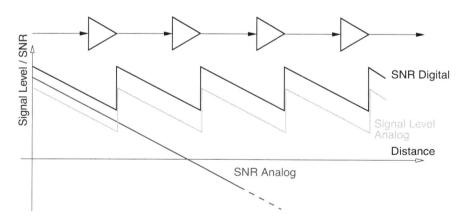

Figure 1.2 Signal level and signal-to-noise ratio over the transmission distance.

- analog transmission
  The intermediate amplifiers can merely increase the signal level. As signal and noise are amplified simultaneously, an improvement in SNR is impossible (even worse, as practical amplifiers are always noisy, the SNR drops (not shown in the figure)).
- digital transmission
  Using digital modulation, *signal regeneration* is possible. This is done by detecting (decoding in case of coded transmission) the binary information sequence. If the regenerators (which are not pure amplifiers but receivers and transmitters) are spaced closely enough, the binary sequence can be detected free of error and retransmission without any noise (the same as the initial transmit signal) is possible.

1.2 Examples of Digital Transmission Systems

Meanwhile, digital communication schemes are wide spread. The schemes are as diverse as the transmission media and requirements are. Some examples of categories of digital transmission systems are listed.

1.2.1 Digital Baseband Transmission over Metallic Wires

To this category belongs data transmission within all circuits for digital processing (e.g., processors). Moreover, transmission over symmetric cables (wire pairs – "plain old telephone lines"), so-called digital subscriber line (DSL) transmission, or coaxial lines is common. DSL schemes support downstream (from the central office to the customer) data rates of 16 Mbit/s over long distances (4 km) and up to 600 Mbit/s over short distances (< 150 m). Over the cable television network, data rates up to 10 Gbit/s are possible (data over cable service interface specification (DOCSIS)).

1.2.2 Optical Transmission Using Light Waves

Fiber-optical transmission offers enormous data rates and builds the backbone of the internet. Currently (2023) plugable transceivers for access/metro/long-haul applications (e.g., quad small form factor plugable double density (QSFP-DD) or octal small form factor plugable (OSFP)) achieve data rates up to 400 GBit/s over a distance of 1000 km or a data rate of 100 GBit/s over a distance of approximately 2500 km. The next generation (announced for 2024) will support 800 GBit/s over up to 3000 km. In the long-haul/ultra long-haul network, line cards are available which transmit more than 180 GBit/s over 10 000 km-long fibers. Current lab experiments have demonstrated more than 270 TBit/s over 1000 km, using a specific fiber that carries a dozen wave modes.

1.2.3 Digital Carrier-Modulated (Wireless) Transmission

Digital bandpass transmission is essential for wireless communications and is widely used in radio and TV broadcasting, but also in mobile phones and directed radio. For example

- digital mobile communications: GSM, UMTS, LTE, 5G
- digital satellite connections: $n \times 64$ kbit/s, $n = 1, \ldots, 2000$
- digital radio, digital audio broadcasting (DAB): 1.5 Mbit/s
- digital television, digital video broadcasting (DVB): 4 Mbit/s up to 30 Mbit/s
- digital directed radio: 140 Mbit/s up to 565 Mbit/s
- coherent optical transmission (free space or wave-guided): data rates up to some terabits per second per fiber have been established
- digital communications over switched telephone line (300 Hz–3.4 kHz)
  - voice-band modems: 300 bit/s (1960) up to 33 600 bit/s (1996)
  - "PCM modems": $\leq 56\,000$ bit/s (1998).

1.2.4 Digital Information Storage

Storage (communication over time) is also an important field of digital communication. This includes (classical) magnetic recording on tapes or disc (hard disc drives) and solid-state memory chips but also systems where the information is stored as "pits" and "lands" (mechanically) which are read out optically, in particular CD, DVD, BluRay.

1.2.5 Driving Forces in the Development

There are two (extreme) applications that have always been at the forefront of the development of digital communications. On the one hand, in the early days of digital communications up to the end of the last century, the development of *voice-band modems* (i.e., the representation of data streams by signals, similar to analog telephone voice in order to use analog switched telephone links for data communication) created very sophisticated technologies (e.g., trellis-coded modulation, signal shaping). Owing to the very low data rates and carrier frequencies, high-complexity algorithms were implementable using digital signal processors. Because of the very limited bandwidth of the telephone voice channel (typically 300 Hz–3.4 kHz, i.e., approximately 3 kHz bandwidth), very high power and bandwidth efficiency are required (30 kbit/s in 3 kHz bandwidth means a spectral efficiency of $\Gamma = 10 \frac{\text{bit/s}}{\text{Hz}}$).

On the other hand, spacecraft and deep-space missions were a field of early application of digital schemes. Here, contrary to the mass-market modems, very high power efficiency is required due to the limited energy resource in spacecraft; bandwidth efficiency is not an issue. Since only single copies were built, the cost was not an issue.

Meanwhile, mobile communications (starting in the 1990s with GSM (2G)) and mobile access to the internet ("ubiquitous wireless communications") are the driving forces in the further development of digital communication systems.

1.3 Basics of Digital Communication Schemes

In order to analyze digital communication systems, some fundamental building blocks and the corresponding nomenclature have to be specified. To that end, we revisit the block diagram of Figure 1.1, repeated as Figure 1.3.

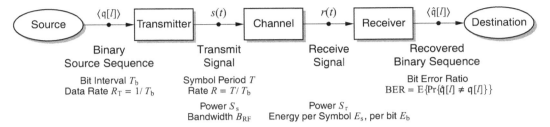

Figure 1.3 Block diagram of digital communication.

1.3.1 Data Source

Without loss of generality, a source data stream represented by an infinite sequence of binary symbols q[l] with 1 bit information per symbol is assumed. This means that the source data are modeled as an i.i.d. sequence with balanced a-priori probabilities.[2] Thus, the sequence ⟨q[l]⟩ is assumed to be free of redundancy (if there is indeed some redundancy, an exploitation of this is not considered here).

Note that throughout this book, without loss of generality, *binary*, *equally probable* symbols from a *memoryless* source (i.i.d. sequence) are assumed.

On average, in any given T_b seconds a source symbol is delivered by the source and the digital communication scheme transmits a data rate

$$R_T = \frac{1}{T_b} \; . \tag{1.1}$$

Definition 1.6 *Elements* and *Sequences*

The notation $x[k]$, $k \in \mathbb{Z}$, is primarily used to specify the kth *element* in a sequence (e.g., discrete-time signal), whereas the expression ⟨$x[k]$⟩ is used to explicitly represent the entire *sequence*.

---

[2] These properties are also beneficial to perform essential tasks at the receiver side like synchronization, equalization, and so on. In order to guarantee this behavior, a scrambler is often applied to the binary source data sequence. For scrambling, a linear, pure IIR system is usually applied over the binary field $\mathbb{F}_2$, together with a pure FIR inverse system for descrambling at the receiver side. The FIR structure guarantees limited error propagation.

1.3.2 Transmitter

The task of the *transmitter* (TX) of a digital communication scheme can be paraphrased as follows.

> **Definition 1.7 *Transmitter***
>
> The *transmitter* performs a mapping of an abstract sequence of binary source symbols $\langle q[l] \rangle$, $l \in \mathbb{Z}$, $q[l] \in \{0, 1\}$, onto a physical transmit signal $s(t)$.

The mapping has to be done in such a way that:

- the signal $s(t)$ does not exceed a predefined bandwidth – this demand is quantified by the *bandwidth efficiency* of the scheme;
- the transmission requires only a little transmit power to be robust against noise – this demand is quantified by the *power efficiency* of the scheme.

Usually, the generation of the continuous-time transmit signal is subdivided into *modulation steps*, where every T seconds transmit symbols are mapped to partial signals, the so-called *signal elements*. T is the *basic period* of the digital communication scheme for its discrete-time representation. This basic period is the time interval between two modulation steps. $1/T$ is the frequency of a *master clock* inside the transmitter. In the respective discrete-time representation, the average information per modulation step or transmit symbol, called the *rate of the digital communication scheme*, is given by

$$R = \frac{T}{T_b} \quad \left[\frac{\text{bit}}{\text{modulation step}}\right] \text{ or } \left[\frac{\text{bit}}{\text{symbol}}\right]. \qquad (1.2)$$

This equation immediately follows from the continuity of information flow throughout a communication scheme: within each building block, information is allowed neither to be lost nor to be generated.

Usually, the mapping of the source sequence to the transmit signal is subdivided into *(channel) coding* and *modulation*, where all features introducing memory into the process of generation of the transmit signal are devoted to coding. In generalization to its meaning in analog communications, the term modulation is used for the mapping of a sequence of symbols in discrete time with respect to the master clock $1/T$ onto the continuous-time transmit signal $s(t)$ independently from step to step (i.e., memoryless w.r.t. the time steps of duration T). This formal separation of coding and modulation into two processes, the introduction of memory on the one hand and memoryless transition from a discrete-time to a continuous-time signal on the other hand, may at first glance look rather artificial. Later, this concept will prove to be very helpful for the derivation of optimum receiver concepts, both in theory and in technical practice. Even if memory within the receive signal is introduced on its way to the receiver by a dispersive channel (e.g., multipath signal propagation or filter effect of lines), the method of modeling the generation of the receiver input signal by an equivalent encoder and memoryless modulation leads to highly efficient receiver implementations, see Chapters 6 and 8.

As a consequence of the splitting of the generation of the signal into coding and modulation, the rate R may be separated into a *code rate* R_c and a *modulation rate* R_m:

$$R = \frac{T}{T_b} = \frac{T_c}{T_b} \cdot \frac{T}{T_c} = R_c \cdot R_m, \qquad (1.3)$$

where T_c is the time interval per *binary* code symbol and

$$R_c = \frac{T_c}{T_b} \quad \left[\frac{\text{bit}}{\text{code symbol}}\right], \qquad (1.4)$$

$$R_m = \frac{T}{T_c} \quad \left[\frac{\text{code symbols}}{\text{modulation step}}\right]. \qquad (1.5)$$

If no channel coding is applied, the code symbols are identical to the binary source symbols ($R_c = 1$) and, thus, $R_m = R$.

1.3.3 Channel

Recall that the *channel* is that part of the communication system that one is either unwilling or unable to change. In this section, we look at the basic modeling of channels.

Block Diagram of the Channel and Its Modeling

In Figure 1.4, the general block diagram of a channel assuming radio-frequency (RF) transmission is given.

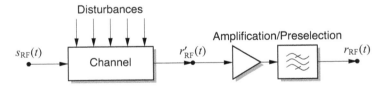

Figure 1.4 Block diagram of a channel (RF transmission).

The transmit signal $s_{RF}(t)$ is fed to the channel. In the channel, the signal is affected by attenuation and disturbances. The actual receive signal $r'_{RF}(t)$ is amplified and filtered (bandpass filter to suppress adjacent channels), leading to the signal $r_{RF}(t)$, which is further processed. Owing to the unavoidable amplifier (consisting of devices working at a temperature > 0 Kelvin and therefore being subject to thermal noise), noise is superimposed. In case we are not able to modify the amplification/preselection device, it belongs to the channel and $s_{RF}(t)$ is the input and $r_{RF}(t)$ the channel output.

> **Box 1.3 Comment on *Signal, Noise, Interference*, and *Disturbance***
>
> In communications literature, the term *signal* is used in two ways. In a general sense, the term denotes the representation of information by means of a physical process. In a specific sense, only the useful part within a signal is named "signal" (i.e., that portion which helps to detect the message at the receiver side in contrast to noise and interference, e.g., in the *signal-to-noise ratio*). Thus, a correct interpretation of the term signal is sometimes tricky and needs context.
>
> A similar problem exists with the term *interference*. On the one hand, interference is what obstructs the detection of the message on the receiver side. On the other hand, the term interference is often used in a more specific way only for that part of the impairment of a receive signal caused by (useful) signals of other communication links (e.g., the signals of other users cause *multi-user interference* in a multi-user communication systems). In this restricted meaning, *interference* has to be distinguished from the impairment *noise* or, more generally, *disturbance*. Noise is an impairment which has its source outside of the communication technology, like thermal noise, manmade noise, and so on.

Attenuation of the transmit signal is caused by resistive losses and reactive filter properties in metallic or optical lines, whereas in wireless communications free-space radiation leads to a dissipation of signal power in directions other than exactly to the small receive antenna and, thus, is the main reason for attenuation. To give an example, the signal power in mobile communications may be attenuated from about 1 W at the transmitter to 10^{-14} W at the receive antenna, that is, by about 140 dB from $10\log_{10}(\frac{1\,\text{W}}{1\,\text{mW}}) = 30$ dBm to $10\log_{10}(\frac{10^{-14}\,\text{W}}{1\,\text{mW}}) = -110$ dBm. The signal attenuation is usually not fully balanced by a receiver input amplifier but we are interested only in the signal-to-noise ratio (SNR) for the receiver input signal $r_{\text{RF}}(t)$, for which the factor of amplification does not matter. Therefore, often a full compensation of attenuation is assumed for simplified analysis.

Equivalent Linearized Model in the Equivalent Complex Baseband Domain

The *signal distortions* can be divided into linear (and dispersive) distortions and nonlinear distortions. Typical sources for linear dispersive properties of the transmission media are:

- linear distortions in metallic wire pairs (inductive/capacitive coupling, skin effect, ...);
- multipath propagation, reflections, diffraction, and scattering in wireless communications;
- modal and wavelength dispersion in wave guides;
- bandlimiting devices (amplifiers, read-out heads, ...).

Nonlinear distortions are typically caused by:

- nonlinear characteristics of power amplifiers or deliberate signal limitation to a foreseen amplitude range (clipping);
- saturation effects in the medium (magnetic recording);
- nonlinear behavior of the medium itself (fiber optics).

Typically, it is assumed that the scheme operates in the linear range (small-signal model), and, thus, nonlinear effects are neglected for the moment. This leads to the equivalent (linearized) model of a communication channel in the equivalent complex baseband (ECB) domain (to be introduced in detail in Chapter 2), given in Figure 1.5.

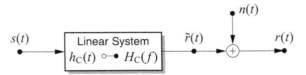

Figure 1.5 Block diagram of the equivalent (linearized) channel model (ECB domain).

This end-to-end model incorporates the actual channel and up/downconversion from baseband to radio frequency and vice versa. Moreover, the compensation for channel attenuation (amplification) has been performed at the receiver. All (linear) dispersive effects are characterized by the *channel transfer function* $H_C(f)$, or equivalently by the *channel impulse response* $h_C(t)$. Hence, a linear, time-invariant (LTI) system is used to model the linear distortions. Finally, all disturbances are combined (summed up) into the *equivalent noise* $n(t)$, which is effective at the receiver input.

The main sources for disturbance or interference are:

- thermal noise, amplified or generated at the receiver input amplifier;
- crosstalk due to other information signals
  - signals in adjacent frequency bands / adjacent tracks
    (adjacent channel interference, ACI)
  - signals in the same frequency band, for example, in cellular systems
    (co-channel interference, CCI)
  - signals in the various pairs of a multi-pair cable
  - non-orthogonal signals in multiplexing
    (multiple-access interference, MAI or multi-user interference, MUI).

White and Colored Gaussian Noise

The simplest and most often used model for noise is *white Gaussian noise*. This means:

- the noise process $n(t)$ is stationary (i.e., its statistical properties do not change over time);

- the probability density function (pdf) of the noise at a particular time instant t is (complex) Gaussian (i.e., "Gaussian" refers to the amplitude characteristics) – noise is typically modeled to be a Gaussian stochastic process (cf. Appendix A);
- the noise has a constant power spectral density (psd) – for real, physical signals this constant, the two-sided psd of white noise, is denoted by

$$\Phi_{n_r n_r}(f) = \frac{N_0}{2}, \qquad \forall f \in \mathbb{R}. \tag{1.6}$$

For signal description in the ECB domain (see Chapter 2), a psd

$$\Phi_{nn}(f) = N_0 = \text{const.} \ \forall f \tag{1.7}$$

results, or the autocorrelation function (acf, inverse Fourier transform of the psd) is

$$\phi_{nn}(\tau) = N_0 \delta(\tau) \tag{1.8}$$

where $\delta(t)$ denotes the Dirac delta function (i.e., "white" refers to the temporal characteristics; there are no dependencies between two time instants).

Note that white noise is a very convenient model but (luckily!) it does not exist in practice. White noise would have infinite power, as the power is given by integrating the psd $P_n = \int_{-\infty}^{\infty} \Phi_{nn}(f) \, df$. It is justified since the receive filter always limits the noise bandwidth; we typically require only that the noise is white within the transmission band used by the communication scheme under consideration.

The theory of Maxwell and Boltzmann says that thermal noise is nearly white up to a frequency limit of about 300 GHz and given by (in case of impedance matching)

$$N_0 = kT_\text{K} \tag{1.9}$$

with k = 1.38×10^{-23} J/K: Boltzmann constant and T_K: absolute temperature (K).

Finally, **information theory** tells us that white Gaussian noise is the "worst-case noise." Since the noise is white, no dependencies over time are present and thus no prediction from the past can be made. Since the noise is Gaussian, for a given power it has maximum entropy (randomness).

Colored Gaussian noise has the same statistical properties as white Gaussian noise, except that the psd is non-constant (non-flat) over the frequencies:

$$\Phi_{nn}(f) \neq \text{const.}, \tag{1.10}$$

or, equivalently, the autocorrelation function does not have a Dirac delta shape.

The denominations "white" and "colored" are due to the analogy to visual light. White light contains all colors (frequency components) with the same intensity. Colored light (e.g., red visual light) contains the respective frequency bands with higher intensity than the others.

In the case of colored noise, a linear filter $H_\text{W}(f)$ may be applied at the receiver frontend, by which colored noise is transformed into white noise (whitening filter). This filter has to be accounted for in the useful signal transfer function, see Chapter 8.

1.3 Basics of Digital Communication Schemes

Without loss of generality, we hence restrict ourselves in the subsequent derivations to white noise. In situations where there are neither (significant) linear signal distortions nor (significant) noise coloring arising within the used frequency band, the simple and very fundamental model of the *additive white Gaussian noise (AWGN) channel* can be applied with satisfactory accuracy. The AWGN channel is given by (ECB domain)

$$H_C(f) = 1, \quad \Phi_{nn}(f) = N_0, \qquad (1.11)$$

and it is visualized in Figure 1.6.

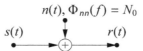

Figure 1.6 Additive white Gaussian noise channel (ECB domain).

In Chapters 3 to 7 we will restrict the analysis to this elementary channel model. Chapters 8 and 9 are devoted to methods for digital communication over linear dispersive channels and/or colored noise.

Note that in the AWGN channel model, a complete compensation for the signal attenuation by pre-amplification is usually subsumed and therefore the model can be formulated at any power level for signal and noise. Signal powers at transmitter and receiver side are modeled to be equal and denoted simply by $S = S_t = S_r$ without subscript. Of course, the noise power density has to be matched to the chosen power level.

1.3.4 Receiver

The task of the *receiver* (RX) of a digital communication scheme can be paraphrased as follows.

Definition 1.8 *Receiver*

The *receiver* performs the inverse mapping of the transmitter, that is, it retrieves the source sequence from the noisy and distorted receive signal.

This recovery of the source sequence from the noisy receive signal should be done such that:

- all information on the source symbols, which is still present in the receive signal, is extracted;

- the probability that the estimated binary symbol $\hat{q}[l]$ is unequal to the transmitted one, that is, the *bit error ratio* ($\Pr\{\cdot\}$: probability; $\mathrm{E}\{\cdot\}$: expectation)

$$\mathrm{BER} \stackrel{\mathrm{def}}{=} \mathrm{E}\{\Pr\{\hat{q}[l] \neq q[l]\}\} \qquad (1.12)$$

is as small as possible;
- the destination (sink) may be informed about the reliability of the detected symbols, that is, by means of an estimate of $\Pr\{\hat{q}[l]\}$.

Definition 1.9 *Energy per bit*

The equivalent average received signal *energy per bit* of transmitted information (in short: energy per bit) is given by

$$E_\mathrm{b} = S_\mathrm{r} \cdot T_\mathrm{b} = \frac{S_\mathrm{r}}{R_\mathrm{T}} = \frac{E_\mathrm{s}}{R}. \qquad (1.13)$$

For discrete-time analysis, the equivalent average received signal *energy per modulation step* (in short: energy per step or energy per symbol)

$$E_\mathrm{s} = S_\mathrm{r} \cdot T = E_\mathrm{b} \cdot R \qquad (1.14)$$

is frequently used.

The power efficiency of a digital communication scheme can be specified in a very general way, quite irrespective of the actually applied transmission method, by the minimum energy per bit E_b of the receive signal that is required to achieve a desired reliability of the received message at a given noise/interference situation (i.e., the bit error ratio, BER should be below the desired threshold for a fixed spectral noise power density of an AWGN channel. For voice communication, BER $< 10^{-5}$ may be sufficient, whereas for pure data streams, thresholds BER $< 10^{-8}$ or BER $< 10^{-15}$ are usual). In this book, only aspects of the basic physical layer of digital communication systems according to the layered ISO/OSI model of communications are considered. Error control mechanisms (e.g., automatic repeat requests) at higher layers are not included when discussing demands on reliability. Also, the additional latency on a data stream caused by higher-layer protocols is beyond the scope of this book.

1.4 Summary of Basic Definitions

Here, we repeat, complete, and collect basic definitions for the analysis of digital communication schemes. To that end, in Figure 1.7, the block diagrams of Figures 1.1 and 1.3 are revisited, now in the simplified form for the AWGN channel model. Up to Chapter 7, only this AWGN model will be further applied.

The blocks of Figures 1.1 and 1.3 can be recognized. Some more details are now shown, and the channel is specialized to the AWGN channel (cf. Figure 1.6), that is, white Gaussian noise is superimposed onto the (RF) transmit signal $s_{RF}(t)$. As can be seen, the transmitter is split into a (channel) coding and mapping block and the modulation. Consequently, the receiver consists of the respective demodulation and (channel) decoding blocks. This separation will be detailed in the following chapters. Moreover, the different symbol intervals are indicated: the source delivers the binary symbols (on average) every T_b, the bit duration. The modulation operates on symbol intervals of duration T.

At the receiver, synchronization has to be performed. The block "synchronization" has to synchronize the receiver-side downconversion (carrier frequency and carrier phase) to the transmitter-side upconversion. Additionally, timing synchronization is required; symbol frequency and symbol phase have to match those of the transmitter.

Optimal carrier frequency/phase and symbol clock synchronization are rather sophisticated fields in digital communications. A sufficient treatment of the topic of receiver synchronization would clearly go beyond the scope of this introductory book. Therefore, we will refrain from addressing the topic here. Specific textbooks devoted to this are available, and we refer the reader to those. For the theoretical analysis, we mostly assume perfect synchronization. At some points (e.g., in the context of non-coherent receivers), we point out the consequences and remedies in case of non-synchronized (or non-perfectly synchronized) receivers.

The digital communication scheme expressed in the ECB (cf. Chapter 2) corresponding to the RF model of Figure 1.7 is shown in Figure 1.8. For brevity, the synchronization part is not shown, despite it being required in any receiver.

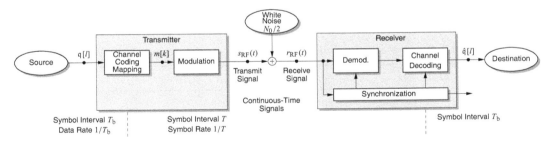

Figure 1.7 Block diagram of a digital communication scheme (real-valued RF signals).

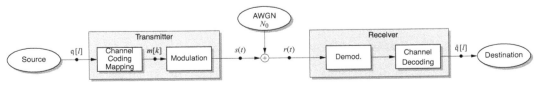

Figure 1.8 Block diagram of a digital communication scheme (ECB signals, synchronization not shown for brevity).

1.4.1 Parameters of a Digital Communication Scheme

Table 1.1 collects the most important parameters of a digital communication scheme. The relations to the other parameters and their units are given.

Table 1.1 Parameters of a digital communication scheme.

Parameter	Symbol	Unit
bit duration (bit interval)	T_b	[s]
symbol interval	T	[s]
transmitted data rate	$R_T = \frac{1}{T_b}$	$\left[\frac{\text{bit}}{\text{s}}\right]$
symbol rate (baud rate)	$\frac{1}{T}$	$\left[\frac{\text{symbols}}{\text{s}}\right]$
average information per transmitted symbol (rate of the digital communication scheme)	$R = \frac{T}{T_b}$	$\left[\frac{\text{bit}}{\text{symbol}}\right]$
code rate (binary code)	$R_c = \frac{T_c}{T_b}$	$\left[\frac{\text{bit}}{\text{code symbol}}\right]$
modulation rate	$R_m = \frac{T}{T_c}$	$\left[\frac{\text{code symbols}}{\text{modulation step}}\right]$
average transmit power	S_s	[W]
average power of the desired (noise-free) part of the receive signal	S_r	[W]
average received signal energy per symbol	$E_s = S_r T$	[Ws]
average received signal energy per bit	$E_b = S_r T_b = \frac{E_s}{R}$	[Ws]
(one-sided) bandwidth of the transmit signal	B_{RF}	[Hz]
bandwidth efficiency (spectral efficiency)	$\Gamma = \frac{R_T}{B_{RF}}$	$\left[\frac{\text{bit/s}}{\text{Hz}}\right]$

When specializing to an AWGN channel, we further consider the quantities collected in Table 1.2.

Table 1.2 Parameters of a digital communication scheme over the AWGN channel.

Parameter	Symbol	Unit
additive white Gaussian noise with two-sided noise power spectral density (RF signal)	$N_0/2$	$\left[\frac{\text{W}}{\text{Hz}}\right]$
two-sided noise power spectral density (ECB signal)	N_0	$\left[\frac{\text{W}}{\text{Hz}}\right]$
power normalization	$S_s = S_r = S$	[W]
energy per bit to noise power spectral density ratio "SNR per bit" (at RX input)	$E_b/N_0 = \frac{E_s/N_0}{R}$ $= S T_b/N_0$	[–]

1.4.2 Optimality Criteria in Digital Communications

We are now ready to define the *optimality criteria*, by which we assess the digital communication schemes studied in this book. The two most important criteria of interest are as follows.

- *Power Efficiency.* Specified by the signal-to-noise ratio

$$10 \log_{10}(E_b/N_0)$$

E_b : (equivalent) received signal energy per information bit
N_0 : one-sided noise power spectral density

minimally required to guarantee a tolerated bit error ratio

$$\text{for example} \quad \text{BER} \leq 10^{-8}$$

(i.e., it is the lower limit on the SNR).
- *Bandwidth (or Spectral) Efficiency.* Specified by the data rate per occupied bandwidth

$$\Gamma \stackrel{\text{def}}{=} \frac{R_T}{B_{RF}} \quad \left[\frac{\text{bit/s}}{\text{Hz}} \right]$$

R_T : data rate to be transmitted in bit/s
B_{RF} : (one-sided) bandwidth of the (RF, physical) transmit signal

In practice, further criteria have relevance, for example:
- *Complexity.* The complexity of TX and RX, that is, the (numerical) complexity of encoding, modulation, detection, receiver synchronization, channel (de)coding, and so on – typically, the receiver requires a much larger complexity than the transmitter.
- *Delay.* The delay (latency) of the data stream caused by the processing chain, which is especially of interest in real-time (interactive) communication scenarios.

Note that the scheme has a

- *high power efficiency* if the required E_b/N_0 (in dB) is *small*
- *high bandwidth efficiency* if Γ is *large*.

1.5 Trade-Off between Power and Bandwidth Efficiency

As in any engineering discipline, the optimality criteria cannot be fulfilled simultaneously. Instead, there exists a trade-off between power and bandwidth efficiency. A system has either a high power efficiency (low required SNR) at a poor bandwidth efficiency, or it has a high bandwidth efficiency (large Γ) at a poor power efficiency.

The trade-off between both quantities (transmission over the AWGN channel) is visualized in the *power/bandwidth plane* given in Figure 1.9.

This diagram is the basis for a comparison of different digital modulation schemes. On the horizontal axis the *minimally required* SNR (expressed as the ratio of (received) energy per information bit E_b and (one-sided) noise power spectral density N_0 of the AWGN channel, i.e., E_b/N_0 in dB) for achieving a desired reliability (e.g., BER = 10^{-5}) is depicted. On the vertical axis the spectral efficiency Γ (in

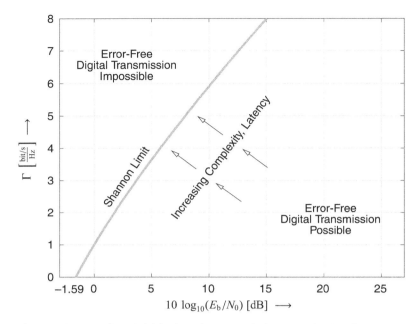

Figure 1.9 Power/bandwidth plane for transmission over the AWGN channel.

$\frac{\text{bit/s}}{\text{Hz}}$) is indicated. Each digital communication format corresponds to a *point* in the power/bandwidth plane.

Information theory provides a fundamental limit for the optimal trade-off between both quantities. This *Shannon limit* is shown as a bold gray line in Figure 1.9. Trade-offs to the right below the line are possible; no (error-free) digital communication schemes can exist to the left above the limit. Coming closer to the limit typically incurs an increase in complexity and latency.

The Shannon limit is obtained from Shannon's famous equation for the capacity of a bandlimited AWGN channel:

$$C_\mathrm{T} = B_\mathrm{RF} \, \log_2\left(1 + \frac{S}{N}\right) \quad \left[\frac{\text{bit}}{\text{s}}\right] \tag{1.15}$$

B_RF: one-sided RF bandwidth
$S = E_\mathrm{b}/T_\mathrm{b}$: signal power (at receiver)
$N = N_0 \cdot B_\mathrm{RF}$: noise power.

In the best case, that is, for an ideal communication scheme, the transmitted data rate $R_\mathrm{T} = 1/T_\mathrm{b}$ equals the channel capacity, hence

$$R_\mathrm{T} \stackrel{!}{=} B_\mathrm{RF} \, \log_2\left(1 + \frac{E_\mathrm{b}}{N_0} \frac{1}{B_\mathrm{RF} T_\mathrm{b}}\right) . \tag{1.16}$$

1.5 Trade-Off between Power and Bandwidth Efficiency

Using the definition of the bandwidth efficiency $\Gamma \stackrel{\text{def}}{=} \frac{R_T}{B_{RF}} = \frac{1}{B_{RF}T_b}$ and solving for $\frac{E_b}{N_0}$, we arrive at

$$\frac{R_T}{B_{RF}} = \log_2\left(1 + \frac{E_b}{N_0}\frac{1}{B_{RF}T_b}\right),$$

$$\Gamma = \log_2\left(1 + \frac{E_b}{N_0}\Gamma\right),$$

$$2^\Gamma = 1 + \frac{E_b}{N_0}\Gamma,$$

$$\frac{E_b}{N_0}\Gamma = 2^\Gamma - 1. \qquad (1.17)$$

Thus

$$\frac{E_b}{N_0} = \frac{1}{\Gamma}\left(2^\Gamma - 1\right). \qquad (1.18)$$

From Figure 1.9 it can be seen that even for an arbitrary small bandwidth efficiency (very small number of transmitted bits per hertz bandwidth), the power efficiency cannot be increased. The minimally required E_b/N_0 is obtained for $\Gamma \to 0$. We have

$$\left(\frac{E_b}{N_0}\right)_{min} = \lim_{\Gamma \to 0} \frac{2^\Gamma - 1}{\Gamma} = \lim_{\Gamma \to 0} \frac{e^{\Gamma \cdot \log_e(2)} - 1}{\Gamma}$$

$$\stackrel{\text{L'Hôpital}}{=} \lim_{\Gamma \to 0} \frac{\log_e(2) e^{\Gamma \cdot \log_e(2)}}{1}$$

$$= \log_e(2) = 0.69, \qquad (1.19)$$

or in dB

$$10 \log_{10}\left(\frac{E_b}{N_0}\right)_{min} = -1.59 \text{ dB}. \qquad (1.20)$$

Below this fundamental limit, no reliable (error-free) transmission is possible at all. Note that this limit has a relation to the second law of thermodynamics (entropy will never decrease within a closed system).

Digital communication schemes with a performance closely approaching the Shannon limit in the power/bandwidth plane are based on very sophisticated channel coding schemes. Here, the noise process is averaged within long codewords close to expectation values, that is, the randomness of noise effective for an entire codeword is greatly reduced. On the one hand, a joint processing of long blocks of data on the transmitter side and of long receive signal intervals at the receiver side introduces high delay/latency for the data stream, which increases linearly with the codelength. This may sometimes not be tolerable, for example, in bidirectional or multilateral communication schemes. On the other hand, usually the complexity for encoding and decoding in powerful coding schemes increases more than linearly with the codelength and becomes unmanageable for extremely long codes. The

Shannon limit, promising full reliability of data (BER = 0), is only valid for codelengths approaching infinity, thus for infinite complexity and latency. There also exist bounds on power and bandwidth efficiency for finite-length codes and some tolerated residual codeword error probability. These bounds show that schemes with a latency of several thousand T_b can exist in principle close to 1 dB away from the Shannon limit, while providing sufficiently high reliability. About 60 years after the derivation of the Shannon limit (published in 1948), schemes were invented that approach this limit in practice (with a high, but manageable, complexity). The sophisticated theory of channel coding, which is the fundament of such powerful digital communication schemes, is not included in this book. Here, only some examples will be addressed.

1.6 Problems

In order to assess your prior knowledge, please carefully answer the following questions. If you have problems answering some of them, please pick up a suitable textbook of your choice and recapitulate the corresponding topics.

1.6.1 Basic Calculus
(a) What are complex numbers? Why and when are they required?
    Give the rules for computing with complex numbers (addition, subtraction, multiplication, and division).
(b) Write down Euler's formula. Describe the Cartesian form and the polar form of complex numbers. How can the forms be converted into each other?
(c) Differentiate (w.r.t. the variable t; $c \in \mathbb{R}$; $j^2 = -1$) the following functions:
    $f(t) = c$, $f(t) = t^2$, $f(t) = \cos(ct)$, $f(t) = e^t$, $f(t) = e^{jct}$.
(d) Integrate (w.r.t. the variable t; $c \in \mathbb{R}$; $j^2 = -1$) the following functions:
    $$f(t) = c, \ f(t) = t^2, \ f(t) = \cos(ct), \ f(t) = e^{jct}, \ f(t) = \begin{cases} 1, & |t| < 1/2 \\ 0, & \text{else} \end{cases}.$$
(e) How is the product of a matrix with a vector defined? What requirements do both have to fulfill, such that the product exists?
(f) How is the scalar product of vectors defined?
(g) What is the Euclidean norm of a vector?
(h) Let a basis (i.e., a set of vectors spanning a space) be given. How can we calculate an orthonormal basis from the given one?
(i) Explain the meaning of eigenvalues and eigenvectors. How are they calculated given a (square) matrix?

1.6.2 Signals and Systems
(a) What are signals and what are systems?
(b) Characterize continuous-time and discrete-time signals!
(c) Under what conditions is a continuous-time signal completely specified by a discrete-time signal?

How is the discrete-time signal obtained from the continuous-time signal? How is the continuous-time signal recovered from the discrete-time signal?
(d) What is the Dirac delta "function"?
(e) Give the definition of the convolution of two signals.
(f) What is the energy and what is the average power of a signal?
(g) What is a linear time-invariant (LTI) system?
How is such a system completely characterized? Given the input signal, how is the output signal calculated?
(h) What is a causal system?
(i) When is a system stable? When is a system minimum phase?
(j) What is an ideal low-pass filter?
(k) How is the Fourier transform of a (continuous-time) signal defined?
What are the properties of the Fourier transform of a (i) real-valued signal, (ii) real-valued even signal, (iii) purely imaginary odd signal?
(l) Let a signal with its corresponding Fourier transform (i.e., its spectrum) be given.
How is the spectrum modified if the time-domain signal is (i) scaled with a constant, (ii) temporarily shifted, (iii) multiplied by e^{jct}, (iv) multiplied by $\cos(ct)$?
(m) What operation in the frequency domain corresponds to convolution in the time domain?
What operation in the time domain corresponds to convolution in the frequency domain?
(n) What is the Hilbert transform of a (continuous-time) signal?
Give the impulse response and the transfer function of the Hilbert transform.
(o) Given a real-valued signal, how is the corresponding analytic signal defined?
What is the main frequency-domain property of such a signal?
(p) How is the z-transform of a discrete-time signal defined?
Given the z-transform of a signal, how is the spectrum (discrete-time Fourier transform) of this signal obtained?
(q) What is the discrete Fourier transform (DFT)? For which signals is the DFT suited?
(r) How are the spectra of a continuous-time signal and a discrete-time signal, generated therefrom via sampling, related?

1.6.3 Random Variables, Stochastic Processes, and Estimation
(a) What is a random variable?
(b) Give the meaning of the cumulative distribution function (cdf) and the probability density function (pdf) of a random variable.
(c) How is the expected value (expectation) of a random variable defined?
What is the intuition behind the expected value?
(d) What are the mean and variance of a random variable?
(e) What is the notion of a stochastic process?
What are the most important quantities characterizing a stochastic process?

(f) What conditions must a process fulfill for it to be called a wide-sense stationary process?
What is an ergodic process?
(g) What is the power spectral density (psd) of a wide-sense stationary stochastic process?
Knowing the psd, how is the power of the process calculated?
(h) Discuss the features and differences of the terms energy-limited signal and power-limited signal. Clarify the difference between an autocorrelation of a deterministic, energy-limited signal and an autocorrelation function (acf) of a power-limited stochastic process.
(i) Let a stochastic process (with given acf and psd) be the input to an LTI system (given by its impulse response/transfer function). How are the acf and psd of the output process related to the quantities of the input process?
(j) A continuous-time stochastic process (with given acf and psd) is sampled. How are the acf and psd of the obtained discrete-time process related to the quantities of the input process?
(k) We want to estimate a random variable x via the (indirect) observation of a random variable y, which depends on x. Explain the maximum-a-posteriori rule and the maximum-likelihood rule for estimating x based on y.

1.6.4 Fundamentals of Communications

(a) What is the meaning of a signal-to-noise ratio?
(b) Express the quotient of power, or energy, of two signals in decibels (dB) for the values 2, 3, 4, 10, 100, and 1000. Give also the quotient of the root mean square (rms) values of two signals in decibels for the values 2, 3, 4, 10, 100, and 1000.
(c) What is meant by pulse-code modulation (PCM)?
(d) An analog signal is converted into a (binary) data stream. How is the data rate related to the sampling frequency and the resolution?
If uniform quantization is used, by how many decibels does the signal-to-noise ratio increase if the number of quantization levels is doubled? By what amount does the data rate increase?
(e) What are the bandwidth and the power of a (transmit) signal?
(f) What is an additive white Gaussian noise (AWGN) channel? Explain the meaning of additive, white, and Gaussian.
(g) What is the main characteristic of an analog transmission scheme?
(h) Let a continuous-time (source) signal be given. How is such a signal transmitted using amplitude modulation (AM)? What are the required operations in the transmitter and receiver?
(i) What are the advantages and disadvantages of using a modulation index < 1 in AM?
(j) How is the bandwidth of the AM radio-frequency transmit signal related to the bandwidth of the source signal to be communicated? Which variant of AM has the smallest bandwidth?

(k) Let a continuous-time (source) signal be given. How is such a signal transmitted using frequency modulation (FM)?
(l) What are the advantages and disadvantages of FM compared to AM?
(m) How is the bandwidth of the FM radio-frequency transmit signal related to the bandwidth of the source signal to be communicated?
(n) What is the fundamental difference of digital transmission schemes compared to analog schemes? What are the main advantages of digital transmission over analog transmission?
(o) What are the main performance measures for any type of communication system? In other words, the utilization of which resources is typically quantified?
(p) What is the main principle of pulse-amplitude modulation (PAM)? How is the (binary) information to be communicated represented in the transmit signal if we consider the simplest form of amplitude-shift keying?
(q) What is the essential difference between baseband transmission and carrier-modulated (bandpass) transmission?
(r) What is a quadrature modulator/demodulator?
(s) What is the optimum receive filter for PAM transmission over an AWGN channel?
(t) Why are digital transmission schemes usually characterized via their bit error ratio (BER)? How (qualitatively) is the BER related to the signal-to-noise ratio?
(u) Given a discrete source with a finite alphabet and known probabilities of the symbols, how is the entropy of such a source calculated and what is its meaning?
(v) What is redundancy and what is the aim of source coding? Give an example of a specific source coding scheme and explain how it works.
(w) What is the meaning of the capacity of a channel?
(x) Let a discrete-time AWGN channel with some signal power and noise power be given. Calculate the capacity of this channel.
(y) Let a continuous-time AWGN channel with some bandwidth, signal power, and noise power spectral density be given. Calculate the capacity of this channel.
(z) What is the aim of channel coding? Give an example of a specific channel coding scheme and how the corresponding decoder works.

2 Equivalent Complex Baseband Signals

In carrier-modulated (digital) communication, the transmit signal has spectral components in a band around a so-called *carrier frequency* f_c. A baseband transmit signal is upconverted to obtain the RF transmit signal and the RF receive signal is downconverted to obtain the baseband receive signal. The processing of transmit and receive signals is done as far as possible in the baseband domain (i.e., around $f = 0$).

Thus, *bandpass signals* are considered, whose spectra only exist for

$$f_0 - \frac{B}{2} < |f| < f_0 + \frac{B}{2}$$

f_0: center frequency (not necessarily the carrier frequency, see Chapter 7)
B: one-sided bandwidth (only the positive part of the frequency axis).

Of course, all physical signals are real-valued. From the field of signals and systems it is well known that real-valued time-domain signals have a spectrum which is symmetric about $f = 0$; the frequency components on the negative part of the frequency axis are a mirrored version of those located on the positive part of the frequency axis. Hence, one side of the spectrum (e.g., that for $f < 0$) is redundant and can be neglected or suppressed. However, when destroying the symmetry of the spectrum, the signal is no longer real-valued but *complex-valued*. This means that instead of using *real* signals, one can equivalently use *analytic signals*.

The aim of this chapter is to mathematically and precisely develop a compact representation of real-valued (bandpass) signals, independent of the actual center frequency f_0 (or carrier frequency f_c). This leads to the concept of *equivalent complex baseband (ECB) signals*, the corresponding systems, and stochastic processes in the ECB domain (a brief introduction to the fundamental concepts of stochastic processes can be found in Appendix A).

PREREQUISITES FOR THIS CHAPTER

- Complex numbers.
- Signals and systems (continuous-time signals, linear time-invariant systems).
- Random variables (pdf, expectation, Gaussian density) and stochastic processes (stationarity, autocorrelation, power spectral density, filtering of processes).

2.1 Equivalent Complex Baseband Signals

2.1.1 Basic Definitions

In communications, in order to represent an information-carrying signal by a physical (RF) signal, all modulation formats (analog and digital) vary a *carrier waveform* suitably. A carrier waveform $c(t)$ is given by

$$c(t) = \sqrt{2}\, A_{\text{rms}} \cdot \cos(2\pi f t + \varphi) \,. \qquad (2.1)$$
$$\phantom{c(t) = \sqrt{2}\,}\underset{A}{\uparrow} \underset{F}{\uparrow}\underset{P}{\uparrow}$$

Each harmonic oscillation is uniquely characterized by the three parameters (root mean square, rms) of amplitude (A), frequency (F), and phase (P). Information is usually represented by modifying these parameters according to the message to be transmitted; this is called *modulation* of the carrier waveform.

Since the *instantaneous* (*momentarily* active) *frequency* $f_M(t)$ and the phase $\varphi(t)$ are related via

$$f_M(t) = \frac{1}{2\pi} \frac{d}{dt} \varphi(t) \,, \qquad (2.2)$$

one parameter (either frequency or phase) is redundant and can be kept fixed. Hence, each modulated (RF) signal $s_{\text{RF}}(t)$ can be written as

$$s_{\text{RF}}(t) = \sqrt{2}\, a_M(t) \cdot \cos(2\pi f_0 t + \varphi_M(t)) \qquad (2.3)$$

f_0: reference frequency
$a_M(t)$: amplitude signal
$\varphi_M(t)$: phase signal.

Note that all signals in (2.3) are *real-valued*. Amplitude and phase signals $a_M(t)$ and $\varphi_M(t)$ are low-pass (baseband) signals; $s_{\text{RF}}(t)$ is typically a bandpass signal.

Example 2.1 Modulation

From the generic (general) form (2.3), specific modulation formats can be obtained:

- amplitude modulation – amplitude $a_s(t)$; $\varphi_s = $ const.
$$s_{\text{RF}}(t) = \sqrt{2}\, a_s(t) \cdot \cos(2\pi f_0 t + \varphi_s) \,;$$

- phase modulation – phase $\varphi_s(t)$; $s_{\text{rms}} = $ const.
$$s_{\text{RF}}(t) = \sqrt{2}\, s_{\text{rms}} \cdot \cos(2\pi f_0 t + \varphi_s(t)) \,;$$

- frequency modulation – frequency $f_s(t)$; $s_{\text{rms}} = \text{const.}$

$$s_{\text{RF}}(t) = \sqrt{2}\, s_{\text{rms}} \cdot \cos\left(2\pi f_0 t + 2\pi \int_{-\infty}^{t} f_s(\tau)\,\mathrm{d}\tau\right);$$

- baseband modulation – amplitude $a_s(t)/\sqrt{2}$; $f_0 = 0$, $\varphi_s = 0$

$$s_{\text{RF}}(t) = \sqrt{2}\, a_s(t)/\sqrt{2} \cdot \cos(2\pi 0 t + 0) = a_s(t).$$

Without loss of generality we can make the following assumptions:

$$a_M(t) \geq 0 \quad \text{and} \quad \varphi_M(t) \in (-\pi, \pi]. \tag{2.4}$$

This can be justified by the following.

- If $a_s(t) < 0$ and $\varphi_s(t)$ are given, using $-\cos(x) = \cos(x + \pi)$, we have

$$\begin{aligned}\sqrt{2}\, a_s(t) \cdot \cos(2\pi f_0 t + \varphi_s(t)) &= -\sqrt{2}\,|a_s(t)| \cdot \cos(2\pi f_0 t + \varphi_s(t)) \\ &= \sqrt{2}\,|a_s(t)| \cdot \cos(2\pi f_0 t + \varphi_s(t) + \pi) \\ &= \sqrt{2}\, a_M(t) \cdot \cos(2\pi f_0 t + \varphi_M(t))\end{aligned}$$

with $a_M(t) = |a_s(t)|$ and $\varphi_M(t) = \varphi_s(t) + \pi$. This means that a negative real-valued signal has a phase of π.

- If $a_s(t) \geq 0$ and $\varphi_s(t) \notin (-\pi, \pi]$ are given, using $\cos(x) = \cos(x + 2\pi z)$, $z \in \mathbb{Z}$, we have

$$\begin{aligned}\sqrt{2}\, a_s(t) \cdot \cos(2\pi f_0 t + \varphi_s(t)) &= \sqrt{2}\, a_s(t) \cdot \cos(2\pi f_0 t + \varphi_s(t) + 2\pi z) \\ &= \sqrt{2}\, a_M(t) \cdot \cos(2\pi f_0 t + \varphi_M(t))\end{aligned}$$

with $a_M(t) = a_s(t)$ and $\varphi_M(t) = \varphi_s(t) + 2\pi z$, with $z \in \mathbb{Z}$, such that $\varphi_M(t) \in (-\pi, \pi]$. This means that the phase (as it is 2π periodic) can always be restricted to the interval $(-\pi, \pi]$ (or, if desired, to $[0, 2\pi)$).

Representation of Modulated Signals

As shown above, all modulated signals can be given according to (2.3), with reference frequency f_0, amplitude signal $a_M(t) \geq 0$, and phase signal $\varphi_M(t) \in (-\pi, \pi]$. Using Euler's formula $\mathrm{e}^{\mathrm{j}x} = \cos(x) + \mathrm{j}\sin(x)$, we can write

$$\begin{aligned}s_{\text{RF}}(t) &= \sqrt{2}\, a_M(t) \cdot \cos(2\pi f_0 t + \varphi_M(t)) \\ &= \sqrt{2}\, a_M(t) \cdot \mathrm{Re}\{\mathrm{e}^{\mathrm{j}(2\pi f_0 t + \varphi_M(t))}\} \\ &= \sqrt{2} \cdot \mathrm{Re}\{a_M(t)\mathrm{e}^{\mathrm{j}\varphi_M(t)} \cdot \mathrm{e}^{\mathrm{j}2\pi f_0 t}\} \\ &= \sqrt{2} \cdot \mathrm{Re}\{s(t) \cdot \mathrm{e}^{\mathrm{j}2\pi f_0 t}\} \end{aligned} \tag{2.5}$$

with

$$s(t) \stackrel{\text{def}}{=} |s(t)|\mathrm{e}^{\mathrm{j}\varphi_s(t)} = a_M(t)\mathrm{e}^{\mathrm{j}\varphi_M(t)}. \tag{2.6}$$

2.1 Equivalent Complex Baseband Signals

The representation (2.5), (2.6) can be interpreted as follows: the *complex-valued* signal $s(t)$, which is denoted as the *equivalent complex baseband* (ECB) signal,[1] compactly represents the *amplitude* and *phase* of the *real-valued* RF signal $s_{RF}(t)$. Or, the amplitude $a_M(t)$ and phase $\varphi_M(t)$ of the RF signal are compactly combined into a complex signal $s(t)$, where $a_M(t)$ gives the amplitude and $\varphi_M(t)$ the phase.

This establishes the desired compact, mathematically equivalent representation of an RF signal independent of the reference (carrier) frequency f_0.

Comparison with Alternating Current Circuit Theory

Alternating current circuit theory is a very useful tool for the steady-state analysis of RLC networks. Thereby, a pure sinusoidal signal with frequency f_0, rms value z_{rms}, and phase φ

$$\sqrt{2} z_{\text{rms}} \cdot \cos(2\pi f_0 t + \varphi) = \sqrt{2} z_{\text{rms}} \cdot \text{Re}\{e^{j(2\pi f_0 t + \varphi)}\} = \sqrt{2} \cdot \text{Re}\{z_{\text{rms}} e^{j\varphi} \cdot e^{j2\pi f_0 t}\} \tag{2.7}$$

is completely characterized by the complex number (rms phasor)

$$z \stackrel{\text{def}}{=} z_{\text{rms}} e^{j\varphi} . \tag{2.8}$$

ECB signals equivalently characterize a real-valued (bandpass) signal $x_{RF}(t)$ by the complex-valued (low-pass) signal $x(t)$

$$x_{RF}(t) = \sqrt{2} \cdot \text{Re}\left\{x(t) \cdot e^{j2\pi f_0 t}\right\} . \tag{2.9}$$

Thus, ECB signals are nothing else than time-varying rms phasors.

Box 2.1 *Analytic Signal*

Let a *real-valued* signal $x(t) \in \mathbb{R}$ be given. The corresponding spectrum

$$x(t) \circ\!\!-\!\!\bullet X(f) \tag{2.10}$$

has the symmetry

$$X(f) = X^*(-f) . \tag{2.11}$$

The corresponding *analytic signal*, which represents the original signal $x(t)$ completely, is defined as

---

[1] The ECB representation of a real-valued signal is sometimes also referred to as its *complex envelope*. Here we avoid this notation because of the different definition of the envelope of a signal, see (2.43).

$$x^+(t) \circ\!\!\!-\!\!\bullet X^+(f) \stackrel{\text{def}}{=} (1 + \text{sgn}(f)) \cdot X(f) = \begin{cases} 2X(f), & f > 0 \\ 0, & f < 0 \end{cases}, \quad (2.12)$$

with

$$\text{sgn}(f) \stackrel{\text{def}}{=} \begin{cases} 1, & f > 0 \\ 0, & f = 0 \\ -1, & f < 0 \end{cases}.$$

Per definition, the analytic signal is complex-valued. Using the correspondence

$$\text{sgn}(f) \bullet\!\!\!-\!\!\circ \text{j}\frac{1}{\pi t} \quad (2.13)$$

of the Fourier transform, we have ($*$ denotes convolution)

$$\begin{aligned} X^+(f) &= (1 + \text{sgn}(f)) \cdot X(f) = X(f) + \text{sgn}(f) \cdot X(f) \\ \updownarrow & \\ x^+(t) &= x(t) + x(t) * \text{j}\frac{1}{\pi t} = x(t) + \text{j}\,\mathcal{H}\{x(t)\} \end{aligned} \quad (2.14)$$

with the *Hilbert transform*

$$\mathcal{H}\{y(t)\} \stackrel{\text{def}}{=} y(t) * \frac{1}{\pi t} = \frac{1}{\pi} \int_{-\infty}^{\infty} \frac{y(\tau)}{t - \tau} \, d\tau. \quad (2.15)$$

A block diagram for the generation of the analytic signal from the original signal is depicted in Figure 2.1.

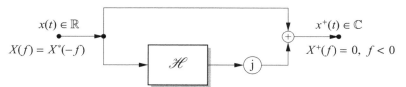

Figure 2.1 Generation of the analytic signal.

Note that the analytic signal $x^+(t)$ has twice the energy of $x(t)$, since

$$\int_0^\infty |2X(f)|^2 \, df = 2\int_{-\infty}^\infty |X(f)|^2 \, df = 2\int_{-\infty}^\infty x^2(t) \, dt = 2E_x. \quad (2.16)$$

2.1.2 ECB Transformation

The above connections between a physical, real-valued (RF) signal and a complex-valued baseband signal establish a *transformation* from the one representation to the other and vice versa.

Let $x_{\text{RF}}(t) \circ\!\!\!-\!\!\bullet X_{\text{RF}}(f)$ be the RF signal and its corresponding spectrum (Fourier transform) and let $x(t) \circ\!\!\!-\!\!\bullet X(f)$ be the ECB signal and its spectrum. In order to have a unique connection between both representations, we demand that (this will become apparent subsequently)

$$|X(f)| = 0, \qquad f < -f_0. \tag{2.17}$$

ECB to RF

Given the ECB signal, the RF signal is obtained as

$$x_{\text{RF}}(t) = \sqrt{2}\,\text{Re}\left\{x(t)\,e^{+j2\pi f_0 t}\right\} = \frac{1}{\sqrt{2}}\left(x(t)\,e^{+j2\pi f_0 t} + x^*(t)\,e^{-j2\pi f_0 t}\right)$$
$$\circ\!\!-\!\!\bullet \tag{2.18}$$
$$X_{\text{RF}}(f) = \frac{1}{\sqrt{2}}\left(X(f - f_0) + X^*(-(f + f_0))\right).$$

The relation for the spectra follows from the properties of the Fourier transform, in particular, $x(t)\,e^{+j2\pi f_0 t} \circ\!\!-\!\!\bullet X(f - f_0)$ and $x^*(t) \circ\!\!-\!\!\bullet X^*(-f)$.

The connection (2.18) can be interpreted as follows:

- modulate the signal – shift the spectrum $X(f)$ upwards by f_0;
- consider only the real part of the signal (suppress the imaginary part);
- scale the signal suitably.

A visualization of this procedure (ECB to RF) is given in Figure 2.2. Thereby, two different signals and their spectra are shown. The one spectrum is shown in light gray; it has a broad spectrum with support starting at $f = -f_0$. The other spectrum is shown in dark gray. It has a narrow spectrum; the spectral components concentrate around $f = 0$. For the light gray case, the spectrum of the RF signal may extend over the entire range of frequencies. The dark gray case corresponds to a bandpass signal; the frequency components of the RF signal are concentrated around $f = f_0$ (and $f = -f_0$).

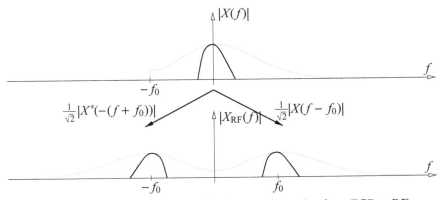

Figure 2.2 Visualization of the spectra for the transformation from ECB to RF domain.

RF to ECB

Since $|X(f)| = 0$, $f < -f_0$, has been presumed, the parts $X(f-f_0)$ and $X^*(-(f+f_0))$ in (2.18) have no spectral overlap. Thus

$$X_{\mathrm{RF}}(f) = \frac{1}{\sqrt{2}}(X(f-f_0) + X^*(-(f+f_0))),$$

which gives

$$\frac{1}{2}(1 + \mathrm{sgn}(f)) X_{\mathrm{RF}}(f) = \frac{1}{\sqrt{2}} X(f-f_0). \tag{2.19}$$

Solving this equation for $X(f)$ leads to the transformation. Given the RF signal, the ECB signal is obtained as

$$X(f) = \frac{1}{\sqrt{2}}(1 + \mathrm{sgn}(f+f_0)) X_{\mathrm{RF}}(f+f_0) = \frac{1}{\sqrt{2}} X_{\mathrm{RF}}^+(f+f_0)$$

$$\updownarrow \tag{2.20}$$

$$x(t) = \frac{1}{\sqrt{2}} x_{\mathrm{RF}}^+(t)\, e^{-\mathrm{j}2\pi f_0 t} = \frac{1}{\sqrt{2}} (x_{\mathrm{RF}}(t) + \mathrm{j}\,\mathscr{H}\{x_{\mathrm{RF}}(t)\})\, e^{-\mathrm{j}2\pi f_0 t}.$$

The connection (2.20) can be interpreted as follows:

- create the *analytic signal* $x_{\mathrm{RF}}^+(t)$ (one-sided spectrum, see Box 2.1);
- scale the signal suitably;
- shift the spectrum by the reference frequency f_0 downwards.

The cut-off of the spectrum at $f < 0$ for the analytic signal guarantees the cut-off of spectra of ECB signals for $f < -f_0$.

A visualization of this procedure (RF to ECB) is given in Figure 2.3. Again, two different spectra are shown; a broad spectrum (light gray) and a narrow one (dark

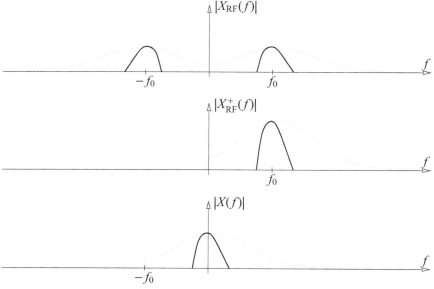

Figure 2.3 Visualization of the spectra for the transformation from RF to ECB domain.

gray). The suppression of the frequency components at $f < 0$ and the frequency translation are visible. In both cases the (relevant part of the) spectrum after the transformation is concentrated around $f = 0$, that is, in the baseband.

Finally, the generation of the ECB signal from the RF signal and vice versa is shown in the block diagrams in Figure 2.4.

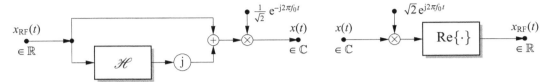

Figure 2.4 Block diagrams for the transformation from RF to ECB and ECB to RF domain.

The definition of the ECB transform is given formally as follows.

Definition 2.1 *Equivalent Complex Baseband (ECB) Signal*

Let the real-valued signal $x_{\mathrm{RF}}(t)$ with spectrum $X_{\mathrm{RF}}(f)$ be given. The corresponding *equivalent complex baseband signal* $x(t)$ (or *complex envelope*) with spectrum $X(f)$ with respect to the *transformation* or *reference frequency* f_0 is calculated as follows:

$$x(t) = \frac{1}{\sqrt{2}} x_{\mathrm{RF}}^+(t) \, e^{-j2\pi f_0 t} = \frac{1}{\sqrt{2}} \left(x_{\mathrm{RF}}(t) + j \mathcal{H}\{ x_{\mathrm{RF}}(t) \} \right) e^{-j2\pi f_0 t}$$

$$\updownarrow \qquad\qquad\qquad\qquad\qquad\qquad\qquad\qquad (2.21)$$

$$X(f) = \frac{1}{\sqrt{2}} X_{\mathrm{RF}}^+(f + f_0) = \frac{1}{\sqrt{2}} \left(1 + \mathrm{sgn}(f + f_0) \right) X_{\mathrm{RF}}(f + f_0) \,.$$

Given the ECB signal $x(t)$, the corresponding real-valued signal is recovered as

$$x_{\mathrm{RF}}(t) = \sqrt{2} \, \mathrm{Re}\left\{ x(t) \, e^{+j2\pi f_0 t} \right\} = \frac{1}{\sqrt{2}} \left(x(t) \, e^{+j2\pi f_0 t} + x^*(t) \, e^{-j2\pi f_0 t} \right)$$

$$\updownarrow \qquad\qquad\qquad\qquad\qquad\qquad\qquad\qquad (2.22)$$

$$X_{\mathrm{RF}}(f) = \frac{1}{\sqrt{2}} \left(X(f - f_0) + X^*(-(f + f_0)) \right) \,.$$

This transformation has the following properties

- The transformation $x_{\mathrm{RF}}(t) \leftrightarrow x(t)$ is linear.
  When transforming a linear combination of signals, the corresponding linear combination is obtained in the other domain.
- By definition, the spectrum of ECB signals is zero for $f < -f_0$.
- ECB signals are the most prominent example of complex-valued signals and their reasonable application. The ECB transform creates an image of a real physical signal in the domain of ECB signal representation.

- The ECB transformation here is defined using the factors $\frac{1}{\sqrt{2}}$ and $\sqrt{2}$, respectively. This guarantees that the real-valued signal $x_{\mathrm{RF}}(t)$ and its equivalent complex baseband signal $x(t)$ have the same energy, as

$$\int_{-\infty}^{\infty} x_{\mathrm{RF}}^2(t)\, \mathrm{d}t = \int_{-\infty}^{\infty} |X_{\mathrm{RF}}(f)|^2\, \mathrm{d}f = \int_{-\infty}^{\infty} |X(f)|^2\, \mathrm{d}f = \int_{-\infty}^{\infty} |x(t)|^2\, \mathrm{d}t \ .$$

Note that in the literature, other factors for normalization are used as well.

2.1.3 The Quadrature Components of ECB Signals

We now turn to an important characterization of ECB signals. As $x(t)$ (corresponding to the real-valued $x_{\mathrm{RF}}(t)$) is a *complex-valued* signal, it can either be given via *magnitude* and *phase* (polar form; see motivation above) or via its *real* and *imaginary part* (Cartesian form).

The complex-valued ECB signal in polar form reads

$$x(t) = |x(t)| \cdot \mathrm{e}^{\mathrm{j}\varphi_x(t)} \tag{2.23}$$

with[2]

$$|x(t)| = \sqrt{x_{\mathrm{I}}^2(t) + x_{\mathrm{Q}}^2(t)} \qquad \textit{magnitude} \text{ or } \textit{envelope}, \tag{2.24}$$

$$\varphi_x(t) = \arg\{x(t)\} \qquad \textit{phase}. \tag{2.25}$$

The envelope $|x(t)|$ is independent of the reference frequency f_0, see Box 2.2 later. Moreover, we have

$$x(t) = \mathrm{Re}\{x(t)\} + \mathrm{j}\,\mathrm{Im}\{x(t)\}$$
$$= x_{\mathrm{I}}(t) + \mathrm{j}\,x_{\mathrm{Q}}(t) \tag{2.26}$$

with

$$x_{\mathrm{I}}(t) \stackrel{\mathrm{def}}{=} \mathrm{Re}\{x(t)\} \qquad \textit{in-phase component}, \tag{2.27}$$

$$x_{\mathrm{Q}}(t) \stackrel{\mathrm{def}}{=} \mathrm{Im}\{x(t)\} \qquad \textit{quadrature component}. \tag{2.28}$$

$x_{\mathrm{I}}(t)$ and $x_{\mathrm{Q}}(t)$ together are denoted as the *quadrature components*.[3]

A complex-valued signal can be displayed in three-dimensional space; the time axis t and the in-phase and quadrature components $x_{\mathrm{I}}(t)$ and $x_{\mathrm{Q}}(t)$. When only looking at the plane defined by $x_{\mathrm{I}}(t)$ and $x_{\mathrm{Q}}(t)$ (i.e., the complex plane), the time axis is not visible anymore. The diagram of $x(t)$ in this (complex) I/Q plane is called a *phasor*, see Figure 2.5.

---

[2] For a remark on the arg function, see Appendix A, Equation (A.7).
[3] Note the different meaning of component and components in this context. Signals in the ECB domain always have to be represented by two real-valued signals, which have to be processed jointly.

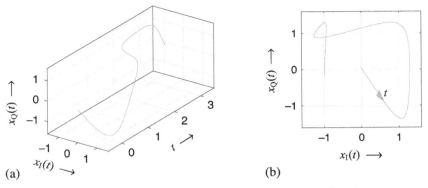

Figure 2.5 (a) Complex-valued signal over time. (b) Corresponding phasor.

Quadrature Components to RF Signal

Given the quadrature components $x_I(t)$ and $x_Q(t)$, the ECB signal $x(t)$ is obviously given by

$$x(t) = x_I(t) + j\, x_Q(t) \,. \tag{2.29}$$

Taking Euler's formula $e^{jx} = \cos(x) + j\sin(x)$ into account, the real-valued RF signal $x_{RF}(t)$ is obtained as

$$\begin{aligned} x_{RF}(t) &= \sqrt{2}\,\mathrm{Re}\{x(t)\,e^{j2\pi f_0 t}\} \\ &= \sqrt{2}\,\mathrm{Re}\{(x_I(t) + j\,x_Q(t))\cdot(\cos(2\pi f_0 t) + j\sin(2\pi f_0 t))\} \\ &= \sqrt{2}\,\bigl(x_I(t)\cdot\cos(2\pi f_0 t) - x_Q(t)\cdot\sin(2\pi f_0 t)\bigr) \,. \end{aligned} \tag{2.30}$$

This equation establishes the *quadrature upconversion* – given the quadrature components, the RF signal is generated. A block diagram of this upconversion (quadrature mixer) is shown in Figure 2.6.

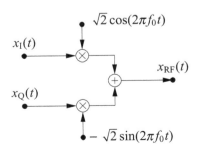

Figure 2.6 Quadrature upconversion.

RF Signal to Quadrature Components

Given the real-valued RF signal $x_{\text{RF}}(t)$, the generation of the quadrature components $x_\text{I}(t)$ and $x_\text{Q}(t)$ is more involved. Per definition, we have

$$x_{\substack{\text{I}\\\text{Q}}}(t) = \substack{\text{Re}\\\text{Im}}\{x(t)\} = \substack{\text{Re}\\\text{Im}}\left\{\frac{1}{\sqrt{2}} x_{\text{RF}}^+(t)\, e^{-j2\pi f_0 t}\right\} \tag{2.31}$$

$$= \frac{1}{\sqrt{2}} \substack{\text{Re}\\\text{Im}}\left\{(x_{\text{RF}}(t) + j\,\mathcal{H}\{x_{\text{RF}}(t)\})(\cos(2\pi f_0 t) - j\sin(2\pi f_0 t))\right\}, \tag{2.32}$$

where the upper index (I) corresponds to the real part (upper operation) and the lower index (Q) to the imaginary part (lower operation). Performing multiplication of the bracketed terms and separating real and imaginary parts, we obtain

$$x_\text{I}(t) = \frac{1}{\sqrt{2}}\left(x_{\text{RF}}(t)\cdot\cos(2\pi f_0 t) + \mathcal{H}\{x_{\text{RF}}(t)\}\cdot\sin(2\pi f_0 t)\right), \tag{2.33}$$

$$x_\text{Q}(t) = \frac{1}{\sqrt{2}}\left(\mathcal{H}\{x_{\text{RF}}(t)\}\cdot\cos(2\pi f_0 t) - x_{\text{RF}}(t)\cdot\sin(2\pi f_0 t)\right). \tag{2.34}$$

These equations establish the *quadrature downconversion* – given the RF signal, the quadrature components are generated. A block diagram of this downconversion is depicted in Figure 2.7.

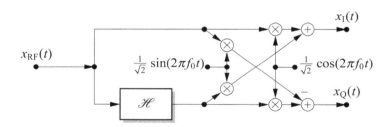

Figure 2.7 Quadrature downconversion (general form).

This form of the downconversion is valid for all RF signals (e.g., the light gray and dark gray spectra in Figures 2.2 and 2.3). However, in (digital) communications we usually deal with *bandpass signals* (dark gray spectrum). For this situation, the downconversion process can be *simplified*.

From (2.20) we have

$$X(f) = \frac{1}{\sqrt{2}}(1 + \text{sgn}(f + f_0))\cdot X_{\text{RF}}(f + f_0). \tag{2.35}$$

If $X_{\text{RF}}(f) = 0$, $|f| > 2f_0$, the (frequency-shifted) Hilbert transform can be replaced by an ideal low-pass filter with cut-off frequency f_0

$$h_{\text{LP}}(t)\ \circ\!\!-\!\!\bullet\ H_{\text{LP}}(f) = \text{rect}\left(\frac{f}{2f_0}\right) = \begin{cases}1, & |f| < f_0\\ 0, & \text{else}\end{cases}, \tag{2.36}$$

as
$$1 + \text{sgn}(f + f_0) = 2 \, \text{rect}\left(\tfrac{f}{2f_0}\right), \quad f < f_0. \tag{2.37}$$

This is visualized in Figure 2.8.

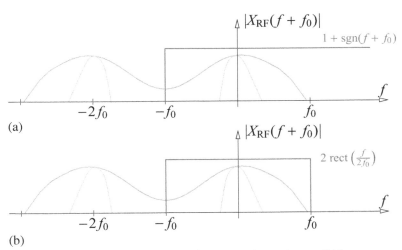

Figure 2.8 (a) (Frequency-shifted) Hilbert transformation vs. (b) low-pass filtering.

This leads to
$$\begin{aligned} X(f) &= \frac{1}{\sqrt{2}} \left(1 + \text{sgn}(f + f_0)\right) \cdot X_{\text{RF}}(f + f_0) \\ &= \sqrt{2} \, \text{rect}\left(\tfrac{f}{2f_0}\right) \cdot X_{\text{RF}}(f + f_0), \quad \text{for } X_{\text{RF}}(f) = 0, \, |f| > 2f_0 \end{aligned}$$
$$\updownarrow \tag{2.38}$$
$$x(t) = \sqrt{2} \, h_{\text{LP}}(t) * \left(x_{\text{RF}}(t) \cdot e^{-j2\pi f_0 t}\right).$$

In summary, given the real-valued RF signal $x_{\text{RF}}(t)$ with $X_{\text{RF}}(f) = 0$ for $|f| > 2f_0$, the quadrature components $x_{\text{I}}(t)$ and $x_{\text{Q}}(t)$ are obtained as

$$x_{\substack{\text{I}\\\text{Q}}}(t) = \substack{\text{Re}\\\text{Im}} \left\{ \sqrt{2} \left(x_{\text{RF}}(t) \cdot e^{-j2\pi f_0 t}\right) * h_{\text{LP}}(t) \right\},$$

thus
$$x_{\text{I}}(t) = \sqrt{2} \left(x_{\text{RF}}(t) \cdot \cos(2\pi f_0 t)\right) * h_{\text{LP}}(t), \tag{2.39}$$
$$x_{\text{Q}}(t) = -\sqrt{2} \left(x_{\text{RF}}(t) \cdot \sin(2\pi f_0 t)\right) * h_{\text{LP}}(t). \tag{2.40}$$

These equations establish the simplified version of quadrature downconversion, valid only for RF signals with $X_{\text{RF}}(f) = 0$, for $|f| > 2f_0$, that is, in particular for bandpass signals where the bandwidth is smaller than twice the center frequency f_0 (i.e., for almost all RF signals in practice). A block diagram of this downconversion is shown in Figure 2.9.

Note the symmetry of the structures for up- and downconversion, Figures 2.6 and 2.9, in this simplified version.

38   2 Equivalent Complex Baseband Signals

Figure 2.9 Quadrature downconversion (simplified form for bandpass signals).

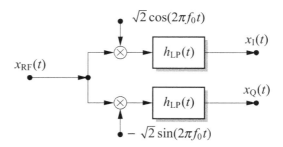

Box 2.2 Envelope of a Signal

Let the RF signal

$$s_{\mathrm{RF}}(t) = \sqrt{2}\,\mathrm{Re}\{s(t)\,e^{j2\pi f_c t}\} \quad (2.41)$$
$$= s_{\mathrm{I}}(t)\,c_{\mathrm{c}}(t) - s_{\mathrm{Q}}(t)\,c_{\mathrm{s}}(t) \quad (2.42)$$

with $\quad c_{\mathrm{c}}(t) \stackrel{\text{def}}{=} \sqrt{2}\cos(2\pi f_c t)$

and $\quad c_{\mathrm{s}}(t) \stackrel{\text{def}}{=} \sqrt{2}\sin(2\pi f_c t)$

be given. The *envelope* of the RF signal is defined as the magnitude of the corresponding ECB signal, that is

$$\mathrm{env}\{s_{\mathrm{RF}}(t)\} \stackrel{\text{def}}{=} |s(t)|\ . \quad (2.43)$$

Both real-valued modulating baseband signals, $s_{\mathrm{I}}(t)$ and $s_{\mathrm{Q}}(t)$, are assumed to have the (two-sided) bandwidth B_{IQ}. As long as $f_c > B_{\mathrm{IQ}}/2$, we have, from the ECB transform (the *Hilbert transform* of the signal $x(t)$ is denoted as $\mathscr{H}\{x(t)\}$, cf. Appendix A)

$$s(t) = s_{\mathrm{I}}(t) + j\,s_{\mathrm{Q}}(t) = \frac{1}{\sqrt{2}}\left(s_{\mathrm{RF}}(t) + j\,\mathscr{H}\{s_{\mathrm{RF}}(t)\}\right) e^{-j2\pi f_c t} \quad (2.44)$$

and thus

$$|s(t)|^2 = s_{\mathrm{I}}^2(t) + s_{\mathrm{Q}}^2(t) = \frac{1}{2}|s_{\mathrm{RF}}(t) + j\,\mathscr{H}\{s_{\mathrm{RF}}(t)\}|^2\,|e^{-j2\pi f_c t}|^2$$
$$= \frac{1}{2}\left(s_{\mathrm{RF}}^2(t) + \mathscr{H}^2\{s_{\mathrm{RF}}(t)\}\right)\ . \quad (2.45)$$

Note that the envelope of a signal is independent of the reference frequency (carrier frequency) used in the ECB transform.

Let $\mathcal{L}\{\cdot\}$ denote low-pass filtering (cut-off frequency $B_{\mathrm{RF}}/2$). Then

$$\mathcal{L}\{s_{\mathrm{RF}}^2(t)\} = \mathcal{L}\{(s_{\mathrm{I}}(t)\,c_{\mathrm{c}}(t) - s_{\mathrm{Q}}(t)\,c_{\mathrm{s}}(t))^2\}$$
$$= \mathcal{L}\{s_{\mathrm{I}}^2(t)\,c_{\mathrm{c}}^2(t) + s_{\mathrm{Q}}^2(t)\,c_{\mathrm{s}}^2(t) - 2s_{\mathrm{I}}(t)\,c_{\mathrm{c}}(t)\,s_{\mathrm{Q}}(t)\,c_{\mathrm{s}}(t)\}$$

with $c_c^2(t) = 1 + \cos(2\pi 2 f_c t)$, $c_s^2(t) = 1 - \cos(2\pi 2 f_c t)$ and $c_c(t)c_s(t) = \sin(2\pi 2 f_c t)$

$$\begin{aligned}
&= \mathcal{L}\{s_I^2(t) + s_Q^2(t) + \text{higher-frequency terms}\} \\
&= s_I^2(t) + s_Q^2(t) \\
&= |s(t)|^2 \,.
\end{aligned} \quad (2.46)$$

This means that the squared envelope can be generated directly from the RF signal by squaring and low-pass filtering (envelope detection).

2.1.4 Up/Downconversion with Frequency and Phase Errors

Figure 2.10 shows the transmitter-side upconversion in cascade with the receiver-side downconversion; the noise-free case is depicted.

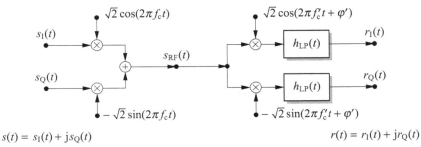

Figure 2.10 Frequency and phase mismatch in up- and downconversion (noise-free channel).

Thereby, the quadrature upconversion according to Figure 2.6 and the simplified quadrature downconversion (valid for bandpass signals) according to Figure 2.9 are employed. At the transmitter side, the carrier frequency f_c is utilized for upconversion and the phase is zero. In contrast, due to non-perfect synchronization at the receiver side, the frequency f_c' is used for downconversion and a phase shift φ' is present.

ECB Model of Frequency and Phase Errors

The ECB receive signal $r(t)$ is given by

$$\begin{aligned}
r(t) &= \mathcal{L}_{\text{ow-pass}}\{s_{\text{RF}}(t) \cdot (\sqrt{2}\cos(2\pi f_c't + \varphi'))\} \\
&\quad + j\,\mathcal{L}_{\text{ow-pass}}\{s_{\text{RF}}(t) \cdot (-\sqrt{2}\sin(2\pi f_c't + \varphi'))\} \\
&= \mathcal{L}_{\text{ow-pass}}\{s_{\text{RF}}(t) \cdot \sqrt{2}e^{-j(2\pi f_c't + \varphi')}\} \\
&= \mathcal{L}_{\text{ow-pass}}\{\sqrt{2}\text{Re}\{s(t)e^{j2\pi f_c t}\} \cdot \sqrt{2}e^{-j(2\pi f_c't + \varphi')}\} \\
&= 2\,\mathcal{L}_{\text{ow-pass}}\{(s_I(t)\cos(2\pi f_c t) - s_Q(t)\sin(2\pi f_c t)) \\
&\quad \cdot (\cos(2\pi f_c't + \varphi') - j\sin(2\pi f_c't + \varphi'))\} \,.
\end{aligned}$$

Using $2\sin(x)\sin(y) = \cos(x-y) - \cos(x+y)$, $2\sin(x)\cos(y) = \sin(x-y) + \sin(x+y)$, and $2\cos(x)\cos(y) = \cos(x-y) + \cos(x+y)$, the definitions $f_\Delta \stackrel{\text{def}}{=} f_c - f_c'$ and $\varphi_\Delta \stackrel{\text{def}}{=} 0 - \varphi'$, as well as the fact that the low-pass filters suppress all frequency components except those in the baseband, we have

$$r(t) = s_I(t) \left(\cos(2\pi f_\Delta t + \varphi_\Delta) + j \sin(2\pi f_\Delta t + \varphi_\Delta) \right)$$
$$+ j s_Q(t) \left(\cos(2\pi f_\Delta t + \varphi_\Delta) + j \sin(2\pi f_\Delta t + \varphi_\Delta) \right)$$
$$= (s_I(t) + j s_Q(t)) \cdot e^{j(2\pi f_\Delta t + \varphi_\Delta)}. \qquad (2.47)$$

Thus, in case of a frequency/phase mismatch between transmitter and receiver, the ECB receive signal is related to the ECB transmit signal by

$$r(t) = s(t) \cdot e^{j(2\pi f_\Delta t + \varphi_\Delta)}, \quad \text{with} \quad \begin{array}{l} \text{frequency error} \quad f_\Delta \stackrel{\text{def}}{=} f_c - f_c' \\ \text{phase error} \quad \varphi_\Delta \stackrel{\text{def}}{=} 0 - \varphi' \end{array}. \qquad (2.48)$$

Looking at (2.48), we see that a frequency error f_Δ causes a multiplication of the transmit signal by $e^{j2\pi f_\Delta t}$. The effect is a *time-dependent* rotation of the receive signal. Visualizing the phasor in three-dimensional space, this corresponds to a twisting or torsion of the signal. If f_Δ is positive, that is, the TX upconversion frequency the is larger than the RX downconversion frequency, a rotation in the mathematically positive the direction (counterclockwise) remains; otherwise ($f_\Delta < 0$), a rotation in the mathematically negative direction (clockwise) is present.

A phase error φ_Δ causes a multiplication of the transmit signal by $e^{j\varphi_\Delta}$. This corresponds to a *time-independent rotation*, a tilting of the phasor.

2.2 LTI Systems and ECB Signals

As explained in Chapter 1, channels are typically modeled as linear time-invariant (LTI) systems. Hence, the filtering of signals by LTI systems is of interest. Since we transmit RF signals, the channel acts on these RF signals. However, for processing at the receiver, that is, after downconversion, a (channel) description in the ECB domain is required.

2.2.1 Equivalent Transfer Function

To that end, we assume a real-valued LTI system with impulse response $h_{\text{RF}}(t) \in \mathbb{R}$ or corresponding transfer function $H_{\text{RF}}(f) = \mathscr{F}\{h_{\text{RF}}(t)\}$. For real-valued (bandpass) signals we have the following relations:

$$y_{\text{RF}}(t) = x_{\text{RF}}(t) * h_{\text{RF}}(t)$$
$$\updownarrow \qquad (2.49)$$
$$Y_{\text{RF}}(f) = X_{\text{RF}}(f) \cdot H_{\text{RF}}(f),$$

where $x_{\text{RF}}(t)$ is the RF input signal and $y_{\text{RF}}(t)$ the corresponding (noise-free) system output signal.

Now, the RF signals are represented by their ECB signals $x(t)$ and $y(t)$, respectively. We want to have a characterization of the system by an impulse response or a transfer function $h(t) \circ\!\!-\!\!\bullet\, H(f)$ in the ECB domain, such that the corresponding relations to (2.49) are valid, that is, the following should hold:

$$y(t) = x(t) * h(t)$$
$$\circ\!\!-\!\!\bullet \qquad (2.50)$$
$$Y(f) = X(f) \cdot H(f) \,.$$

Such an equivalent ECB system is obtained by the following considerations. We have

$$X_{\text{RF}}(f) = \frac{1}{\sqrt{2}} \Big(X(f - f_0) + X^*(-(f + f_0)) \Big) \qquad (2.51)$$

and

$$Y_{\text{RF}}(f) = \frac{1}{\sqrt{2}} \Big(Y(f - f_0) + Y^*(-(f + f_0)) \Big) \,. \qquad (2.52)$$

Assume that the same (except for the scaling c_H) holds for systems as well. Then we have

$$H_{\text{RF}}(f) = c_H \Big(H(f - f_0) + H^*(-(f + f_0)) \Big) \,. \qquad (2.53)$$

Using this ansatz, and considering that the spectra of ECB signals vanish for $f < -f_0$ (hence, the spectrum $X(f - f_0)$ vanishes for $f < 0$), we arrive at

$$Y_{\text{RF}}(f) \stackrel{!}{=} X_{\text{RF}}(f) \cdot H_{\text{RF}}(f) \qquad (2.54)$$
$$= \frac{1}{\sqrt{2}} \Big(X(f - f_0) + X^*(-(f + f_0)) \Big) \cdot c_H \Big(H(f - f_0) + H^*(-(f + f_0)) \Big)$$
$$= \frac{c_H}{\sqrt{2}} \Big(X(f - f_0) \cdot H(f - f_0) + X^*(-(f + f_0)) \cdot H^*(-(f + f_0))$$
$$+ \underbrace{X(f - f_0) \cdot H^*(-(f + f_0))}_{=0,\ f<0\ \ \ \ =0,\ f>0} + \underbrace{X^*(-(f + f_0)) \cdot H(f - f_0)}_{=0,\ f>0\ \ \ \ =0,\ f<0} \Big)$$
$$= \frac{c_H}{\sqrt{2}} \Big(Y(f - f_0) + Y^*(-(f + f_0)) \Big) \,.$$

Comparing with (2.52), this demand is obviously fulfilled for $c_H = 1$.

Hence, we can state the definition of ECB *systems* (note that the same procedure as for *signals*, except for the scaling, is present).

Definition 2.2 *Equivalent Complex Baseband (ECB) Systems*

Let the real-valued system $h_{\text{RF}}(t) \circ\!\!-\!\!\bullet\, H_{\text{RF}}(f)$ be given. The corresponding *equivalent complex baseband system* $h(t)$ with transfer function $H(f)$ is calculated as follows (transformation frequency f_0):

$$h(t) = \frac{1}{2}\left(h_{\mathrm{RF}}(t) + \mathrm{j}\,\mathscr{H}\{h_{\mathrm{RF}}(t)\}\right) \mathrm{e}^{-\mathrm{j}2\pi f_0 t}$$

$$\updownarrow \qquad\qquad (2.55)$$

$$H(f) = \frac{1}{2}\left(1 + \mathrm{sgn}(f + f_0)\right) H_{\mathrm{RF}}(f + f_0)\,.$$

Given the ECB system $h(t)$, the corresponding real-valued system is recovered as

$$h_{\mathrm{RF}}(t) = 2\,\mathrm{Re}\left\{h(t)\,\mathrm{e}^{+\mathrm{j}2\pi f_0 t}\right\}$$

$$\updownarrow \qquad\qquad (2.56)$$

$$H_{\mathrm{RF}}(f) = H(f - f_0) + H^*(-(f + f_0))\,.$$

Caution: Other scaling factors (1/2 and 2) as for signals!

2.2.2 ECB Systems and Quadrature Components

Similar to ECB signals, the quadrature components of systems, that is, the real and imaginary parts of the impulse response, are of interest. We have

$$h(t) = h_{\mathrm{I}}(t) + \mathrm{j}\,h_{\mathrm{Q}}(t) \qquad (2.57)$$

with

$$h_{\mathrm{I}}(t) \stackrel{\mathrm{def}}{=} \mathrm{Re}\{h(t)\}\,, \qquad (2.58)$$
$$h_{\mathrm{Q}}(t) \stackrel{\mathrm{def}}{=} \mathrm{Im}\{h(t)\}\,. \qquad (2.59)$$

Since

$$\begin{aligned} y(t) &= x(t) * h(t) \\ &= \left(x_{\mathrm{I}}(t) + \mathrm{j}\,x_{\mathrm{Q}}(t)\right) * \left(h_{\mathrm{I}}(t) + \mathrm{j}\,h_{\mathrm{Q}}(t)\right) \\ &= \underbrace{\left(x_{\mathrm{I}}(t) * h_{\mathrm{I}}(t) - x_{\mathrm{Q}}(t) * h_{\mathrm{Q}}(t)\right)}_{y_{\mathrm{I}}(t)} + \mathrm{j}\underbrace{\left(x_{\mathrm{I}}(t) * h_{\mathrm{Q}}(t) + x_{\mathrm{Q}}(t) * h_{\mathrm{I}}(t)\right)}_{y_{\mathrm{Q}}(t)}, \quad (2.60) \end{aligned}$$

an ECB LTI system with complex-valued impulse response $h(t)$ incorporates four real-valued LTI systems with impulse responses $h_{\mathrm{I}}(t)$ and $h_{\mathrm{Q}}(t)$, respectively. This is visualized in Figure 2.11.

For other operations than linear filtering and up/downconversion, the correspondences between real-valued RF signals and ECB signals are not immediate. In particular, the ECB equivalent of a product of real-valued RF signals is not the product of the corresponding ECB signals!

Finally, we note that the definition of ECB systems allows some degree of freedom. As, per definition, the spectrum $X(f)$ of any ECB signal vanishes for $f < -f_0$,

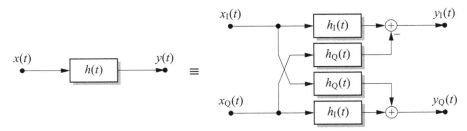

Figure 2.11 ECB LTI system and real-valued description (quadrature components).

the restriction $H(f) = 0$, $f < -f_0$, is not required for ECB systems (this part of the transfer function is irrelevant). In some situations, it is more convenient to define a *substitute system*, specifically

$$\tilde{H}(f) = H_{\text{RF}}(f + f_0) \,. \tag{2.61}$$

This system leads to exactly the same output signal $y(t)$ as the actual ECB system $H(f) = \frac{1}{2}(1 + \text{sgn}(f + f_0))H_{\text{RF}}(f + f_0)$. This fact can conveniently be used to find simpler ECB descriptions. This is shown in the following (important) example.

Example 2.2 Delay Element

As shown in Chapter 1, in mobile communications, delayed versions of signals arrive at the receiver. Hence, a delay element is an important characteristic to represent a channel.

The input/output relation of a delay element (delay t_0) with respect to the real-valued signals is given by

$$h_{\text{RF}}(t) = \delta(t - t_0)$$
$$\updownarrow$$
$$H_{\text{RF}}(f) = e^{-j2\pi f t_0} \,.$$

Using the definition of ECB systems, this leads to the intricate expression

$$H(f) = \frac{1}{2}(1 + \text{sgn}(f + f_0)) \cdot e^{-j2\pi(f+f_0)t_0}$$
$$\updownarrow$$
$$h(t) = \frac{1}{2}\left(\delta(t - t_0) + \frac{j}{\pi(t - t_0)}\right) \cdot e^{-j2\pi f_0 t_0} \,.$$

However, a simple substitute system can be given by

$$\tilde{H}(f) = H_{\text{RF}}(f + f_0) = e^{-j2\pi(f+f_0)t_0}$$
$$\updownarrow$$
$$\tilde{h}(t) = \mathcal{F}^{-1}\{\tilde{H}(f)\} = \delta(t - t_0) \cdot e^{-j2\pi f_0 t_0} \,.$$

The last equation shows that a delay acting on the RF signal causes a *delay* (t_0) and a *phase offset* ($2\pi f_0 t_0$; the delay t_0 translates into this phase shift compared to the reference oscillation) on the ECB signal.

2.3 Stochastic Processes in the Equivalent Complex Baseband

Let a (wide-sense) stationary real-valued stochastic process $x_{RF}(t)$ with *mean* $m_{x_{RF}}$ and *autocorrelation function (acf)* $\phi_{x_{RF}x_{RF}}(\tau)$ be given. We are interested in the parameters m_x and $\phi_{xx}(\tau)$ of the corresponding ECB process $x(t)$. To that end, we first have to look at how $x(t)$ is obtained from $x_{RF}(t)$. This can be done by considering Figure 2.12, which repeats Figure 2.4. Each sample function of $x_{RF}(t)$ corresponds to a sample function of $x(t)$. By doing the transformation (conceptionally) for all sample functions, the ECB process is obtained. Hence, we can write (the same way as for signals)

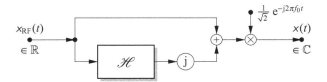

Figure 2.12 ECB transformation (RF to ECB) of a stochastic process $x_{RF}(t)$.

$$x(t) = \frac{1}{\sqrt{2}} \left(x_{RF}(t) + j\mathcal{H}\{x_{RF}(t)\} \right) e^{-j2\pi f_0 t} \qquad (2.62)$$

$$= \frac{1}{\sqrt{2}} \left(\chi(t) + j\tilde{\chi}(t) \right) e^{-j2\pi f_0 t}, \qquad (2.63)$$

with the definition for a compact notation

$$\chi(t) \stackrel{\text{def}}{=} x_{RF}(t), \qquad (2.64)$$

$$\tilde{\chi}(t) \stackrel{\text{def}}{=} \mathcal{H}\{x_{RF}(t)\}. \qquad (2.65)$$

From $x(t)$, the mean and autocorrelation function/power spectral density can be calculated and the respective consequences can be derived.

Mean. As the Hilbert transform eliminates any mean ($H_{\mathcal{H}}(f = 0) = -j\,\text{sgn}(f = 0) = 0$), the mean of the ECB process calculates as

$$m_x(t) \stackrel{\text{def}}{=} \mathrm{E}\{x(t)\} = \frac{1}{\sqrt{2}} \left(\mathrm{E}\{x_{RF}(t)\} + j\mathrm{E}\{\mathcal{H}\{x_{RF}(t)\}\} \right) e^{-j2\pi f_0 t}$$

$$= \frac{1}{\sqrt{2}} m_{x_{RF}} \cdot e^{-j2\pi f_0 t}. \qquad (2.66)$$

As can be seen from (2.66), in general, the mean $m_x(t)$ is a function of time. Only if $m_{x_{RF}} = 0$ do we have $m_x = 0$ (for all times). **Hence, the ECB process can only be stationary if the RF process is zero mean.**

2.3 Stochastic Processes in the Equivalent Complex Baseband   45

Autocorrelation function. The acf of the ECB process is calculated as follows:

$$\phi_{xx}(\tau) \stackrel{\text{def}}{=} \mathrm{E}\{x(t+\tau) \cdot x^*(t)\}$$

$$= \frac{1}{2}\mathrm{E}\{(\chi(t+\tau) + \mathrm{j}\tilde{\chi}(t+\tau)) \cdot (\chi(t) - \mathrm{j}\tilde{\chi}(t))\} \cdot \mathrm{e}^{-\mathrm{j}2\pi f_0(t+\tau)} \cdot \mathrm{e}^{+\mathrm{j}2\pi f_0 t}$$

$$= \frac{1}{2}(\phi_{\chi\chi}(\tau) + \phi_{\tilde{\chi}\tilde{\chi}}(\tau) + \mathrm{j}\phi_{\tilde{\chi}\chi}(\tau) - \mathrm{j}\phi_{\chi\tilde{\chi}}(\tau))\,\mathrm{e}^{-\mathrm{j}2\pi f_0 \tau} \;. \tag{2.67}$$

Since $\tilde{\chi}(t) = \mathcal{H}\{\chi(t)\}$, we have for its power spectral density (psd):

(A) $\Phi_{\tilde{\chi}\tilde{\chi}}(f) = \Phi_{\chi\chi}(f) \cdot |-\mathrm{j} \cdot \mathrm{sgn}(f)|^2 = \Phi_{\chi\chi}(f)$

and thus

$$\phi_{\chi\chi}(\tau) = \phi_{\tilde{\chi}\tilde{\chi}}(\tau) \;; \tag{2.68}$$

(B) $\Phi_{\chi\tilde{\chi}}(f) = \Phi_{\chi\chi}(f) \cdot (-\mathrm{j} \cdot \mathrm{sgn}(f))^* = \mathrm{j}\Phi_{\chi\chi}(f)\,\mathrm{sgn}(f)$

which is purely imaginary and has symmetry $\Phi_{\chi\tilde{\chi}}(-f) = -\Phi_{\chi\tilde{\chi}}(f)$, thus

$$\phi_{\chi\tilde{\chi}}(\tau) = -\phi_{\chi\tilde{\chi}}(-\tau) = -\phi_{\tilde{\chi}\chi}(\tau) \;; \tag{2.69}$$

(C) $\mathcal{H}\{\phi_{\chi\chi}(\tau)\} \mathrel{\circ\!\!-\!\!\bullet} \Phi_{\chi\chi}(f) \cdot (-\mathrm{j} \cdot \mathrm{sgn}(f)) = \Phi_{\tilde{\chi}\chi}(f)$

so

$$\mathcal{H}\{\phi_{\chi\chi}(\tau)\} = \phi_{\tilde{\chi}\chi}(\tau) \;. \tag{2.70}$$

Combining all results, we arrive at

$$\phi_{xx}(\tau) = \frac{1}{2}(\phi_{\chi\chi}(\tau) + \phi_{\tilde{\chi}\tilde{\chi}}(\tau) + \mathrm{j}\phi_{\tilde{\chi}\chi}(\tau) - \mathrm{j}\phi_{\chi\tilde{\chi}}(\tau))\,\mathrm{e}^{-\mathrm{j}2\pi f_0 \tau}$$

$$\stackrel{A,\,B}{=} 1\,(\phi_{\chi\chi}(\tau) + \mathrm{j}\phi_{\tilde{\chi}\chi}(\tau))\,\mathrm{e}^{-\mathrm{j}2\pi f_0 \tau}$$

$$\stackrel{C}{=} 1\,(\phi_{\chi\chi}(\tau) + \mathrm{j}\mathcal{H}\{\phi_{\chi\chi}(\tau)\})\,\mathrm{e}^{-\mathrm{j}2\pi f_0 \tau}$$

$$= 1\,(\phi_{x_{\mathrm{RF}}x_{\mathrm{RF}}}(\tau) + \mathrm{j}\mathcal{H}\{\phi_{x_{\mathrm{RF}}x_{\mathrm{RF}}}(\tau)\})\,\mathrm{e}^{-\mathrm{j}2\pi f_0 \tau}$$

and thus for the psd

$$\Phi_{xx}(f) = 1\,(1 + \mathrm{sgn}(f+f_0))\Phi_{x_{\mathrm{RF}}x_{\mathrm{RF}}}(f+f_0) \;. \tag{2.71}$$

Hence, we can define the ECB transformation of the acf/psd as follows.

Definition 2.3 *Equivalent Complex Baseband (ECB) Random Processes*

Let the real-valued wide-sense stationary random process $x_{\mathrm{RF}}(t)$ with zero mean and autocorrelation function $\phi_{x_{\mathrm{RF}}x_{\mathrm{RF}}}(\tau)$ or power spectral density

$\Phi_{x_{RF}x_{RF}}(f)$ be given. The corresponding *equivalent complex baseband random process* $x(t)$ has acf and psd

$$\phi_{xx}(\tau) = 1\,(\phi_{x_{RF}x_{RF}}(\tau) + j\,\mathcal{H}\{\phi_{x_{RF}x_{RF}}(\tau)\})\,e^{-j2\pi f_0 \tau}$$

$$\circ\!\!-\!\!\bullet \qquad\qquad\qquad (2.72)$$

$$\Phi_{xx}(f) = 1\,(1 + \operatorname{sgn}(f+f_0))\,\Phi_{x_{RF}x_{RF}}(f+f_0)\,.$$

Given the ECB process $x(t)$, the corresponding real-valued process $x_{RF}(t)$ has acf and psd

$$\phi_{x_{RF}x_{RF}}(\tau) = 1\,\operatorname{Re}\{\phi_{xx}(\tau)\,e^{+j2\pi f_0 \tau}\}$$

$$\circ\!\!-\!\!\bullet \qquad\qquad\qquad (2.73)$$

$$\Phi_{x_{RF}x_{RF}}(f) = \frac{1}{2}\,(\Phi_{xx}(f-f_0) + \Phi_{xx}(-(f+f_0)))\,.$$

Caution: Other scaling factors (1 and 1/2) as for signals or systems.[4]

The transformation of a real-valued, wide-sense stationary RF process gives the corresponding ECB process directly with its specific properties. Conversely, there are properties an ECB process has to meet, such that the (inverse) transformation results in an RF process that is real-valued and wide-sense stationary. Only then is the complex-valued process a valid representation of a stationary real-valued RF process.

Looking at the derivations above, this is summarized as follows. A complex-valued wide-sense stationary process $x(t)$ is only equivalent to a real-valued wide-sense stationary process $x_{RF}(t)$ if:

- the mean is zero

$$m_x \equiv 0\,; \qquad\qquad (2.74)$$

- the real and imaginary parts have the same acf

$$\phi_{x_I x_I}(\tau) = \phi_{x_Q x_Q}(\tau)\,; \qquad\qquad (2.75)$$

- the ccf between the real and the imaginary part is an odd function

$$\phi_{x_I x_Q}(\tau) = -\phi_{x_I x_Q}(-\tau)\,. \qquad\qquad (2.76)$$

In summary, these constraints (derived by T. Grettenberg in 1965) guarantee that the complex process is *rotationally invariant*, that is, $x(t)$ and $x(t)e^{j\alpha}$, $\alpha \in \mathbb{R}$, have the same statistical properties with respect to first- and second-order moments.

---

[4] Note that the scaling factors $\sqrt{2}$ or $1/\sqrt{2}$, respectively, in the ECB transform of *signals*, which aimed to conserve the signal power, now lead to the factor 1 for the acf, which represents *power*. Demanding a power conservation at all transformations (albeit other approaches are present in the literature), the use of three different scaling factors for signals, systems, and acf/psd is inevitable.

2.3.1 ECB Model of White Noise

An important noise model is *white noise*. A real-valued, zero-mean stationary process $n(t)$ is white if the psd is constant over all frequencies (cf. Section 1.3), that is,

$$\Phi_{nn}(f) = \frac{N_0}{2} \quad \forall f \in \mathbb{R}, \qquad \phi_{nn}(\tau) = \frac{N_0}{2} \delta(\tau) \qquad (2.77)$$

(cf. Figure 2.13). Note that we demand the psd be constant; the pdf can be arbitrary. If, additionally, the pdf is Gaussian, we have white Gaussian noise, which is the most important noise model.

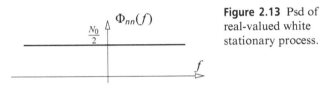

Figure 2.13 Psd of a real-valued white stationary process.

For historical reasons (use of a one-sided spectra), we have

N_0:   *one*-sided psd (historical definition for the real-valued process)
$N_0/2$:   *two*-sided psd.

As already stated in Chapter 1, white noise does not exist, as its power would be infinite; $\sigma_n^2 \to \infty$. However, it is a very useful model; in particular, it is the most random noise among all types of noise.

Contrary to continuous-time white noise, *discrete-time* (zero-mean) white noise $n[k]$ exists and is well defined by

$$\phi_{nn}[\kappa] = \sigma_n^2 \delta[\kappa], \qquad \Phi_{nn}\left(e^{j2\pi fT}\right) = \sigma_n^2 \qquad (2.78)$$

with finite variance σ_n^2.

Let a real-valued (physical) white-noise process $n_{\text{RF}}(t)$ with psd $\Phi_{n_{\text{RF}} n_{\text{RF}}}(f) = N_0/2$ be given. The corresponding ECB model is

$$\tilde{n}(t) = \tilde{n}_{\text{I}}(t) + j\,\tilde{n}_{\text{Q}}(t) \qquad (2.79)$$

with $\qquad \Phi_{\tilde{n}\tilde{n}}(f) = \frac{N_0}{2}(1 + \text{sgn}(f + f_0)) = \begin{cases} N_0, & f \geq -f_0 \\ 0, & f < -f_0 \end{cases}. \qquad (2.80)$

The psd is visualized in Figure 2.14. As the psd is not constant over *all* frequencies, this noise model is not white anymore.

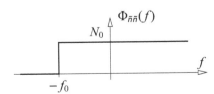

Figure 2.14 Psd of the ECB model corresponding to white RF noise.

However, similar to systems, we may define a substitute model that has the same effect. Remember that white noise is only a useful *model* and some band-limiting receiver input filter $H_{\text{RF}}(f)$ always has to be present to limit the noise power. For the corresponding ECB transfer function, we have

$$H(f) = 0 \qquad \text{for } f < -f_0 \,. \tag{2.81}$$

Hence, we can *define* a substitute random process where the psd has an arbitrary value for $f < -f_0$. In particular, we may define the ECB model $n(t)$ corresponding to white noise as

$$\Phi_{nn}(f) = N_0 \quad \forall f, \qquad \phi_{nn}(\tau) = N_0 \delta(\tau) \,. \tag{2.82}$$

This psd is visualized in Figure 2.15.

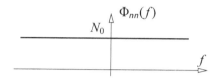

Figure 2.15 Psd of the ECB substitute model corresponding to white RF noise.

Since we demand rotational invariance, we define, for the (zero-mean) quadrature components of $n(t) = n_{\text{I}}(t) + j\, n_{\text{Q}}(t)$

$$\phi_{n_{\text{I}} n_{\text{I}}}(\tau) = \phi_{n_{\text{Q}} n_{\text{Q}}}(\tau) = \frac{N_0}{2}\delta(\tau) \,, \tag{2.83}$$

$$\Phi_{n_{\text{I}} n_{\text{I}}}(f) = \Phi_{n_{\text{Q}} n_{\text{Q}}}(f) = \frac{N_0}{2} \,, \tag{2.84}$$

$$\phi_{n_{\text{I}} n_{\text{Q}}}(\tau) = 0 \,. \tag{2.85}$$

2.3.2 Filtering of White Noise in the ECB Domain

Finally, we consider the filtering of white complex-valued stationary noise $n(t)$. It is the input to an ECB system with impulse response/transfer function (cf. Figure 2.16, which repeats Figure 2.11)

$$h(t) = h_{\text{I}}(t) + j\, h_{\text{Q}}(t) \circ\!\!-\!\!\bullet H(f) \,. \tag{2.86}$$

If the input process $n(t)$ has parameters

$$\Phi_{nn}(f) = N_0 \,, \tag{2.87}$$

$$\phi_{n_{\text{I}} n_{\text{I}}}(\tau) = \phi_{n_{\text{Q}} n_{\text{Q}}}(\tau) = \frac{N_0}{2}\delta(\tau) \,, \qquad \phi_{n_{\text{I}} n_{\text{Q}}}(\tau) = 0 \,, \tag{2.88}$$

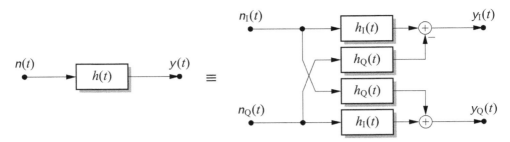

Figure 2.16 Filtering of white complex-valued stationary noise.

the output process $y(t) = n(t) * h(t)$ has parameters

$$\Phi_{yy}(f) = N_0 |H(f)|^2, \tag{2.89}$$

$$\phi_{y_I y_I}(\tau) = \phi_{y_Q y_Q}(\tau) = \frac{N_0}{2} \left(\varphi_{h_I h_I}(\tau) + \varphi_{h_Q h_Q}(\tau) \right), \tag{2.90}$$

$$\phi_{y_I y_Q}(\tau) = \frac{N_0}{2} \left(\varphi_{h_I h_Q}(\tau) - \varphi_{h_I h_Q}(-\tau) \right), \tag{2.91}$$

with $\varphi_{h_\square h_\blacksquare}(\tau) = h_\square(\tau) * h_\blacksquare^*(-\tau)$, $\square, \blacksquare \in \{I, Q\}$.

Since the input process is assumed to be rotationally invariant, the output process is stationary and rotationally invariant, too. If the input process is complex Gaussian, the output is also complex Gaussian.

2.4 Problems

2.4.1 Equivalent Complex Baseband

Let the radio-frequency signal

$$x_{\mathrm{RF}}(t) = \sqrt{2} \cos(2\pi f_1 t)$$

be given (unmodulated carrier).
(a) Calculate the equivalent complex baseband (ECB) signal $x(t)$ corresponding to $x_{\mathrm{RF}}(t)$. Use f_0 (not necessarily equal to f_1) as transformation frequency.
(b) Sketch the phasor, that is, the diagram of $x(t)$ in the (complex) I/Q plane.
(c) Characterize $x(t)$ in case of $f_1 > f_0$ and in case of $f_1 < f_0$.
(d) How does the result simplify if $f_0 = f_1$? Interpret the result.

2.4.2 Equivalent Complex Baseband

Let the radio-frequency signal

$$x_{\mathrm{RF}}(t) = \sqrt{2} \cos(2\pi f_1 t) + \sqrt{2} \cos(2\pi f_2 t)$$

with $f_1 = f_c + \Delta f$ and $f_2 = f_c - 3\Delta f$ ($\Delta f > 0$) be given.
(a) Calculate the equivalent complex baseband (ECB) signal $x(t)$ corresponding to $x_{RF}(t)$. Use f_c as transformation frequency.
(b) Sketch the magnitude of the spectrum $X(f)$ corresponding to $x(t)$.
(c) Is the ECB signal real-valued?
(d) How should the transformation frequency f_0 be chosen, such that the ECB signal is real-valued? Give $x(t)$ for this choice.

2.4.3 Equivalent Complex Baseband
We consider the radio-frequency signal
$$x_{RF}(t) = \sqrt{2}\, q(t)\, \cos(2\pi f_c t),$$
where $q(t)$ is a real-valued analog (source) signal whose bandwidth is much smaller then f_c.
(a) Calculate the equivalent complex baseband (ECB) signal $x(t)$ corresponding to $x_{RF}(t)$. Use f_c as transformation frequency.
(b) What does the phasor of $x(t)$ look like in principle?
(c) Repeat the problems for $x_{RF}(t) = -\sqrt{2}\, q(t)\, \sin(2\pi f_c t)$.

2.4.4 Equivalent Complex Baseband
The radio-frequency signal
$$x_{RF}(t) = \text{rect}\,(t/T)\, \cos(2\pi f_c t)$$
is given.
(a) Determine the equivalent complex baseband (ECB) signal $x(t)$ corresponding to $x_{RF}(t)$; use f_c as transformation frequency f_0.
(b) How does the result simplify if we assume $f_c \gg 1/T$?

The signal $x_{RF}(t)$ is passed through an LTI system with transfer function
$$H_{RF}(f) = \text{rect}\left(\frac{f - f_c}{\Delta f}\right) + \text{rect}\left(\frac{f + f_c}{\Delta f}\right)$$
where $\Delta f \ll f_c$ and $\Delta f \ll 1/T$.
(c) Give the spectrum $Y_{RF}(f)$ of the corresponding system output signal $y_{RF}(t)$ via
- direct calculation in the radio-frequency domain
- calculation using ECB signals.

Sketch $Y_{RF}(f)$ graphically.
(d) Give an approximation for $y_{RF}(t)$ in case of $f_c \gg 1/T$.

2.4.5 Random Variables
Let $x \in \mathbb{R}$ be a random variable with probability density function (pdf)
$$f_x(x) = b \cdot \text{rect}\left(\frac{x}{2a}\right), \qquad a, b \in \mathbb{R}.$$
(a) Give b as a function of a, such that $f_x(x)$ is a valid pdf.
(b) Sketch the cumulative distribution function (cdf) $F_x(x)$ corresponding to $f_x(x)$.

Now, a constant $c \in \mathbb{R}, c \neq 0$, is added to the random variable, that is, a new variable y is created, with

$$y = x + c.$$

(c) Calculate and sketch the pdf $f_y(y)$ and the cdf $F_y(y)$ of y.

Finally, we consider the random variable z with

$$z = 2x.$$

(d) Calculate and sketch the pdf $f_z(z)$ and the cdf $F_z(z)$ of z.

2.4.6 Filtering of White Noise

Real-valued white Gaussian noise $n(t)$ with (two-sided) power spectral density $N_0/2$ is the input to the following LR system:

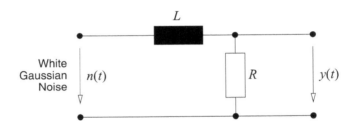

(a) Give the probability density function of the output process $y(t)$.
(b) Calculate the power spectral density $\Phi_{yy}(f)$ of $y(t)$.
(c) Calculate the power of the input process $n(t)$ and that of the output process $y(t)$, respectively.
(d) Give the autocorrelation functions of the input and output processes $x(t)$ and $y(t)$.

2.4.7 SNR at the Receiver

An RF signal with transmit power 40 W, carrier frequency 1 GHz, and bandwidth 1 MHz is radiated. On the way to the receiver, the signal is attenuated by 130 dB and the noisy receiver input amplifier is modeled by the addition of white noise (AWGN) with two-sided psd 4×10^{-19} W/Hz to the receiver input signal. The amplifier is implemented as a small-band amplifier around the carrier frequency with a quasi-rectangularly shaped transfer function.

(a) Give the (theoretic) SNR at the receiver input.
(b) Determine the SNR_r at the output of the receiver input amplifier/filter.
(c) Sketch the psds of signal and noise of the amplifier output in the ECB domain with respect to the carrier frequency.
(d) Determine again the SNR_r, now using the ECB representation of signal and noise.

2.4.8 Complexity-Efficient I/Q Conversion

An implementation of the transform of an RF signal $x_{RF}(t)$ into its quadrature components in the ECB domain (and back) by means of digital signal processing is considered. The power spectral density (psd) of the RF signal $x_{RF}(t)$ is concentrated within a frequency band of one-sided width B_x, which is located symmetrically around the carrier frequency f_c. Usually f_c is much higher than B_x, that is, $f_c \gg B_x$. In order to simplify the subsequent digital signal processing, the RF signal is often downconverted to a signal $x_{IF}(t)$ around an *intermediate frequency* f_{IF} by amplitude modulation (see the figure below). This intermediate frequency f_{IF} again denotes the (one-sided) middle of the down-shifted spectral band of width B_x. For complexity-efficient digitization of $x_{IF}(t)$, a small sampling frequency f_s is desirable. Thus, the intermediate frequency f_{IF} should be chosen rather small.

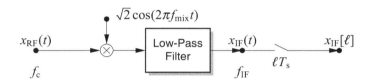

(a) Give the required mixing frequency f_{mix} for a desired intermediate frequency f_{IF}.
(b) Specify the smallest intermediate frequency f_{IF} for which no overlap of mixing products in the spectrum of $x_{IF}(t)$ occurs.

The intermediate carrier frequency f_{IF} is chosen as reference frequency f_0 for the ECB transform, that is, $f_0 = f_{IF}$.

(c) Specify the smallest intermediate carrier frequency f_{IF} for which the simplified version of the RF (here: IF) to ECB transform with low-pass filters can be applied instead of the general version with a Hilbert transform. Compare this frequency with the result of part (b).

An intermediate carrier frequency is chosen for which both conditions of parts (b) and (c) are satisfied. For digitization, the signal $x_{IF}(t)$ is sampled at a sampling frequency

$$f_s = 4 f_{IF} = 4 f_0 .$$

(d) Is the sampling theorem satisfied for the signal $x_{IF}(t)$ and this sampling frequency?
(e) For quadrature downconversion, the signal has to be multiplied by (scaled versions of) $\cos(2\pi f_0 t)$ and $-\sin(2\pi f_0 t)$. Show that these multiplications are very simple when operating only on the samples $x_{IF}[\ell]$ and the specific sampling frequency is used.

Hint: Calculate the discrete-time sequences of sampled cosine and sine functions.

(f) Draw a block diagram of this quadrature downconversion starting with $x_{\text{IF}}[\ell]$. In what situations is it possible to reduce the sampling frequency (decimation) after suitable processing with low-pass filters?

(g) Is an equivalent simple digital implementation also possible for the quadrature upconversion? If yes, draw the corresponding block diagram.

3 Digital Pulse-Amplitude Modulation

As explained in Chapter 1, the task of the transmitter in a digital transmission scheme is the mapping of a binary information sequence onto a physical transmit signal. The most basic and widely used form of this mapping is *digital pulse-amplitude modulation (PAM)*. As the name suggests, here the information is carried by the (complex-valued) amplitude of a basic pulse. Since we deal with real-valued and complex-valued amplitude coefficients in a unified manner, we use the denomination PAM for both baseband and carrier-modulated signal formats. PAM is the simplest form of digital modulation but establishes the basis for enhanced variants (see Chapter 4) and signal-space concepts (see Chapter 6).

In this chapter, we study mainly *uncoded transmission*; no channel coding is considered. In other words, when looking at Figures 1.7 and 1.8, we deal with the *modulation* and *demodulation* operations. The interface between the source and the modulation reduces to a symbol-by-symbol mapping of blocks of binary source symbols to signal points. The strategies for joint demodulation and decoding are explained where appropriate.

PREREQUISITES FOR THIS CHAPTER
- Equivalent complex baseband signals and systems.
- Random variables and probability; Gaussian processes and power spectral density.

3.1 Basic Principles

To define the transmit signal in digital PAM schemes, some basic "ingredients" are required.

- *Basic Pulse Shape*
  A basic pulse shape, we denote it by $g(t)$, has to be defined. The constraints on this pulse will be discussed subsequently. The pulse can also be viewed as the impulse response of a transmit filter. The pulse has spectrum and energy

spectrum of the pulse $\quad G(f) = \mathcal{F}\{g(t)\} = \int_{-\infty}^{\infty} g(t) e^{-j2\pi f t} \, dt$

energy of the pulse $\quad E_g = \int_{-\infty}^{\infty} |g(t)|^2 \, dt = \int_{-\infty}^{\infty} |G(f)|^2 \, df$

- *Signal Constellation*
  A signal constellation, denoted by $\mathcal{A}$, has to be specified. This constellation $\mathcal{A} \subset \mathbb{C}$ is the set of *amplitude coefficients* or *signal points* $a \in \mathcal{A}$. The parameters connected to the constellation are

  cardinality $\quad M \stackrel{\text{def}}{=} |\mathcal{A}|$

  rate of the modulation $\quad R_m = \log_2(M)$

  rate (no channel coding) $\quad R = R_m \left[\frac{\text{bit}}{\text{modulation step}} \right]$

  For a constellation $\mathcal{A}$, a *mapping* $\mathcal{M}$ has to be defined. It specifies which R_m-tuple of binary symbols is mapped to which signal point.

- *Symbol Period*
  The symbol period or *duration of a modulation interval*, denoted by T, has to be given. It is tightly connected to the basic pulse shape $g(t)$. Owing to the continuity of the information flow symbol period, bit duration, and rate are related by

  relation to bit duration $\quad T = T_b \cdot R$

3.1.1 Modulation

The operation of modulation can be specified as follows. All operations are carried out per time step of duration T, the symbol period. The *discrete time index* counting the time steps (modulation intervals) is always denoted as k; we have $k \in \mathbb{Z}$.

In each modulation interval, $R_m = \log_2(M)$ binary symbols are collected and mapped onto an *amplitude coefficient* or a *signal point* $a[k]$ drawn from the constellation, that is, the set $\mathcal{A}$. Thus, $a[k] \in \mathcal{A}$. Hence, in the *mapping* step, the sequence of binary source symbols $\langle q[l] \rangle$ (with discrete time index l and bit duration T_b; or, in case of channel coding, the sequence of code symbols with duration T_c) is transformed into the sequence $\langle a[k] \rangle$ of amplitude coefficients (with discrete time index k and symbol duration T). This mapping step establishes the transition from abstract (finite-field) elements to real/complex numbers.

Even in *digital* communications, the transmit signal has to be a physically realizable and thus a *continuous-time* (analog) signal. To that end, the *modulation* step has to transform the sequence of amplitude coefficients $a[k]$ into the transmit signal $s(t)$. This is done by weighting (scaling) the pulse shape $g(t)$ by the coefficient $a[k]$ in each modulation step k. The weighted pulses at the time instants kT are finally superimposed to form the transmit signal $s(t)$.

This weighting can also be viewed as *filtering* the sequence $\langle a[k] \rangle$ with an LTI system with impulse response $g(t)$. At the input, a discrete-time sequence is present and – since the impulse response is continuous – at the output, a continuous-time signal is obtained. This filtering is given by a "mixed" form of convolution, between continuous- and discrete-time convolution. Note the following variants of convolution that exist:

continuous $\quad y(t) = \int x(\tau) h(t-\tau) \, d\tau$,
"mixed" $\quad\quad y(t) = \sum_k x[k] h(t - kT)$,
discrete $\quad\quad y[k] = \sum_\kappa x[\kappa] h[k - \kappa]$.

That is, for each T, the signal $a[k]\delta(t - kT)$ is input, the impulse response scaled by $a[k]$ is triggered, and $a[k]g(t - kT)$ is produced at the output. Owing to the superposition principle of LTI systems, the input $\sum_{k=-\infty}^{\infty} a[k]\,\delta(t-kT)$ produces the output $\sum_{k=-\infty}^{\infty} a[k]\,g(t-kT)$. Hence, the discrete-time sequence $\langle a[k] \rangle$ is represented as a continuous-time signal with weighted Dirac functions; this signal is filtered by an LTI system with impulse response $g(t)$.

The transmit signal of a PAM scheme is thus given by

$$s(t) = \sum_{k=-\infty}^{\infty} a[k]\, g(t - kT) \,, \qquad a[k] \in \mathcal{A} \,. \tag{3.1}$$

The respective *block diagram* of the PAM transmitter for uncoded transmission is depicted in Figure 3.1. The blocks for the two steps described above are clearly visible: the (redundancy-free) mapping of the binary source symbols $q[l]$, $l \in \mathbb{Z}$, onto the sequence of amplitude coefficients $a[k]$, $k \in \mathbb{Z}$, and the filtering of the sequence $\langle a[k] \rangle$ with the pulse shape $g(t)$ (modulation).

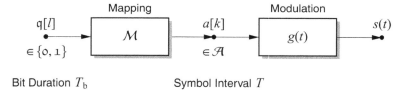

Figure 3.1 Block diagram of the PAM transmitter.

3.1.2 Baseband and Carrier-Modulated Transmission

In *baseband transmission* the signal $s(t)$ from (3.1) is used directly for transmission over the medium. No upconversion to some carrier frequency takes place. Such modulation forms are suited for low-pass transmission media (e.g., cables). As the transmit signal $s(t)$ has to be real-valued, both the pulse $g(t)$ and the amplitude coefficients $a[k]$ have to be real-valued.

In *carrier-modulated transmission* the signal $s(t)$ is the ECB signal corresponding to the actual RF transmit signal. Such modulation forms are suited for bandpass channels (e.g., in radio communications). As the transmit signal $s(t)$ may be complex-valued, both the pulse $g(t)$ and the amplitude coefficients $a[k]$ may be complex. However, in practice, the basic pulse shape $g(t)$ is mostly chosen to be real-valued.

The upconversion to the carrier frequency follows the concept introduced in Chapter 2. In PAM schemes, the carrier frequency f_c is used as reference frequency in the ECB transformation f_0. Note that this upconversion is often called "modulation." In digital communications, especially when dealing with ECB signals where this upconversion is not visible, the term "modulation" is used to denote the step from the discrete-time abstract representation of information to the physical signal. Modulation is the process of varying the parameters of a carrier signal according to the source signal – in the ECB domain this reduces to the generation of the complex baseband signal.

The RF transmit signal is thus given by

$$s_{\mathrm{RF}}(t) = \sqrt{2}\,\mathrm{Re}\{s(t) \cdot e^{j2\pi f_c t}\}, \qquad (3.2)$$

with the ECB transmit signal

$$s(t) = \sum_{k=-\infty}^{\infty} a[k]\, g(t - kT). \qquad (3.3)$$

Assuming a real-valued pulse $g(t)$, and writing the amplitude coefficients as $a[k] = a_{\mathrm{I}}[k] + j a_{\mathrm{Q}}[k]$, where $a_{\mathrm{I}}[k] = \mathrm{Re}\{a[k]\}$ and $a_{\mathrm{Q}}[k] = \mathrm{Im}\{a[k]\}$, the quadrature components of the transmit signal calculate as

$$\begin{aligned}
s_{\mathrm{RF}}(t) &= \sqrt{2}\,\mathrm{Re}\left\{\left(\sum_{k=-\infty}^{\infty}(a_{\mathrm{I}}[k]+ja_{\mathrm{Q}}[k])\,g(t-kT)\right)\bigl(\cos(2\pi f_c t)+j\sin(2\pi f_c t)\bigr)\right\} \\
&= \sqrt{2}\left(\sum_{k=-\infty}^{\infty} a_{\mathrm{I}}[k]\,g(t-kT)\right)\cdot \cos(2\pi f_c t) \qquad \text{in-phase component} \\
&\quad - \sqrt{2}\left(\sum_{k=-\infty}^{\infty} a_{\mathrm{Q}}[k]\,g(t-kT)\right)\cdot \sin(2\pi f_c t) \qquad \text{quadrature component.}
\end{aligned} \qquad (3.4)$$

The following figures show block diagrams of the PAM transmitter. In Figure 3.2, the generation of the RF transmit signal is shown. Note that this diagram is only valid if the pulse shape $g(t)$ is real-valued; in case of a complex-valued pulse, the parallel filters (in-phase and quadrature branch) have to be replaced by the cross-coupled structure of Figure 2.11.

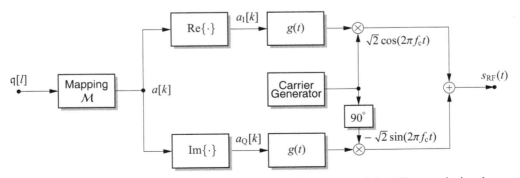

Figure 3.2 Block diagram of the PAM transmitter. Generation of the RF transmit signal. The pulse shape $g(t)$ is assumed to be real-valued.

Figure 3.3 Block diagram of the PAM transmitter. ECB model; the pulse shape $g(t)$ can be real- or complex-valued.

In Figure 3.3, the corresponding ECB model is shown. As the upconversion is absorbed, only the mapping and modulation (pulse-shaping) blocks remain. It is noteworthy that the ECB model is valid no matter whether the pulse shape is real- or complex-valued.

3.2 Average Power Spectral Density of PAM Signals

We are always interested in the *power* and *bandwidth* efficiency of a particular digital transmission scheme. In order to quantify the bandwidth efficiency, the statistical properties of the transmit signal have to be known, specifically the (average) power spectral density, which determines the bandwidth.

We now calculate the average power spectral density of PAM schemes. To that end, the pulse shape $g(t)$, the symbol period T, and the constellation $\mathcal{A}$ (the "ingredients") are assumed to be given. Moreover, as the source sequence is drawn from a (wide-sense stationary) random process, the sequence $\langle a[k] \rangle$ of amplitude coefficients is also drawn from a (wide-sense stationary) discrete-time random process. Hence, the sequence is characterized by its autocorrelation sequence and power spectral density

$$\phi_{aa}[\kappa] = \mathrm{E}\{a[k+\kappa]a^*[k]\},$$

$$\Phi_{aa}\left(\mathrm{e}^{\mathrm{j}2\pi fT}\right) \stackrel{\text{def}}{=} \sum_{\kappa=-\infty}^{\infty} \phi_{aa}[\kappa]\, \mathrm{e}^{-\mathrm{j}2\pi fT\kappa}. \tag{3.5}$$

As explained above, the generated continuous-time PAM signal (ECB domain) reads

$$s(t) = \sum_{k=-\infty}^{\infty} a[k]\, g(t-kT). \tag{3.6}$$

3.2.1 Average Autocorrelation Function of $s(t)$

Owing to the inherent clocking (pulses are produced every T), the PAM transmit signal $s(t)$ is drawn from a *cyclo-stationary* process; its parameters (mean, variance, etc.) vary over time with period T (see Section A.4 of Appendix A). Hence, we consider *average* parameters (mean power, mean acf, etc.) obtained by averaging the parameters over one period of duration T.

3.2 Average Power Spectral Density of PAM Signals

Of particular interest is the *average autocorrelation function* of the transmit signal, defined as

$$\bar{\phi}_{ss}(\tau) \stackrel{\text{def}}{=} \frac{1}{T} \int_0^T E\{s(t+\tau)s^*(t)\} \, dt \, . \tag{3.7}$$

Using (3.1), the average acf can be written as

$$\bar{\phi}_{ss}(\tau) = \frac{1}{T} \int_0^T E\left\{\left(\sum_k a[k]g(t+\tau-kT)\right)\left(\sum_l a[l]g(t-lT)\right)^*\right\} dt$$

$$= \frac{1}{T} \int_0^T E\left\{\sum_k \sum_l a[k]a^*[l] g(t+\tau-kT)g^*(t-lT)\right\} dt$$

since $g(t)$ is deterministic and due to the linearity of expectation

$$= \frac{1}{T} \int_0^T \sum_k \sum_l E\{a[k]a^*[l]\} g(t+\tau-kT)g^*(t-lT) \, dt$$

defining $\kappa = k - l$

$$= \frac{1}{T} \int_0^T \sum_k \sum_\kappa E\{a[k]a^*[k-\kappa]\} g(t+\tau-kT)g^*(t-(k-\kappa)T) \, dt$$

considering $\phi_{aa}[\kappa] = E\{a[k+\kappa]a^*[k]\}$ and interchanging integration and summation

$$= \frac{1}{T} \sum_\kappa \phi_{aa}[\kappa] \sum_k \int_0^T g(t+\tau-kT)g^*(t-(k-\kappa)T) \, dt$$

using $\sum_k \int_0^T f(t-kT) \, dt = \int_{-\infty}^{\infty} f(t) \, dt$

$$= \frac{1}{T} \sum_\kappa \phi_{aa}[\kappa] \int_{-\infty}^{\infty} g(t+\tau)g^*(t+\kappa T) \, dt \, .$$

With the definition of the *autocorrelation* of the basic pulse $g(t)$ (cf. also (A.16))

$$\varphi_{gg}(\tau) \stackrel{\text{def}}{=} g(\tau) * g^*(-\tau) = \int_{-\infty}^{\infty} g(t+\tau)g^*(t) \, dt \tag{3.8}$$

we arrive at

$$\bar{\phi}_{ss}(\tau) = \frac{1}{T} \sum_\kappa \phi_{aa}[\kappa]\varphi_{gg}(\tau - \kappa T) \, . \tag{3.9}$$

This result again is a mixed convolution analog to the signal generation in (3.1), now of the autocorrelation sequence of a discrete-time stochastic process and the autocorrelation of the continuous-time pulse (and with an additional normalization by $1/T$).

3.2.2 Average Power Spectral Density of $s(t)$

Taking the Fourier transform of the average acf $\bar{\phi}_{ss}(\tau)$, the *average power spectral density* of the transmit signal is obtained as (transformation w.r.t. τ)

$$\bar{\Phi}_{ss}(f) \stackrel{\text{def}}{=} \mathscr{F}_\tau\{\bar{\phi}_{ss}(\tau)\}$$

$$= \mathscr{F}_\tau\left\{\frac{1}{T}\sum_\kappa \phi_{aa}[\kappa]\varphi_{gg}(\tau - \kappa T)\right\}$$

due to the linearity of the Fourier transform

$$= \frac{1}{T}\sum_\kappa \phi_{aa}[\kappa] \cdot \mathscr{F}_\tau\{\varphi_{gg}(\tau - \kappa T)\}$$

shifting in time domain corresponds to modulation of the spectrum

$$= \frac{1}{T}\sum_\kappa \phi_{aa}[\kappa] \cdot \mathscr{F}_\tau\{\varphi_{gg}(\tau)\} \cdot e^{-j2\pi f T\kappa}$$

using $\varphi_{gg}(\tau) \circ\!\!-\!\!\bullet |G(f)|^2$

$$= \frac{1}{T}\sum_\kappa \phi_{aa}[\kappa] \cdot |G(f)|^2 \cdot e^{-j2\pi f T\kappa}$$

and taking the definition (3.5) of the psd of the sequence of amplitude coefficients into account, we arrive at

$$\bar{\Phi}_{ss}(f) = \frac{1}{T}\Phi_{aa}(e^{j2\pi f T})|G(f)|^2 \ .$$

Thus, the average power spectral density of PAM signals is given by

$$\boxed{\bar{\Phi}_{ss}(f) = \Phi_{aa}(e^{j2\pi f T})\frac{|G(f)|^2}{T} \ .} \qquad (3.10)$$

This equation gives rise to some remarks. First, in (3.10), the three "ingredients" are noticeable again. The statistical properties of the amplitude coefficients (via the psd $\Phi_{aa}(e^{j2\pi f T})$), the basic pulse shape (via its squared spectrum $|G(f)|^2$, since we deal with powers), and the symbol timing T together determine the average psd of the PAM transmit signal.

Moreover, since $\langle a[k]\rangle$ is a discrete-time sequence (T-spaced), due to the properties of the Fourier transform, its spectrum is periodic over frequency with period $1/T$ – a discrete-time sequence has a spectrum which is infinitely broad! Via the pulse-shaping step (modulation), the periodic spectrum is multiplied by $|G(f)|^2$. Since the spectrum $G(f)$ of the pulse shape is concentrated in the baseband, the respective part of the periodic spectrum $\Phi_{aa}\left(e^{j2\pi f T}\right)$ is cut out; the other spectral replicas are suppressed and a finite bandwidth can be achieved. This is visualized in Figure 3.4.

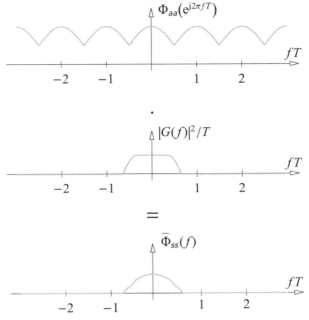

Figure 3.4 Generation of the PAM transmit signal. Visualization in the frequency domain. For illustration purposes only, the psd $\Phi_{aa}\left(e^{j2\pi fT}\right)$ of the amplitude coefficients is shown as a garland shape (usually it is flat).

However, in order not to lose information, at least one period of the psd $\Phi_{aa}\left(e^{j2\pi fT}\right)$ has to be cut out. Since one period of the psd has width $1/T$, the (average) power spectral density of the PAM transmit signal has as least bandwidth $1/T$. PAM schemes[1] cannot have a smaller bandwidth and thus a higher bandwidth efficiency than R_m.

Finally, if the basic pulse shape is real-valued, $|G(f)|$ is symmetric about $f = 0$. The average power spectral density of the RF transmit signal is then symmetric about the carrier frequency f_c. Only in rare cases is a non-symmetric psd required or used. Hence, in most systems, the basic pulse shape is real-valued.

3.2.3 Uncoded Transmission

The above derivation is equally valid for coded and uncoded transmission. However, if we restrict ourselves to uncoded transmission, the psd $\Phi_{aa}\left(e^{j2\pi fT}\right)$ can be given easily.

If the source symbols q[l] are independent, uniformly distributed, and no channel coding is present, the amplitude coefficients $a[k]$ are also independent and

---

[1] This fundamental statement holds for uncoded schemes. By means of special channel coding methods, the bandwidth efficiency may be increased. For details, see Chapter 6.

uniformly distributed and have an autocorrelation sequence

$$\phi_{aa}[\kappa] = \mathrm{E}\{a[k+\kappa]a^*[k]\} = \begin{cases} \sigma_a^2 + |m_a|^2, & \kappa = 0 \\ |m_a|^2, & \kappa \in \mathbb{Z} \setminus \{0\} \end{cases}, \quad (3.11)$$

with mean of $\mathcal{A}$
$$m_a \stackrel{\text{def}}{=} \mathrm{E}\{a\} = \frac{1}{M} \sum_{\forall a \in \mathcal{A}} a \quad (3.12)$$

and variance of $\mathcal{A}$
$$\sigma_a^2 \stackrel{\text{def}}{=} \mathrm{E}\{|a - m_a|^2\} = \frac{1}{M} \sum_{\forall a \in \mathcal{A}} |a|^2 - |m_a|^2. \quad (3.13)$$

This autocorrelation sequence $\phi_{aa}[\kappa]$ is visualized in Figure 3.5.

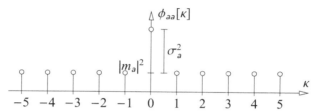

Figure 3.5 Autocorrelation sequence $\phi_{aa}[\kappa]$ for independent and uniformly distributed amplitude coefficients $a[k]$. The constellation $\mathcal{A}$ has mean m_a and variance σ_a^2.

Using the identity $\sum_{k=-\infty}^{\infty} e^{j2\pi fTk} = \frac{1}{T} \sum_{\ell=-\infty}^{\infty} \delta(f - \frac{\ell}{T})$, the psd of the sequence of amplitude coefficients is given by

$$\Phi_{aa}\left(e^{j2\pi fT}\right) = \sum_{k=-\infty}^{\infty} \phi_{aa}[k] \cdot e^{-j2\pi fTk}$$

$$= \sigma_a^2 + |m_a|^2 \cdot \sum_{k=-\infty}^{\infty} e^{-j2\pi fTk}$$

$$= \sigma_a^2 + \frac{|m_a|^2}{T} \cdot \sum_{\ell=-\infty}^{\infty} \delta\left(f - \frac{\ell}{T}\right). \quad (3.14)$$

Using this psd in (3.10), the average psd of uncoded PAM is given by

$$\bar{\Phi}_{ss}(f) = \sigma_a^2 \frac{|G(f)|^2}{T} + |m_a|^2 \frac{|G(f)|^2}{T^2} \sum_{\ell=-\infty}^{\infty} \delta\left(f - \frac{\ell}{T}\right). \quad (3.15)$$

When looking at (3.15), one can see that if the signal constellation has a nonzero mean ($m_a \neq 0$), the psd contains Dirac pulses at all frequency positions that are multiples of $1/T$ (i.e., at all multiples of the symbol frequency). Only if $G\left(\frac{\ell}{T}\right) = 0$, the respective Dirac pulse is suppressed.

This effect is known from analog modulation: in analog amplitude modulation with carrier (AM with carrier), a carrier component is present. This can be achieved by adding a DC offset to the (analog) source signal and feeding this signal to the modulator. As will be seen later (and is known from analog modulation formats), carrier components lower the power efficiency of a communication system; most of the transmit power is "wasted" in the carrier. However, typically a (much) simpler demodulation is enabled (cf. envelope detection of AM). We will come back to this issue in Chapter 4.

Average Transmit Power and Average Energy per bit

Having the (average) psd of the transmit signal, the (average) *transmit power* can be calculated by integrating the psd over all frequencies.

In case of *zero-mean* amplitude coefficients ($m_a = 0$), we have $\bar{\Phi}_{ss}(f) = \sigma_a^2 \frac{|G(f)|^2}{T}$ and thus $\int \bar{\Phi}_{ss}(f) \, df = \sigma_a^2 \frac{\int |G(f)|^2 \, df}{T}$. Since $E_g = \int |G(f)|^2 \, df$, we arrive at

$$\text{average transmit power} \qquad S_s = \sigma_a^2 \frac{E_g}{T} \, . \qquad (3.16)$$

In a pure additive noise channel model (i.e., a channel where the channel gain is 1, achieved by scaling at the receiver such that the signal attenuation is fully compensated for), the transmit power S_s and receive signal power S_r are equal: $S \stackrel{\text{def}}{=} S_s = S_r$. Here we have

$$\text{average energy per symbol} \qquad E_s = S \cdot T = \sigma_a^2 \, E_g \, , \qquad (3.17)$$

$$\text{average energy per bit} \qquad E_b = \frac{E_s}{R} \stackrel{\text{uncoded}}{=} \frac{\sigma_a^2 E_g}{\log_2(M)} \, . \qquad (3.18)$$

For $m_a \neq 0$ the situation is more involved. It can be shown that for $g(t)$ a so-called $\sqrt{\text{Nyquist}}$ pulse (see Section 3.4), we have

$$\text{average transmit power} \qquad S_s = \left(\sigma_a^2 + |m_a|^2\right) \frac{E_g}{T} \, . \qquad (3.19)$$

3.3 Choice of the Signal Constellation

The signal constellation $\mathcal{A}$ is one of the main "ingredients" in digital PAM schemes. The choice of the set of signal points (amplitude coefficients) influences the transmission rate, the transmit power, and the error ratio. Depending on the situation and demands, particular constellations are of interest; three main "families" of PAM schemes can be distinguished. In general, regularly spaced arrangements are of interest. Usually, the cardinality M of the constellation is a power of 2 (i.e., $M = 2^m$) such that m binary symbols are represented per modulation interval.

3.3.1 Amplitude-Shift Keying (ASK)

In *baseband transmission*, the signal constellation has to be real-valued (i.e., $\mathcal{A} \subset \mathbb{R}$). This gives rise to *amplitude-shift keying (ASK)*.[2] Two basic variants of regular ASK constellations exist.

In *bipolar ASK*, the signal points are uniformly spaced and symmetrical with respect to the origin. The spacing of adjacent points is chosen[3] to be 2. For an even cardinality M of the constellation (M-ary constellation; MASK), we have

$$\mathcal{A} = \{\pm 1, \pm 3, \pm 5, \ldots, \pm(M-1)\}. \tag{3.20}$$

For example, 2ASK then has the constellation $\mathcal{A}_{2\text{ASK}} = \{-1, +1\}$.

Such a bipolar ASK constellation ($M = 8$) is depicted in Figure 3.6.

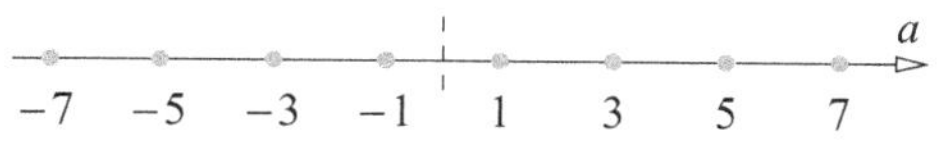

Figure 3.6 Bipolar ASK signal constellation ($M = 8$).

The mean and variance of the constellation (3.20) are calculated as follows:

$$m_a = 0, \qquad \sigma_a^2 = \frac{M^2 - 1}{3}. \tag{3.21}$$

In contrast, in *unipolar ASK*, the signal points are uniformly spaced but only on the non-negative part of the real axis. An M-ary constellation is given by

$$\mathcal{A} = \{0, 2, 4, \ldots, (2M - 2)\}. \tag{3.22}$$

Such a constellation ($M = 8$) is depicted in Figure 3.7.

Figure 3.7 Unipolar ASK signal constellation ($M = 8$).

Here, the mean and variance of the constellation are given by

$$m_a = M - 1, \qquad \sigma_a^2 = \frac{M^2 - 1}{3} \tag{3.23}$$

(i.e., it has a nonzero mean).

3.3.2 Phase-Shift Keying (PSK)

In *bandpass transmission*, the signal constellation may be complex-valued. In some situations it is useful to choose the constellation points such that all have the same magnitude. The information is then represented purely in the phase, leading to

---

[2] Note that many authors call this PAM but it is more reasonable to denote the basic principle as PAM, which then includes the three families (subtypes) given here as special PAM schemes.

[3] The chosen normalization of the spacing of the signal point will prove to be convenient when subsequently calculating the error ratio. Of course, any other normalization is also possible.

phase-shift keying (PSK). In the context of digital communications and description in the ECB domain, it is noteworthy that phase-shift keying belongs to the family of digital "amplitude" modulation.

Spacing the points uniformly on the unit circle, for M-ary PSK we have

$$\mathcal{A} = \{e^{j2\pi \frac{m-1}{M}} \mid m \in \{1, 2, \ldots, M\}\}, \quad (3.24)$$

visualized in Figure 3.8 for different cardinalities.

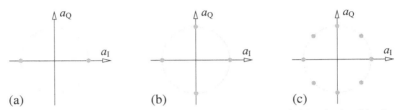

Figure 3.8 PSK signal constellations. (a) $M = 2$ (2PSK is identical to bipolar 2ASK); (b) $M = 4$ (4PSK is equivalent to 4QAM); and (c) $M = 8$.

The mean and variance of PSK constellations are given by

$$m_a = 0, \qquad \sigma_a^2 = 1, \quad \forall M \in \mathbb{N}. \quad (3.25)$$

3.3.3 Digital Quadrature-Amplitude Modulation (QAM)

Finally, in bandpass transmission, the signal points can be spaced regularly in the complex plane, leading to *quadrature-amplitude modulation (QAM)* constellations. In QAM, the in-phase and quadrature components of the signal points are usually chosen from a bipolar ASK constellation.

If $M = 2^m$ is a square number ($M = 4, 16, 64, \ldots$), the boundary of the constellation is a square and $\left(\sqrt{M} = 2^{(m/2)}\right)$-ary ASK (representing $m/2$ bits) is used independently per component. These are the so-called *square constellations*. If $M = 2^m$ is not a square number ($M = 32, 128, \ldots$), the boundary is typically chosen (for minimum average transmit power) to be cross shaped, leading to the so-called *cross constellations*. Here, the in-phase and quadrature components can no longer be chosen independently.

Some QAM constellations are depicted in Figure 3.9.

For square constellations, the mean and variance are given by

$$m_a = 0, \qquad \sigma_a^2 = \frac{2(M-1)}{3}. \quad (3.26)$$

Table 3.1 gives the exact numbers for the cross constellations.

Table 3.1 Variances of QAM constellations.

M	4	16	32 cross	64	128 cross	256
σ_a^2	2	10	20	42	82	170

Figure 3.9 QAM signal constellations ($M = 4, 16, 32, 64, 128,$ and 256).

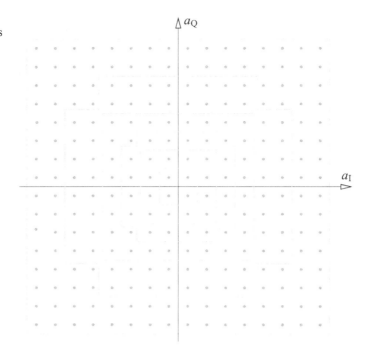

Besides these regular constellations, freely designed constellations may be used for special purposes.

3.4 Receiver Structure and Choice of the Basic Pulse Shape

As stated in the introduction to this chapter, one of the main items in digital PAM is the basic pulse shape $g(t)$. In this section, we derive the constraints on $g(t)$; these constraints emerge from the design of an optimum receiver. Hence, we now turn to the study of the receiver.

3.4.1 Optimum Receive Filter for PAM

When assuming digital PAM over an AWGN channel, the receive signal (ECB domain) is given by

$$r(t) = s(t) + n(t) = \sum_k a[k]\, g(t - kT) + n(t), \qquad (3.27)$$

where $n(t)$ is complex-valued white Gaussian noise with noise power spectral density N_0 ($N_0/2$ per quadrature component). The natural question to be answered is: What is the optimum receiver for this setting? For that, the end-to-end model of PAM is shown in Figure 3.10.

3.4 Receiver Structure and Choice of the Basic Pulse Shape

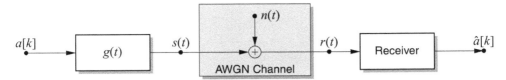

Figure 3.10 End-to-end model of digital PAM over an AWGN channel.

Since in PAM the information is represented in symbols $a[k]$ transmitted every T, the respective operations at the receiver are also done based on this timing. Moreover, for each operation at the transmitter a corresponding operation at the receiver has to be performed. Hence, any receiver for PAM will carry out the following steps.

- The receive signal $r(t)$ is filtered with a linear filter with impulse response $h_R(t)$ or corresponding transfer function $H_R(f)$; the *receive filter*. This is the operation corresponding to the transmit filter (pulse shaping) and is essential for the limitation of noise and interference.
- The filtered signal is *sampled* in order to obtain *decision variables* $d[k]$. Sampling is the corresponding operation to the T-spaced excitement of the transmit filter (TX: transition from a T-spaced discrete-time sequence to a continuous-time TX signal; RX: transition from a continuous-time RX signal to a T-spaced discrete-time sequence).
- Decisions are taken based on $d[k]$; estimates $\hat{a}[k]$ on the amplitude coefficients are generated (performed by a slicer). This step can be seen as corresponding to the mapping operation at the transmitter.

In order to derive the *optimum receive filter* $h_R(t)$ (transfer function $H_R(f)$) for digital PAM, which is the transmission of a *pulse train*, we first consider the transmission of *a single pulse* $g(t)$ over the AWGN channel. The task is then the optimal detection of a pulse $g(t)$ with finite energy E_g, which is disturbed by additive white Gaussian noise. Notice that in contrast to analog communications, the pulse shape is known to the receiver. This situation is visualized in Figure 3.11. Following the above procedure, the receiver performs filtering with $h_R(t)$ and sampling at the detection instant $t = T_d$.

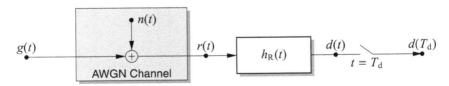

Figure 3.11 Detection of a single pulse $g(t)$ transmitted over the AWGN channel.

However, we still have to define what we mean by "optimal," that is, an optimization criterion has to be specified. In communications, a suitable measure is the *signal-to-noise ratio (SNR)*, which is linked to the symbol error ratio. Hence, we consider the SNR at the detection time T_d, which is given by

$$\text{SNR}_d \stackrel{\text{def}}{=} \frac{|\tilde{d}(T_d)|^2}{\sigma^2_{n_{1,d}}}, \qquad (3.28)$$

where the *useful signal* is given by

$$\tilde{d}(T_d) = g(t) * h_R(t)\Big|_{t=T_d} = \int_{-\infty}^{\infty} G(f)\, H_R(f)\, e^{j2\pi f T_d}\, df \qquad (3.29)$$

and the *noise power* (per quadrature component) calculates as

$$\sigma^2_{n_{1,d}} = \frac{N_0}{2} \int_{-\infty}^{\infty} |H_R(f)|^2\, df\,. \qquad (3.30)$$

Before solving the optimization problem for $H_R(f)$, we first consider an illustrative example.

Example 3.1 Action of the Receive Filter

One can imagine two extreme cases for the receive filter. On the one hand, the receive filter may have a *large* bandwidth. For a rectangular pulse shape, the situation may look as shown in Figure 3.12.

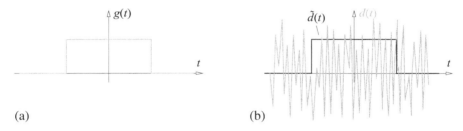

Figure 3.12 (a) Transmitted pulse and (b) noise-free and noisy detected pulse. Receive filter with large bandwidth.

Here, the transmit pulse is almost undistorted by filtering but the AWGN on the channel is not sufficiently bandlimited. If the bandwidth of the receive filter tends to infinity, the noise power tends to infinity and thus $\text{SNR}_d \to 0$ (the denominator is infinite).

If, on the other hand, the receive filter has a *small* bandwidth, the situation may look as shown in Figure 3.13.

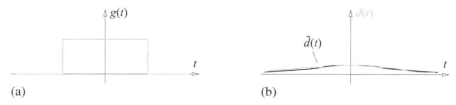

Figure 3.13 (a) Transmitted pulse and (b) noise-free and noisy detected pulse. Receive filter with small bandwidth.

Now, the AWGN is limited to obtain a finite (small) noise power at the filter output. However, the useful signal is heavily distorted; almost no useful power is present. If the bandwidth of the receive filter tends to zero, we finally have $\text{SNR}_d \to 0$ (the numerator is zero).

Consequently, neither a very large nor a very small bandwidth is optimal. Hence, a *compromise* with respect to signal distortion and noise limitation has to be found, as illustrated in Figure 3.14.

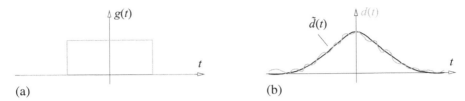

Figure 3.14 (a) Transmitted pulse and (b) noise-free and noisy detected pulse. Receive filter with optimized bandwidth.

Here, the maximum signal-to-noise ratio (i.e., $\text{SNR}_d \to \max$) may be achieved.

Upper Bound on the SNR and Matched Filter

In order to find the optimum receive filter together with the optimum sampling instant T_d for detection, we derive an *upper bound* on the achievable SNR. A receive filter achieving this bound is then optimal.

The SNR at the detection instant

$$\text{SNR}_d = \frac{|\tilde{d}(T_d)|^2}{\sigma_{n_{\text{I},d}}^2} = \frac{\left| \int_{-\infty}^{\infty} G(f) H_R(f) e^{j2\pi f T_d} df \right|^2}{\frac{N_0}{2} \int_{-\infty}^{\infty} |H_R(f)|^2 df} \qquad (3.31)$$

can be written, using $\bar{H}_R^*(f) \stackrel{\text{def}}{=} H_R(f) e^{j2\pi f T_d}$ and considering $|\bar{H}_R(f)| = |H_R(f)|$, as

$$= \frac{\left| \int_{-\infty}^{\infty} G(f) \bar{H}_R^*(f) df \right|^2}{\frac{N_0}{2} \int_{-\infty}^{\infty} |\bar{H}_R(f)|^2 df} . \qquad (3.32)$$

The numerator can be upper bounded by the *Cauchy–Schwarz inequality* (see Box 3.1) $\left| \int_{-\infty}^{\infty} G(f) \bar{H}_R^*(f) df \right|^2 \leq \left(\int_{-\infty}^{\infty} |G(f)|^2 df \right) \cdot \left(\int_{-\infty}^{\infty} |\bar{H}_R(f)|^2 df \right)$, leading to

$$\text{SNR}_d \leq \frac{\int_{-\infty}^{\infty} |G(f)|^2 df \cdot \int_{-\infty}^{\infty} |\bar{H}_R(f)|^2 df}{\frac{N_0}{2} \int_{-\infty}^{\infty} |\bar{H}_R(f)|^2 df} \qquad (3.33)$$

$$= \frac{\int_{-\infty}^{\infty} |G(f)|^2 df}{N_0/2} = \frac{E_g}{N_0/2} . \qquad (3.34)$$

Box 3.1 The Cauchy–Schwarz Inequality

Euclidean Space $\mathbb{R}^n$
For all vectors $a, b \in \mathbb{R}^n$ we have

$$a^\top \cdot b = |a| \cdot |b| \cdot \cos(\phi_{ab}) \leq |a| \cdot |b|$$

since $|\cos(\phi)| \leq 1$. With a_i, b_i denoting the components of the vectors, this gives

$$\left(\sum_i a_i \cdot b_i\right)^2 \leq \left(\sum_i a_i^2\right) \cdot \left(\sum_i b_i^2\right).$$

Equality holds if and only if $a_i = c \cdot b_i$, $c \in \mathbb{R}$, that is, $a = cb$ (collinear vectors).
The relations are visualized for $n = 2$ in Figure 3.15.

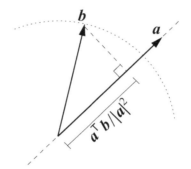

Figure 3.15 Visualization of the scalar product in two-dimensional space.

Complex-Valued Vectors
For vectors $a, b \in \mathbb{C}^n$ we have

$$\left|\sum_i a_i \cdot b_i^*\right|^2 \leq \left(\sum_i |a_i|^2\right) \cdot \left(\sum_i |b_i|^2\right).$$

Equality holds if and only if $a = cb$, $c \in \mathbb{C}$.

Complex-Valued Functions
The relation can be extended to (complex-valued) functions ($a_i \to A(x)$, $b_i \to B(x)$, $\sum_i \to \int \cdot \, dx$). Here, for all functions $A(x), B(x) \in \mathbb{C}$, we have

$$\left|\int A(x) \cdot B^*(x) \, dx\right|^2 \leq \left(\int |A(x)|^2 \, dx\right) \cdot \left(\int |B(x)|^2 \, dx\right).$$

Again, equality holds if and only if $A(x) = cB(x)$, $c \in \mathbb{C}$.

Hence, regardless of the pulse shape $g(t)$ – only the energy E_g matters – we have

$$\text{SNR}_d \leq \frac{E_g}{N_0/2}. \tag{3.35}$$

From the Cauchy–Schwarz inequality we know that equality in the above derivation of the upper bound (3.35) is obtained if (const is a real-valued constant)

$$G(f) \stackrel{!}{=} \text{const} \cdot \bar{H}_R(f) . \qquad (3.36)$$

Taking the definition $\bar{H}_R(f) = \left(H_R(f) e^{j2\pi f T_d}\right)^*$ into account and solving the equation for $H_R(f)$ gives

$$\text{const} \cdot \left(H_R(f) e^{j2\pi f T_d}\right)^* \stackrel{!}{=} G(f)$$

$$\left(H_R(f) e^{j2\pi f T_d}\right)^* \stackrel{!}{=} \frac{1}{\text{const}} G(f)$$

$$H_R(f) e^{j2\pi f T_d} \stackrel{!}{=} \frac{1}{\text{const}} G^*(f)$$

$$H_R(f) \stackrel{!}{=} \gamma \, G^*(f) e^{-j2\pi f T_d} ,$$

where $\gamma = \frac{1}{\text{const}}$ is a real-valued constant. Using the properties of the Fourier transform (see Section A.1 of Appendix A), the time-domain description (receiver impulse response) is then

$$h_R(t) = \gamma \, g^*(T_d - t) .$$

In summary, the optimum receive filter for the finite-energy pulse $g(t)$ in additive white noise is thus given by the so-called *matched filter (MF)*

$$H_R(f) = \gamma \, G^*(f) \, e^{-j2\pi f T_d} \qquad \circ\!\!-\!\!\bullet \qquad h_R(t) = \gamma \, g^*(T_d - t) . \qquad (3.37)$$

The matched filter corresponding to $g(t)$ is thus given by the *time-reversed and delayed (by T_d) conjugate-complex impulse $g(t)$*. Employing the matched filter, at the sampling instant T_d the highest possible SNR, $\text{SNR}_{d,\text{MF}} = \frac{E_g}{N_0/2}$, is achieved – this maximum SNR is independent of the shape of $g(t)$.

Detection Signal and Detection Instant

Applying the matched filter, the useful part of the detection signal (the output of the matched filter) is given by

$$\begin{aligned}\tilde{d}(t) &= g(t) * h_R(t) \\ &= g(t) * \gamma \, g^*(T_d - t) \\ &= \gamma \, \varphi_{gg}(t - T_d) , \end{aligned} \qquad (3.38)$$

where $\varphi_{gg}(t) \stackrel{\text{def}}{=} g(t) * g^*(-t)$ denotes the autocorrelation of the pulse $g(t)$ (cf. (3.8)).

Hence, a shifted and scaled version of the autocorrelation $\varphi_{gg}(t)$ is present. In the sampling (detection) instant, we have

$$\tilde{d}(T_d) = \gamma \, \varphi_{gg}(T_d - T_d) = \gamma \, \varphi_{gg}(0)$$
$$= \gamma \int_{-\infty}^{\infty} |g(\tau)|^2 \, d\tau$$
$$= \gamma \, E_g \in \mathbb{R} \, . \tag{3.39}$$

Since the autocorrelation function of the AWGN (ECB signal) is $N_0 \, \delta(\tau)$, the autocorrelation function of the noise after the matched filter calculates as

$$\phi_{n_d n_d}(\tau) = N_0 \, \delta(\tau) * h_R(\tau) * h_R^*(-\tau)$$
$$= N_0 \, \gamma \, g^*(T_d - \tau) * \gamma \, g(T_d + \tau)$$
$$= N_0 \, \gamma^2 \, \varphi_{gg}(\tau) \, , \tag{3.40}$$

and thus the noise power is

$$\sigma_{n_d}^2 = \phi_{n_d n_d}(0)$$
$$= N_0 \, \gamma^2 \, \varphi_{gg}(0) \, . \tag{3.41}$$

A fundamental property of the matched filter can be recognized: at its output, the useful signal $\tilde{d}(t)$ and the acf $\phi_{n_d n_d}(\tau)$ of the noise are proportional to each other and proportional to the autocorrelation $\varphi_{gg}(\tau)$ of the pulse $g(t)$.

Still the optimum detection instant has to be determined. As an autocorrelation function has its maximum at $\tau = 0$, sampling has to be done at that time instant where $\tilde{d}(t)$ equals $\gamma \varphi_{gg}(0) = \gamma E_g$ – this is exactly at $t = T_d$. However, when looking at the definition (3.37) of the matched filter, T_d is a parameter – it may be chosen arbitrarily. When varying T_d, the delay of the matched filter impulse response changes and thus the delay in the detection signal changes. Hence, as long as the sampling instant is adapted (matched) to the delay in the matched filter, optimum performance is achieved.

In most subsequent theoretical derivations, we will therefore use $T_d = 0$ – the discrete-time sequences $a[k]$ at the transmitter and $d[k] = d(kT)$ are taken to be perfectly simultaneous. However, one has to take into account that the time reversal of $g(t)$ (assuming the basic pulse shape to be causal) leads to a *non-causal* impulse response of the receive filter. In calculations, this fact can be ignored. Of course, non-causal filters cannot be implemented in practice. Consequently, a *delay* has to be introduced in order to enable a causal impulse response $h_R(t)$. The sampling instants then have to be shifted by the same delay. Hence, the parameter T_d is required when implementing systems; it may be ignored in the theoretical analysis.

Example 3.2 Pulse Shape, Matched Filter, Detection Signal

Assume that the basic pulse shape $g(t)$ has the given (real-valued) triangular form with causal support in $t \in [0, \theta]$. Then, if no delay is introduced, the corresponding matched filter would be strictly non-causal and the optimum detection instant would simply be $T_d = 0$. This is shown in Figure 3.16.

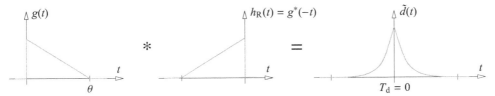

Figure 3.16 Pulse shape $g(t)$ and corresponding matched filter. $T_d = 0$.

If a delay of at least θ is introduced, the matched filter can be implemented in a causal way. Choosing $T_d = \theta$, the situation is shown in Figure 3.17. Now, the optimum detection instant is $t = T_d = \theta$.

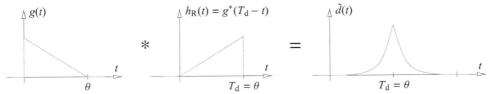

Figure 3.17 Pulse shape $g(t)$ and corresponding matched filter. $T_d = \theta$.

3.4.2 Intersymbol Interference-Free Transmission – the Nyquist Criterion

Up to now we have considered the optimal reception of a single pulse only. However, in PAM schemes, a pulse train is transmitted. Owing to linearity of filtering and sampling, the result can be directly transferred to the PAM situation. For illustration, a block diagram of the PAM transmission over the AWGN channel including matched filtering, sampling, and decision is depicted in Figure 3.18.

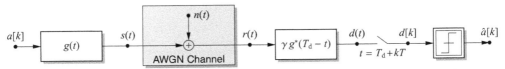

Figure 3.18 Block diagram of the PAM transmission over the AWGN channel including receiver-side processing.

Now, the useful signal (without noise) after the matched filter is given by

$$\begin{aligned}
\tilde{d}(t) &= s(t) * \gamma\, g^*(T_d - t) \\
&= \left(\sum_\kappa a[\kappa]\, g(t - \kappa T) \right) * \gamma\, g^*(T_d - t) \\
&= \gamma \sum_\kappa a[\kappa]\, (g(t - \kappa T) * g^*(T_d - t)) \\
&= \gamma \sum_\kappa a[\kappa]\, \varphi_{gg}(t - \kappa T - T_d) \quad (3.42)
\end{aligned}$$

and after sampling we arrive at

$$\tilde{d}[k] \overset{\text{def}}{=} \tilde{d}(T_{\text{d}} + kT)$$
$$= \gamma \sum_\kappa a[\kappa]\, \varphi_{gg}(T_{\text{d}} + kT - T_{\text{d}} - \kappa T)$$
$$= \gamma \sum_\kappa a[\kappa]\, \varphi_{gg}((k - \kappa)T)\,. \qquad (3.43)$$

Note that the useful detection signal $\tilde{d}(t)$ is again a PAM signal. Instead of the basic pulse shape $g(t)$, due to receiver-side filtering, the pulse is now given by the (time-shifted) autocorrelation $\varphi_{gg}(t)$ of the basic pulse $g(t)$.

Based on the samples $d[k]$, *decisions* on the amplitude coefficients should be taken – estimates $\hat{a}[k]$ should be produced. It is an obvious demand that at time step k (detection time $T_{\text{d}} + kT$) the estimate on the current amplitude coefficient $a[k]$ should be generated independently of previously and subsequently transmitted amplitude coefficients. Hence, the question is as follows.

> *What are the constraints on the pulse shape $g(t)$ and, thus, on its autocorrelation $\varphi_{gg}(t)$, such that symbol-by-symbol decisions on the coefficients $a[k]$ are possible without interference from prior/posterior transmitted symbols, that is, without intersymbol interference (ISI)?*

Intersymbol-interference-free decision instants are present, if the sum in (3.43) vanishes and, thus, if $\tilde{d}[k]$ is only dependent on a single amplitude coefficient, that is, if

$$\tilde{d}[k] = \tilde{d}(kT + T_{\text{d}}) \overset{!}{=} a[k]\,. \qquad (3.44)$$

Comparing this with

$$\tilde{d}[k] = \gamma \sum_\kappa a[\kappa]\, \varphi_{gg}((k - \kappa)T)\,,$$

this is only possible for

$$\varphi_{gg}(\lambda T) = \begin{cases} \frac{1}{\gamma}, & \lambda = 0 \\ 0, & \lambda \in \mathbb{Z} \setminus \{0\} \end{cases}\,. \qquad (3.45)$$

Hence, the acf $\varphi_{gg}(\tau)$ of the pulse $g(t)$ has to have equidistant zeros. Functions with this property, that is, zeros at λT, with $\lambda \in \mathbb{Z} \setminus \{0\}$, are called *Nyquist functions*.[4]

For ISI-free detection, the *autocorrelation* $\varphi_{gg}(\tau)$ of the pulse $g(t)$ has to fulfill the (first) *Nyquist criterion*

$$\varphi_{gg}(\lambda T) = \int_{-\infty}^{\infty} g(\tau + \lambda T)\, g^*(\tau)\, \mathrm{d}\tau \overset{!}{=} \begin{cases} E_g, & \lambda = 0 \\ 0, & \lambda \in \mathbb{Z} \setminus \{0\} \end{cases}\,. \qquad (3.46)$$

---

[4] In honor of Harry Nyquist (1889–1976). The theorem was formulated in 1928. His name is Swedish and is pronounced [ny:kvist].

3.4 Receiver Structure and Choice of the Basic Pulse Shape

Comparing (3.43) and (3.44), it turns out that for unit-gain end-to-end transmission the scaling factor γ of the matched filter has to be

$$\gamma = \frac{1}{E_g} . \quad (3.47)$$

The *autocorrelation* $\varphi_{gg}(\tau)$ of the *pulse* $g(t)$ has to be a Nyquist function – not the pulse itself! The *pulse* $g(t)$ cascaded with its corresponding *matched filter* $g^*(-t)$ has to meet the Nyquist criterion, that is, $\varphi_{gg}(t) = g(t) * g^*(-t)$.

A pulse $g(t)$ whose autocorrelation $\varphi_{gg}(\tau)$ is a Nyquist function is a *square-root Nyquist function*, that is, a $\sqrt{\text{Nyquist}}$ *pulse*.

> For optimum intersymbol-interference-free PAM transmission over the AWGN channel, the basic pulse shape $g(t)$ has to be a $\sqrt{\text{Nyquist}}$ pulse.

Since

$$\varphi_{gg}(\lambda T) = \int g(\tau + \lambda T) \, g^*(\tau) \, d\tau \stackrel{!}{=} 0 , \quad \lambda \in \mathbb{Z} \setminus \{0\} , \quad (3.48)$$

the pulse $g(t)$ has to be *orthogonal* to temporarily shifted versions of itself. The denominations "square-root Nyquist" and "temporal orthogonal" are thus synonymous.

An example of a Nyquist function is plotted in Figure 3.19.

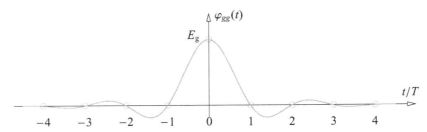

Figure 3.19 Exemplary Nyquist function. Equidistant zeros at λT, with $\lambda \in \mathbb{Z} \setminus \{0\}$.

Discrete-Time End-to-End Model

At the input of the pulse-shaping filter at the transmitter side, the discrete-time sequence $\langle a[k] \rangle$ is present. After matched filtering and sampling, again a discrete-time sequence, now the detection sequence $\langle d[k] \rangle$, is obtained. Hence, from end to end, a *discrete-time channel model* is established (see Figure 3.20).

Owing to the square-root Nyquist property of the basic pulse shape, the end-to-end input/output relation reads ($\gamma = 1/E_g$)

$$d[k] = a[k] + n[k] . \quad (3.49)$$

3 Digital Pulse-Amplitude Modulation

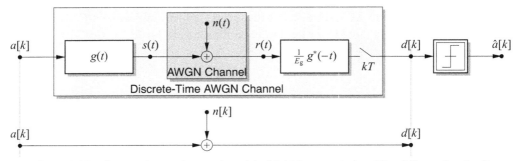

Figure 3.20 Discrete-time end-to-end model of PAM transmission. The different levels of abstraction (inner continuous-time AWGN channel, outer discrete-time AWGN channel) are indicated.

Considering (3.40), the autocorrelation sequence of the additive Gaussian noise $n[k]$ is given by

$$\phi_{nn}[\kappa] = \phi_{n_d n_d}(\kappa T) \tag{3.50}$$

$$= N_0 \gamma^2 \varphi_{gg}(\kappa T) = \begin{cases} \frac{N_0}{E_g}, & \kappa = 0 \\ 0, & \kappa \in \mathbb{Z} \setminus \{0\} \end{cases}. \tag{3.51}$$

This means that the discrete-time noise $n[k]$ is *white* with variance $\sigma_n^2 = \frac{N_0}{E_g}$. In summary:

> *When employing a square-root Nyquist basic pulse shape $g(t)$ in PAM schemes over the AWGN channel, a discrete-time AWGN end-to-end channel is present. Using pulses with energy E_g and having a noise power spectral density (ECB domain) N_0, for unit-gain transmission the noise power $\sigma_n^2 = \frac{N_0}{E_g}$ is active.*

The Nyquist Criterion in the Frequency Domain

Up to now, the Nyquist criterion has been stated in the time domain. Using the properties of the Fourier transform, a respective criterion can be given in the frequency domain.

The orthogonality condition (3.48) in the time domain is

$$\int_{-\infty}^{\infty} g(t + \lambda T) \, g^*(t) \, \mathrm{d}t = E_g \, \delta[\lambda], \quad \lambda \in \mathbb{Z} \tag{3.52}$$

or compactly

$$g(t) * g^*(-t) \Big|_{t=\lambda T} = E_g \, \delta[\lambda] \tag{3.53}$$

($\delta[\cdot]$ is the discrete-time unit pulse).

3.4 Receiver Structure and Choice of the Basic Pulse Shape

Using the relations of the Fourier transform (cf. Appendix A)

$$g(t) \circ\!\!-\!\!\bullet G(f) \qquad g(t) * g^*(-t) \circ\!\!-\!\!\bullet |G(f)|^2 , \qquad (3.54)$$

$$x(t) \circ\!\!-\!\!\bullet X(f) \qquad x[k] \stackrel{\text{def}}{=} x(kT) \circ\!\!-\!\!\bullet \tfrac{1}{T} \sum_\mu X\left(f - \tfrac{\mu}{T}\right) , \qquad (3.55)$$

we have the Nyquist criterion in the frequency domain:

The periodic continuation of the spectrum $|G(f)|^2$ of a time function $\varphi_{gg}(t)$, which fulfills the first Nyquist criterion, sums up to a constant, that is

$$\sum_{\mu=-\infty}^{\infty} |G\left(f - \tfrac{\mu}{T}\right)|^2 = E_g T . \qquad (3.56)$$

Examples of √Nyquist and Nyquist Functions

Rectangular Pulse in the Time Domain A rectangular pulse in the time domain of duration (at most) T (this situation is called *hard keying*) is a √Nyquist pulse. Denoting the pulse amplitude by $\hat{g}$, the pulse is given by (non-causal description; for definition of rect, si, and tri, see Section A.1 of Appendix A)

$$g(t) = \hat{g}\, \text{rect}(t/T)$$
$$\circ\!\!-\!\!\bullet$$
$$G(f) = \hat{g}\, T\, \text{si}(\pi f T) . \qquad (3.57)$$

The pulse energy is

$$E_g = \hat{g}^2\, T \qquad (3.58)$$

and the autocorrelation is a Nyquist pulse, given by

$$\varphi_{gg}(t) = E_g\, \text{tri}\left(\tfrac{t}{T}\right)$$
$$\circ\!\!-\!\!\bullet$$
$$\Phi_{gg}(f) = E_g T\, \text{si}^2(\pi f T) . \qquad (3.59)$$

Here we have

$$\sum_\mu E_g T\, \text{si}^2\left(\pi \left(f - \tfrac{\mu}{T}\right) T\right) = E_g T , \qquad (3.60)$$

as shown in Figure 3.21.

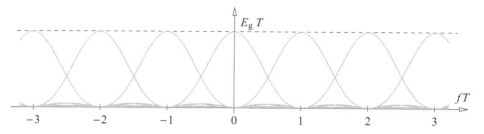

Figure 3.21 Superposition of shifted si² functions to a constant (dashed).

Time-Limited Pulse In general, any pulse $g(t)$ that is time limited to the symbol interval T, specifically

$$g(t) = 0 \ , \qquad t \notin \left(-\frac{T}{2}, \frac{T}{2}\right) , \tag{3.61}$$

has an autocorrelation $\varphi_{gg}(t)$ that is limited to $(-T, T)$. Hence, $g(t)$ is a $\sqrt{\text{Nyquist}}$ pulse.

Rectangular Pulse in the Frequency Domain The counterpart to the time-domain rectangular pulse is a rectangular pulse in the frequency domain. Again, we see that $1/T$ is the smallest possible bandwidth of a basic pulse and therefore for a PAM signal with ISI-free detection. For smaller bandwidth, a spectral continuation, which results in a constant, cannot exist. The $\sqrt{\text{Nyquist}}$ pulse is given by (non-causal description)

$$\begin{aligned} g(t) &= \hat{g} \ \text{si}(\pi t/T) \\ &\circ\!\!-\!\!\bullet \\ G(f) &= \hat{g} T \ \text{rect}(fT) \end{aligned} \tag{3.62}$$

with pulse energy

$$E_g = \hat{g}^2 \ T \tag{3.63}$$

and corresponding Nyquist pulse

$$\begin{aligned} \varphi_{gg}(t) &= E_g \ \text{si}(\pi t/T) \\ &\circ\!\!-\!\!\bullet \\ \Phi_{gg}(f) &= E_g T \ \text{rect}(fT) \ . \end{aligned} \tag{3.64}$$

Obviously, we have

$$\sum_\mu E_g T \ \text{rect}(fT - \mu) = E_g T \ , \qquad \forall f , \tag{3.65}$$

as shown in in Figure 3.22.

Nyquist Pulse with Trapezoidal Shape in the Frequency Domain A pulse $g(t)$, whose squared magnitude spectrum has the trapezoidal shape depicted in Figure 3.23, is a $\sqrt{\text{Nyquist}}$ pulse.

3.4 Receiver Structure and Choice of the Basic Pulse Shape

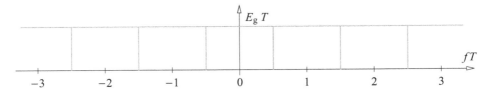

Figure 3.22 Superposition of shifted rect functions to a constant.

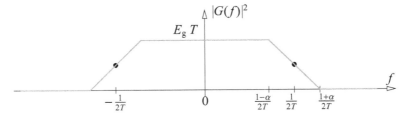

Figure 3.23 Trapezoidal shape in the frequency domain.

The parameter $\alpha \in [0, 1]$ is called the *roll-off factor* or *bandwidth-excess factor* and enables a trade-off between

- $\alpha = 0$: rectangular shape in the frequency domain
- $\alpha = 1$: triangular shape in the frequency domain.

As the RF bandwidth is given by

$$B_{\mathrm{RF}} = (1 + \alpha) \frac{1}{T} \tag{3.66}$$

via the parameter α, the relative extension of the bandwidth over its absolute minimum $1/T$ can be controlled.

Pulses with Point-Symmetrical Slopes in the Frequency Domain In generalization of the trapezoidal shape, any spectrum whose squared magnitude has slopes symmetric about the points

$$\left(f = \pm \frac{1}{2T}, \; |G(f)|^2 = \frac{E_g T}{2} \right) \tag{3.67}$$

is a spectrum of a $\sqrt{\text{Nyquist}}$ pulse. Such a symmetric roll-off is called the *Nyquist slope of a spectrum*.

Cosine-Roll-Off Nyquist Pulses An often-employed Nyquist slope has a cosine shape. A cosine-roll-off Nyquist pulse in the frequency domain (roll-off factor α) is given as

$$|G(f)|^2 = E_g T \cdot \begin{cases} 1, & |f| < \frac{1-\alpha}{2T} \\ \frac{1}{2}\left(1 - \sin\left(\frac{\pi T}{\alpha}(|f| - \frac{1}{2T})\right)\right), & \frac{1-\alpha}{2T} \leq |f| \leq \frac{1+\alpha}{2T} \\ 0, & |f| > \frac{1+\alpha}{2T} \end{cases} \tag{3.68}$$

and visualized in Figure 3.24.

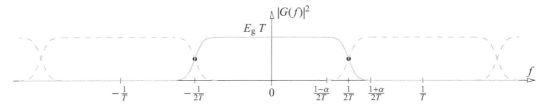

Figure 3.24 Cosine-roll-off Nyquist pulse (and periodic repetitions) in the frequency domain.

Assuming a linear phase, the corresponding symmetric √Nyquist pulse in the time domain reads

$$g(t) = \sqrt{\frac{E_g}{T}} \frac{4\alpha t \cos(\pi(1+\alpha)t/T) + T\sin(\pi(1-\alpha)t/T)}{\pi t(1-(4\alpha t/T)^2)}. \quad (3.69)$$

Note that L'Hôpital's rule gives for $t = 0$, $g(0) = \frac{4\alpha + \pi(1-\alpha)}{\pi}$ and for $|t| = \frac{T}{4\alpha}$ we have $g\left(\pm\frac{T}{4\alpha}\right) = \left(4\alpha \cos\left(\pi\frac{(1+\alpha)}{(4\alpha)}\right) - \pi(1+\alpha)\sin\left(\pi\frac{(1+\alpha)}{(4\alpha)}\right) + \pi(1-\alpha)\cos\left(\pi\frac{(1-\alpha)}{(4\alpha)}\right)\right)/(-2\pi)$.

Example 3.3 Cosine-Roll-Off √Nyquist and Nyquist Pulse: $\alpha = 1/3$

Figure 3.25 shows a cosine-roll-off √Nyquist (a) and the corresponding Nyquist pulse (b) together with their spectra.

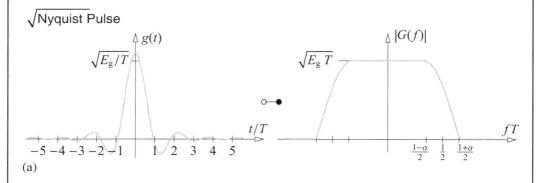

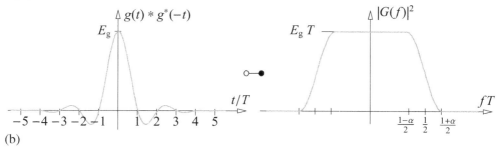

Figure 3.25 (a) Cosine-roll-off √Nyquist and (b) corresponding Nyquist pulse together with their spectra.

Observe that the √Nyquist pulse does not have equidistant zeros at kT, as the Nyquist pulse does. Only the shifted versions of the spectrum $|G(f)|^2$ of the Nyquist pulse (b, right-hand image) add up to a constant.

Bandwidth of PAM Signals Using √Nyquist Functions

Now, the bandwidth and thus the spectral efficiency of PAM signals using a √Nyquist pulse with spectral roll-off can be stated. For that, we have to distinguish between baseband and carrier-modulated transmission. With $R_T = R/T$ and $\Gamma = R_T/B_{RF}$, we get

baseband transmission:

$$B_{RF} = \frac{1}{2T}(1+\alpha), \qquad \Gamma = \frac{2R}{1+\alpha}; \qquad (3.70)$$

carrier-modulated transmission:

$$B_{RF} = \frac{1}{T}(1+\alpha), \qquad \Gamma = \frac{R}{1+\alpha}. \qquad (3.71)$$

Figure 3.26 compares the psds of PAM ($\sigma_a^2 = 1$) when using hard keying ($g(t)$ is a rect pulse) and a cosine-roll-off √Nyquist pulse with $\alpha = 0.5$. The average psd of hard keying has significant side lobes that decay only with $1/f$; the bandwidth is unlimited. Using a cosine-roll-off √Nyquist pulse, that is, *soft keying*, a compact, bandlimited psd is obtained (however, the time-domain pulse has infinite duration). This illustrates the effect of soft keying on the bandwidth efficiency.

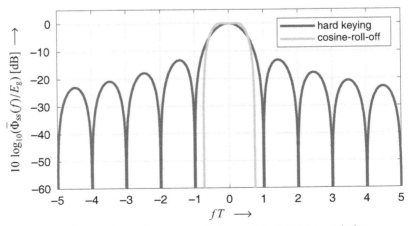

Figure 3.26 Comparison of average psds (log scale). PAM transmission ($\sigma_a^2 = 1$) employing hard keying and cosine-roll-off √Nyquist pulse ($\alpha = 0.5$).

Example 3.4 Phasor Demo

For visualization, examples of phasors are available on the companion web page. The demos run directly in the web browser. The phasor can be rotated with the mouse, moved (Ctrl + mouse), and zoomed (Shift + mouse).

The phasors are plotted in three-dimensional space; the real and imaginary parts of the complex baseband signal $s(t)$ over time (normalized to the symbol duration T). A number of possible signals (corresponding to different sequences of amplitude coefficients) are plotted on top of each other. One of them is plotted in red. Rotate the phasor such that the real part (in-phase component) or the imaginary part (quadrature component) is plotted over time (eye pattern, cf. next subsection), and rotate the phasor such that the complex plane (imaginary vs. real part) is visible.

The following phasors of signals using 4QAM are available.

Phasor 1: Transmit signal $s(t)$ for rectangular basic pulse shape.
Phasor 3: Detection signal $d(t)$ (after matched filter) for rectangular basic pulse shape.
Phasor 4: Transmit signal $s(t)$ for $\sqrt{\text{Nyquist}}$ cosine-roll-off ($\alpha = 0.5$) basic pulse shape.
Phasor 5: Detection signal $d(t)$ for $\sqrt{\text{Nyquist}}$ cosine-roll-off ($\alpha = 0.5$) basic pulse shape (hence, cosine-roll-off Nyquist end-to-end pulse).

3.4.3 Eye Pattern

We come back to the detection signal, that is, the signal after the matched filter. Looking at this signal, the sensitivity to noise and errors in timing synchronization can be assessed. Figure 3.27 shows the block diagram of the PAM receiver (w.r.t. RF signals and non-causal matched filters). The basic pulse shape is assumed to be real-valued, that is, $g(t) \in \mathbb{R}$.

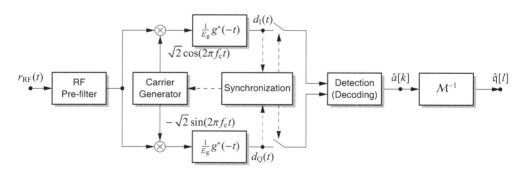

Figure 3.27 Block diagram of the PAM receiver (RF signals).

A depiction of *all* sample functions of a cyclic-stationary stochastic process in a *single diagram* creates a periodic pattern with period T, the so-called *eye pattern*. Thereby, the detection signal $d(t)$, that is, $d_I(t) = \text{Re}\{d(t)\}$ and $d_Q(t) = \text{Im}\{d(t)\}$, is

3.4 Receiver Structure and Choice of the Basic Pulse Shape

visualized for all possible sequences $a[k]$ of amplitude coefficients. In practice (e.g., in labs when measuring specific transmission systems), the eye pattern is obtained using an oscilloscope with large fade-out time if it is triggered by the symbol clock.

In the subsequent example (Figure 3.28), typical eye patterns are shown.

Example 3.5 Eye Pattern of ASK Transmission

The eye pattern of 2ASK with cosine-roll-off $\sqrt{\text{Nyquist}}$ pulse $g(t)$ (hence, end-to-end Nyquist pulse $\varphi_{gg}(t)$) is shown in Figure 3.28. Different roll-off factors α are used. When using 4QAM, the same diagrams are obtained per quadrature component.

As can be seen, increasing α, the *horizontal eye opening* becomes larger. This leads to a higher robustness against timing errors (wrong sampling instants, called symbol-timing *clock jitter*). In case of perfect sampling at $t = kT$, in all cases the correct values ± 1 are obtained. Since for larger α the spectral efficiency is lower, a trade-off between bandwidth efficiency and robustness against symbol-timing clock jitter is enabled.

Figures 3.29 and 3.30 show the eye patterns of 4ASK and 8ASK, respectively, with cosine-roll-off $\sqrt{\text{Nyquist}}$ pulse $g(t)$ with roll-off factor $\alpha = 0.33$.

As can be seen by a comparison of the figures, when increasing M, the cardinality of the signal constellation, the horizontal eye opening decreases. The larger the signal constellation, the more sensitive is the timing.

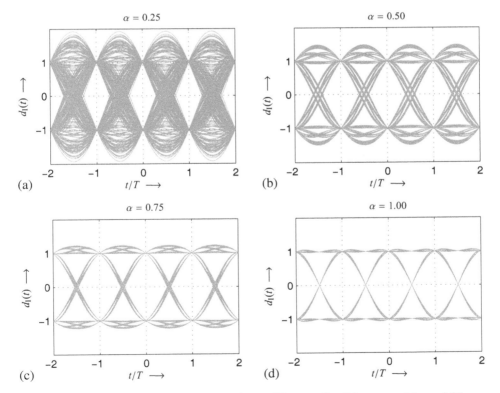

Figure 3.28 Eye pattern of 2ASK transmission. Different roll-off factors α: (a) $\alpha = 0.25$; (b) $\alpha = 0.50$; (c) $\alpha = 0.75$; (d) $\alpha = 1.00$.

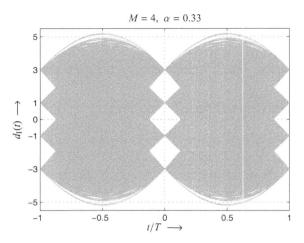

Figure 3.29 Eye pattern of 4ASK transmission. Roll-off factor $\alpha = 0.33$.

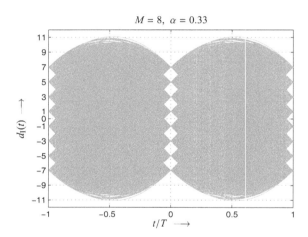

Figure 3.30 Eye pattern of 8ASK transmission. Roll-off factor $\alpha = 0.33$.

3.4.4 Crest Factor of PAM Transmit Signals

The "peakiness" of a signal can be quantified by the ratio between the maximum signal amplitude that can occur and its root mean square (rms) value. This so-called *crest factor* is an important parameter concerning practical implementation of power amplifiers. The crest factor of the ECB transmit signal $s(t)$ with average power S_s is defined as

$$\zeta \stackrel{\text{def}}{=} \frac{\max_{\forall t} |s(t)|}{\sqrt{S_s}} \, . \tag{3.72}$$

Equivalently, the *peak-to-average power ratio (PAPR)* can be given. This is related to the crest factor by

$$\text{PAPR} \overset{\text{def}}{=} \frac{\max_{\forall t} |s(t)|^2}{S_s} = \zeta^2 . \tag{3.73}$$

Note that due to the multiplication with sin/cos carriers in the upconversion, the crest factor of the RF signal is higher by a factor of $\sqrt{2}$ (for PAPR, by a factor of 2) than the crest factor of the ECB signal. The quantities in the RF domain are thus 3 dB worse than in the ECB domain. Here, we always consider the ECB signals.

The power amplification of a continuous-time transmit signal with a high dynamic range may be rather inefficient as a corresponding high-output back-off (see (4.3)) has to be chosen in order to avoid serious nonlinear signal distortions. Usually, the power consumption of a high-power amplifier (HPA) is dominated by its nominal output peak power, not by the average power of its output signal. Thus, the PAPR of the transmit signal of a digital communication scheme influences its power efficiency in a direct way.[5]

Taking (3.1) into account, the crest factor of a PAM transmit signal is then given by

$$\zeta = \frac{\max_{\forall t, \langle a[k] \rangle} \left| \sum_k a[k] g(t - kT) \right|}{\sqrt{\mathrm{E}\{|a|^2\} \cdot E_g/T}} . \tag{3.74}$$

Using $|x \cdot y| = |x| \cdot |y|$ and the triangle inequality for complex numbers $|x + y| \leq |x| + |y|$, $x, y \in \mathbb{C}$, we have

$$\max_{\forall t, \langle a[k] \rangle} \left| \sum_k a[k] g(t - kT) \right| \leq \max_{a \in \mathcal{A}} |a| \cdot \max_{\forall t} \sum_k |g(t - kT)| , \tag{3.75}$$

and thus

$$\zeta \leq \zeta_a \cdot \zeta_g , \tag{3.76}$$

with

$$\zeta_a \overset{\text{def}}{=} \frac{\max_{a \in \mathcal{A}} |a|}{\sqrt{\mathrm{E}\{|a|^2\}}} \qquad \text{crest factor of signal constellation } \mathcal{A}, \tag{3.77}$$

$$\zeta_g \overset{\text{def}}{=} \frac{\max_{\forall t} \sum_k |g(t - kT)|}{\sqrt{E_g/T}} \qquad \text{crest factor of basic pulse } g(t). \tag{3.78}$$

Crest Factor ζ_a of the PAM Constellation $\mathcal{A}$

Tables 3.2–3.5 (later) on the parameters of commonly used PAM signal constellations also include the crest factor ζ_a according to (3.77). As can be seen, when

---

[5] It should be mentioned that there exist methods to control the supply voltage for the HPA according to a short-time average of its upcoming input signal by a sophisticated dynamic power control unit with high efficiency. By this, a high PAPR problem can be mitigated to some extent.

increasing the cardinality, the crest factor usually increases too. Since in PSK all signal points have magnitude one, peak power and average power coincide; the crest factor is $\zeta_a = 1$. However, this is only the factor concerning the constellation; the actual crest factor ζ of the continuous-time transmit signal may be higher.

Crest Factor ζ_g of the Basic Pulse Shape $g(t)$

The second factor in (3.76) is the contribution of the basic pulse shape $g(t)$ according to (3.78). Typically, this factor has to be determined numerically. For the important example of cosine-roll-off $\sqrt{\text{Nyquist}}$ pulses according to (3.69), the crest factor ζ_g of the basic pulse is given by

$$\zeta_g = \max_{\forall t \in [-T/2, T/2]} \sum_k \left| \frac{4\alpha(t-kT)\cos(\pi(1+\alpha)(t-kT)/T) + T\sin(\pi(1-\alpha)(t-kT)/T)}{\pi(t-kT)(1-(4\alpha(t-kT)/T)^2)} \right|, \quad (3.79)$$

which means that $|g(t)|$ is aliased (folded) modulo T and the maximum over one period is searched. This is visualized in Figure 3.31. In part (a) $\alpha = 0.1$ is used, and in part (b) $\alpha = 0.4$. The pulse $g(t)$ is shown in dark gray, the shifted versions $|g(t-kT)|$ in light gray, and the aliased version in medium gray.

For smaller α, the side lobes of the pulse are more pronounced, leading to a larger absolute sum and thus to a larger crest factor. Note that the maximum peak in (3.79) typically occurs at $t/T = -1/2$.

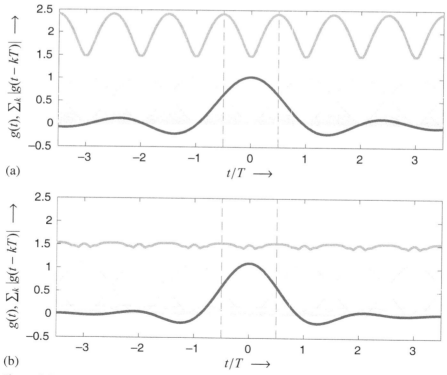

Figure 3.31 Visualization of the sum in (3.79). (a) $\alpha = 0.1$ ($\zeta_g = 2.43$ or $20\log_{10}(\zeta_g) = 7.69$ dB); (b) $\alpha = 0.4$ ($\zeta_g = 1.54$ or $20\log_{10}(\zeta_g) = 3.74$ dB).

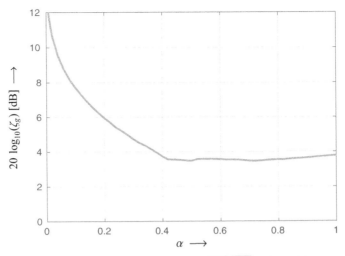

Figure 3.32 Crest factor ζ_g (in dB) of $\sqrt{\text{Nyquist}}$ pulses with cosine-roll-off spectral shape over the roll-off factor α.

The crest factor ζ_g of $\sqrt{\text{Nyquist}}$ pulses with square-root cosine-roll-off spectral shape is shown in Figure 3.32 over the roll-off factor α.

As long as α is not smaller than approximately 0.33, a moderate crest factor (around 4 dB) is obtained. If α is getting smaller, the crest factor increases. However, the bandwidth of the pulse decreases. Hence, a trade-off between bandwidth efficiency and power efficiency, here in the sense of how efficiently power amplification can be performed, is present. Note that for $\alpha \to 0$ the bandwidth is the smallest possible one ($1/T$, brick-wall spectrum) and the basic pulse shape is given by the sinc function. Unfortunately, a sinc function corresponds to an infinite crest factor of the PAM signal due to the divergence of the harmonic series in the superimposed $|\operatorname{si}(\cdot)|$ functions.

3.4.5 Remarks on the Roll-Off Factor and the Nyquist Bandwidth

As demonstrated in Figures 3.28 through 3.30, the roll-off factor α of Nyquist pulses has a great impact on the horizontal opening of the eye pattern, especially for higher-order modulation. Thus, in technical practice, it is difficult to implement schemes with $\alpha < \frac{1}{3}$ when no adaptive equalization is applied or with $\alpha < \frac{1}{10}$, respectively, for receiver frontends that employ digital signal processing together with control loops for adaptive equalization in order to guarantee ISI-free sampling.

Note that a very low roll-off factor (i.e., $\alpha \to 0$) is practically impossible. Some reasons for this restriction (besides others) are as follows.

- The horizontal eye opening tends to zero, that is, the sensitivity against symbol-timing clock jitter increases to infinity.
- The (useful) signal loses its property to be cyclo-stationary and becomes stationary (at least w.r.t. the first- and second-order moments). As a consequence,

symbol-clock extraction and/or symbol-clock phase synchronization will be extremely difficult or even impossible.
- The peak-to-average power ratio of the transmit signal $s(t)$ grows significantly (and tends to infinity for $\alpha \to 0$). Thus, the high-power amplifier in the transmitter frontend has to operate at large back-offs, leading to a very poor power efficiency (for details, see Section 4.1).
- Filters with extremely steep slopes are very difficult to implement and cause very high latency.

3.5 Detection of the PAM Signal Points

We now know how to filter the PAM receive signal and when to sample. The end-to-end discrete-time channel has been derived. The next block in the receiver processing chain is the detector. Based on the detection sample $d[k]$ symbol-by-symbol estimates $\hat{a}[k]$ for the amplitude coefficients and thus the transmitted information have to be generated.

The situation we are faced with can be stated as follows: PAM transmission with some given constellation $\mathcal{A}$ is considered. As shown above, employing a $\sqrt{\text{Nyquist}}$ pulse $g(t)$ at the transmitter and the corresponding matched filter at the receiver and assuming a continuous-time AWGN channel, regardless of the actual pulse shape, an end-to-end discrete-time AWGN channel model is present. Hence, the channel input/output relation reads

$$d[k] = a[k] + n[k], \quad \text{with} \quad a[k] \in \mathcal{A}. \tag{3.80}$$

This means that $a[k]$ is transmitted and $d[k]$ (a noisy version of $a[k]$) is observed.

The question now is how to generate the desired estimate $\hat{a}[k]$ based on $d[k]$. As this can be performed independently symbol by symbol for schemes without channel coding, in the following we may drop the discrete time index k.

3.5.1 Maximum-a-Posteriori (MAP) Strategy

Since both data a and noise n are random quantities, we have to apply an optimum decision rule from statistics. The main idea is to calculate the *probability* for each amplitude coefficient $a \in \mathcal{A}$ given the observation d, that is, the *conditional probability* $\Pr\{a \mid d\}$ (a-posteriori probability), and to decide in favor of the signal point that has the largest probability. This is stated in the following.

> **Maximum-a-Posteriori (MAP) Strategy**
> *Decide in favor of the signal point $a \in \mathcal{A}$ that has the largest probability given (after, a-posteriori) the observation of d:*
>
> $$\hat{a} = \underset{a \in \mathcal{A}}{\text{argmax}} \; \Pr\{a \mid d\}.$$

> In other words, the MAP strategy seeks the maximum probability when looking backward from the observation d to the message a.
>
> The MAP strategy leads to the smallest possible probability of symbol errors.

Box 3.2 argmax and argmin

Let $f(x)$ be a real-valued function of the argument x. The operator "argmax" returns the *maximizer* of the function, that is, the argument $x_{\max}$ for which the function is maximized. In contrast, the operator "max" returns the *maximum*, that is, the function value at the optimal point. Thus we have

$$x_{\max} = \underset{x}{\mathrm{argmax}}\, f(x) \quad \Leftrightarrow \quad \underset{x}{\max}\, f(x) = f(x_{\max}).$$

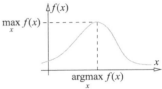

Figure 3.33 argmax and max.

This is visualized in Figure 3.33.

The same holds for argmin and min, where the *minimizer* of the function and the *minimum*, respectively, are returned.

3.5.2 Maximum-Likelihood (ML) Strategy

Using Bayes' rule, and since the detection variable d is a continuous random variable and thus has to be characterized by its pdf rather than a probability, we can write

$$\Pr\{a \mid d\} = f_d(d \mid a) \cdot \frac{\Pr\{a\}}{f_d(d)}, \tag{3.81}$$

where $f_d(d \mid a)$ is the pdf for the receive symbol d given (assuming) the transmitted data symbol a. This pdf fully characterizes the action of the channel.

When looking at (3.81), we see that the denominator $f_d(d)$ does not depend on a and is thus irrelevant for the maximizer. We thus have

$$\hat{a} = \underset{a \in \mathcal{A}}{\mathrm{argmax}}\, \Pr\{a \mid d\} = \underset{a \in \mathcal{A}}{\mathrm{argmax}}\, f_d(d \mid a)\, \Pr\{a\}. \tag{3.82}$$

Typically, the signal points are drawn *uniformly* from the constellation $\mathcal{A}$ (equiprobable signaling). Then, $\Pr\{a\} = 1/M$, $\forall a$, and consequently this term is also irrelevant for maximization.

This leads to the following decision rule, which is optimum for *equiprobable* signal points:

> **Maximum-Likelihood (ML) Strategy**
>
> *Decide in favor of the signal point $a \in \mathcal{A}$ for which the observation d has the largest probability (density)*
>
> $$\hat{a} = \underset{a \in \mathcal{A}}{\operatorname{argmax}}\ f_d(d \mid a) .$$
>
> In other words, the ML strategy seeks the amplitude coefficient a, from which the observation d emerges with highest probability (density). It seeks the maximum probability when looking forward from message a to observation d.

As already stated, the ML strategy is only equivalent to the MAP strategy when all messages have the same a-priori probabilities. For unequal a-priori probabilities, the ML strategy is sub-optimal. Detection methods implementing the ML strategy can be derived in a direct way from the channel statistics and, therefore, are sometimes easier to establish than the MAP strategy.

ML Detection on the AWGN Channel

The above decision rules are valid in general, for all types of channels. However, we are particularly interested in the ML rule specialized to the AWGN channel.

Since for the AWGN channel the noise has a Gaussian pdf, the conditional pdf of the observation $d = a + n$ (ECB signals) is given by

$$f_d(d \mid a) = f_n(d - a) = \frac{1}{\pi \sigma_n^2} e^{-|d-a|^2/\sigma_n^2} . \tag{3.83}$$

This leads to

ML rule

$$\hat{a} = \underset{a \in \mathcal{A}}{\operatorname{argmax}}\ f_d(d \mid a)$$

pdf assuming AWGN channel

$$= \underset{a \in \mathcal{A}}{\operatorname{argmax}} \left(\frac{1}{\pi \sigma_n^2} e^{-|d-a|^2/\sigma_n^2} \right)$$

$\log(\cdot)$ is strictly increasing

$$= \underset{a \in \mathcal{A}}{\operatorname{argmax}} \left(\log \left(\frac{1}{\pi \sigma_n^2} e^{-|d-a|^2/\sigma_n^2} \right) \right)$$

$\log(ab) = \log(a) + \log(b)$

$$= \underset{a \in \mathcal{A}}{\operatorname{argmax}} \left(\log \left(\frac{1}{\pi \sigma_n^2} \right) + \log \left(e^{-|d-a|^2/\sigma_n^2} \right) \right)$$

constants are irrelevant

$$= \underset{a \in \mathcal{A}}{\operatorname{argmax}} \left(-|d-a|^2/\sigma_n^2 \right)$$

3.5 Detection of the PAM Signal Points

$$\operatorname{argmax}(-x) = \operatorname{argmin}(x)$$

$$= \underset{a \in \mathcal{A}}{\operatorname{argmin}} |d - a|^2 . \tag{3.84}$$

Note that $|d - a|^2$ is the squared Euclidean distance between observation d and hypothesis a in the complex plane. Hence, rather than calculating probabilities, squared Euclidean distances have to be compared on the AWGN channel (since the Gaussian density is proportional to the exponential of the squared distance). In summary, for the AWGN channel, the ML rule is simply given by

> *ML detection on the AWGN channel is accomplished by minimizing the (squared) **Euclidean distance** between observation d and hypothesis a.*

Note that this basic principle is also true if channel coding is applied. Here, distances in a high-dimensional space have to be evaluated. For details, see Chapter 6.

Decision Regions for ML Decoding

Using ECB signals, the detection variable d is a complex number, a point in the complex plane. Applying the ML rule, one unique point a from the signal constellation $\mathcal{A}$ is assigned to each point $d \in \mathbb{C}$ (ties are resolved arbitrarily). Hence, given an M-ary signal constellation, the ML decision rule induces M (disjoint) *decision regions* $\mathcal{G}_a$. For each signal point a, a corresponding decision region exists. All points within the region $\mathcal{G}_a$ are closer (w.r.t. Euclidean distance) to the corresponding signal point a than to all other signal points. The union of all decision regions gives the entire complex plane.

The boundaries of the decision regions are given by the *perpendicular bisectors* between pairs of signal points. For the situation of two signal points, Figure 3.34 shows the *decision boundary*. Points located exactly on the boundary have equal distance to the signal points; one can choose freely to which signal point the boundary is assigned.

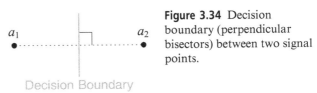

Figure 3.34 Decision boundary (perpendicular bisectors) between two signal points.

Hence we arrive at a *geometrical interpretation* of digital transmission:

- points (in some space) are transmitted;
- the noise shifts the transmitted point randomly;
- the receiver measures Euclidean distances and decides in favor of the point closest to the received point.

Again (see Chapter 6), this geometric interpretation holds in the same way for coded transmission in a corresponding high-dimensional signal space.

Example 3.6 Decision Regions

Figures 3.35 to 3.38 show examples for the decision regions for ASK, PSK, and QAM constellations. The decision boundaries are the dashed lines.

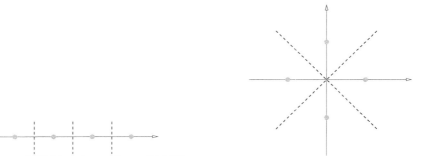

Figure 3.35 Decision regions of 4ASK.

Figure 3.36 Decision regions of 4PSK.

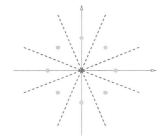

Figure 3.37 Decision regions of 8PSK.

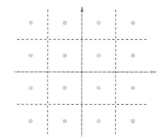

Figure 3.38 Decision regions of 16QAM.

Starting from (3.82) and (3.83), the decision boundaries for the MAP decision rule can be derived in an analogous way as the ML rule for the AWGN channel. If two signal points a_1 and a_2 have different but known a-priori probabilities $\Pr\{a_1\}$ and $\Pr\{a_2\}$, the perpendicular bisector between the corresponding decision regions simply has to be shifted by

$$\frac{\sigma_n^2}{2|a_1 - a_2|} \cdot \log\left(\frac{\Pr\{a_1\}}{\Pr\{a_2\}}\right) \tag{3.85}$$

in direction towards the point with lower a-priori probability.

3.6 Bit and Symbol Error Ratio

After having derived the end-to-end transmission model and the optimum detection strategy, we are ready to calculate the error ratio of digital PAM transmission. Note that in popular science there is a misconception that, in contrast to *analog*, *digital* always means perfect and error-free. This is not the case. Even in digital transmission the transmit signal is analog and due to channel noise, detection errors will occur.

When looking at the last subsection and the construction of the decision regions, it is obvious that when transmitting a particular signal point a, a *symbol error* occurs if the decision variable d lies outside the decision region corresponding to this signal point a. The distribution of d given a is characterized by the pdf $f_d(d \mid a)$. When integrating the pdf over the entire complex plane except for the correct decision region $\mathcal{G}_a$, the probability that the decision variable lies outside $\mathcal{G}_a$ is obtained. At the same time, this quantity is the probability that a symbol error occurs. Hence, the *symbol error ratio (SER)*, assuming that a specific signal point a has been transmitted, is given by[6]

$$\Pr\{\hat{a} \neq a\} = \iint_{d \notin \mathcal{G}_a} f_d(d \mid a)\, \mathrm{d}d \;=\; \iint_{d \notin \mathcal{G}_a} f_n(d - a)\, \mathrm{d}d\,. \tag{3.86}$$

The *average symbol error ratio* is then obtained by averaging (taking the expected value) over the contributions of all possible transmit symbols $a \in \mathcal{A}$. This is required if the decision regions are not all congruent. Then we have (in general and for equiprobable signal points, respectively)

$$\mathrm{SER} \stackrel{\mathrm{def}}{=} \mathrm{E}\{\Pr\{\hat{a} \neq a\}\} \stackrel{\mathrm{equiprob.}}{=} \frac{1}{M} \sum_{a \in \mathcal{A}} \Pr\{\hat{a} \neq a\}\,. \tag{3.87}$$

Assuming an AWGN channel, the pdf of the Gaussian noise is given by

baseband transmission: $\quad f_n(n) = \dfrac{1}{\sqrt{2\pi\sigma_{n,\mathrm{I}}^2}}\, \mathrm{e}^{-n^2/(2\sigma_{n,\mathrm{I}}^2)}\,, \qquad \sigma_{n,\mathrm{I}}^2 = N_0/(2E_\mathrm{g})\,,$

ECB signals: $\quad f_n(n) = \dfrac{1}{\pi\sigma_n^2}\, \mathrm{e}^{-|n|^2/\sigma_n^2}\,, \qquad \sigma_n^2 = N_0/E_\mathrm{g}\,.$

Example 3.7 Decision Regions and Error Ratio Calculation

Figures 3.39 and 3.40 depict the regions (shown in light gray) over which the conditional pdf has to be integrated. The white regions are the decision regions; the light gray regions the complement thereof. In Figure 3.39, the situation for the signal point $1 + \mathrm{j}$ is shown. In Figure 3.40, the situation for the point $3 - \mathrm{j}$. Note that the decision regions are not congruent.

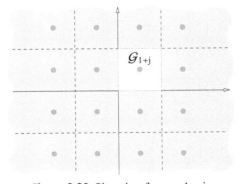

Figure 3.39 Situation for $a = 1 + \mathrm{j}$.

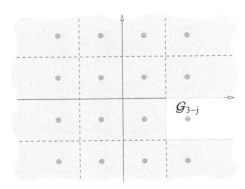

Figure 3.40 Situation for $a = 3 - \mathrm{j}$.

---

[6] In general, d is complex-valued and integration is done in the two-dimensional complex plane; this is symbolized by the double integrals.

3.6.1 Symbol Error Ratio of ASK

We start with the calculation of the symbol error ratio of ASK, that is, baseband transmission, since all signals are real-valued.

The situation is visualized in Figure 3.41. Assume that the signal point a has been transmitted. Owing to the additive Gaussian noise, the probability density for the received detection variable d is given by the Gaussian bell shape centered at the signal point a, that is, $f_n(d - a)$. Since we assume the spacing of the signal points to be 2, the decision region (here a one-dimensional interval) is $\mathcal{G}_a = [a - 1, a + 1]$. If d lies inside $\mathcal{G}_a$, a correct decision is obtained. Otherwise, if d lies outside $\mathcal{G}_a$, a symbol error occurs. The probability for this event is given by integrating the pdf $f_n(d - a)$ over all d not in $\mathcal{G}_a$ (i.e., from $-\infty$ to $a - 1$ and from $a + 1$ to ∞). The value of the integral is given by the shaded parts below the Gaussian function.

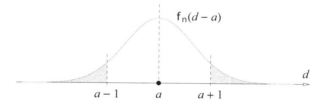

Figure 3.41 Superimposed Gaussian noise and probability of error (shaded parts).

A so-called "inner point," a point not at the edge of the constellation, has two neighboring signal points (as shown in Figure 3.41). Here, the symbol error ratio calculates as

$$\Pr\{\hat{a} \neq a\} = \int_{-\infty}^{a-1} f_n(d - a) \, \mathrm{d}d + \int_{a+1}^{\infty} f_n(d - a) \, \mathrm{d}d$$

substituting $n = d - a$

$$= \int_{-\infty}^{-1} f_n(n) \, \mathrm{d}n + \int_{+1}^{\infty} f_n(n) \, \mathrm{d}n$$

$f_n(n)$ is symmetric

$$= 2 \int_{1}^{\infty} f_n(n) \, \mathrm{d}n$$

$f_n(n)$ is Gaussian

$$= 2 \int_{1}^{\infty} \frac{1}{\sqrt{2\pi\sigma_{n,\mathrm{I}}^2}} e^{-n^2/(2\sigma_{n,\mathrm{I}}^2)} \, \mathrm{d}n$$

using (3.90)

$$= 2\,\mathrm{Q}\left(\frac{1}{\sigma_{n,\mathrm{I}}}\right)$$

and since $\sigma_{n,\mathrm{I}}^2 = N_0/(2E_\mathrm{g})$

$$= 2\,\mathrm{Q}\left(\sqrt{\frac{2E_\mathrm{g}}{N_0}}\right). \tag{3.88}$$

A so-called "outer point," a point at the edge of the constellation, has only a single nearest neighbor; errors occur only in one "direction." Here, the symbol error ratio calculates as

3.6 Bit and Symbol Error Ratio

$$\Pr\{\hat{a} \neq a\} = 1 \, Q\left(\sqrt{\frac{2E_g}{N_0}}\right). \quad (3.89)$$

Note that

$$Q(x) \stackrel{\text{def}}{=} \int_x^\infty \frac{1}{\sqrt{2\pi}} e^{-y^2/2} \, dy \quad (3.90)$$

is the so-called *complementary Gaussian integral function*, see Box 3.3.

An M-ary ASK constellation has $M - 2$ inner and 2 outer points. Assuming equiprobable signal points, the average symbol error ratio is given by

$$\begin{aligned}
\text{SER} &= (M - 2) \frac{1}{M} 2 \, Q\left(\sqrt{\frac{2E_g}{N_0}}\right) + 2 \frac{1}{M} 1 \, Q\left(\sqrt{\frac{2E_g}{N_0}}\right) \\
&= \left(\frac{2(M - 2)}{M} + \frac{2}{M}\right) Q\left(\sqrt{\frac{2E_g}{N_0}}\right) \\
&= \frac{2M - 2}{M} Q\left(\sqrt{\frac{2E_g}{N_0}}\right). \quad (3.91)
\end{aligned}$$

Box 3.3 *Complementary Gaussian Integral Function*

The *complementary Gaussian integral function* is defined as

$$Q(x) \stackrel{\text{def}}{=} \int_x^\infty \frac{1}{\sqrt{2\pi}} e^{-y^2/2} \, dy. \quad (3.92)$$

Unfortunately, there is no closed-form expression for this integral; hence, the abbreviation $Q(x)$ is common.

The function is plotted in Figure 3.42: (a) on a linear scale and (b) on a double-logarithmic scale.

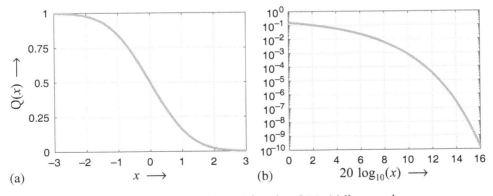

Figure 3.42 Complementary Gaussian integral function $Q(x)$: (a) linear scale; (b) double-logarithmic scale.

Note that the Q-function has the following relations to the *error function* erf(x) and the *complementary error function* erfc(x) :

$$\text{erf}(x) \stackrel{\text{def}}{=} \frac{2}{\sqrt{\pi}} \int_0^x e^{-t^2} dt , \qquad (3.93)$$

$$\text{erfc}(x) \stackrel{\text{def}}{=} 1 - \text{erf}(x) , \qquad (3.94)$$

$$\text{erfc}(x) = 2\, \text{Q}\left(\sqrt{2}x\right) , \qquad (3.95)$$

$$\text{Q}(x) = \frac{1}{2} \text{erfc}\left(\frac{x}{\sqrt{2}}\right) . \qquad (3.96)$$

An upper bound on the Q-function is given by

$$\text{Q}(x) \leq \frac{1}{2} e^{-x^2/2}, \qquad x \geq 0 . \qquad (3.97)$$

In most applications we are interested in an expression of the error ratio depending on the *energy per bit in relation to the noise power spectral density*, E_b/N_0. To that end, we note that for a general constellation the minimum spacing might not be 2 (or distance to decision boundary 1) but is given by $d_{E,\min} = \min_{a_i, a_j \in \mathcal{A},\, a_i \neq a_j} |a_i - a_j|$. Then, following the above derivation, we have ($N = 2$ for inner, $N = 1$ for outer points)

$$\Pr\{\hat{a} \neq a\} = N\, \text{Q}\left(\frac{\frac{1}{2}d_{E,\min}}{\sigma_{n,I}}\right) = N\, \text{Q}\left(\sqrt{\frac{E_g}{2N_0} d_{E,\min}^2}\right) \stackrel{!}{=} N\, \text{Q}\left(\sqrt{d_{\min}^2 \frac{E_b}{N_0}}\right) , \qquad (3.98)$$

where, considering $E_b = \frac{E_s}{\log_2(M)} = \frac{\sigma_a^2 E_g}{\log_2(M)}$, we have used the following definition:[7]

Definition 3.1 *Normalized Minimum Squared Euclidean Distance* of a Signal Constellation

The *normalized minimum squared Euclidean distance* of a signal constellation $\mathcal{A}$ (with arbitrary spacing of the signal points) is defined as

$$d_{\min}^2 \stackrel{\text{def}}{=} \frac{E_g}{2E_b} \min_{\substack{a_i, a_j \in \mathcal{A} \\ a_i \neq a_j}} |a_i - a_j|^2 = \frac{\log_2(M)}{2\sigma_a^2} \min_{\substack{a_i, a_j \in \mathcal{A} \\ a_i \neq a_j}} |a_i - a_j|^2 . \qquad (3.99)$$

Then, the (average) symbol error ratio of bipolar ASK can be written as

$$\text{SER} = \frac{2M - 2}{M} \text{Q}\left(\sqrt{d_{\min}^2 \frac{E_b}{N_0}}\right) . \qquad (3.100)$$

---

[7] The normalization by two times the energy per information bit is due to the historical definition of $N_0/2$ for the two-sided noise psd. It may look strange but simplifies notation in many equations similar to (3.100).

3.6.2 Symbol Error Ratio of General Constellations

The same derivations can be conducted for PSK and QAM constellations; thereby, the one-dimensional integration over the real line has to be replaced by a two-dimensional integration over the complex plane. However, in most situations (tight) approximations or upper bounds are sufficient. Such upper bounds can be derived by employing a *union-bound technique*.

For a 4QAM constellation this is visualized in Figure 3.43. In order to calculate the symbol error ratio assuming that $a = 1 + j$ has been transmitted, one has to integrate the pdf $f_d(d \mid a)$ over the gray-shaded area depicted on the left figure. An upper bound is obtained by integrating over the gray area in the middle *plus* integrating over the gray area on the right. By this, the contributions in the third quadrant (real and imaginary parts negative) are counted twice; hence, we obtain a strict upper bound.

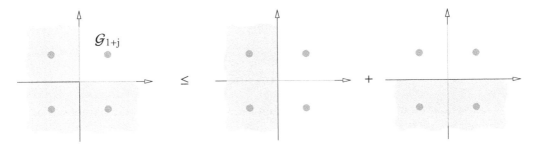

Figure 3.43 Upper bound on the symbol error ratio derived by the union bound.

Denoting the signal points by $a^{(1)} = 1+j$, $a^{(2)} = -1+j$, $a^{(3)} = -1-j$, and $a^{(4)} = 1-j$, mathematically, the error probability (assuming $a^{(1)}$ has been transmitted but due to symmetry equally valid for all points) can be written as

$$
\begin{aligned}
&\Pr\{\hat{a} \neq a^{(1)}\} \\
&= \iint_{d \notin \mathcal{G}_{a^{(1)}}} f_d(d \mid a^{(1)}) \, dd = \iint_{d \in (\mathcal{G}_{a^{(2)}} \cup \mathcal{G}_{a^{(3)}} \cup \mathcal{G}_{a^{(4)}})} f_d(d \mid a^{(1)}) \, dd \\
&= \iint_{d \in \mathcal{G}_{a^{(2)}}} f_d(d \mid a^{(1)}) \, dd + \iint_{d \in \mathcal{G}_{a^{(3)}}} f_d(d \mid a^{(1)}) \, dd + \iint_{d \in \mathcal{G}_{a^{(4)}}} f_d(d \mid a^{(1)}) \, dd \\
&= \iint_{d \in (\mathcal{G}_{a^{(2)}} \cup \mathcal{G}_{a^{(3)}})} f_d(d \mid a^{(1)}) \, dd + \iint_{d \in (\mathcal{G}_{a^{(3)}} \cup \mathcal{G}_{a^{(4)}})} f_d(d \mid a^{(1)}) \, dd - \iint_{d \in \mathcal{G}_{a^{(3)}}} f_d(d \mid a^{(1)}) \, dd \\
&= \iint_{\mathrm{Re}\{d\}<0} f_d(d \mid a^{(1)}) \, dd + \iint_{\mathrm{Im}\{d\}<0} f_d(d \mid a^{(1)}) \, dd - \iint_{\mathrm{Re}\{d\}<0 \wedge \mathrm{Im}\{d\}<0} f_d(d \mid a^{(1)}) \, dd \\
&\leq \iint_{\mathrm{Re}\{d\}<0} f_d(d \mid a^{(1)}) \, dd + \iint_{\mathrm{Im}\{d\}<0} f_d(d \mid a^{(1)}) \, dd \, . \tag{3.101}
\end{aligned}
$$

Each of the last two integrals considers only the transmitted signal point and one competing nearest neighbor. Hence, they reduce to a one-dimensional integral, simply given by

$$\iint_{\mathrm{Re}\{d\}<0} f_d(d \mid a^{(1)}) \, \mathrm{d}d = \iint_{\mathrm{Im}\{d\}<0} f_d(d \mid a^{(1)}) \, \mathrm{d}d$$

$$= \int_1^\infty \frac{1}{\sqrt{2\pi\sigma_{n,\mathrm{I}}^2}} e^{-n^2/(2\sigma_{n,\mathrm{I}}^2)} \, \mathrm{d}n = \mathrm{Q}\left(\sqrt{\frac{2E_g}{N_0}}\right). \quad (3.102)$$

In other words, the error ratio is bounded by

$$\Pr\{\hat{a} \neq a\} \leq 2\,\mathrm{Q}\left(\sqrt{\frac{2E_g}{N_0}}\right) \quad (3.103)$$

whereas the correct value in the present case of 4QAM would be

$$\Pr\{\hat{a} \neq a\} = 2\,\mathrm{Q}\left(\sqrt{\frac{2E_g}{N_0}}\right) - \left(\mathrm{Q}\left(\sqrt{\frac{2E_g}{N_0}}\right)\right)^2. \quad (3.104)$$

The factor 2 in front of the Q-function is nothing other than the number of nearest neighbors.

Using the same procedure, the error probability can be bounded for the regular PSK and QAM constellations introduced in Section 3.3. The average error ratio then depends on the *average number of nearest-neighbor signal points* and the (normalized) distance between the neighboring signal points, expressed as d_{min}^2.

Figure 3.44 shows the double-counted regions (dark gray) for (a) 16QAM and (b) 8PSK. The nearest neighbors of the considered signal point are circled. The multiple-counted areas exhibit greater distances to the correct signal point than the boundaries to the nearest neighbors. Because of the steep descent of the Q-function, the difference of the union bound from a correct result is small in the regime of

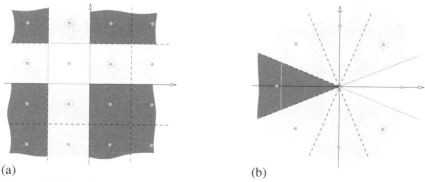

Figure 3.44 Double-counted regions (dark gray) for (a) 16QAM and (b) 8PSK. Circled points: nearest neighbors.

high SNR or low error ratio, respectively. Here, the bound (3.105) is quite tight and valuable for system analysis.

In summary:

> An *upper bound* on the *symbol error ratio* of digital PAM transmission is given by
>
> $$\text{SER} \leq N_{\min} \, Q\left(\sqrt{d_{\min}^2 \frac{E_b}{N_0}}\right) \qquad (3.105)$$
>
> $N_{\min}$: average number of nearest-neighbor signal points
> E_b/N_0: energy per bit to noise power spectral density ratio
> $d_{\min}^2$: normalized squared Euclidean distance of the scheme.

Note that for non-regular signal constellations, (3.105) may indeed be an approximation. An upper bound is only present when the nearest neighbors determine entirely the decision regions of all signal points (as can be seen from Figure 3.44, the decision boundaries are the perpendicular bisectors between the signal point of interest and its nearest neighbors). If signal points at higher distances also affect the shape of the decision region, using the nearest-neighbor approximation, some regions are not counted as errors.

3.6.3 Approximation of the Bit Error Ratio

Up to now, we have calculated the *symbol error ratio*, that is, the probability that a transmitted symbol is detected erroneously. However, each symbol may represent a number of bits. In some situations we are more interested in the probability of a *bit error*.

To derive an approximation of the *bit error ratio (BER)* we assume that:

- the system is operated at medium to high SNRs – here, decision errors in favor of nearest neighbors dominate;
- the signal points are labeled using *Gray labeling* – the binary labels of nearest-neighbor points differ in a single position.

Example 3.8 Gray Labeling

Gray labeling of binary vectors to signal points of a constellation is defined as follows: for any pair of points that have a distance equal to the minimum Euclidean distance, the corresponding binary vectors differ in exactly one digit (have Hamming distance 1). Gray labeling is not unique. It has been shown that Gray labeling exists for any ASK, PSK, or QAM constellation in one or two dimensions, but it does not exist in general for constellations defined in a higher-dimensional signal space.

In Figures 3.45 and 3.46 a possible Gray labeling is shown for an 8ASK and a 16QAM constellation, respectively.

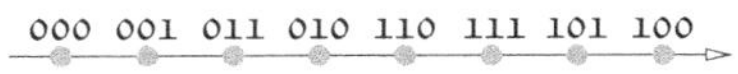

Figure 3.45 Gray labeling for 8ASK.

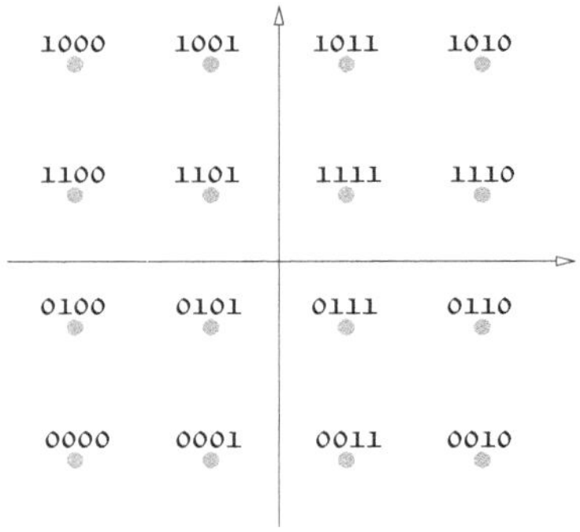

Figure 3.46 Gray labeling for 16QAM.

Since each symbol error causes a single bit error and the label of each signal point contains $\log_2(M)$ bits, we arrive at

$$\text{BER} \approx \frac{N_{\min}}{\log_2(M)} \, Q\left(\sqrt{d_{\min}^2 \frac{E_b}{N_0}}\right). \tag{3.106}$$

3.6.4 Normalized Minimum Squared Euclidean Distance of PAM Constellations

Tables 3.2 to 3.5 collect the normalized squared Euclidean distances $d_{\min}^2$, average number of nearest neighbors $N_{\min}$, variance σ_a^2 and mean m_a of important families of PAM constellations.

M-ary bipolar ASK

$$d_{\min}^2 = \frac{6 \log_2(M)}{M^2 - 1}. \tag{3.107}$$

Table 3.2 Parameters of M-ary bipolar ASK ($m_a = 0$).

M	2	4	8	16	32
$d_{\min}^2$	2	0.8	0.286	0.094	0.0293
$N_{\min}$	1	1.5	1.75	1.875	1.9375
σ_a^2	1	5	21	85	341
ζ_a	1	$\sqrt{9/5}$	$\sqrt{7/3}$	$\sqrt{45/17}$	$\sqrt{31/11}$
	0 dB	2.55 dB	3.68 dB	4.23 dB	4.50 dB

M-ary unipolar ASK

$$d_{\min}^2 = \frac{6 \log_2(M)}{4M^2 - 6M + 2}. \tag{3.108}$$

Table 3.3 Parameters of *M*-ary unipolar ASK.

M	2	4	8	16	32
d_{min}^2	1	0.286	0.0857	0.0258	0.0077
N_{min}	1	1.5	1.75	1.875	1.9375
σ_a^2	1	5	21	85	341
m_a	1	3	7	15	31
ζ_a	$\sqrt{2}$	$\sqrt{18/7}$	$\sqrt{14/5}$	$\sqrt{90/31}$	$\sqrt{1922/651}$
	3 dB	4.10 dB	4.47 dB	4.63 dB	4.70 dB

M-ary PSK

$$d_{min}^2 = 2 \log_2(M) \cdot \sin^2\left(\frac{\pi}{M}\right) . \tag{3.109}$$

Table 3.4 Parameters of *M*-ary PSK ($m_a = 0$).

M	2	4	8	16	32
d_{min}^2	2	2	0.88	0.305	0.096
N_{min}	1	2	2	2	2
σ_a^2	1	1	1	1	1
ζ_a	1	1	1	1	1
	0 dB	0 dB	0 dB	0 dB	0 dB

M-ary QAM

$$d_{min}^2 = \frac{3 \log_2(M)}{M - 1} \quad \text{(square constellations).} \tag{3.110}$$

Table 3.5 Parameters of *M*-ary QAM ($m_a = 0$).

M	2	4	16	32 (cross)	64	128 (cross)	256
d_{min}^2	2	2	0.8	0.5	0.286	0.171	0.094
N_{min}	1	2	3	3.25	3.5	3.625	3.75
σ_a^2	1	2	10	20	42	82	170
ζ_a	1	1	$\sqrt{9/5}$	$\sqrt{17/10}$	$\sqrt{7/3}$	$\sqrt{85/41}$	$\sqrt{45/17}$
	0 dB	0 dB	2.55 dB	2.30 dB	3.68 dB	3.17 dB	4.23 dB

3.6.5 Bit Error Ratio of Selected PAM Schemes

Figure 3.47 shows the approximation (3.106) of the bit error ratios of selected PAM schemes (using Gray labeling) over the signal-to-noise ratio expressed as E_b/N_0 (in dB).

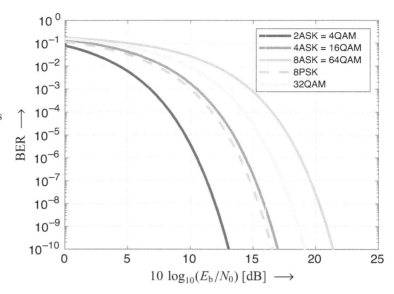

Figure 3.47 Bit error ratios of selected PAM schemes (using Gray labeling) over the signal-to-noise ratio expressed as E_b/N_0 (in dB).

Note that M-ary ASK and M^2-ary QAM constellations have the same (approximations for the) bit error ratio; the doubled number of nearest neighbors cancels the doubled number of binary symbols. Moreover, $d_{\min, M\text{ASK}} = d_{\min, M^2\text{QAM}}$; see (3.107) and (3.110).

The error ratio curves are characterized by the shape of the Q-function. As the signal-to-noise ratio increases, the bit error ratios decrease. From Figure 3.47, the exchange between power and bandwidth efficiency is visible. 2ASK carries a single bit per symbol. Here, the lowest SNR is required to achieve a certain reliability (e.g., BER = 10^{-5}). Going to larger cardinalities, more bits per symbol are transmitted; the bandwidth efficiency increases. In turn, a larger SNR is required to guarantee a desired reliability; the power efficiency decreases.

3.7 Assessment of PAM Schemes in the Power/Rate and Power/Bandwidth Plane

In Chapter 1 we introduced the power/bandwidth plane for transmission over the AWGN channel (see Figure 1.9). We now characterize the power/bandwidth trade-off of digital PAM schemes.

Recall the criteria for assessing (digital) modulation formats.

- The *power efficiency* is specified by the signal-to-noise ratio (AWGN channel)

$$10 \log_{10}(E_b/N_0) \,, \qquad (3.111)$$

which is minimally required in order to guarantee a tolerated bit error ratio (e.g., BER $\leq 10^{-5}$).
- The *bandwidth efficiency* is given by the data rate per occupied bandwidth. When using a $\sqrt{\text{Nyquist}}$ pulse with roll-off factor α, it reads

3.7 Assessment of PAM Schemes in the Power/Bandwidth Plane

$$\Gamma \stackrel{\text{def}}{=} \frac{R_T}{B_{RF}} = \frac{\log_2(M)}{1+\alpha}. \tag{3.112}$$

- The *rate* of a digital communication scheme is the counterpart to the bandwidth efficiency for an analysis in the discrete-time domain (symbol duration T). For uncoded PAM, the rate $R = \log_2(M)$ exactly corresponds to the bandwidth efficiency if the roll-off factor $\alpha = 0$, that is, if no bandwidth excess over the minimal required bandwidth is present.

For selected digital PAM schemes, the *power/bandwidth plane* is depicted in Figure 3.48. Thereby, the roll-off factor $\alpha = 0$ is assumed that gives the smallest bandwidth of the transmit signal and, hence, the best bandwidth efficiency and thus the *power/rate plane*.

Each modulation format (each constellation) corresponds to a point in the power/bandwidth plane. This point characterizes the following.

- Power efficiency – on the horizontal axis the value of the *minimally required* E_b/N_0 is marked; for SNRs larger than the value, the actual bit error ratio is below the tolerated bit error ratio (BER $\leq 10^{-5}$ or BER $\leq 10^{-8}$). The SNR limit can be read from Figure 3.47.
- Bandwidth efficiency – on the vertical axis the value of Γ or R, respectively, is marked. Note that for $\alpha = 0$ we have $\Gamma = R = \log_2(M)$; the bandwidth efficiency here is simply given by the rate, that is, the number of bits each symbol carries!

As a reference point, one can remember that for BPSK (2ASK, 2PSK) that carries a single bit, for a tolerated bit error ratio BER $\leq 10^{-5}$, an SNR of at least $10 \log_{10}(E_b/N_0) = 9.6$ dB is required.

Looking at *ASK transmission*, doubling the number of signal points increases the number of carried bits per symbol and thus the bandwidth efficiency by 1. However, as

$$d_{\min}^2 = \frac{6 \log_2(M)}{M^2 - 1} \approx \frac{6 \log_2(M)}{M^2} \tag{3.113}$$

for large M, doubling M decreases $d_{\min}^2$ to

$$d_{\min, 2M}^2 = \frac{6 \log_2(2M)}{(2M)^2} = \frac{6(1 + \log_2(M))}{4M^2} \approx \frac{1}{4}\frac{6 \log_2(M)}{M^2} \approx \frac{1}{4} d_{\min, M}^2. \tag{3.114}$$

Hence, the normalized squared Euclidean distance is decreased by a factor of 4. For the same error ratio (same argument $d_{\min}^2 E_b/N_0$ of the Q-function), a four-times higher SNR is required; this corresponds to 6 dB. In summary, for ASK transmission, an extra bit[8] costs 6 dB.

The situation is similar in *PSK transmission*. As the signal points are restricted to the unit circle, basically a one-dimensional arrangement is present. When doubling the number of signal points (one bit extra), the distance between neighboring signal

---

[8] This relation coincides with a well-known fact from pulse-code modulation (quantization). There, an extra bit in the resolution of the quantizer brings an increase in SNR by 6 dB.

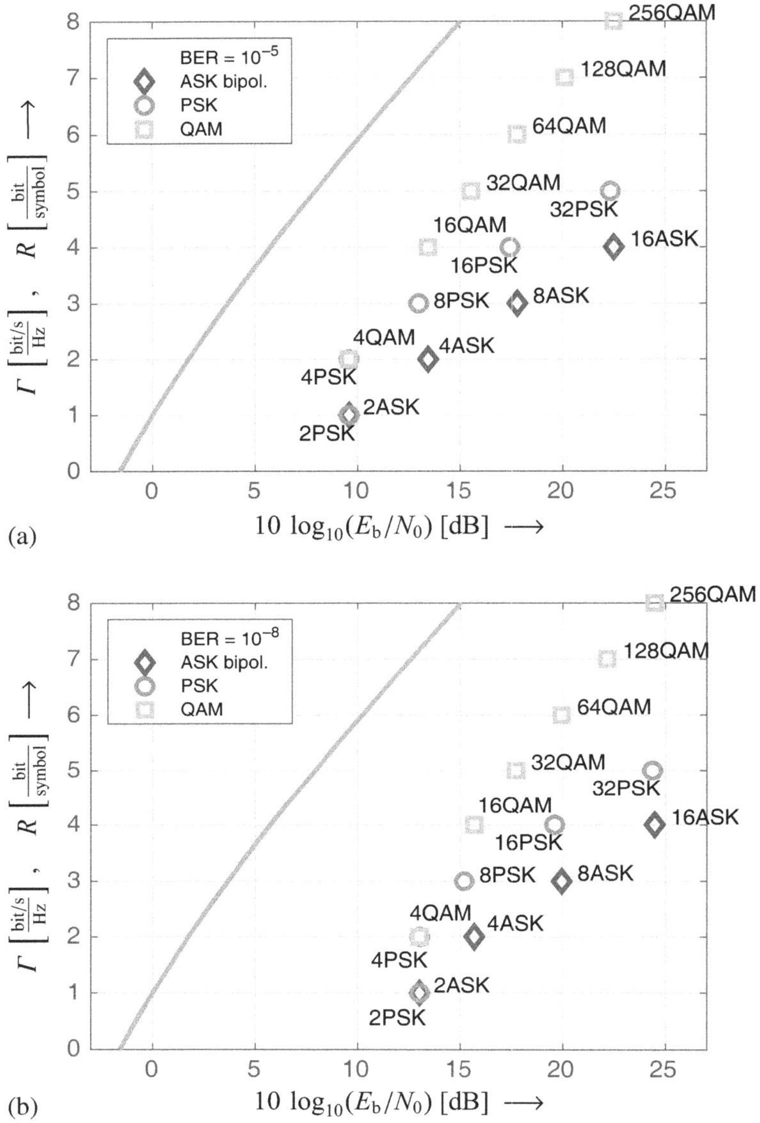

Figure 3.48 Power/bandwidth plane for digital PAM schemes. $\alpha = 0$ (i.e., power/rate plane). (a) Tolerated bit error ratio BER $\leq 10^{-5}$. (b) BER $\leq 10^{-8}$.

points is (almost, large M) halved; the (normalized) squared distance is reduced by a factor of 4 as in ASK. Here, an extra bit costs 6 dB too.

In *QAM signaling*, where both in-phase and quadrature components are used and a true two-dimensional constellation is present, the situation changes. Doubling the number of signal points gives

$$d^2_{\min, 2M} = \frac{3 \log_2(2M)}{(2M) - 1} = \frac{3(1 + \log_2(M))}{2M - 1} \approx \frac{1}{2} \frac{3 \log_2(M)}{M} \approx \frac{1}{2} d^2_{\min, M}. \quad (3.115)$$

Hence, in QAM an extra bit costs 3 dB (two extra bits correspond to an extra bit per dimension, which costs 2×3 dB $= 6$ dB). It should be noted that the Shannon

3.7 Assessment of PAM Schemes in the Power/Bandwidth Plane

limit (gray solid line) also has (for large SNRs) a slope of 1 bit per 3 dB as, from (1.18), $E_b/N_0 = (2^\Gamma - 1)/\Gamma$ and thus

$$E_b/N_0|_{\Gamma+1} = (2^{\Gamma+1}-1)/(\Gamma+1) \approx (2 \cdot 2^\Gamma)/\Gamma \approx 2 \cdot E_b/N_0|_\Gamma . \qquad (3.116)$$

Lowering the tolerated bit error ratio from BER = 10^{-5} (Figure 3.48a) to BER = 10^{-8} (Figure 3.48b) leads to a poorer power efficiency; higher SNRs are required and the points are shifted to the right.

Finally, in Figure 3.49, digital PAM schemes are assessed in the power/bandwidth plane under the variation of the roll-off factor α. The top cross of each vertical bar corresponds to $\alpha = 0$; the bottom cross to $\alpha = 1$. Note that for $\alpha = 1$

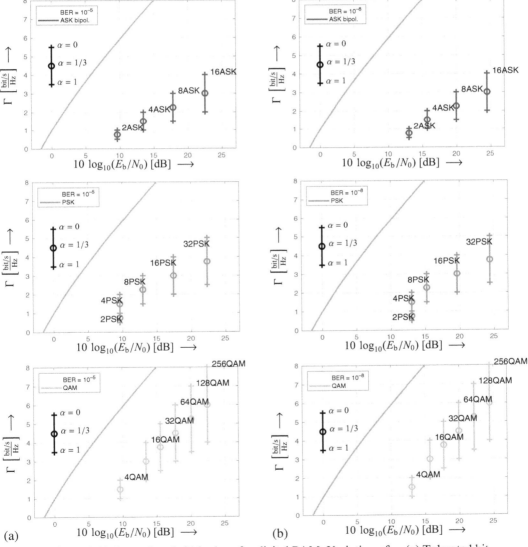

Figure 3.49 Power/bandwidth plane for digital PAM. Variation of α. (a) Tolerated bit error ratio BER $\leq 10^{-5}$. (b) BER $\leq 10^{-8}$.

the bandwidth is doubled compared to the case of $\alpha = 0$, and thus the bandwidth efficiency is halved. The middle circles correspond to $\alpha = 1/3$; in view of (3.112), only three-quarters of the maximal bandwidth efficiency is achieved.

3.8 Problems

3.8.1 Error Probability in Case of Gaussian Noise

We consider digital transmission over the complex-valued discrete-time AWGN channel. The detection variable (one fixed modulation step) is given by $d = a + n$, where n is complex-valued Gaussian noise with variance σ_n^2. Two PAM signal constellations are considered: $\mathcal{A}_A = \{+1, -1\}$ and $\mathcal{A}_B = \{+1+j, -1-j\}$.

(a) What are the optimum decision boundaries for $\mathcal{A}_A$ and $\mathcal{A}_B$, respectively, according to the maximum-likelihood criterion?

(b) We consider $\mathcal{A}_A$. Let $a = 1$ be transmitted. Calculate $\Pr\{\hat{a} = -1\}$, that is, the probability that a (wrong) decision in favor of $a = -1$ is taken. Interpret the result.

(c) We now consider $\mathcal{A}_B$. Let $a = 1+j$ be transmitted. Calculate $\Pr\{\hat{a} = -1-j\}$.

(d) Express both results with respect to the quotient energy per bit and one-sided noise power spectral density for specification of the signal-to-noise ratio. Compare the resulting formula.

3.8.2 Amplitude-Shift Keying

Pulse-amplitude modulation schemes using bipolar amplitude-shift keying (ASK) constellations are considered. The signal constellation with M points (M even) is defined as

$$\mathcal{A} = \{\pm 1, \pm 3, \pm 5, \ldots, \pm(M-1)\}.$$

We assume that all signal points $a \in \mathcal{A}$ are used equiprobably, that is, $\Pr\{a\} = 1/M$. Determine the parameters

(a) rate of the modulation scheme R_m
(b) average energy per symbol E_s
(c) average energy per bit E_b
(d) normalized squared Euclidean distance d_{min}^2
(e) average number of nearest neighbors N_{min}
(f) crest factor of the constellation ζ_a.

Hint: You may use the sum formulas $\sum\limits_{i=1}^{n} i = \frac{n(n+1)}{2}$ and $\sum\limits_{i=1}^{n} i^2 = \frac{n(n+1)(2n+1)}{6}$.

3.8.3 PAM Transmission Scheme I

We want to design a digital carrier-modulated pulse-amplitude modulation (PAM) transmission scheme (without channel coding) operating over an additive white Gaussian noise (AWGN) channel with (two-sided) noise power spectral density $N_0/2$. We preferably consider equivalent complex baseband signals.

(a) Why should the basic pulse shape $g(t)$ be a square-root Nyquist pulse?

Subsequently we use the pulse $g(t)$, whose (real-valued) spectrum reads

$$g(t) \circ\!\!-\!\!\bullet\; G(f) = 10^{-6}\,\text{Vs} \cdot \begin{cases} 1, & |f| < 0.8\ \text{MHz} \\ \sqrt{\frac{1.2\ \text{MHz} - |f|}{0.4\ \text{MHz}}}, & 0.8\ \text{MHz} \le |f| \le 1.2\ \text{MHz} \\ 0, & |f| > 1.2\ \text{MHz}. \end{cases}$$

(b) Show (e.g., graphically) that $g(t)$ is a $\sqrt{\text{Nyquist}}$ pulse for a suitable choice of the symbol duration T. How should T be chosen?

Now, additionally, a 4QAM signal constellation

$$\mathcal{A} = \{+1+j,\ +1-j,\ -1+j,\ -1-j\}$$

with normalized minimum squared Euclidean distance $d_{\min}^2 = 2$ is employed.

(c) What data rate is supported? Give the spectral efficiency of the transmission scheme.
(d) Calculate the energy E_g of the pulse $g(t)$. How large is E_b, the transmitted energy per information bit?
(e) Assume that the optimum receiver is used. How large may the noise power spectral density $N_0/2$ be such that a symbol error ratio less than or equal to 8×10^{-6} is still guaranteed?
Hint: Below, a plot of the complementary Gaussian integral function is given.

Supplementary Material: Complementary Gaussian integral function

$$Q(x) \stackrel{\text{def}}{=} \int_x^\infty \frac{1}{\sqrt{2\pi}}\, e^{-y^2/2}\, dy.$$

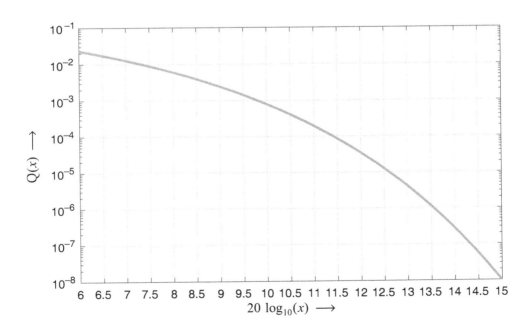

3.8.4 PAM Transmission Scheme II
Binary bipolar pulse-amplitude modulation (PAM) is used for transmission of a data rate of 20 Mbit/s over an additive white Gaussian noise (AWGN) channel with attenuation D. The block diagram for ECB signals and systems is shown in the following figure:

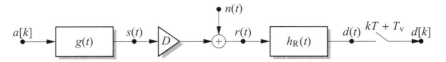

The amplitude coefficients are taken from $a[k] \in \mathcal{A} = \{+1, -1\}$ and the basic pulse shape is given by

$$g(t) = 1\,\text{V} \cdot \text{rect}\left(\frac{t}{T}\right).$$

(a) Sketch the basic pulse shape.
(b) Determine the duration T of the modulation interval.
(c) Plot the (ECB) transmit signal $s(t)$ for the following section of the data sequence:

$$\langle a[-1], a[0], a[1], \ldots, a[5] \rangle = \langle -1, -1, 1, -1, -1, -1, 1 \rangle.$$

(d) Determine the impulse response $h_R(t)$ and the transfer function $H_R(f)$ of the causal "optimum" receive filter. In what sense is this filter optimal?
(e) Sketch the pulse at the output of this filter, that is, the so-called detection pulse shape. Does intersymbol interference occur?
(f) Sketch the useful signal at the output of the receive filter in the time interval $0 \leq t \leq 5T$. Use the data sequence given in part (c). Give the optimal detection/sampling time instants.
(g) Sketch the eye pattern.
(h) Determine the autocorrelation function of the noise at the output of the receive filter.

On the way through the channel, the signal is attenuated by 120 dB.
(i) Calculate the received energy per bit.

The one-sided power spectral density of the noise $n(t)$ is assumed to be $3.33 \times 10^{-21}\,\text{V}^2\,\text{s}$.
(j) Calculate the bit error probability.
(k) Give the discrete-time representation of the PAM transmission.
(l) Why is "soft keying" preferred over "hard keying" in practical digital communication systems? Give advantages and disadvantages of hard and soft keying, respectively.

3.8.5 PAM Transmission Scheme III
The basic pulse shape

$$g(t) = c \cdot \text{si}(\pi t/T) \qquad \text{with } c = 1\,\text{V and } T = 1\,\mu\text{s}$$

is used for binary bipolar ($\mathcal{A} = \{\pm 1\}$) digital PAM transmission over an AWGN channel. The power spectral density of the additive white noise (ECB model) is 10^{-7} V^2/Hz. In each case, the optimum receive filter is used.

(a) Give the noise power at the input and at the output of the optimum receive filter, respectively.
(b) Calculate the signal-to-noise ratio (expressed as E_b/N_0) at the optimum sampling instants. Give the bit error ratio BER assuming that an optimum threshold decision is used.

Assume that (e.g., due to an implementation error) the decision boundary of the threshold device is shifted to 0.2. Owing to this mismatch, the error probability will increase. This may be compensated for by increasing the transmit power adequately.

(c) How much more transmit power (in %) is necessary to achieve the same bit error ratio as in part (b)? Give the new value for the amplitude c of the basic pulse shape.

3.8.6 Decision Threshold

We derive Equation (3.85).

Hint: Because of rotational invariance of complex Gaussian noise, without loss of generality, it is sufficient to look only at two signal points on the real axis. Consequently, we consider two points $a_1, a_2 \in \mathbb{R}$ with probabilities Pr$\{a_1\}$ and Pr$\{a_2\}$, respectively.

(a) Recapitulate the MAP strategy and write it in a form where the conditional probability $f_d(d \mid a)$ appears.
(b) For the setting of two signal points, find the boundary between the two corresponding decision regions.

3.8.7 On–Off Keying

Digital transmission using binary unipolar ASK is also called "on–off keying" (OOK). We consider OOK over the AWGN channel using the basic pulse shape

$$g(t) = 1 \text{ V} \cdot \text{rect}(t/T) .$$

(a) Give a possible mapping from the binary source symbols to the amplitude coefficients.
(b) Plot the ECB transmit signal $s(t)$ and the RF transmit signal $s_{\text{RF}}(t)$ (using $f_c = 3/T$), respectively, for the following section of the source sequence:

$$\langle\, q[-1], q[0], q[1], \ldots, q[5]\,\rangle = \langle 0, 0, 1, 0, 1, 1, 0 \rangle.$$

(c) Why is this form of modulation called "on–off keying"?
(d) Sketch the eye pattern assuming that the optimum receive filter is used.
(e) What are the advantages and disadvantages of OOK over binary bipolar transmission?

3.8.8 Design of a Digital Directed Radio Link (Microwave Transmission)

A data stream with a bit rate of 128 Mbit/s has to be transmitted over a distance of 80 km by means of a directed radio link. A frequency band of 28 MHz bandwidth, symmetric around a carrier frequency of 13 GHz, is available. The microwave directed radio channel can be modeled as an AWGN channel with 117 dB signal attenuation from transmitter to receiver (including antenna gains at both sides) and noise with two-sided noise power spectral density of 4×10^{-21} W/Hz at the receiver side.

We consider digital QAM schemes with 2^{R_m} signal points where the rate of the modulation, R_m, has to be an integer.

(a) What bandwidth (spectral) efficiency is required for the transmission scheme?
(b) What is the smallest regular QAM constellation with rate $R_m \in \mathbb{N}$ for which this bandwidth efficiency is achievable. Give the modulation rate, symbol rate (baud rate), and duration of the symbol interval for this QAM scheme.
(c) What maximum roll-off factor is possible for the basic square-root Nyquist PAM pulse without violation of the bandwidth limitation. The QAM scheme of part (b) is applied.
(d) Calculate the minimally required received signal energy per information bit at the receiver frontend in order to achieve a bit error ratio less than 10^{-10}. Gray labeling is assumed. What minimum average transmit signal power results?
(e) Give the crest factors or peak-to-average power ratio (PAPR) with respect to the signal constellation and with respect to the basic PAM pulse. Calculate an upper bound on the peak transmit power of the RF signal. For what nominal peak power has the high-power amplifier to be designed?

For the design according to part (b), a very low spectral roll-off factor for the basic PAM pulse has to be chosen (see result of part (c)).

(f) Explain why a very small roll-off factor should be avoided in digital PAM transmission schemes in principle. Give at least four reasons.

In order to mitigate the problem concerning the roll-off factor, a QAM scheme with a doubled number of signal points compared to part (b) is considered.

(g) Specify the size of the QAM constellation, its modulation rate, symbol rate, and the maximal possible roll-off factor.
(h) Calculate the minimally required received signal energy per information bit at the receiver frontend in order to achieve a bit error ratio less than 10^{-10}. Gray labeling is assumed. What minimum average transmit signal power now results?
(i) Give the crest factors or the peak-to-average power ratios (PAPR) with respect to the signal constellation and with respect to the basic PAM pulse for this setting. Calculate an upper bound on the peak transmit power of the RF signal. For what nominal peak power has the high-power amplifier to be designed in this scenario? Compare the result with that of part (e).

Now the additional application of an error-correcting code is considered. In particular, a binary block code with codelength 504 and 432 information symbols. By this (shortened binary BCH) code, all error patterns with up to eight errors per codeword are correctable. The required data reliability of BER $< 10^{-10}$ (after decoding) is achieved when the error ratio for binary code symbols before decoding is 6×10^{-4} or less.

(j) Specify the code rate R_c of this binary block code!

Of course, the transmitted information data rate of 128 Mbit/s has to be guaranteed. In order to transmit the additional binary check symbols without violating the demands on the signal bandwidth and the roll-off factor, the size of the QAM constellation has to be doubled again.

(k) Specify the size of this QAM constellation, its modulation rate, symbol rate, and the maximal possible roll-off factor. Give also the overall rate [bit/modulation step] of this coded scheme.

(l) Calculate the minimally required received signal energy per binary code symbol at the receiver frontend in order to achieve a bit error ratio less than 6×10^{-4}. Gray labeling is assumed. What minimum average transmit signal power now results?

(m) Give the gain in reduction of average and peak signal power due to the application of the error-correcting code compared to the results of part (h).
Hint: Take into account that the symbol duration of the binary code symbols is less than that of the information bits.

By sophisticated channel-coding schemes with optimum combining of signal shaping, coding, and modulation, together with soft-decision decoding, a power efficiency is achievable within 1 dB of the limits given by information theory (Shannon bound).

(n) What minimum average transmit signal power results for such a sophisticated transmission scheme? What minimum average transmit signal is necessary if the demand on the roll-off factor of part (g) still has to be satisfied? What gains with respect to power efficiency are achievable by sophisticated coding over uncoded transmission or over an application of simple error-correcting codes?

4 Variants of PAM Transmission Schemes

In the last chapter, we dealt with the basic principles of digital PAM transmission. In some situations it is convenient to do some modifications to the transmitter, in order to achieve some desired properties of the transmit signal, and/or to the receiver, in order to vary the demodulation process.

In this chapter we consider two aspects that are addressed by variants of PAM transmission schemes. First, have a look at the *crest factor* or *peak-to-average power ratio* of the transmit signal. For efficient power amplification this parameter should be as low as possible. We show that the peak-to-average power ratio can be lowered by a simple modification of the transmitter, leading to *offset QAM*. A further restriction finally leads to *minimum-shift keying*, which even has a constant envelope. In the last step, the reduced bandwidth efficiency is tackled by the variant *Gaussian minimum-shift keying*.

The second aspect studied in this chapter is the replacement of the coherent I/Q demodulator by different principles. First, "carrierless" amplitude and phase modulation is treated as an alternative approach to PAM. Here, no explicit mixing of the pulse-shaped continuous-time baseband signal to the RF domain is required; the RF signal is generated directly by pulse shaping. Second, in some cases (e.g., fiber-optics transmission) coherent reception is possible in principle but very costly. Here it is desired that even when demodulating without phase information (i.e., conducting *energy detection*) a performance (w.r.t. data rate and error ratio) close to a coherent receiver is enabled. We study in detail an advanced scheme, called the *Kramers–Kronig coherent receiver*, which meets this aim by performing more complex operations in the digital part.

PREREQUISITES FOR THIS CHAPTER

- PAM transmission (Chapter 3).
- Causal and minimum-phase systems (Section A.3 of Appendix A for Section 4.2.2).

4.1 Schemes with Reduced Peak-to-Average Power Ratio

All practical systems have a limited amplitude range (e.g., for all active electronic circuits there is a lower and an upper limit on the signal amplitude that can be tolerated or processed (almost) linearly). Since each transmit signal has to be amplified before it can be fed to the transmission medium (the physical representation of information requires energy), the *dynamic range* of the signal has to be limited. Large peaks should be avoided in the transmit signal. Owing to pulse shaping (filtering with $g(t)$), the PAM transmit signal may exhibit such large peaks. However, by a simple modification called *offset QAM* (or *staggered QAM*), these peaks can be reduced.

4.1.1 Offset QAM

Characterization of Power Amplifiers

As already stated, high-power amplifiers (HPA) at the transmitter frontend are not ideal (simply scaling the signal by the desired factor) but exhibit *nonlinear and dispersive* characteristics. This means that these systems introduce amplitude and phase distortions. As the power amplifiers act on the RF signal, the distortions are caused in the RF domain.

However, we are more interested in the effect on the ECB signal, as shown in Figure 4.1. For most types of power amplifiers (e.g., traveling-wave tubes, GaAs-FET amplifiers etc.), adequate baseband models are available. Typically, these ECB models are *nonlinear* but fortunately, the dispersive behavior can be well approximated here by a simple phase shift, and thus by a *non-dispersive* model.

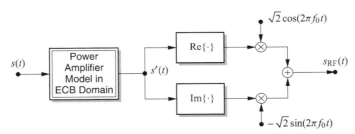

Figure 4.1 Transmitter frontend (upconversion) taking an ECB model of an RF power amplifier into account.

Let the undistorted ECB signal be

$$s(t) = |s(t)| \cdot e^{j \arg(s(t))}, \qquad (4.1)$$

then a common nonlinear, non-dispersive model is given by *AM/AM and AM/PM conversion*. Here, the distorted ECB signal $s'(t)$ (cf. Figure 4.1) is related to the undistorted one, $s(t)$, by

$$s'(t) = A\left(\frac{|s(t)|}{s_0}\right) \cdot e^{j\left(\arg(s(t)) + P\left(\frac{|s(t)|}{s_0}\right)\right)} \qquad (4.2)$$

s_0: reference amplitude
$A(\cdot)$: AM/AM characteristics
$P(\cdot)$: AM/PM characteristics.

The *AM/AM characteristics* $A(\cdot)$ give the output amplitude as a function of the input amplitude (normalized to s_0); the *AM/PM characteristics* give the phase deviation as a function of the input amplitude (i.e., a variable delay of the RF signal depending on its amplitude).

In order to avoid severe signal distortions by the HPA and, especially, the generation of out-of-band radiation, the average transmit power S_s has to be reduced compared to the nominal power of the HPA, S_{nominal}. This reduction is called (output) *power back-off* and defined as

$$\text{PBO} \stackrel{\text{def}}{=} 10 \log_{10} \left(\frac{S_s}{S_{\text{nominal}}} \right) \text{ [dB] .} \quad (4.3)$$

The back-off has to be chosen according to the peak-to-average power ratio (PAPR) of the transmit signal. Typically, the back-off causes a loss in the power efficiency of communication schemes as the power consumption of the HPA is related to its nominal power.

Example 4.1 AM/AM and AM/PM Characteristics of Traveling-Wave Tube

Figure 4.2 shows the AM/AM and AM/PM characteristics of a typical traveling-wave tube power amplifier (i.e., a vacuum tube suited for amplifying microwave RF signals, $\approx$300 MHz to 50 GHz).

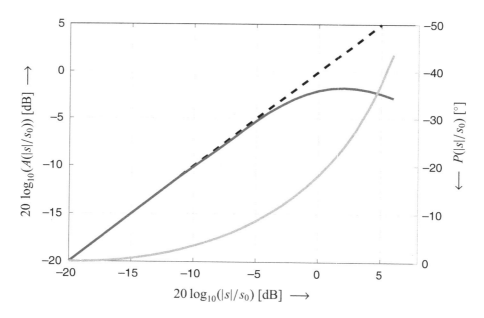

Figure 4.2 AM/AM (dark gray) and AM/PM (light gray) characteristics of a traveling-wave tube power amplifier (amplification gain normalized to one).

As long as the input amplitude is small, the output amplitude follows the input amplitude and (almost) no phase change is caused. For large input amplitudes, saturation effects are caused; the output amplitude is limited and may even drop for very large inputs. In addition, a phase deviation is caused; the higher the input amplitude, the larger the phase deviation.

Principle of Offset QAM

Recall the generation of the RF PAM signal by pulse shaping and upconversion with a quadrature mixer. The essential parts of Figure 3.2 are repeated in Figure 4.3.

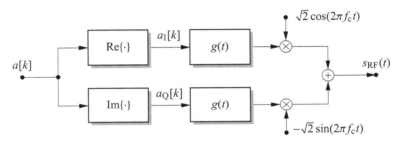

Figure 4.3 Block diagram of a PAM transmitter (real-valued pulse $g(t)$).

A slight modification leads to the transmitter for *offset QAM (OQAM)*: the signal in the quadrature branch is delayed by half a symbol period (i.e., by $T/2$). The transmitter is shown in Figure 4.4.

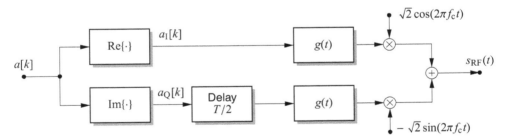

Figure 4.4 Block diagram of an offset QAM (OQAM) transmitter (real-valued pulse $g(t)$).

The purpose of this modification is to lower the fluctuations in the envelope of the RF transmit signal, and hence to lower the crest factor ζ_g contributed by the pulse. For OQAM it is given by

$$\zeta_g \stackrel{\text{def}}{=} \frac{\max_{\forall t} \sqrt{\left(\sum_k |g(t-kT)|\right)^2 + \left(\sum_k |g(t-kT-T/2)|\right)^2}}{\sqrt{2E_g/T}}. \tag{4.4}$$

As a result, the transmit signal exhibits less amplitude modulation (AM) – thus, less AM/AM and AM/PM conversion is caused. To have the same level of tolerable distortions, the power amplifier can be operated with less back-off, and thus more efficiently.

Note that the delay has no influence on the average psd of the transmit signal. A delay in the time domain corresponds to a phase ramp in the frequency domain; the magnitude remains unchanged.

The effect of the temporal offset to the ECB signal is best explained using "hard keying," that is, $g(t) = \sqrt{E_g/T} \, \mathrm{rect}(t/T)$. Moreover, a 4QAM signal constellation is used. In Figure 4.5 the situation is visualized for (a) (conventional) QAM and (b) OQAM. On the left, the *phasor* (i.e., the trajectory of the ECB signal in the complex plane) is shown. On the right, the quadrature components over time are shown.

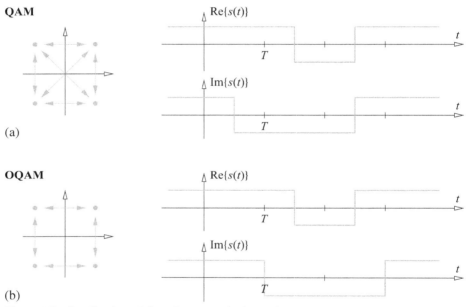

Figure 4.5 Visualization of the ECB transmit signals for (a) (conventional) QAM and (b) OQAM. Left: phasor. Right: quadrature components over time. 4QAM signal constellation and rectangular (in the time domain) basic pulse shape.

In conventional QAM the symbol transitions for in-phase and quadrature components occur at the same time. If, for $k = 2$ (as shown), the amplitude coefficient $a[2] = -1 - \mathrm{j}$ has to be sent and at $k = 3$ the amplitude coefficient $a[3] = 1 + \mathrm{j}$, at $t = 2.5T$ a transition between these two points in the complex plane takes place. Both in-phase and quadrature signals change from $-1 \cdot \sqrt{E_g/T}$ to $+1 \cdot \sqrt{E_g/T}$. The phasor (momentarily) crosses the origin; the amplitude $|s(t)|$ fluctuates from $\sqrt{2}\sqrt{E_g/T}$ to 0 and back to $\sqrt{2}\sqrt{E_g/T}$ again.

Using OQAM, these fluctuations are significantly lowered. Now, the in-phase or the quadrature signal changes; the changes alternate. Hence, the phasor changes either in the direction of the real or the imaginary part; crossings of the origin cannot occur.

Example 4.2 QAM / OQAM Phasors and Crest Factor of the Pulse Shape

The phasors of QAM and OQAM for the practically more relevant case of cosine-roll-off $\sqrt{\text{Nyquist}}$ pulses ("soft keying") are shown in Figures 4.6 and 4.7, respectively. The roll-off factor is $\alpha = 0.5$ and the constellation 4QAM.

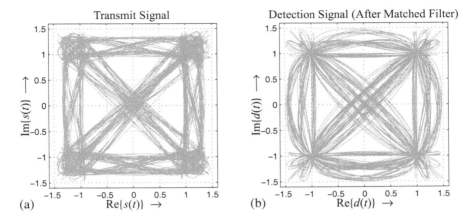

Figure 4.6 Phasors of (a) the QAM transmit signal $s(t)$ with cosine-roll-off $\sqrt{\text{Nyquist}}$ pulses and (b) the corresponding detection signal $d(t)$.

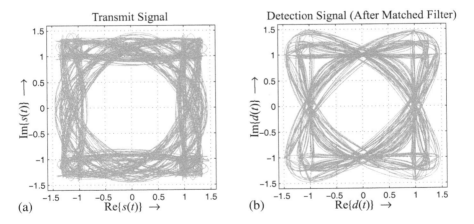

Figure 4.7 Phasors of (a) the OQAM transmit signal $s(t)$ with cosine-roll-off $\sqrt{\text{Nyquist}}$ pulses and (b) the corresponding detection signal $d(t)$.

As for hard keying, OQAM avoids the zero crossings of the phasor; the amplitude is more stable (but not constant).

Note that in PAM, $\sqrt{\text{Nyquist}}$ pulses $g(t)$ should be used. Hence, the phasors shown for $s(t)$ do not have sharp points at $\pm 1 \pm j$. Only when looking at the detection signals $d(t)$, where the end-to-end pulse $\varphi_{gg}(t) = g(t) * g^*(-t)$ is present (which is then a

Nyquist pulse), does the phasor exactly crosses the points defined by the constellation $\mathcal{A}$ at the detection instants kT. Since in OQAM the quadrature component is delayed by $T/2$, detection is alternately done in the real part ($t = kT$) and in the imaginary part ($t = kT + T/2$). At these instants, either the real part or the imaginary part of $d(t)$ takes on the values ± 1.

The crest factor ζ_g of OQAM when using $\sqrt{\text{Nyquist}}$ pulses with square-root cosine-roll-off spectral shape is shown in Figure 4.8 over the roll-off factor α. For comparison, the crest factor ζ_g of conventional QAM is repeated from Figure 3.32.

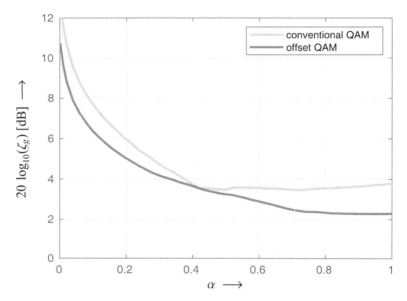

Figure 4.8 Crest factor ζ_g of OQAM with $\sqrt{\text{Nyquist}}$ pulses with square-root cosine-roll-off spectral shape over the roll-off factor α (dark gray). The light gray curve shows the crest factor of conventional QAM.

The reduction in the crest factor may not seem so dramatic – peak power is reduced only moderately. The additional effect of avoiding zero crossings, and thus reducing the dynamic range of the power amplifier input signal and, in turn, the dynamic distortions, is not captured in this figure.

Example 4.3 Phasor Demo

For visualization, examples of phasors are available on the companion web page. The demos run directly in the web browser. The phasor can be rotated using the mouse, moved (Ctrl + mouse), and zoomed (Shift + mouse).

The phasors are plotted in three-dimensional space; the real and imaginary part of the complex baseband signal $s(t)$ over time (normalized to the symbol duration T). A number of possible signals (corresponding to different sequences of amplitude coefficients) are plotted on top of each other. One of them is plotted in red. Rotate the phasor such that the real part (in-phase component) or the imaginary part (quadrature component) is plotted over time (eye pattern), and rotate the phasor such that the complex plane (imaginary vs. real part) is visible.

The following phasors of signals using 4QAM are available.

Phasor 1: Transmit signal $s(t)$ for rectangular basic pulse shape (cf. Figure 4.5a).

Phasor 2: Transmit signal $s(t)$ for rectangular basic pulse shape, offset QAM (cf. Figure 4.5b).

Phasor 3: Detection signal $d(t)$ (after matched filter) for rectangular basic pulse shape.

Phasor 4: Transmit signal $s(t)$ for $\sqrt{\text{Nyquist}}$ cosine-roll-off ($\alpha = 0.5$) basic pulse shape (cf. Figure 4.6a).

Phasor 5: Detection signal $d(t)$ for $\sqrt{\text{Nyquist}}$ cosine-roll-off ($\alpha = 0.5$) basic pulse shape (hence, cosine-roll-off Nyquist end-to-end pulse) (cf. Figure 4.6b).

Phasor 6: Transmit signal $s(t)$ for $\sqrt{\text{Nyquist}}$ cosine-roll-off basic pulse shape ($\alpha = 0.5$), offset QAM (cf. Figure 4.7a).

Phasor 7: Detection signal $d(t)$ for $\sqrt{\text{Nyquist}}$ cosine-roll-off ($\alpha = 0.5$) basic pulse shape (hence, cosine-roll-off Nyquist end-to-end pulse), offset QAM (cf. Figure 4.7b).

4.1.2 Minimum-Shift Keying (MSK)

Using offset QAM, the variations in the envelope of the transmit signal can be lowered, in particular, zero crossings can be avoided, but still some amplitude modulation is present.

The fluctuations in the magnitude of the ECB transmit signal can be avoided completely if the following basic pulse shape is employed in offset 4QAM:

$$g(t) = \sqrt{\frac{2E_g}{T}} \cos(\pi t/T) \cdot \text{rect}(t/T) . \tag{4.5}$$

The prefactor guarantees that the pulse has energy E_g. The pulse together with its $T/2$-delayed version is shown in Figure 4.9. The resulting modulation format is known as *minimum-shift keying (MSK)*.

Instead of delaying the signal that is input to the pulse shaping in the quadrature branch (as drawn in Figure 4.4), equivalently, the pulse shape in this branch can be shifted by $T/2$. In this sense, the in-phase branch of MSK uses the pulse shape shown in the top part of Figure 4.9. According to (4.5), it is half a cycle of a cosine wave; limited to the finite duration $t \in [-T/2, T/2]$. The quadrature branch of MSK uses the pulse shape shown in the bottom part of Figure 4.9. Since shifting a cosine wave by a quarter of a period results in a sine wave, the pulse in the quadrature

Figure 4.9 Basic pulse shape $g(t)$ used in minimum-shift keying (and $T/2$-shifted version).

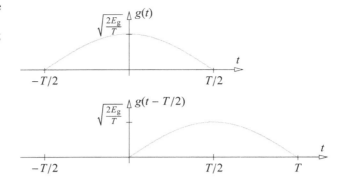

Figure 4.10 Phasor of minimum-shift keying.

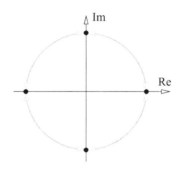

branch is a sine function limited to the finite duration $t \in [0, T]$. Since

$$s(t) = \pm \cos(\pi t/T) \pm \mathrm{j}\sin(\pi t/T) = \pm \mathrm{e}^{\mp \mathrm{j}\pi t/T}, \qquad (4.6)$$

we have $|s(t)| = 1$, $\forall t$. A constant envelope is present, as in Figure 4.10.

Note that a constant envelope is only obtained by the combination of 4QAM ($a = \pm 1 \pm \mathrm{j}$) and the specific pulse shape (4.5). For other signal constellations and/or other pulses, fluctuations in the magnitude will still occur.

In MSK – although it is a pulse-*amplitude* modulation format (complex amplitude!) – pure phase modulation is present (only the phase changes over time); the crest factor is $\zeta = 1$. Very efficient power amplification is possible, as no AM/AM and no (time-dependent) AM/PM conversion occurs – the power amplifier is driven at constant operating point, which may be the nominal power of the HPA (i.e., no back-off is necessary).

Power and Bandwidth Efficiency of MSK

As discussed in Chapter 3, the power efficiency of digital PAM can be characterized by the *normalized squared Euclidean distance* $d_{\min}^2$. As in PAM, this parameter depends only on the signal constellation and is independent of the basic pulse shape. MSK has the $d_{\min}^2$ of 4QAM; it is the same as for conventional 4QAM and given by (cf. (3.110))

4.1 Schemes with Reduced Peak-to-Average Power Ratio

$$d_{\min}^2 = 2 \ . \tag{4.7}$$

The bandwidth efficiency is characterized by the average power spectral density. For uncoded PAM with a zero-mean signal constellation, the average psd is given by (cf. (3.15))

$$\bar{\Phi}_{ss}(f) = \sigma_a^2 \cdot \frac{|G(f)|^2}{T} \ . \tag{4.8}$$

Using $\sigma_a^2 = 2$ for 4QAM and the correspondence of the Fourier transform

$$g(t) = \sqrt{\frac{2E_g}{T}} \cos(\pi t/T) \cdot \text{rect}(t/T)$$
$$ \circ\!\!-\!\!\bullet \tag{4.9}$$
$$G(f) = \sqrt{\frac{2E_g}{T}} \frac{2T}{\pi} \cdot \frac{\cos(\pi fT)}{1 - (2fT)^2} \ ,$$

we arrive at (for $f = \pm\frac{1}{2T}$ the continuous extension has to be considered)

$$\bar{\Phi}_{ss}(f) = \frac{16E_g}{\pi^2} \cdot \frac{\cos^2(\pi fT)}{(1 - (2fT)^2)^2} \ . \tag{4.10}$$

Since the time-domain pulse $g(t)$ is time-limited, according to the properties of the Fourier transform, the corresponding spectrum $G(f)$ cannot be strictly bandlimited; hence, the average psd of the MSK transmit signal has infinite bandwidth. The broad main lobe and the distinct side lobes are visible in Figure 4.11.

Consequently, instead of specifying the bandwidth where the entire spectrum is contained, an $X\%$ bandwidth $B_{X\%}$ has to be given. $X\%$ of the transmit power is contained within the bandwidth $B_{X\%}$, that is

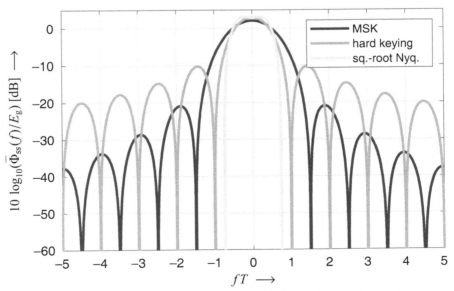

Figure 4.11 Comparison of (average) psds (log scale). MSK, hard keying, and $\sqrt{\text{Nyquist}}$ cosine-roll-off pulse ($\alpha = 0.5$).

$$\int_{-B_{X\%}/2}^{+B_{X\%}/2} \bar{\Phi}_{ss}(f) \, \mathrm{d}f \stackrel{\text{def}}{=} \frac{X}{100} \int_{-\infty}^{\infty} \bar{\Phi}_{ss}(f) \, \mathrm{d}f \, . \tag{4.11}$$

For MSK we have ($R_\mathrm{T} = 1/T_\mathrm{b}$)

$$B_{99\%} = 1.18 \, R_\mathrm{T} \, , \qquad \Gamma_{99\%} = \frac{R_\mathrm{T}}{B_{99\%}} = 0.85 \left[\frac{\text{bit/s}}{\text{Hz}}\right] . \tag{4.12}$$

Note that this spectral efficiency has to be compared to that of 4QAM using $\sqrt{\text{Nyquist}}$ pulses. For the best case of a roll-off factor $\alpha = 0$, we would have $\Gamma = \log_2(M) = 2$; hence, MSK sacrifices approximately only half of the possible spectral efficiency. This is the price to be paid for having a constant envelope. As (almost) always in engineering: "There is no free lunch."

Finally, Figure 4.11 compares the psds of MSK with that of hard keying ($g(t)$ is a rect pulse) and that when using a $\sqrt{\text{Nyquist}}$ cosine-roll-off pulse with $\alpha = 0.5$.

MSK has a broad main lobe and distinct side lobes, which, however, decay faster (with $1/f^2$, since $g(t)$ is continuous but not differentiable at $t = \pm T/2$) than the side lobes of the psd for the rect pulse, which is given by a sinc function (this psd decays with $1/f$, since the rect pulse is discontinuous). Using a $\sqrt{\text{Nyquist}}$ cosine-roll-off pulse, a compact (bandlimited) psd is obtained.

Equivalence between MSK and Binary Frequency Modulation

In PAM, the information is carried in the amplitude coefficients $a[k]$. Using 4QAM, per modulation step k of duration T, two bits are transmitted ($R = 2$); or one bit per duration $T/2 = T_\mathrm{b}$ (the bit duration).

As desired, no amplitude modulation is left in MSK. Hence, the information is purely represented in the phase. As *phase* and *frequency* are related via differentiation, one can also argue that the information is carried in the instantaneous (momentary) frequency $f_\mathrm{M}(t)$.

In order to derive the connection between the representation of the information via the amplitude coefficient $a[k]$ and that via the frequency $f_\mathrm{M}(t)$, in Figure 4.12, an exemplary MSK transmit signal is plotted.

In the two top plots, in-phase (lighter gray) and quadrature component (darker gray) are shown for the given sequence of amplitude coefficients $a[k]$. The phasor ($s(t)$ in the complex plane) is shown in the third row. During an interval of $T/2$, the signal moves with constant speed along the indicated quarter circle. At $t = kT$ the signal $s(t)$ is real-valued, it lies on the real axis; for $t = kT + T/2$ it is purely imaginary, it lies on the imaginary axis. The corresponding phase signal $\varphi(t)$ is shown in the fourth row. The phase either increases linearly by $\pi/2$ or decreases linearly by $\pi/2$ during an interval of duration $T/2$. Note that the phase is displayed over the interval $[-\pi, \pi]$; the discontinuity at $t = 3T$ is only due to 2π phase wrapping in this interval. As $f_\mathrm{M}(t) = \frac{1}{2\pi} \frac{\mathrm{d}\varphi(t)}{\mathrm{d}t}$, the linear phase change corresponds to an instantaneous frequency that is constant over each interval of duration $T/2$, shown in the bottom row of the figure.

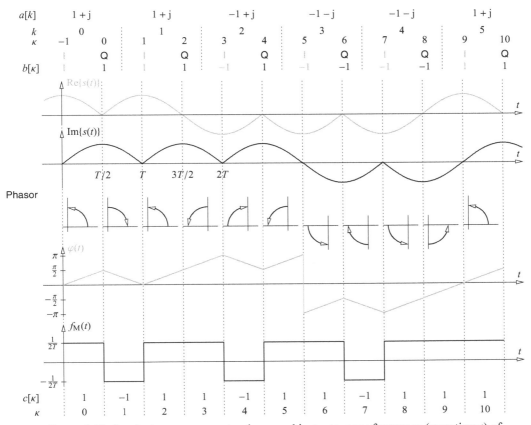

Figure 4.12 Quadrature components, phase, and instantaneous frequency (over time t) of an exemplary MSK transmit signal.

A *binary frequency modulation* is present; the information is represented as a positive or negative instantaneous frequency of the transmit signal (in the ECB domain; hence, upward or downward tuning from the carrier frequency).

Using the *serialized description* $b[\kappa]$, with $b[\kappa] \in \{\pm 1\}$, $\kappa \in \mathbb{Z}$, such that

$$a[k] = b[2k - 1] + j\, b[2k], \qquad k \in \mathbb{Z} \tag{4.13}$$

(in-phase and quadrature components are demultiplexed into a binary data sequence of twice the speed), a thorough study of Figure 4.12 reveals that each pair $(b[\kappa - 1], b[\kappa])$ determines the *instantaneous frequency* $f_\mathrm{M}(t)$ within a time interval of bit duration $T_\mathrm{b} = T/2$. Defining $c[\kappa] \in \{\pm 1\}$ as

$$c[\kappa] = (-1)^\kappa \cdot b[\kappa] \cdot b[\kappa - 1], \tag{4.14}$$

we have

$$f_\mathrm{M}(t) = \frac{1}{2\pi}\frac{\mathrm{d}\varphi(t)}{\mathrm{d}t} = \pm\frac{1}{4T_\mathrm{b}} c[\kappa], \quad \text{for}\quad t \in [\kappa T_\mathrm{b}, (\kappa + 1) T_\mathrm{b}). \tag{4.15}$$

Hence, $c[\kappa]$ can be interpreted as the input symbol (time step κ) to an equivalent *frequency modulator*, called binary *frequency-shift keying (FSK)* (see Chapter 7).

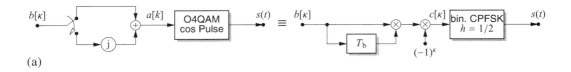

(a)

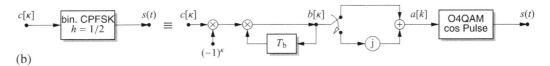

(b)

Figure 4.13 Equivalent MSK transmitter structures: (a) $b[\kappa]$ given; (b) $c[\kappa]$ given.

This equivalence leads to the block diagrams of an MSK transmitter shown in Figure 4.13. In Figure 4.13(a) it is assumed that the binary sequence $\langle b[\kappa]\rangle$ (QAM view) is given. It can either be demultiplexed into the amplitude coefficients $a[k]$, which are fed into the offset QAM transmitter employing the cosine pulse (4.5) (left part), or using (4.14) the binary sequence $\langle c[\kappa]\rangle$ is calculated, which is fed into a binary frequency modulator that guarantees continuous phase (hence CPFSK) (right part). Note that FSK modulators are usually characterized by their *modulation index h*, which gives the separation of the instantaneous frequencies relative to the symbol rate of CPFSK. Since $f_\mathrm{M}(t) = \pm\frac{1}{4T_\mathrm{b}}$ we have $h = 1/2$. A CPFSK modulator can be implemented by means of a *voltage-controlled oscillator (VCO)* (or, in digital implementations, by a numerically controlled oscillator (NCO)).

In Figure 4.13(b) it is assumed that the binary sequence $\langle c[\kappa]\rangle$, directly representing the instantaneous frequency, is given (FSK view). It can either be fed into the CPFSK modulator or, inverting (4.14), the binary sequence $\langle b[\kappa]\rangle$ can be calculated and fed into the offset QAM modulator.

Since a particular variant of offset QAM is present and offset QAM (except considering the delay) has the same receiver as conventional QAM, MSK can be demodulated using the well-known frontend structure (matched filtering and sampling) and a threshold decision device as discussed in Chapter 3.

Note that MSK is the only modulation format that can be viewed either as a PAM or FSK scheme. The denomination "minimum" will be explained in Chapter 7.

Example 4.4 Phasor Demo

Again, for visualization, the phasor of MSK is available online.

Phasor 8: Transmit signal $s(t)$ of MSK.

To visualize that even when using PSK no constant envelope is achieved, phasors for transmit signals employing 8PSK and rectangular or $\sqrt{\text{Nyquist}}$ cosine-roll-off basic pulse shape are available.

Phasor 9: Transmit signal $s(t)$ for 8PSK rectangular basic pulse shape.

Phasor 10: Transmit signal $s(t)$ for 8PSK $\sqrt{\text{Nyquist}}$ cosine-roll-off ($\alpha = 0.5$) basic pulse shape.

4.1.3 Gaussian Minimum-Shift Keying (GMSK)

MSK guarantees a constant envelope of the transmit signal, and thus is well suited for power amplification. However, as we have seen, the power spectral density is very broad, leading to poor spectral efficiency. A natural question is whether a constant envelope can be achieved with a more compact spectrum.

A basic principle of the Fourier transform is that quick changes correspond to large bandwidth; discontinuities ("jumps") in the time-domain signal lead to spectra that only decay with $1/f$. The MSK transmit signal is continuous but not differentiable; sudden changes in the "direction" occur. Or, when looking at Figure 4.12, the instantaneous frequency is discontinuous. This signal, which is not the transmit signal itself but the frequency thereof, looks like a hard-keyed BPSK signal.

The main idea of *Gaussian minimum-shift keying (GMSK)*, which is the modulation format in the *Global System for Mobile Communications (GSM)*, is to "smoothen" the instantaneous frequency. In order to enlighten this procedure, in Figure 4.14 the block diagrams of (a) the MSK and (b) the GMSK transmitter are contrasted.

In both schemes, a continuous-time instantaneous frequency signal is generated in the same way as a BPSK transmit signal would be generated. When starting with the serialized O4QAM description, the coefficients $b[\kappa]$ are first converted into the FSK description, that is, the binary frequency symbols $c[\kappa]$. This sequence is filtered (pulse shaped) with the rectangular pulse of duration T_b.

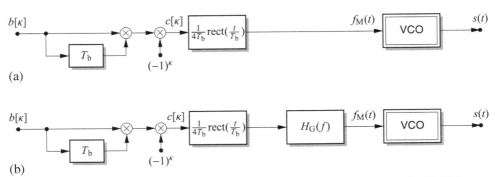

Figure 4.14 (a) Block diagram of the MSK transmitter. (b) Block diagram of the GMSK transmitter.

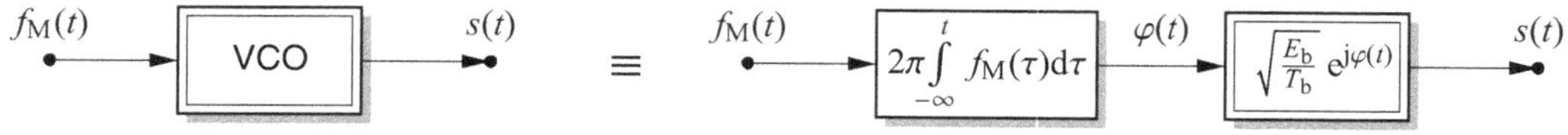

4 Variants of PAM Transmission Schemes

$f_M(t) \longrightarrow \boxed{\text{VCO}} \longrightarrow s(t) \quad \equiv \quad f_M(t) \longrightarrow \boxed{2\pi \int_{-\infty}^{t} f_M(\tau) d\tau} \xrightarrow{\varphi(t)} \boxed{\sqrt{\frac{E_b}{T_b}} e^{j\varphi(t)}} \longrightarrow s(t)$

Figure 4.15 Block diagram of a voltage-controlled oscillator (VCO) in the ECB domain.

In MSK, this hard-keyed signal is directly fed to the voltage-controlled oscillator (VCO). Note that in the ECB domain, a VCO is given as depicted in Figure 4.15 – the instantaneous frequency $f_M(t)$ is integrated to obtain the phase function $\varphi(t)$. A phase modulator produces the ECB transmit signal (E_b: energy per bit)

$$s(t) = \sqrt{\frac{E_b}{T_b}} e^{j\varphi(t)} = \sqrt{\frac{E_b}{T_b}} e^{j2\pi \int_{-\infty}^{t} f_M(\tau) d\tau} . \tag{4.16}$$

In GMSK, the hard-keyed signal is filtered with an LTI system with transfer function $H_G(f)$ (corresponding impulse response $h_G(t)$). This signal, $f_M(t)$, where the jumps have been smoothed, is then fed to the VCO. In GMSK, the low-pass filter is chosen to have a *Gaussian* (hence the denomination) transfer function, and thus it also has a Gaussian impulse response, specifically

$$H_G(f) = e^{-(f/B_b)^2 \cdot \log(2)/2}$$
$$\updownarrow \tag{4.17}$$
$$h_G(t) = \frac{1}{\sqrt{2\pi T_b^2 \tau^2}} e^{-(t/T_b)^2/(2\tau^2)} , \qquad \tau \stackrel{\text{def}}{=} \frac{\sqrt{\log(2)}}{2\pi B_b T_b} .$$

Thereby, B_b is a bandwidth parameter that can be tuned. In the GSM system, $B_b T_b = 0.3$ is selected.

Since rectangular pulse shaping and Gaussian filtering are done in cascade, the entire *frequency pulse* reads

$$g_F(t) \stackrel{\text{def}}{=} \frac{1}{4T_b} \text{rect}\left(\frac{t}{T_b}\right) * h_G(t)$$

$$= \frac{1}{4T_b} \int_{t-T_b/2}^{t+T_b/2} \frac{1}{\sqrt{2\pi T_b^2 \tau^2}} e^{-(\tau/T_b)^2/(2\tau^2)} d\tau$$

$$= \frac{1}{4T_b} \int_{(t-T_b/2)/(\tau T_b)}^{(t+T_b/2)/(\tau T_b)} \frac{1}{\sqrt{2\pi}} e^{-\tau^2/2} d\tau$$

$$= \frac{1}{4T_b} \left(Q\left(\frac{t-T_b/2}{\tau T_b}\right) - Q\left(\frac{t+T_b/2}{\tau T_b}\right) \right) , \tag{4.18}$$

where we have used the complementary Gaussian integral function $Q(x)$ (cf. page 95). In Figure 4.16, this frequency pulse is depicted for different choices of B_b. For very large B_b, almost no filtering takes place ($H_G(f) \to 1$) and the rectangular frequency pulse of MSK is obtained. For small B_b, a narrowband filtering is performed and the frequency pulse is broadened, leading to a stronger smearing of the jumps.

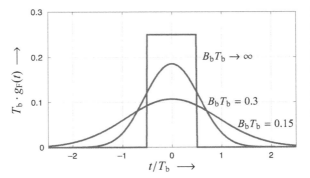

Figure 4.16 GMSK frequency pulse for different choices of $B_b T_b$. Non-causal description.

Note the difference between pulse shaping in PAM and shaping of the instantaneous frequency in GMSK. In PAM, the transmit signal itself is affected. In GMSK, the signal "instantaneous frequency" is treated; the transmit signal is obtained after the nonlinear operation e^{jx} (the VCO). Thus, GMSK is a *nonlinear* modulation format (in contrast to the linear format PAM).

Example 4.5 Instantaneous Frequency over Time

Exemplary signals of the instantaneous frequency over time (for the same binary data sequence) are given for MSK and GMSK in Figures 4.17 and 4.18.

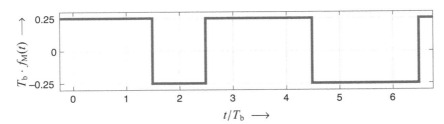

Figure 4.17 Example for the instantaneous frequency of MSK.

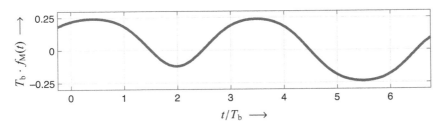

Figure 4.18 Example for the instantaneous frequency of GMSK.

The smooth (continuous) course in case of GMSK is clearly visible.

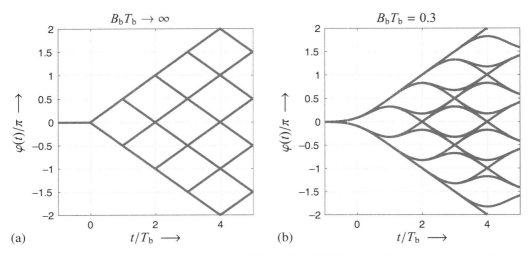

Figure 4.19 Phase trajectories for (a) MSK and (b) GMSK. Start of transmission at $t = 0$.

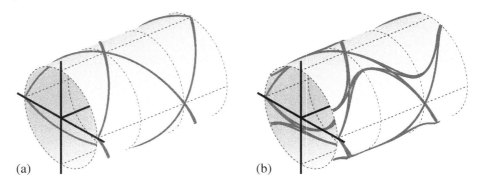

Figure 4.20 Phase cylinders for (a) MSK and (b) GMSK ($B_b T_b = 0.3$).

As the magnitude is constant in MSK and GMSK, we may look solely at the instantaneous frequency, or equivalently the phase. Integrating the rectangular course of the MSK frequency signal gives a linearly increasing or decreasing phase (cf. Figure 4.12). For GMSK, smooth transitions of the phase are present.

Similar to the eye pattern in PAM, the phases of the transmit signals for all possible binary data signals can be plotted on top of each other, leading to the so-called *phase trajectories*. These diagrams are shown for (a) MSK and (b) GMSK in Figure 4.19.

Taking (i) the constant magnitude of the (G)MSK transmit signals and (ii) that the phase is ambiguous with respect to 2π into account, the phase trajectories can be wrapped around a cylinder. These *phase cylinders* give the ECB transmit signals in three-dimensional space, as shown in Figure 4.20.

Power and Bandwidth Efficiency of GMSK

We now discuss the power and bandwidth efficiency of GMSK. We again emphasize that GMSK is a nonlinear modulation format. Consequently, the calculation of the average psd $\bar{\Phi}_{ss}(f)$ and of the normalized squared Euclidean distance $d^2_{\min}$ of Chapter 3 cannot be applied here.

Instead, the (average) psd of GMSK may be obtained from numerical (Monte Carlo) simulations. To that end, for a random binary data sequence, a sufficiently long part of a GMSK transmit signal is generated. This signal is used to deduce the average psd, applying methods from spectral estimation (e.g., averaged periodograms). The psds generated in this way are depicted in Figure 4.21.

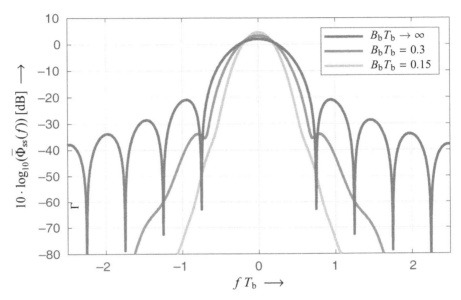

Figure 4.21 Average power spectral density of GMSK. Different choices of $B_b T_b$.

For $B_b T_b \to \infty$, the broad spectrum of MSK is obtained. When decreasing $B_b T_b$, a stronger smoothing is active and the spectrum becomes more compact. For $B_b T_b = 0.15$, an even more compact psd is obtained in comparison to $B_b T_b = 0.3$.

As we will see in Chapter 6, the power efficiency of any digital modulation scheme can be characterized by the normalized minimum squared Euclidean distance $d^2_{\min}$. For the general case, this parameter is defined as the minimum energy of the difference of any pair of signals representing different messages (cf. (6.73))

$$d^2_{\min} = \frac{1}{2E_b} \cdot \min_{\substack{s^{(i)}(t), s^{(j)}(t) \\ s^{(i)}(t) \neq s^{(j)}(t)}} \int_{-\infty}^{\infty} |s^{(i)}(t) - s^{(j)}(t)|^2 \, dt \ . \tag{4.19}$$

Basically, $d^2_{\min}$ quantifies how well two transmit sequences can be distinguished. Note that this is a parameter of the modulation scheme – it compares *transmit* signals. To make use of the properties of the modulation scheme, a suitable optimum receiver has to be used. The design of the optimum GMSK receiver will be discussed briefly in Section 7.4 in the context of the more general class of continuous-phase modulation. However, in practice, often a sub-optimum (but low-complexity) receiver is used.

As GMSK is derived from MSK and MSK is an offset QAM scheme where the conventional PAM receiver applies, GMSK may be demodulated as if it were a PAM scheme. The block diagram of this sub-optimum receiver is shown in Figure 4.22. It follows the well-known concept of filtering, sampling, and threshold decision. Owing to the offset principle, decisions in the real and imaginary parts are interlaced.

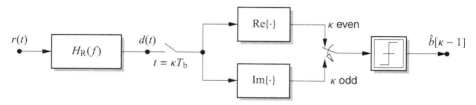

Figure 4.22 Sub-optimum receiver for GMSK.

As GMSK is a nonlinear modulation format, no matched filter can be given in the strict sense. As a compromise, the receive signal $r(t)$ is filtered with a Gaussian low-pass filter, that is

$$H_R(f) = \sqrt{\frac{1}{E_b B_R}} \sqrt{\frac{\log(2)}{\pi}} e^{-(f/B_R)^2 \cdot \log(2)/2}$$

$$h_R(t) = \sqrt{\frac{2}{E_b}} B_R \sqrt{\frac{\pi}{\log(2)}} e^{-(t B_R \pi)^2 \cdot 2/\log(2)} , \qquad (4.20)$$

where the filter bandwidth B_R (3 dB cut-off frequency) has to be optimized to the transmitter-side bandwidth B_b and to the actual signal-to-noise ratio on the channel. For the GSM system, $B_R T_b \approx 0.29$ is a convenient choice (in the literature, this filter is sometimes specified in the RF domain, consequently with twice the bandwidth). The noise variance after the low-pass filter is $\sigma_n^2 = \frac{N_0}{E_b}$; compared to PAM with a matched filter (since $E_g = E_b$), the same noise power is present.

Example 4.6 Eye Pattern of GMSK with Sub-optimum Receiver

When approximating the receiver for GMSK by a linear receiver, as in PAM, eye patterns can be studied. These patterns are shown in Figure 4.23 for different bandwidths B_b and $B_R T_b = 0.29$.

For $B_b T_b < 0.5$, the effect of ISI is clearly visible; there are no more sharp points in the eye pattern at the sampling instants. When decreasing B_b, the vertical eye opening decreases, that is, the power efficiency declines as the symbol error ratio is dominated by the smallest distance to the decision threshold. Note that, even for $B_b T_b \to \infty$ (MSK), the vertical eye opening is clearly smaller than 2 as the matched filter is not used and the low-pass filter causes significant attenuation.

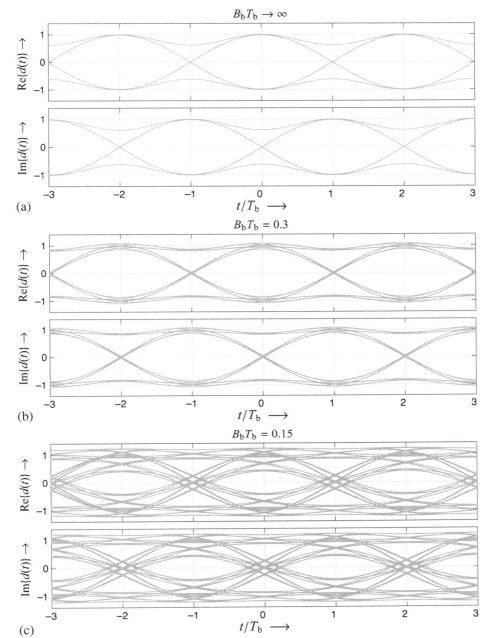

Figure 4.23 Eye pattern of GMSK with sub-optimum receiver according to Figure 4.22; $B_R T_b = 0.29$. (a) $B_b T_b \to \infty$; (b) $B_b T_b = 0.3$; and (c) and $B_b T_b = 0.15$.

However, as the filtering gets stronger (see above), a more compact spectrum is obtained. Hence, via $B_b T_b$ a trade-off between power and bandwidth efficiency is enabled.

The offset between the in-phase and quadrature components is also visible; alternatingly (each $T_b = T/2$), decisions are taken in the real and imaginary part, respectively.

Finally, in Figure 4.24, the normalized squared Euclidean distance of GMSK is plotted over the bandwidth B_b.

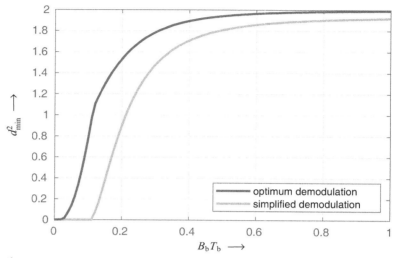

Figure 4.24 Normalized squared Euclidean distance of GMSK over the normalized bandwidth $B_b T_b$. Optimum and sub-optimum ($B_R T_b = 0.29$) demodulation.

The dark gray line corresponds to a (hypothetical, very complex) optimum receiver, the light gray line to the discussed concept of approximate filtering and sampling. For $B_b T_b \to \infty$, MSK is obtained and the optimum demodulation strategy achieves $d_{min}^2 = 2$ of MSK (of 4QAM). As $B_b T_b$ decreases, d_{min}^2 decreases too, but the bandwidth efficiency increases. For $B_b T_b$ down to approximately 0.3, the loss is not significant. However, for even smaller $B_b T_b$, the Euclidean distance drops significantly. From this diagram, the choice $B_b T_b = 0.3$ in GSM seems reasonable.

For the approximate demodulation, the loss in the vertical eye opening is expressed as an equivalent decrease in d_{min}^2. The light gray curve shows that for sufficiently large $B_b T_b$, only a small loss is caused compared to the theoretical optimum receiver.

Note that the idea of generating a suitable (frequency or) phase signal, which represents the information (the amplitude remains constant), can be generalized. In the literature there is the broad field of *continuous-phase modulation (CPM)*. A short introduction to this class of digital modulation schemes will be given in Section 7.4.

4.2 Schemes without Explicit I/Q Demodulation

In Chapter 3, digital PAM was treated in detail. We now study two alternative approaches, which avoid the explicit mixing of the received signal from the RF to the baseband domain. The shown variants complement each other: the one is particularly suited for low carrier frequencies, since all processing can be done in the digital domain; the other is interesting for very high carrier frequencies, since the analog receiver frontend can be implemented with less effort.

4.2.1 "Carrierless" Amplitude and Phase Modulation

We first address an alternative approach to PAM transmission, which avoids (i) the explicit mixing of the pulse-shaped continuous-time baseband signal to the RF domain by directly generating the RF signal and (ii) the explicit mixing of the receive signal to the baseband by simply performing sampling. This is of particular interest if the carrier frequency is not too high. This form of modulation is, for example, standardized for high-rate transmission over twisted-pair lines (asymmetric digital subscriber line, ADSL).

Basic Concept

To highlight the similarities and differences of so-called *carrierless amplitude and phase modulation (CAP)* to conventional PAM, in Figure 4.25 the transmission model of PAM (employing a real-valued pulse shape $g(t)$) is repeated. The pulse shaping in the I and Q branches and the upconversion to the carrier frequency by mixing with sine and cosine waves are visible. At the receiver side, the corresponding operations are applied; downmixing followed by matched filtering and sampling. Combining both real-valued signals establishes the complex-valued detection signal.

In Figure 4.26, the transmission model of CAP is shown. Here, two different pulses, $g_I(t)$ and $g_Q(t)$, are employed. No explicit mixing in the RF domain is performed. Consequently, at the receiver, without downconversion only filtering using the matched filters for $g_I(t)$ and $g_Q(t)$, respectively, is done.

This approach only works if we impose some constraints on the pulses $g_I(t)$ and $g_Q(t)$. Specifically, we have to demand the following.

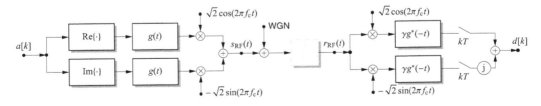

Figure 4.25 Block diagram of PAM transmission (real-valued pulse $g(t)$).

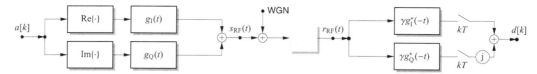

Figure 4.26 Block diagram of CAP transmission. $g_I(t)$ and $g_Q(t)$ are real-valued pulses.

- When considering the in-phase or the quadrature branch separately, we see directly that in both branches real-valued PAM transmission is present. In order to have intersymbol-interference-free sampling instants (when assuming an AWGN channel), both $g_I(t)$ and $g_Q(t)$ have to be real-valued $\sqrt{Nyquist}$ *pulses* (both with the same energy E_g), that is

$$\int_{-\infty}^{\infty} g_I(t+kT) \cdot g_I(t)\, dt = E_g \cdot \delta[k], \qquad k \in \mathbb{Z}, \tag{4.21}$$

$$\int_{-\infty}^{\infty} g_Q(t+kT) \cdot g_Q(t)\, dt = E_g \cdot \delta[k], \qquad k \in \mathbb{Z}. \tag{4.22}$$

- When considering both branches together, we see that parallel PAM transmission is present. In order to have no crosstalk at the sampling instants, $g_I(t)$ and $g_Q(t)$ have to be *mutually orthogonal*, that is

$$\int_{-\infty}^{\infty} g_I(t+kT) \cdot g_Q(t)\, dt = 0, \qquad k \in \mathbb{Z}. \tag{4.23}$$

- Finally, at the output of the pulse-shaping filters, an RF signal should be present directly. This is achieved if $g_I(t)$ and $g_Q(t)$ have *bandpass spectra* centered at the carrier frequency f_c.

Note that, owing to the parallel, individual use of the two branches, CAP can be seen as a *multiplexing scheme*. Two signals are combined at the transmitter – because of the constraints, the signals can be separated at the receiver. Moreover, the approach is an instance of the (much more) general version of the *signal-space concept* of digital transmission, which will be studied in detail in Chapter 6.

Choice of the Pulses

Basically, any pair of pulses $g_I(t)$ and $g_Q(t)$ which fulfills the above-stated demands can be used in CAP. However, usually the pair is derived from a given pulse $g(t)$, which may be used in conventional PAM.

To that end, a real-valued $\sqrt{Nyquist}$ (baseband) pulse $g(t)$ is selected. Its spectrum has to be bandlimited; we denote the maximum frequency as f_g. Mathematically, we demand

$$\int_{-\infty}^{\infty} g(t+kT) \cdot g(t)\, dt = E_g \cdot \delta[k], \qquad k \in \mathbb{Z}, \tag{4.24}$$

with
$$G(f) = \mathcal{F}\{g(t)\} = 0, \qquad |f| > f_g. \tag{4.25}$$

Given $g(t)$ and the carrier frequency $f_c > f_g$, the pulses are obtained as
$$g_I(t) = \sqrt{2}g(t) \cdot \cos(2\pi f_c t), \tag{4.26}$$
$$g_Q(t) = -\sqrt{2}g(t) \cdot \sin(2\pi f_c t). \tag{4.27}$$

Note that for this construction, $g_I(t)$ and $g_Q(t)$ constitute a *Hilbert pair*, in the sense that the one pulse is the Hilbert transform of the other. This can be seen from the following (the scaling by $\sqrt{2}$ is ignored for brevity):

$$
\begin{array}{rcl}
\mathcal{H}\{-g(t)\sin(2\pi f_c t)\} & = & g(t)\cos(2\pi f_c t) \\
\updownarrow & & \updownarrow \\
\frac{1}{2j}(-G(f-f_c)+G(f+f_c)) \cdot (-j \cdot \mathrm{sgn}(f)) & = & \frac{1}{2}(G(f-f_c)+G(f+f_c)).
\end{array}
\tag{4.28}
$$

We start with the upper-left part and transform the signal to the frequency domain (lower-left part). Modulation (multiplication with $\sin(2\pi f_c t)$) generates spectra shifted to $\pm f_c$; the Hilbert transform is equal to a multiplication of the spectrum with the $-j \cdot \mathrm{sgn}(f)$ function. Since $f_c > f_g$, the parts of the spectrum, $G(f \pm f_c)$, do not overlap and are individually affected by the sign function. This gives the expression in the lower-right part. After an inverse Fourier transform, the signal in the upper-right part is obtained and the relation is proved.

It still has to be shown that this construction results in pulses $g_I(t)$ and $g_Q(t)$ that fulfill the demands. To that end, we consider the *autocorrelations* $\varphi_{g_I g_I}(\tau)$ and $\varphi_{g_Q g_Q}(\tau)$ and the *crosscorrelation* $\varphi_{g_I g_Q}(\tau)$. The demands are equivalently written as

$$\varphi_{g_I g_I}(kT) = E_g \cdot \delta[k], \qquad \varphi_{g_Q g_Q}(kT) = E_g \cdot \delta[k], \qquad \varphi_{g_I g_Q}(kT) = 0. \tag{4.29}$$

We have
$$\varphi_{g_I g_I}(kT) = \int_{-\infty}^{\infty} g_I(t+kT) \cdot g_I(t)\, dt = (g_I(t) * g_I(-t))\big|_{t=kT}$$
$$\updownarrow \tag{4.30}$$
$$\Phi_{g_I g_I}(e^{j2\pi fT}) = \sum_{\mu} G_I\left(f - \tfrac{\mu}{T}\right) \cdot G_I^*\left(f - \tfrac{\mu}{T}\right)$$

definition of the pulse
$$= \sum_{\mu} \sqrt{2}\tfrac{1}{2}\left(G\left(f - f_c - \tfrac{\mu}{T}\right) + G\left(f + f_c - \tfrac{\mu}{T}\right)\right)$$
$$\cdot \sqrt{2}\tfrac{1}{2}\left(G^*\left(f - f_c - \tfrac{\mu}{T}\right) + G^*\left(f + f_c - \tfrac{\mu}{T}\right)\right)$$

"$+f_c$" and "$-f_c$" parts do not overlap
$$= \frac{1}{2}\sum_{\mu}\left(\left|G\left(f - f_c - \tfrac{\mu}{T}\right)\right|^2 + \left|G\left(f + f_c - \tfrac{\mu}{T}\right)\right|^2\right)$$

$g(t)$ is a $\sqrt{\text{Nyquist}}$ pulse

$$= \frac{1}{2}(E_g + E_g) = E_g$$

$$\circ\!\!-\!\!\bullet$$

$$\varphi_{g_I g_I}(kT) = E_g \cdot \delta[k] .$$

(4.31)

In the same way, the property of $g_Q(t)$ can be shown.

For the crosscorrelation, we have[1]

$$\varphi_{g_I g_Q}(kT) = \int_{-\infty}^{\infty} g_I(t+kT) \cdot g_Q(t)\, dt = \left. (g_I(t) * g_Q(-t)) \right|_{t=kT}$$

$$\bullet\!\!-\!\!\circ$$

$$\Phi_{g_I g_Q}(e^{j2\pi fT}) = \sum_\mu G_I\left(f - \tfrac{\mu}{T}\right) \cdot G_Q^*\left(f - \tfrac{\mu}{T}\right)$$

(4.32)

definition of the pulses

$$= \sum_\mu \sqrt{2}\tfrac{1}{2}\left(G\left(f - f_c - \tfrac{\mu}{T}\right) + G\left(f + f_c - \tfrac{\mu}{T}\right)\right)$$
$$\cdot \sqrt{2}\tfrac{1}{2j}\left(G^*\left(f - f_c - \tfrac{\mu}{T}\right) - G^*\left(f + f_c - \tfrac{\mu}{T}\right)\right)$$

"$+f_c$" and "$-f_c$" parts do not overlap

$$= \frac{1}{2j} \sum_\mu \left(\left|G\left(f - f_c - \tfrac{\mu}{T}\right)\right|^2 - \left|G(f + f_c - \tfrac{\mu}{T})\right|^2 \right)$$

$g(t)$ is a $\sqrt{\text{Nyquist}}$ pulse

$$= \frac{1}{2j}(E_g - E_g) = 0$$

$$\bullet\!\!-\!\!\circ$$

$$\varphi_{g_I g_Q}(kT) = 0$$

(4.33)

Example 4.7 Mutually Orthogonal $\sqrt{\text{Nyquist}}$ Pulses

We chose $g(t)$ as a $\sqrt{\text{Nyquist}}$ cosine-roll-off pulse with roll-off factor $\alpha = 0.3$. The pulse is depicted in the top row of Figure 4.27. Choosing the carrier frequency to be $f_c = 1.8/T$, the pulses $g_I(t)$ and $g_Q(t)$ drawn in the next two rows are obtained. The autocorrelations $\varphi_{g_I g_I}(kT)$ and $\varphi_{g_Q g_Q}(kT)$ and the crosscorrelation $\varphi_{g_I g_Q}(kT)$ are depicted in the final three rows. The zeros at the sampling instants are clearly visible.

---

[1] This can also be deduced from the property of the Hilbert transform: a signal and its Hilbert transform are orthogonal to each other. Since $g(t)$ is a $\sqrt{\text{Nyquist}}$ pulse, this orthogonality also holds for shifted (by kT) versions of the pulses.

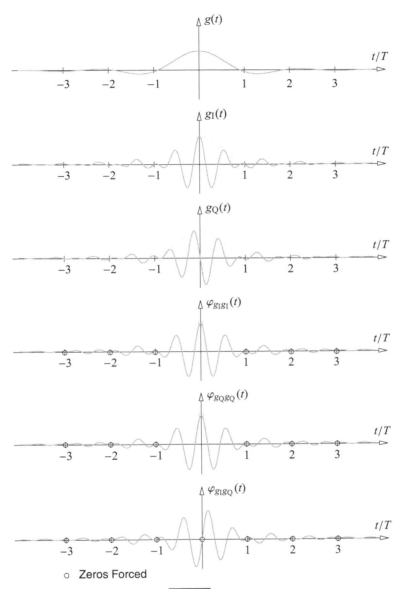

Figure 4.27 Mutually orthogonal $\sqrt{\text{Nyquist}}$ pulses ($\alpha = 0.3$, $f_c = 1.8/T$), their autocorrelations, and their crosscorrelation.

Interpretation and Discussion

For an interpretation of the differences in conventional PAM and CAP, spectra of the transmitter-side signal are shown in Figures 4.28 and 4.29. For illustration purposes, $G(f)$ and the spectra $A_I\left(e^{j2\pi fT}\right) = \mathscr{F}_*\{\text{Re}\{a[k]\}\}$ and $A_Q\left(e^{j2\pi fT}\right) = \mathscr{F}_*\{\text{Im}\{a[k]\}\}$ of the I and Q sequences are assumed to be real-valued. Even though the psd of the sequence of amplitude coefficients is usually flat, in order to visualize

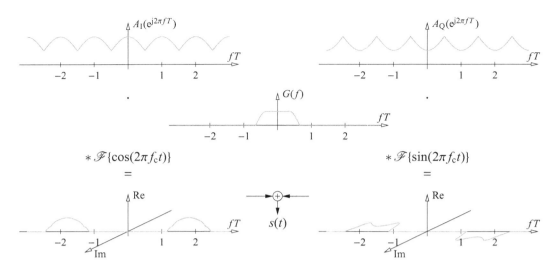

Figure 4.28 Generation of the PAM transmit signal. Visualization in the frequency domain.

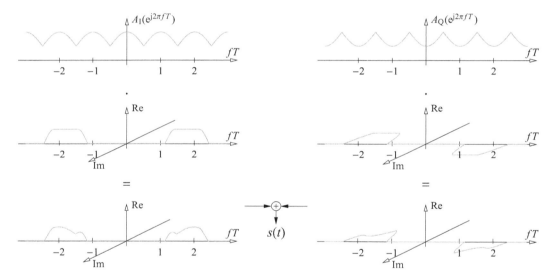

Figure 4.29 Generation of the CAP transmit signal. Visualization in the frequency domain.

the differences, the periodic spectra of the discrete-time sequences are shown as a garland shape.

In conventional PAM (cf. also Figure 3.4), the *baseband period* (plus roll-off) of the periodic spectrum of the sequences of amplitude coefficients is cut out and then shifted to the carrier frequency (here $f_c = 1.8/T$ is used and $g(t)$ is a $\sqrt{\text{Nyquist}}$ cosine-roll-off pulse with roll-off factor $\alpha = 0.3$). Adding up the signals in the in-phase and quadrature parts, the transmit signal $s(t)$ is obtained.

In contrast, in CAP, the spectrum of the pulse is frequency shifted; using this shifted filter, a *bandpass period* (around f_c) is cut out. Adding up both parts again gives the transmit signal $s(t)$.

4.2 Schemes without Explicit I/Q Demodulation

Mathematically, when looking only at the in-phase branch, we have

$$s_{\mathrm{I,PAM}}(t) = \left(\sum_k a_{\mathrm{I}}[k]\, g(t - kT)\right) \cdot \sqrt{2}\cos(2\pi f_c t) \qquad (4.34)$$

and

$$\begin{aligned} s_{\mathrm{I,CAP}}(t) &= \sum_k a_{\mathrm{I}}[k]\, g_{\mathrm{I}}(t - kT) \\ &= \sum_k a_{\mathrm{I}}[k]\, \left(\sqrt{2} g(t - kT)\cos(2\pi f_c(t - kT))\right). \end{aligned} \qquad (4.35)$$

Note that, when $f_c = n/T$, $n \in \mathbb{N}$, we have $\cos(2\pi f_c(t - kT)) = \cos(2\pi f_c t - 2\pi n k) = \cos(2\pi f_c t)$ and can write

$$\begin{aligned} s_{\mathrm{I,CAP}}(t) &= \sum_k a_{\mathrm{I}}[k]\, \sqrt{2} g(t - kT)\cos(2\pi f_c t) \\ &= s_{\mathrm{I,PAM}}(t). \end{aligned} \qquad (4.36)$$

If the carrier frequency f_c is an integer multiple of the symbol frequency $1/T$, the order of the operations "filtering" (cutting out of spectra) and "modulation" (frequency shift) can be interchanged. Under this condition, conventional PAM and CAP produce the same transmit signal.

We now look at the receiver-side signals in conventional PAM and CAP, see Figures 4.30 and 4.31. In the PAM receiver, the receive signal (here no noise is assumed) is downconverted and (matched) filtered in the *baseband*. Sampling produces the discrete-time sequence with periodic spectrum. In contrast, in the CAP receiver, filtering is done in the *bandpass domain*. The downconversion is inherently done via sampling.

Both approaches are again identical if $f_c = n/T$, $n \in \mathbb{N}$. This means that if the carrier frequency and the symbol frequency are coupled, a PAM signal can also be demodulated by the CAP receiver (and vice versa).

In contrast to conventional PAM, CAP requires no explicit mixing of the RF band at the transmitter. Typically, low carrier frequencies f_c are used and the complete RF signal generation can be implemented using digital signal processing.

Moreover, CAP requires no explicit downconversion at the receiver side. At first glance, this seems to be a big advantage as, in contrast to conventional PAM, no carrier frequency and carrier phase synchronization have to be implemented. However, still the symbol frequency and the symbol phase have to be synchronized, as has to be done in PAM. A look at the example in Figure 4.27 reveals the disadvantage of CAP: the autocorrelations and crosscorrelations – which determine the eye diagram – have very steep transitions. The horizontal eye opening will be very small; CAP is rather sensitive to jitters in the symbol timing. The effort that can be saved by not implementing a frequency synchronization has to be shifted to the timing synchronization. Nevertheless, CAP is an attractive alternative to conventional PAM for $f_c \approx \frac{1}{2T} \ldots \frac{2}{T}$.

140 4 Variants of PAM Transmission Schemes

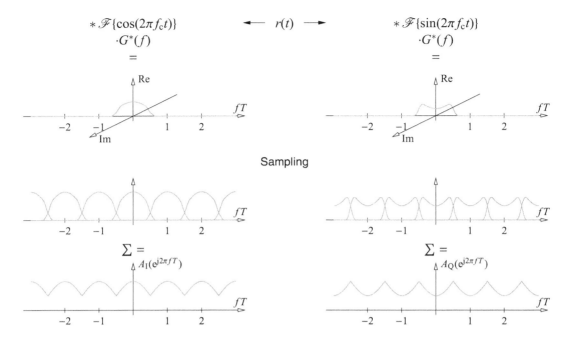

Figure 4.30 Signals at the PAM receiver. Visualization in the frequency domain.

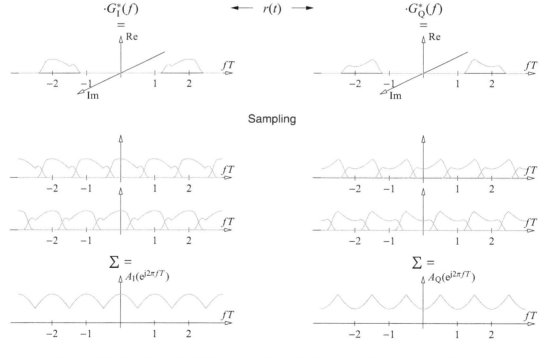

Figure 4.31 Signals at the CAP receiver. Visualization in the frequency domain.

4.2.2 Coherent Reception by Energy Detection

If the carrier frequency is very large, a coherent reception is possible in principle but – due to technological constraints – very costly. This is particularly the case in fiber-optics transmission. Here it is desired that even when demodulating without phase information (i.e., employing *energy detection* with a single photo diode), the joint modulation of in-phase and quadrature components is possible.

This section explains the concept of the *Kramers–Kronig coherent receiver* that recently became popular in fiber optics. We first review some general modulation/demodulation concepts, that is, the up- and downconversion of signals. Then, the underlying basic concept is studied in detail. The basic principles utilized are the subject of many textbooks on signals and systems. However, the required properties of causal signals and minimum-phase systems are briefly summarized in Section A.3 in the Appendix.

Coherent Demodulation

First, we recapitulate *classical I/Q modulation with coherent demodulation*, that is, the generation of an RF signal from an ECB representation and vice versa. Both real-valued modulating baseband signals, $s_I(t)$ and $s_Q(t)$, are assumed to have the (two-sided) bandwidth B_{IQ}; hence the complex signal

$$s(t) = s_I(t) + j\, s_Q(t) \tag{4.37}$$

also has the (two-sided) bandwidth B_{IQ}. After I/Q modulation to a carrier frequency f_c, the real-valued RF signal

$$s_{RF}(t) = \sqrt{2}\,\mathrm{Re}\left\{s(t)\,e^{j2\pi f_c t}\right\} \tag{4.38}$$

has the (one-sided) bandwidth $B_{RF} = B_{IQ}$, see Figure 4.32 and compare with Chapter 2. When using coherent I/Q demodulation, the quadrature signals $s_I(t)$ and $s_Q(t)$ can be chosen independently; both may carry individual data.

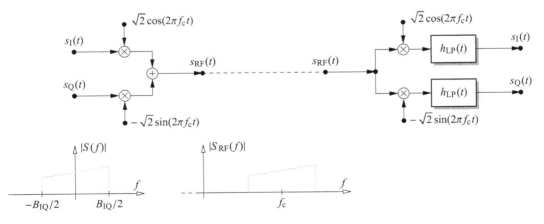

Figure 4.32 I/Q modulation and coherent demodulation.

Non-coherent Demodulation of AM with Carrier

Non-coherent (amplitude-only) demodulation can be applied if the envelope of the time-domain signal has no zero crossings. To that end, the modulating baseband signal has to be real-valued (i.e., only $s_\mathrm{I}(t)$ can be used) and a sufficiently large carrier component (equivalent to a DC component in the modulating signal) has to be added. Hence, the baseband signal reads

$$s(t) = A + s_\mathrm{I}(t) \in \mathbb{R}^+ \,. \tag{4.39}$$

If $s_\mathrm{I}(t)$ has the (two-sided) bandwidth B_IQ, the real-valued RF signal has the (one-sided) bandwidth $B_\mathrm{RF} = B_\mathrm{IQ}$, see Figure 4.33. Compared to the coherent reception above, only half of the data rate can be transmitted as the quadrature component is not usable.

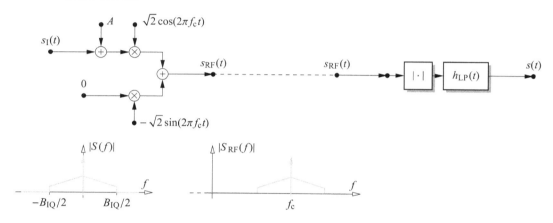

Figure 4.33 Modulation and non-coherent demodulation.

In case of modulation with a (sufficiently large) carrier, the data-carrying signal can be detected by squaring the RF signal (energy detection), low-pass filtering, and taking the square root. Alternatively, the absolute value of the RF signal can be used, followed by low-pass filtering. In each case, the DC offset A in (4.39) has to be removed by a high-pass filter.

Single-Sideband Modulation

Given a real-valued baseband signal $s_\mathrm{I}(t)$ with the (two-sided) bandwidth B_IQ, the *analytic signal*

$$s(t) = s_\mathrm{I}(t) + \mathrm{j}\mathcal{H}\{s_\mathrm{I}(t)\} \tag{4.40}$$

may be formed. The real-valued RF signal of this *single-sideband modulation* then has (one-sided) bandwidth $B_\mathrm{RF} = B_\mathrm{IQ}/2$, see Figure 4.34. Compared to I/Q modulation with coherent reception, the data rate is halved, however, the *spectral efficiency*

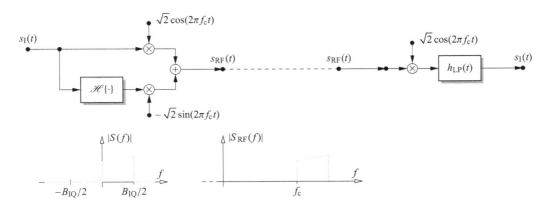

Figure 4.34 Single-sideband modulation (upper side band) and coherent demodulation.

is the same. Single-sideband signals have to be demodulated coherently – since only the in-phase branch is modulated, only this branch has to be down converted.

Compatible Single-Sideband Modulation

Single-sideband signals cannot be demodulated non-coherently as the envelope does not directly reflect the modulating signal. However, a so-called *compatible single-sideband modulation* exists.

If $x(t) = x_I(t) + j\,x_Q(t) = x_I(t) + j\mathcal{H}\{x_I(t)\}$ is an analytic signal (a signal with one-sided spectrum), then any power thereof (i.e., $x^l(t)$, $l = 1, 2, \ldots$) is also an analytic signal as $\mathcal{F}\{x^l(t)\} = X(f) * \cdots * X(f)$. Moreover, any function $\text{fct}(\cdot) = \sum_{l=0}^{\infty} a_l\, x^l(t)$ of $x(t)$, expressed as a Taylor series, is also an analytic signal. Hence, if $x(t)$ is an analytic signal, then

$$s(t) = e^{x(t)} = e^{x_I(t)} \cdot e^{j\mathcal{H}\{x_I(t)\}} \tag{4.41}$$

is also an analytic signal.

If we demand that the envelope of $s(t)$, that is, $|s(t)| = e^{x_I(t)}$, carries a signal, say $q(t) \in \mathbb{R}$, we have to choose $x_I(t) = \log(A + q(t))$ with A large enough such that $A + q(t) > 0$, $\forall t$. The transmit signal (ECB) is then given by

$$\begin{aligned}
s(t) &= (A + q(t))e^{j\mathcal{H}\{\log(A+q(t))\}} \\
&= (A + q(t))\left(\cos(\mathcal{H}\{\log(A + q(t))\}) + j\sin(\mathcal{H}\{\log(A + q(t))\})\right) \\
&= \underbrace{(A + q(t))\cos(\mathcal{H}\{\log(A + q(t))\})}_{s_I(t)} + j\underbrace{(A + q(t))\sin(\mathcal{H}\{\log(A + q(t))\})}_{s_Q(t)},
\end{aligned} \tag{4.42}$$

and the corresponding RF transmit signal reads

$$s_{\text{RF}}(t) = \sqrt{2}(A + q(t))\cos(2\pi f_c t + \mathcal{H}\{\log(A + q(t))\}). \tag{4.43}$$

The respective modulation scheme is depicted in Figure 4.35. Even though the transmit signal is single-sided, the receive signal can be envelope-detected as in

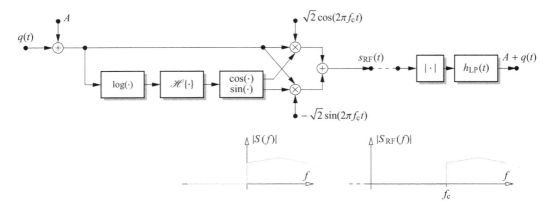

Figure 4.35 Compatible single-sideband modulation and non-coherent demodulation.

Figure 4.33. However, it should be noted that, in general, the bandwidth of the transmit signal is much larger (in theory infinite) compared to that of the modulating signal. This scheme has to be understood from the context of the history of analog modulation.

Single-Sideband Modulation with Carrier

The transmit signal suitable for the *Kramers–Kronig coherent receiver* can be termed *single-sideband signal with carrier* – it can be seen as a generalization of amplitude modulation with carrier. In contrast to the classical case, now both quadrature components are utilized independently. The real-valued modulating baseband signals, $s_\mathrm{I}(t)$ and $s_\mathrm{Q}(t)$, have (two-sided) bandwidth B_IQ; a carrier component at $f = -B_\mathrm{IQ}/2$ is added. The real-valued RF signal has (one-sided) bandwidth $B_\mathrm{RF} = B_\mathrm{IQ}$, see Figure 4.36.

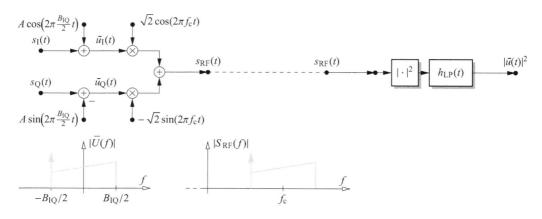

Figure 4.36 Single-sideband modulation with carrier.

4.2 Schemes without Explicit I/Q Demodulation

The baseband signal is given by

$$\bar{u}(t) = s_I(t) + j\, s_Q(t) + A e^{-j2\pi \frac{B_{IQ}}{2} t}$$
$$= \left(s_I(t) + A\cos\left(2\pi \frac{B_{IQ}}{2} t\right)\right) + j\left(s_Q(t) - A\sin\left(2\pi \frac{B_{IQ}}{2} t\right)\right)$$
$$= \bar{u}_I(t) + j\, \bar{u}_Q(t). \qquad (4.44)$$

In the next section, we will show how the modulating baseband signals, $s_I(t)$ and $s_Q(t)$, can be recovered at the receiver based on *energy detection*, that is, evaluating only the *squared envelope* of the receive signal.

Receiver for Single-Sideband Modulation with Carrier

Let a (complex-valued) signal $s(t)$ with spectrum $S(f)$ of bandwidth B_{RF}

$$S(f) = 0, \qquad |f| > B_{IQ}/2 \qquad (4.45)$$

be given, see Figure 4.37.

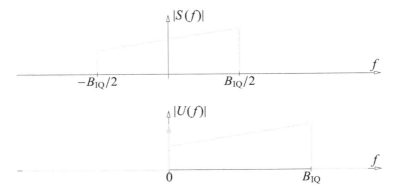

Figure 4.37 Visualization of the spectra of the signals $s(t)$ and $u(t)$.

The signal

$$u(t) \stackrel{\text{def}}{=} A + s(t)\, e^{j2\pi \frac{B_{IQ}}{2} t}$$
$$\circ\!\!-\!\!\bullet$$
$$U(f) = A\delta(f) + S\left(f - \frac{B_{IQ}}{2}\right) \qquad (4.46)$$

then has a *single-sided spectrum*, that is, $U(f) = 0$, $f < 0$ ("causal spectrum"), which can be made *minimum-phase* by choosing A large enough ($|A| > |s(t)|$, $\forall t$, is sufficient), see Section A.3.2 of Appendix A.

Then, the magnitude and phase of $u(t) = |u(t)|\, e^{j\phi_u(t)}$ are related by (sgn($\cdot$): sign function)

$$\phi_u(t) = \mathcal{H}\{\log(|u(t)|)\}$$
$$\circ\!\!-\!\!\bullet$$
$$\Phi_u(f) = -j\,\text{sgn}(f)\, \mathcal{F}\{\log(|u(t)|)\}. \qquad (4.47)$$

Note that $(C, F \in \mathbb{R})$

$$\bar{u}(t) \stackrel{\text{def}}{=} A\,e^{j2\pi(C+Ft)} + s(t)\,e^{j2\pi\left(\left(\frac{B_{IQ}}{2}+F\right)t+C\right)}$$

$$\bar{U}(f) = A\,e^{j2\pi C}\,\delta(f-F) + e^{j2\pi C}\,S\!\left(f - \left(\frac{B_{IQ}}{2}+F\right)\right) \qquad (4.48)$$

has the same magnitude as $u(t)$, since

$$|\bar{u}(t)| = \left| A\,e^{j2\pi(C+Ft)} + s(t)\,e^{j2\pi\left(\left(\frac{B_{IQ}}{2}+F\right)t+C\right)} \right|$$

$$= \left| e^{j2\pi(C+Ft)} \left(A + s(t)\,e^{j2\pi\frac{B_{IQ}}{2}t} \right) \right|$$

$$= \left| A + s(t)\,e^{j2\pi\frac{B_{IQ}}{2}t} \right| = |u(t)|\,. \qquad (4.49)$$

Hence, the phase $\phi(t)$ of $u(t)$ can be reconstructed from $|u(t)|$ only up to a constant (C) corresponding to a phase shift and a linear term (Ft) corresponding to a frequency shift.

Note that, in principle, the signal $u(t)$ can also be reconstructed if only the phase is measured at the receiver.

In summary, the reconstruction of the quadrature components ($s_I(t)$ and $s_Q(t)$ in Figure 4.36) according to the Kramers–Kronig concept is depicted in Figure 4.38. A preferable version is shown; see the discussion below.

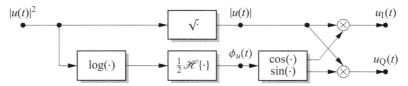

Figure 4.38 Reconstruction of the quadrature components for single-sideband modulation with carrier.

The receiver implements the calculation of

$$\phi_u(t) = \arg\{u(t)\}$$
$$= \mathcal{H}\{\log(|u(t)|)\} = \frac{1}{2}\mathcal{H}\{\log(|u(t)|^2)\}\,, \qquad (4.50)$$

followed by

$$u_I(t) = |u(t)|\cos(\phi_u(t))\,, \qquad (4.51)$$
$$u_Q(t) = |u(t)|\sin(\phi_u(t))\,, \qquad (4.52)$$

or more compactly

$$u(t) = |u(t)|\,e^{j\phi_u(t)}\,. \qquad (4.53)$$

From this, $s_I(t)$ and $s_Q(t)$ can readily be deduced by $s(t) = (u(t) - A)\,e^{-j2\pi\frac{B_{IQ}}{2}t}$.

Discussion

After the analog receiver frontend, the signals are typically A/D converted (sampled) and further manipulated by means of digital signal processing. There, the sampling frequency has to be larger than $1/T$; some oversampling has to be done. Only after the matched filter is applied, is downsampling at the symbol duration T possible.

In the coherent receiver (cf. Figure 4.32), the signals in the I and Q branch both have (two-sided) bandwidth B_{IQ}; hence, two A/D converters are required, each with sampling rate at least B_{IQ}. In the approach of single-sideband modulation with carrier, employing energy detection, a single signal, $|u(t)|^2$, has to be sampled. Since

$$|u(t)|^2 = u(t) \cdot u^*(t) \quad \circ\!\!-\!\!\bullet \quad U(f) * U^*(-f), \quad (4.54)$$

the signal $|u(t)|^2$ has doubled bandwidth compared to $u(t)$; hence, $2B_{IQ}$. Thus, sampling has to be done at a sampling rate of at least $2B_{IQ}$, but only a single (real-valued) signal has to be acquired.

In comparison, both receiver structures (coherent and non-coherent for single-side-band modulation with carrier) require $2B_{IQ}$ samples per second; either two ADCs, each operating at a sampling frequency B_{IQ}, or one ADC operating at a sampling frequency $2B_{IQ}$.

We finally note that the signal $|u(t)|^2$ should be sampled (*energy detection* should be used), not the signal $|u(t)|$ that would be obtained by *envelope detection*. This signal has a (much) larger bandwidth; taking the square root of a signal is a nonlinear operation and broadens the spectrum.

Alternative Receiver Structure

Instead of adding the carrier component at the transmitter side, and thus wasting transmit power, the carrier can be added at the receiver. Since the receiver with energy detection at its frontend is anyway blind to absolute phases, no coherency to the transmitter-side carrier is required. Figure 4.39 shows the modulation/demodulation scheme and Figure 4.40 compares the Kramers–Kronig receiver

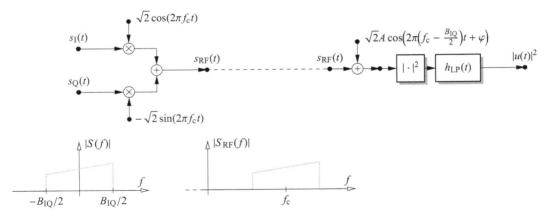

Figure 4.39 Single-sideband modulation with carrier addition at the receiver.

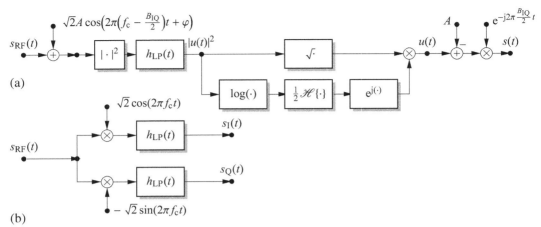

Figure 4.40 Comparison of (a) the Kramers–Kronig concept with (b) the conventional coherent I/Q demodulator.

concept with the conventional coherent I/Q demodulator. Under the conditions given above, both versions perform the same. The non-coherent version requires no synchronization at the frontend (an arbitrary phase φ can be present; a remaining phase offset in $s(t)$ may be compensated for in the digital part) – the carrier is simply added (no multiplication has to be carried out as in the I/Q demodulator) and only energy detection (e.g., via a photo diode) has to be conducted. However, more effort (numerical calculations) is required in the subsequent digital signal processing part (cf. Figure 4.38) to recover the quadrature components $s_I(t)$ and $s_I(t)$ from $|u(t)|^2$. There, phase and frequency shift have to be compensated.

Example 4.8 Kramers–Kronig Receiver

In Figure 4.41, this example shows exemplary signals present in a Kramers–Kronig receiver.

In the top row, the quadrature components (solid in-phase, dashed quadrature component) of the complex baseband signal are plotted. Here, 4QAM signaling with square-root cosine-roll-off pulse (roll-off factor $\alpha = 0.3$) is assumed. The second row displays the corresponding RF signal.

In the third row, the signal after energy detection and its low-pass filtered version, $|u(t)|^2$, are plotted. From this signal, the signals in the next two rows are calculated – the first one by taking the square root, the second by calculating the Hilbert transform of the logarithm of the given signal. These two real-valued signals give the complex-valued signal $u(t)$ in polar form.

Finally, by removing the DC offset and removing the spectral shift by half the signal bandwidth, the recovered signal $\hat{s}(t)$ is identical to the original signal.

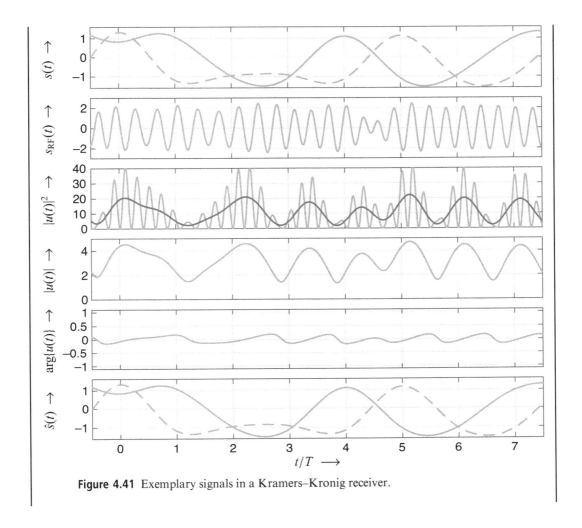

Figure 4.41 Exemplary signals in a Kramers–Kronig receiver.

4.3 Problems

4.3.1 Offset Quadrature-Amplitude Modulation

Offset quadrature-amplitude modulation (offset QAM, or OQAM) with a 4QAM signal constellation is considered.
(a) What is meant by "offset QAM"?
(b) Sketch a block diagram showing the generation of the transmit signal in the ECB (equivalent complex baseband) domain. Thereby, the ECB transformation frequency should be equal to the carrier frequency, that is, $f_0 = f_c$.
(c) Describe the advantages and disadvantages of offset QAM compared to conventional QAM.
(d) Sketch the phasors (I/Q plot) of conventional and offset QAM assuming $g(t) = \mathrm{rect}(t/T)$.

We now assume $g(t) = c \cdot \cos(\pi t/T) \cdot \text{rect}(t/T)$.

(e) Determine the constant c such that $g(t)$ has energy E_g.
(f) Show that for this choice of the basic pulse shape, the resulting OQAM transmit signal has a constant envelope.
(g) Sketch the phasor of the corresponding signal.
(h) How many different signal elements exist in one time interval T? How many different signal transitions exist within the interval $[0, T/2]$?

4.3.2 "Carrierless" Modulation

We consider a variant to carrier modulation using binary amplitude modulation. The block diagram of the conventional approach is plotted in the following figure. The amplitude coefficients $a[k]$ are drawn from the set $\mathcal{A} = \{-1, +1\}$. The carrier frequency is denoted by f_c and the basic pulse shape by $g(t)$.

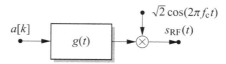

(a) Plot the radio-frequency transmit signal $s_{RF}(t)$ for the following section of the data sequence:

$$\langle a[-1], a[0], a[1], \ldots, a[5] \rangle = \langle -1, -1, 1, -1, -1, -1, 1 \rangle.$$

Use $f_c = 2.5/T$ and $g(t) = \sqrt{E_g/T}\, \text{rect}(t/T)$.

We now turn to an alternative approach, given in the following figure. Thereby, the impulse response of the transmit filter is given by

$$g_I(t) = \sqrt{2}\, g(t) \cos(2\pi f_c t),$$

where $g(t)$ is the same pulse as in the above setting.

(b) Plot the radio-frequency transmit signal $s_{RF}(t)$ for the following section of the data sequence:

$$\langle a[-1], a[0], a[1], \ldots, a[5] \rangle = \langle -1, -1, 1, -1, -1, -1, 1 \rangle.$$

Use again $f_c = 2.5/T$ and $g(t) = \sqrt{E_g/T}\, \text{rect}(t/T)$.

(c) Calculate the average power spectral density of the transmit signal $s_{RF}(t)$.
(d) Sketch the block diagram of the respective optimum receiver when transmitting over the additive white Gaussian noise (AWGN) channel.
(e) What condition does the carrier frequency f_c have to fulfill, such that both approaches generate identical transmit signals?
(f) Give an alternative receiver for carrier-modulated binary amplitude-shift keying if the carrier frequency f_c meets the requirements of part (e).

4.3.3 A-PSK and High-Power Amplification

Under peak-power limitation, multi-amplitude PSK signal constellations (A-PSK) offer for $M \gg 2$ signal points (i.e., in the regime of high bandwidth efficiency), a higher power efficiency than pure PSK.

For modulation rate 4, the following 16-ary A-PSK constellation is popular: 12 points are located equidistantly on an outer circle of radius 1 and 4 points on an inner circle with radius 0.4 (see the figure).

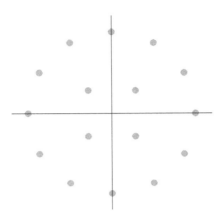

(a) Determine the average energy per symbol of 16 A-PSK.
(b) Determine the minimum normalized squared Euclidean distance and the number of nearest-neighbor error events.
Compare these numbers to the usual 16-ary QAM and 16-ary PSK.
(c) Determine the peak-to-average power ratio (PAPR) of 16-ary A-PSK and the quotient minimum squared Euclidean distance to peak power.
Give the gain in power efficiency of 16-ary A-PSK over the usual 16-ary QAM and 16-ary PSK under a peak-power limitation.

We now discuss the amplification of a 16-ary A-PSK transmit signal by a traveling-wave tube (TWT) high-power amplifier (HPA). For the specification of the TWT-HPA, the characteristics of AM/AM and AM/PM conversion of Figure 4.2 are used. The nominal maximum output power s_0^2/R of the HPA is 1 kW on an antenna impedance of $R = 75\,\Omega$.

(d) Explain why the non-dispersive specification of a HPA by AM/AM and AM/PM characteristics in the ECB domain corresponds to a dispersive non-linear distortion of the radio frequency (RF) real signal, which is amplified.

For 16-ary A-PSK a $\sqrt{\text{Nyquist}}$ pulse with cosine-roll-off factor 0.4 is used as the basic PAM pulse. In order to avoid serious out-of-band power radiation due to nonlinear distortions by the HPA, an (output) back-off of 5 dB is applied.

(e) Determine the average power and the peak power of the output signal of the TWT-HPA, that is, of the transmit signal.

(f) Determine the radii of the outer and inner circle of signal points relative to reference amplitude s_0 of the HPA. Give the AM/AM and AM/PM conversions for inner and outer signal points.
(g) Draw a rough sketch of the signal constellation that will approximately be generated by the samples at the output of the matched filter at the receiver (AWGN channel, noise-free signal).
Specify the loss in data reliability when the signal distortion due to the TWT-HPA is not taken into account, either at the transmitter or at the receiver side.
(h) Propose measures against the distortions caused by the TWT-HPA at the transmitter or at the receiver side.

4.3.4 GMSK

Gaussian minimum-shift keying (GMSK) together with the sub-optimum receiver of Figure 4.22 is analyzed.
(a) Express the amplitude of the MSK transmit signal (ECB domain) in terms of:
  (i) the energy per symbol and the time per symbol for the 4OQAM representation of MSK;
  (ii) the energy per bit and the time per bit.
(b) Explain why, despite smoothing of the phase signal, the same expressions also hold for GMSK.

In the sub-optimum receiver structure of Figure 4.22, a compromise Gaussian low-pass filter with transfer function (ECB domain) given in (4.20) is applied to limit the noise bandwidth when transmitting over an AWGN channel. The normalized 3 dB cut-off frequency B_R is chosen, for $B_R T_b = 0.29$. Eye patterns for the real and imaginary parts of the detection signals, resulting from phase smoothing at the transmitter side and the employed receiver-input filter, are shown in Figure 4.23.
(c) Show that by means of the normalization factor used in the compromise receive filter, the detection signal (cf. the noise-free eye pattern in Figure 4.23) is dimensionless.
(d) Calculate the end-to-end gain factor at DC ($f = 0$) from the discrete-time amplitude coefficients, over the AWGN channel to the output of the compromise receive filter.
(e) Determine the attenuation (in dB) of the receiver-input filter at the Nyquist frequency with respect to binary signaling per quadrature component.
(f) Discuss the linear distortion of the received signal due to the receiver-input filter.
(g) Calculate the noise variance at the output of the receiver-input filter for AWGN in terms of the one-sided noise power spectral density (real signals), that is, the two-sided noise power spectral density in the ECB domain.
(h) Extract the vertical eye opening at optimum decision instants for $B_b T_b \to \infty$, $B_b T_b = 0.3$, and $B_b T_b = 0.15$ from Figure 4.23.
(i) Give the worst-case bit error ratio according to the vertical eye opening for the three examples of smoothed phase function with normalized bandwidths $B_b T_b$ in Figure 4.23.

(j) Express the power efficiency of the considered GMSK schemes with suboptimum receiver by means of an equivalent minimum normalized squared Euclidean distance. Compare the results with the curves in Figure 4.24.

Finally, we approximately optimize the normalized bandwidth $B_R T_b$ of the compromise receive filter. To that end we expect that the worst-case eye opening (sampling instances) for MSK is caused by the constant sequence $c[\kappa] = \langle \ldots 1, 1, 1, 1, \ldots \rangle$ (which corresponds to alternating sequences for the quadrature components, see Figures 4.12 and 4.14). The transmit signal is then

$$s(t) = \sqrt{\frac{E_b}{T_b}} \, e^{j 2\pi t/(4T_b)}.$$

(k) Give the detection signal $d(t)$ when employing the above compromise Gaussian low-pass filter $H_R(f)$ and the samples of the real part at $t = kT = 2kT_b$ that determine the eye opening.

(l) Give the quotient of the squared eye opening (derived in part (k)) and the noise variance (derived in part (g)) which determines the error ratio.

(m) Define $x = \frac{1}{B_R T_b}$ and introduce suitable constants. Express the above quotient conveniently, dependent on the free parameter x. Find x such that the quotient is maximized.

5 Non-coherent Demodulation

Up to now we have dealt with coherent reception. This means that we have (implicitly) assumed that the receiver is (perfectly) synchronized to the transmitter. This synchronicity has to be achieved by some form of synchronization unit. However, in practice, this requirement may not hold: some (small) residual errors are always present. But, even worse, in some scenarios synchronization is not possible at all; in particular, when we have very fast-varying channel conditions (e.g., due to fast-moving transmitters or receivers). Even if perfect frequency generators were used at TX and RX, in mobile scenarios synchronicity is destroyed due to the Doppler effect.

We now study digital transmission schemes that are robust to non-perfect *carrier frequency* and *carrier phase* synchronization. To that end we consider *differential PSK*, which can tolerate phase errors and, to some amount, frequency errors. Then, schemes that require no phase (and frequency) synchronization at all, so-called *non-coherent demodulation schemes*.

PREREQUISITES FOR THIS CHAPTER

- PAM transmission (Chapter 3).

5.1 Phase and Frequency Errors in PAM Transmission

Figure 5.1 shows the transmitter-side upconversion in cascade with an AWGN channel and the receiver-side downconversion (valid for bandpass signals).

Since the transmitter uses the carrier frequency f_c (and zero phase) but the receiver employs carrier frequency f_c' with a phase shift φ' for downconversion, the following errors are present:

$$\begin{aligned} \text{frequency error} \quad & f_\Delta \stackrel{\text{def}}{=} f_c - f_c' \, ; \\ \text{phase error} \quad & \varphi_\Delta \stackrel{\text{def}}{=} 0 - \varphi' \, . \end{aligned} \qquad (5.1)$$

5.1 Phase and Frequency Errors in PAM Transmission

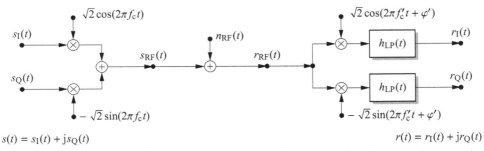

Figure 5.1 Frequency and phase mismatch in up- and downconversion. Transmission over the AWGN channel.

As was shown in Section 2.1.4, the ECB receive signal is related to the ECB transmit signal by

$$r(t) = s(t) \cdot e^{j(2\pi f_\Delta t + \varphi_\Delta)} + n(t), \qquad (5.2)$$

where $n(t)$ is complex, rotationally invariant white Gaussian noise.

Equation (5.2) shows that a frequency error f_Δ causes a multiplication of the transmit signal by $e^{j2\pi f_\Delta t}$, which causes a *time-dependent rotation* of the receive signal. The phasor is not stable but rotates with the frequency error f_Δ. A phase error φ_Δ causes a multiplication of the transmit signal by $e^{j\varphi_\Delta}$, which causes a *time-independent rotation* of the receive signal. The phasor is not aligned with the coordinate axis but tilted according to the phase error φ_Δ.

Example 5.1 Phasors with Phase and Frequency Errors

As before, for visualization, phasors are available on the companion web page. In all cases, 4QAM and a rectangular basic pulse shape are assumed.

Phasor 1: Signal $r(t) = s(t)$ for ideal synchronization.
Phasor 11: Signal $r(t)$ in case of a phase error ($\varphi_\Delta = +20°$).
Phasor 12: Signal $r(t)$ in case of a frequency error ($f_\Delta T = +0.02$).

5.1.1 PAM Scheme with Phase Error

In Figure 5.2, the main parts of a PAM scheme over a channel with phase error only ($f_\Delta \approx 0$) are depicted.

Owing to the linearity of the pulse-shaping filter, we have

$$\begin{aligned}
r(t) &= s(t) \cdot e^{j\varphi_\Delta} + n(t) \\
&= \left(\sum_k a[k]\, g(t - kT)\right) \cdot e^{j\varphi_\Delta} + n(t) \\
&= \sum_k e^{j\varphi_\Delta}\, a[k]\, g(t - kT) + n(t),
\end{aligned} \qquad (5.3)$$

5 Non-coherent Demodulation

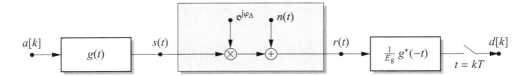

Figure 5.2 PAM scheme over a channel with phase error.

and after matched filtering and sampling, the end-to-end discrete-time description reads

$$d[k] = e^{j\varphi_\Delta} a[k] + n[k] \,. \tag{5.4}$$

Hence, the phase error between transmitter- and receiver-side mixers amounts to a rotation of the ECB signal by the phase difference (error).

In perfect *coherent demodulation*, no phase error is present (i.e., $\varphi_\Delta = 0$).

If carrier phase synchronization is performed but does not work perfectly, the carrier phase error φ_Δ is a random variable. If some statistics are available at the receiver, this knowledge can be exploited for an adapted detection strategy that is called *partially coherent reception*.

If no phase synchronization takes place/is possible, and thus no knowledge of φ_Δ is available, the worst case has to be taken into account, where the phase is *uniformly* distributed over the interval $[0, 2\pi)$. The task of *non-coherent demodulation* is the demodulation (reception) without (any) knowledge of the carrier phase, that is, the reception even if the useful signal is rotated by a random (uniformly distributed) phase offset.

5.1.2 Phase Ambiguities

When the channel adds a random phase change and no synchronization takes place, the phase cannot be used for information transmission. However, even if phase synchronization is performed, there is still a problem. This is visualized in Figure 5.3.

For simplicity, we first consider BPSK. The (noise-free) situation after the channel has rotated the signal is shown on the left of the figure. The task of the synchronization device is to rotate the signal such that the signal points are back on the real axis. Unfortunately, there is an ambiguity: there are two possible angles

Figure 5.3 Synchronization of BPSK.

(i.e., hypotheses on the phase rotation of the channel) such that the alignment is achieved. When looking only at the constellation diagram, this ambiguity cannot be resolved. This is due to the fact that the receiver has no *absolute* phase reference, that is, the receiver has no knowledge where 0° is. For BPSK, an incorrect locking of the carrier phase synchronization leads to an inversion of all symbols.

This effect also occurs for the other families of PAM signal constellations. For bipolar MASK there is a twofold ambiguity (0° vs. 180°), in MPSK an M-fold ambiguity is present, and for MQAM a fourfold ambiguity has to be taken into account (the constellation rotated by an integer multiple of 90° coincides with the original constellation).

Basically, there are two ways to overcome this problem.

- *Synchronization Sequences*
  The data stream may be preceded by a synchronization sequence. This sequence has to be chosen such that any rotation of the sequence by the ambiguous phases can be detected. In case of time-varying channels, synchronization sequences have to be inserted periodically. Their use leads to a (small) loss in data rate.
- *Differential Encoding*
  Even though the *absolute* phase is not known, still *phase changes* can be detected at the receiver. Hence, instead of representing the information in the absolute phase, the information can be represented in *phase differences*, that is, the change in phase of subsequent symbols carries the information.[1]

5.2 Differential Phase-Shift Keying (DPSK)

5.2.1 Differential M-ary PSK

We now consider *differential PSK (DPSK)* using an M-ary constellation. Here, in modulation step k, the information bits are first mapped onto an M-ary *information symbol*

$$b[k] \in \{0, 1, 2, \ldots, (M-1)\} \,. \tag{5.5}$$

An obvious strategy would be to take blocks of $\log_2(M)$ bits in the mapping step (Figure 3.1) and read this binary string as an integer. The transmit symbols (which are taken from an M-ary PSK constellation) are then generated according to

$$a[k] = a[k-1] \cdot e^{j2\pi b[k]/M} \,, \quad \text{with} \quad a[-1] = 1 \,. \tag{5.6}$$

We define the normalized phase of $a[k]$ as (here, $\arg(\cdot) \in [0, 2\pi)$ is assumed, cf. Appendix A)

$$m[k] \stackrel{\text{def}}{=} \frac{M}{2\pi} \arg\{a[k]\} \in \{0, 1, 2, \ldots, (M-1)\} \,, \tag{5.7}$$

---

[1] Even though the differentiation of the phase gives the instantaneous frequency, differential phase modulation is usually not counted in the family of frequency modulation.

that is, $a[k] = e^{j2\pi m[k]/M}$ is the mapping of the *signal number* $m[k]$ to the amplitude coefficient $a[k]$ that is drawn from an M-ary PSK constellation $\mathcal{A} = \{a = e^{j2\pi m/M} \mid m \in \{0, 1, \ldots, M-1\}\}$. Then, the encoding (5.6) can equivalently be given as

$$m[k] = (m[k-1] + b[k]) \bmod M, \quad \text{with} \quad m[-1] = 0. \tag{5.8}$$

This means that, since the phase is 2π periodic, the signal numbers $m[k]$ are accumulated (integrated) modulo M.

The block diagram of an M-ary DPSK transmitter is shown in Figure 5.4. In part (a), the structure corresponding to (5.6) is depicted; in part (b), the structure corresponding to (5.8). Both variants operate exactly the same; given the same input sequence (and the respective starting conditions $a[-1] = 1$ and $m[-1] = 0$), the same output sequence will be generated.

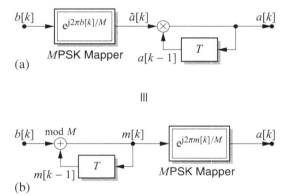

Figure 5.4 Block diagram of the transmitter in M-ary DPSK. Both variants (a) and (b) operate in exactly the same way.

The simplified block diagram for differential BPSK ($M = 2$) is given in Figure 5.5. The addition modulo 2 of the binary symbols can be seen as an XOR combination.

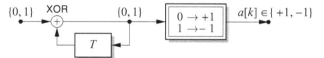

Figure 5.5 Block diagram of the transmitter in differential BPSK.

The two possible (and equivalent) transmitter structures give rise directly to two possible (but not equivalent!) receiver structures. These variants are compared in Figure 5.6.

5.2.2 Differential Coherent Demodulation (DCPSK)

The first receiver concept is based on (5.8) and repeated in Figure 5.7. Here, a *coherent* PSK receiver detects the symbols $m[k]$ (estimates $\hat{m}[k]$); the accumulation done prior to the mapping at the transmitter is compensated for by a *differentiation*

5.2 Differential Phase-Shift Keying (DPSK)

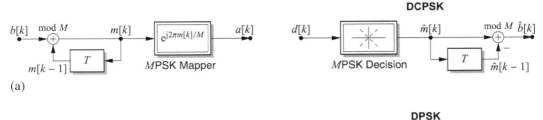

(a)

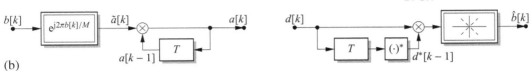

(b)

Figure 5.6 Block diagrams of the DPSK transmitter and possible receiver concepts. (a) DCPSK; (b) DPSK.

Figure 5.7 Block diagram of a DCPSK receiver.

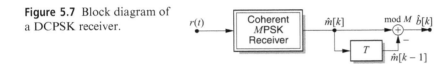

modulo M of the estimates $\hat{m}[k]$. Thus, a coherent PSK scheme is augmented by differential encoding (integration) at the transmitter and differential decoding at the receiver.

The purpose of this strategy is to eliminate the phase ambiguities of the synchronization. Still synchronization has to be performed but it may lock at any of the M possible (ambiguous) phases. Hence, this receiver concept is denoted as *differential coherent PSK (DCPSK)*. In other words, any constant phase offset caused by false locking of the synchronization is eliminated by the final differentiation.

In order to calculate the error probability for DCPSK, we first consider *binary* transmission. Let BER_{PSK} denote the bit error ratio of the inner coherent BPSK scheme. The final estimate is in error if either $\hat{m}[k]$ is erroneous and $\hat{m}[k-1]$ is correct, or the other way round. If either both are correct or both are erroneous, a correct decision is obtained. Hence, the bit error probability of binary DCPSK is given by

$$\begin{aligned}
\text{BER}_{\text{DCPSK}} &= \text{BER}_{\text{PSK}} \cdot (1 - \text{BER}_{\text{PSK}}) + (1 - \text{BER}_{\text{PSK}}) \cdot \text{BER}_{\text{PSK}} \\
&= 2\,\text{BER}_{\text{PSK}} \cdot (1 - \text{BER}_{\text{PSK}}) \\
&= 2\,\text{BER}_{\text{PSK}} - 2\,\text{BER}_{\text{PSK}}^2 \,.
\end{aligned} \qquad (5.9)$$

For non-binary transmission, we look at the symbol error ratio. Here, it is more unlikely (compared to the binary case) that subsequent errors compensate for each other. A good approximation is given by

$$\text{SER}_{\text{DCPSK}} \approx 2\,\text{SER}_{\text{PSK}} - 1.5\,\text{SER}_{\text{PSK}}^2 \,. \qquad (5.10)$$

As for small BER or SER (e.g., 10^{-3}), the squared quantity is much smaller (e.g., 10^{-6}) and can be ignored; the error ratio of DCPSK is approximately doubled compared to a coherent scheme. Owing to the dispersive, non-recursive (FIR) system at the receiver, any detection error is doubled. In DCPSK, symbol errors occur in pairs.

5.2.3 Differential Demodulation (DPSK)

The second receiver concept is based on (5.6) and repeated in more detail in Figure 5.8. After matched filtering and sampling, first a *differential demodulation* is performed. Then, via an MPSK threshold device, estimates $\hat{b}[k]$ of the data symbols are generated. This reversal of the order of differentiation and the nonlinear operation of detection (threshold decision) leads to a different behavior (and, thus, performance) compared to the strategy studied above.

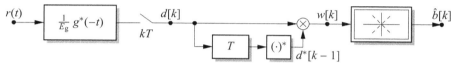

Figure 5.8 Block diagram of a DPSK receiver (ECB signals).

Note that, as in each time step two consecutive detection symbols $d[k]$ are combined, this strategy only works if the carrier phase offset φ_Δ remains (approximately) constant over (at least) two symbol intervals.

The multiplication of the current detection symbol $d[k]$ with the conjugate-complex version of the preceding one can be seen as the simplest version of synchronization – the phase of the preceding detection symbol (i.e., $d[k-1]$) is used as phase reference. Multiplication with a conjugate-complex number causes a subtraction of the phase; a back rotation by the respective angle. However, since instantaneous quantities are used, the carrier phase estimate is as noisy as the signal point itself.

Example 5.2 Differential Detection

An idea of the different behavior of coherent and differential demodulation is given by Figure 5.9. Binary (D)PSK is assumed.

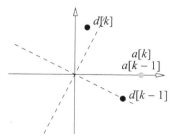

Figure 5.9 Signal points in the complex plane. Light gray: transmitted; dark gray: received.

5.2 Differential Phase-Shift Keying (DPSK)

We assume that at both time instants $k-1$ and k the signal point $a = 1$ has been sent. The respective detection symbols at the receiver are drawn in dark gray.

In case of coherent reception, that is, if the receiver (perfectly) knows where $0°$ is, decisions are taken with respect to the original coordinate system (solid). Since both receive points lie in the right half of the complex plane, in both time steps a (correct) decision in favor of $a = 1$ is made.

When using differential detection, the point $d[k-1]$ gives the phase reference, that is, it defines where (the receiver assumes that) $0°$ is. The rotated dashed coordinate system results. Now, at time step k a decision is made in favor of $a = -1$; an error occurs.

The increased knowledge in case of coherent reception leads to a better performance but at the expense of an increased complexity (if synchronization is possible at all).

In order to calculate the error ratio of DPSK employing differential detection, we have to study the statistics of the final decision variable $w[k]$, given by

$$w[k] \stackrel{\text{def}}{=} d[k] \cdot d^*[k-1]$$
$$= |d[k]| \cdot |d[k-1]| \cdot e^{j(\arg\{d[k]\} - \arg\{d[k-1]\})} . \quad (5.11)$$

We first assume *binary DPSK* and, without loss of generality, $b[k] = 0$, that is, no phase change has been transmitted. (The subsequent derivation holds equally for $b[k] = 1$.) Let $\psi = \varphi_\Delta + \arg\{a[k]\}$ be the carrier phase plus the symbol phase, which is constant over the two symbol intervals k and $k-1$. The decision variable $w[k]$ (cf. Figure 5.8) reads

$$w[k] = (e^{j\psi} + n[k]) \cdot (e^{j\psi} + n[k-1])^* , \quad (5.12)$$

and a bit error occurs if $w[k]$ is such that it would be decided in favor of $a = -1$. Hence

$$\text{BER} = \Pr\{\pi/2 < \arg\{w[k]\} < 3\pi/2\} = \Pr\{\text{Re}\{w[k]\} < 0\} , \quad (5.13)$$

and using $\text{Re}\{z_1 \cdot z_2^*\} = \left|\frac{z_1+z_2}{2}\right|^2 - \left|\frac{z_1-z_2}{2}\right|^2$ we obtain

$$\text{BER} = \Pr\left\{\frac{|Y_1|^2}{4} < \frac{|Y_2|^2}{4}\right\} = \Pr\{|Y_1| < |Y_2|\} . \quad (5.14)$$

Thereby

- $Y_1 \stackrel{\text{def}}{=} 2e^{j\psi} + n[k] + n[k-1]$ is a complex Gaussian random variable with mean $2e^{j\psi}$ and variance $2\sigma_n^2$; since $n[k]$ and $n[k-1]$ are uncorrelated, the variances add up. Thus, $|Y_1|$ has a *Rice distribution* (see Box 5.1) with the parameters $2\sigma_n^2 = 2N_0/E_g$ and $|\eta| = |2e^{j\psi}| = 2$.
- $Y_2 \stackrel{\text{def}}{=} n[k] - n[k-1]$ is a complex, zero-mean Gaussian random variable with variance $2\sigma_n^2$. Thus, $|Y_2|$ has a *Rayleigh distribution* with parameter $2\sigma_n^2$.
- The random variables Y_1 and Y_2 are uncorrelated, because

$$\mathrm{E}\{Y_1 \cdot Y_2^*\} = 2\,\mathrm{e}^{\mathrm{j}\psi} \cdot \mathrm{E}\{n^*[k] - n^*[k-1]\} + \mathrm{E}\{|n[k]|^2\}$$
$$- \mathrm{E}\{n[k] \cdot n^*[k-1]\} + \mathrm{E}\{n^*[k]n[k-1]\} - \mathrm{E}\{|n[k-1]|^2\}$$
$$= 0 + \sigma_n^2 - 0 + 0 - \sigma_n^2 = 0. \tag{5.15}$$

Hence, as they are Gaussian, they are statistically independent.

Combining the results, the error probability (5.14) is equal to the probability that a Rayleigh-distributed random variable exceeds a Ricean-distributed one uncorrelated with it and is given by

$$\mathrm{BER} = \Pr\{|Y_1| < |Y_2|\}$$
$$= \iint_{y_1 < y_2} f_{|Y_1|,|Y_2|}(y_1, y_2)\,\mathrm{d}y_1 \mathrm{d}y_2$$

Y_1 and Y_2 independent

$$= \int_0^\infty \int_{y_1}^\infty f_{\mathrm{Rice}}(y_1) \cdot f_{\mathrm{Rayleigh}}(y_2)\,\mathrm{d}y_2 \mathrm{d}y_1$$
$$= \int_0^\infty f_{\mathrm{Rice}}(y_1) \int_{y_1}^\infty f_{\mathrm{Rayleigh}}(y_2)\,\mathrm{d}y_2 \mathrm{d}y_1. \tag{5.16}$$

See Box 5.1 on the Rayleigh, exponential, and Rice distributions. Since

$$\int_{y_1}^\infty f_{\mathrm{Rayleigh}}(y_2)\,\mathrm{d}y_2 = \int_{y_1}^\infty \frac{y_2}{\sigma_n^2} \cdot \mathrm{e}^{-y_2^2/(2\sigma_n^2)}\,\mathrm{d}y_2 = \mathrm{e}^{-y_1^2/(2\sigma_n^2)}, \tag{5.17}$$

we have

$$\mathrm{BER} = \int_0^\infty \frac{y_1}{\sigma_n^2} \mathrm{e}^{-(y_1^2+\eta^2)/(2\sigma_n^2)} \mathrm{I}_0(y_1 \eta/\sigma_n^2) \cdot \mathrm{e}^{-y_1^2/(2\sigma_n^2)}\,\mathrm{d}y_1$$

substituting $z^2 = 2 y_1^2$, $\mathrm{d}y_1 = \frac{1}{\sqrt{2}}\mathrm{d}z$, and $b^2 = \eta^2/2 = 2$

$$= \int_0^\infty \frac{z}{\sqrt{2}\sigma_n^2} \mathrm{e}^{-(2b^2+z^2)/(2\sigma_n^2)} \mathrm{I}_0(bz/\sigma_n^2) \frac{1}{\sqrt{2}}\,\mathrm{d}z$$
$$= \frac{1}{2}\mathrm{e}^{-b^2/(2\sigma_n^2)} \cdot \int_0^\infty f_{\mathrm{Rice}}(z)\,\mathrm{d}z = \frac{1}{2}\mathrm{e}^{-\frac{2}{2N_0/E_g}}. \tag{5.18}$$

Box 5.1 Rayleigh, Exponential, and Rice Distributions

Rayleigh Distribution

Consider a zero-mean complex Gaussian random variable $n_\mathrm{G} = n_\mathrm{I} + \mathrm{j}n_\mathrm{Q}$ with variance σ_n^2. Its pdf is denoted as

$$f_{n_G}(n_I, n_Q) = \frac{1}{\pi \sigma_n^2} e^{-(n_I^2 + n_Q^2)/\sigma_n^2} . \tag{5.19}$$

The *cdf of the magnitude* of n_G is obtained as

$$F_{|n_G|}(\rho) = \Pr\{|n_G| \leq \rho\} = \iint\limits_{n_I^2 + n_Q^2 \leq \rho^2} f_{n_G}(n_I, n_Q) \, dn_I dn_Q . \tag{5.20}$$

To solve the integral, we change from Cartesian to polar coordinates:

$$n_I = r \cdot \cos \alpha , \quad n_Q = r \cdot \sin \alpha ,$$

with

$$r = |n_G| = \sqrt{n_I^2 + n_Q^2} , \quad \alpha = \arg\{n_G\} , \quad dn_I dn_Q = r \, dr d\alpha . \tag{5.21}$$

Hence we arrive at

$$F_{|n_G|}(\rho) = \frac{1}{\pi \sigma_n^2} \int_0^\rho \int_0^{2\pi} e^{-r^2/\sigma_n^2} r \, dr d\alpha = \frac{2}{\sigma_n^2} \int_0^\rho r \, e^{-r^2/\sigma_n^2} \, dr$$

$$= 1 - e^{-\rho^2/\sigma_n^2} \tag{5.22}$$

using $\int x e^{-x^2} dx = -\frac{1}{2} e^{-x^2} + \text{const}$.

Note that, this cdf of the magnitude of a zero-mean Gaussian random variable is known as the *Rayleigh distribution*, that is, the cdf is given by

$$F_{|n_G|}(\rho) = 1 - e^{-\rho^2/\sigma_n^2} , \quad \rho \geq 0 \tag{5.23}$$

and the pdf reads

$$f_{|n_G|}(\rho) = \frac{d}{d\rho} F_{|n_G|}(\rho) = \begin{cases} \frac{2\rho}{\sigma_n^2} e^{-\rho^2/\sigma_n^2}, & \rho \geq 0 \\ 0, & \rho < 0 \end{cases} . \tag{5.24}$$

Exponential Distribution
In a number of situations we are interested in the distribution of the *squared magnitude* (i.e., the instantaneous power) of a zero-mean complex Gaussian random variable n_G with variance σ_n^2. Following the above derivations, the cdf is obtained as

$$F_{|n_G|^2}(\rho) = \Pr\{|n_G|^2 \leq \rho\} = \Pr\{|n_G| \leq \sqrt{\rho}\}$$

$$= \frac{1}{\pi \sigma_n^2} \int_0^{\sqrt{\rho}} \int_0^{2\pi} e^{-r^2/\sigma_n^2} r \, dr d\alpha = 1 - e^{-\rho/\sigma_n^2} . \tag{5.25}$$

Thus, the squared magnitude follows an *exponential distribution*, that is, the cdf is given by

$$F_{|n_G|^2}(\rho) = 1 - e^{-\rho/\sigma_n^2} , \quad \rho \geq 0 \tag{5.26}$$

and the pdf reads

$$f_{|n_G|^2}(\rho) = \frac{d}{d\rho} F_{|n_G|^2}(\rho) = \begin{cases} \frac{1}{\sigma_n^2} e^{-\rho/\sigma_n^2}, & \rho \geq 0 \\ 0, & \rho < 0 \end{cases}. \quad (5.27)$$

Rice Distribution

Consider a complex Gaussian random variable $n_{G,\eta}$ with variance σ_n^2 and mean value $\eta = \eta_I + j\eta_Q$. Its pdf is denoted as $f_{n_{G,\eta}}(n_I, n_Q)$. The *cdf of the magnitude* of $n_{G,\eta}$ is obtained as

$$F_{|n_{G,\eta}|}(\rho) = \Pr\{|n_{G,\eta}| \leq \rho\}$$
$$= \iint\limits_{n_I^2+n_Q^2 \leq \rho^2} f_{n_{G,\eta}}(n_I, n_Q) \, dn_I dn_Q ,$$

which can be written as

$$= \iint\limits_{n_I^2+n_Q^2 \leq \rho^2} f_{n_G}(n_I - \eta_I, n_Q - \eta_Q) \, dn_I dn_Q . \quad (5.28)$$

As above, we change from Cartesian to polar coordinates and arrive at

$$F_{|n_{G,\eta}|}(\rho) = \frac{1}{\pi \sigma_n^2} \iint\limits_{n_I^2+n_Q^2 \leq \rho^2} e^{-((n_I-\eta_I)^2+(n_Q-\eta_Q)^2)/\sigma_n^2} \, dn_I dn_Q$$

$$= \frac{2}{\sigma_n^2} \int_0^\rho r \, e^{-(r^2+|\eta|^2)/\sigma_n^2} \cdot \underbrace{\frac{1}{2\pi} \int_0^{2\pi} e^{(2r|\eta|\cos(\alpha+\arg(\eta)))/\sigma_n^2} \, d\alpha}_{} \, dr$$

$$= \frac{2}{\sigma_n^2} \int_0^\rho r \, e^{-(r^2+|\eta|^2)/\sigma_n^2} \cdot I_0\left(\frac{2r|\eta|}{\sigma_n^2}\right) dr$$

substituting $r' = \frac{\sqrt{2}}{\sigma_n} r$

$$= \int_0^{\rho \sqrt{2}/\sigma_n} r' \, e^{-(r'^2+2|\eta|^2/\sigma_n^2)/2} \cdot I_0\left(\frac{|\eta|\sqrt{2}}{\sigma_n} r'\right) dr'$$

$$= 1 - Q_M\left(\frac{|\eta|\sqrt{2}}{\sigma_n}, \frac{\rho\sqrt{2}}{\sigma_n}\right). \quad (5.29)$$

Since the integrals cannot be solved analytically, we have used the definitions of the *zeroth-order modified Bessel function of the first kind*

$$I_0(x) \stackrel{\text{def}}{=} \frac{1}{2\pi} \int_0^{2\pi} e^{x \cos(\alpha+\beta)} \, d\alpha \,, \qquad (5.30)$$

where β is arbitrary, and the *Marcum-Q-function* (which can be seen as a generalization of the Q-function)

$$Q_M(y,x) \stackrel{\text{def}}{=} \int_x^\infty r \, e^{-(r^2+y^2)/2} \cdot I_0(yr) \, dr \,. \qquad (5.31)$$

Note that this cdf of the magnitude of a Gaussian random variable with mean η is known as the *Rice distribution*, that is, the cdf is given by

$$F_{|n_{G,\eta}|}(\rho) = 1 - Q_M\left(\frac{|\eta|\sqrt{2}}{\sigma_n}, \frac{\rho\sqrt{2}}{\sigma_n}\right), \qquad \rho \geq 0 \qquad (5.32)$$

and the pdf reads

$$f_{|n_{G,\eta}|}(\rho) = \frac{d}{d\rho} F_{|n_{G,\eta}|}(\rho) = \begin{cases} \frac{2\rho}{\sigma_n^2} e^{-(\rho^2+|\eta|^2)/\sigma_n^2} I_0\left(\frac{2|\eta|\rho}{\sigma_n^2}\right), & \rho \geq 0 \\ 0, & \rho < 0 \end{cases}. \qquad (5.33)$$

As can be seen, the phase of the mean η is immaterial; only its magnitude is relevant.

The Rice pdf is visualized in Figure 5.10 for different mean values; it converges to the Rayleigh distribution for $\eta \to 0$. Conversely, for large η, the pdf is well approximated by a one-dimensional Gaussian pdf.

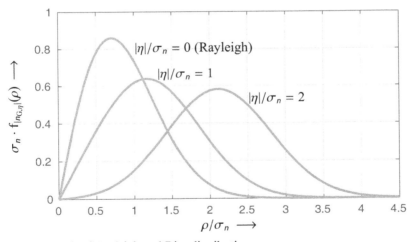

Figure 5.10 Pdfs of Rayleigh and Rice distributions.

Finally, since for $M = 2$ we have $E_b = E_g$, we arrive at

$$\text{BER}_{\text{DPSK}} = \frac{1}{2} e^{-E_b/N_0} \,. \qquad (5.34)$$

This result has to be compared with that of coherent BPSK. Using the upper bound (3.97) on the Q-function, we have

$$\text{BER}_{\text{coherent}} = Q\left(\sqrt{2E_b/N_0}\right) \leq \frac{1}{2} e^{-E_b/N_0}. \qquad (5.35)$$

Hence, the error ratio of binary DPSK coincides with the upper bound for coherent BPSK. This means that at high SNR a *binary* DPSK has almost the same power efficiency as coherent BPSK.

For non-binary DPSK ($M > 2$) the derivation of the symbol error ratio follows the same idea but is more involved. It calculates as

$$\text{SER}_{\text{DPSK}} < 1 + Q_M(X_1, X_2) - Q_M(X_2, X_1) \qquad (5.36)$$

with

$$X_1 = \sqrt{\log_2(M)\left(1 - \sin\left(\frac{\pi}{M}\right)\right) \frac{E_b}{N_0}}, \qquad X_2 = \sqrt{\log_2(M)\left(1 + \sin\left(\frac{\pi}{M}\right)\right) \frac{E_b}{N_0}}.$$

Thereby, $Q_M(y, x)$ is the Marcum-Q-function, given in (5.31).

For M and E_b/N_0 sufficiently large, the following approximation can be used:

$$\text{SER}_{\text{DPSK}} \approx 2\, Q\left(\sqrt{d_{\min}^2 \frac{E_b}{N_0}/2}\right), \qquad (5.37)$$

which (due to the factor $1/2$ in the argument compared to (3.105)) corresponds to a *loss of 3 dB* compared to coherent PSK. This loss can be justified as the phase reference is as noisy as the detection variable itself – owing to the differential detector, two noisy variables are combined, leading to twice the noise variance.[2]

A comparison of the symbol error ratio of differential and coherent PSK schemes over the signal-to-noise ratio E_b/N_0 (in dB) is displayed in Figure 5.11. A comparison of the exact calculation (5.36) and the approximation (5.37) is given in Figure 5.12.

In case of binary transmission ($M = 2$), all variants perform approximately the same. For $M \geq 4$, the different behavior of DCPSK and DPSK when compared to coherent reception is visible. Using DCPSK (dashed curves), the error ratio is approximately doubled. The error ratio curve of coherent PSK (solid curves) is *vertically* shifted by a factor of 2; this is almost invisible for small SER. For non-coherent reception of DPSK (dashed-dotted curves), almost 3 dB loss occurs. The error ratio curves are *horizontally* shifted by 3 dB.

For completeness, the power/bandwidth planes for PSK and DPSK schemes are given in Figure 5.13. The roll-off factor of the $\sqrt{\text{Nyquist}}$ pulse is chosen to be $\alpha = 0$, that is, the bandwidth efficiency corresponds to the rate R. As can be seen, the non-coherent schemes perform up to 3 dB worse (have a poorer power efficiency) than the coherent schemes. In the binary case ($\Gamma = 1$), only a very small degradation is caused.

---

[2] In the binary case, this doubling of the noise power is eliminated since only the in-phase component is considered in the decision process.

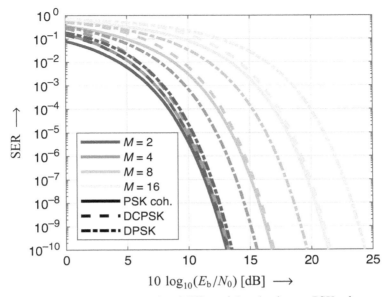

Figure 5.11 Symbol error ratio of differential and coherent PSK schemes over the signal-to-noise ratio E_b/N_0 (in dB).

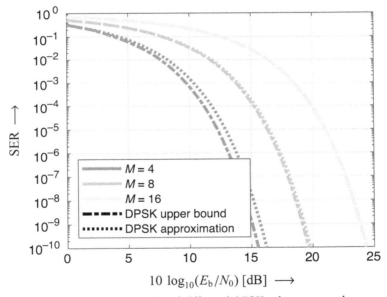

Figure 5.12 Symbol error ratio of differential PSK schemes over the signal-to-noise ratio E_b/N_0 (in dB).

Various proposals can be found in the literature for the improvement of DPSK for $M > 2$ in order to overcome the 3 dB gap from coherent PSK. One approach is to extend the observation interval for decisions over more than two symbols, similar to stabilizing the estimation of the reference phase by long-term averaging. These methods are known under the term *multiple-symbol differential detection (MSDD)*. In Chapter 6, we will give a short introduction to MSDD.

168 5 Non-coherent Demodulation

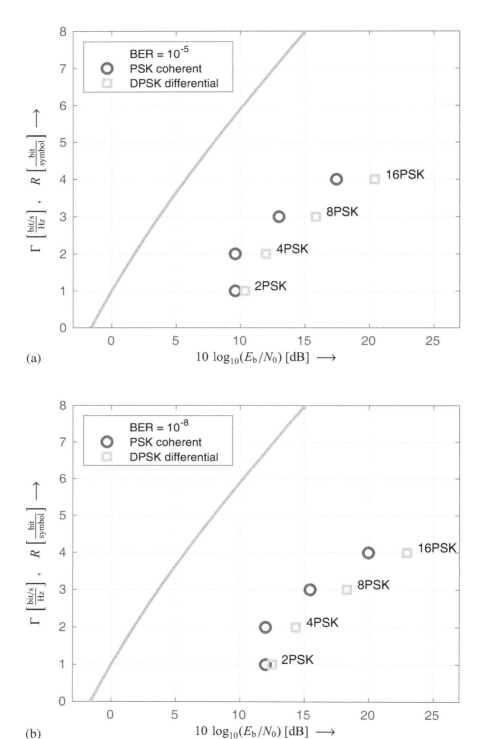

Figure 5.13 Power/bandwidth plane for PSK and DPSK schemes. $\alpha = 0$, that is, power/rate plane. (a) Tolerated bit error ratio BER $\leq 10^{-5}$; (b) BER $\leq 10^{-8}$.

5.3 Non-coherent PAM Schemes

We now assess schemes where no knowledge at all on the phase is required but still reliable transmission is enabled. Specifically, we consider the *non-coherent demodulation of unipolar ASK*. Note that "non-coherent" is always a property of the *receiver* – the receiver-side downconversion is not aligned with the transmitter-side upconversion. However, usually some care has to be taken at the transmitter, such that the receiver can do the demodulation without performing carrier synchronization (but it can also do (better) with synchronization).

5.3.1 Non-coherent Demodulation of Unipolar ASK

When using bipolar ASK constellations, the phase is relevant and represents information. For example, for $\mathcal{A} = \{-1, +1\}$, the phases 0 and π have to be distinguished. If the channel rotates the transmitted signal by π, the entire data sequence will be inverted. Note that this effect also occurs in baseband transmission: if a (twisted-pair) cable is connected the wrong way round, the polarity is inverted.

If the phase cannot be estimated reliably (e.g., in fast-changing environments), the phase cannot be used to carry information – still the amplitude remains. Hence, if the channel introduces a random (uniformly distributed) phase rotation and non-coherent demodulation has to be applied, only *unipolar ASK* constellations can be employed, where information is solely represented in the amplitude.

The simplest variant is obtained when using binary ($M = 2$) unipolar ASK transmission – since one of the two signal points is $a = 0$, this variant is called *on–off keying (OOK)*.

For binary unipolar ASK and *coherent reception*, the decision boundary would be the line at $d = 1$ parallel to the imaginary axis. Since the channel rotates the signal constellation by an arbitrary angle, in *non-coherent reception* the decision can only be based on the magnitude. Hence, it has to be classified whether $|d|$ is larger or smaller than a (to be determined) threshold ρ. The decision boundary for non-coherent detection of unipolar 2ASK is visualized in Figure 5.14.

Error Probability of OOK

We are, of course, interested in the error probability of OOK, that is, the error ratio of non-coherent detection of unipolar 2ASK. Note that for the binary transmission

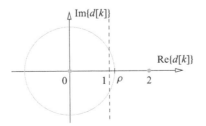

Figure 5.14 Decision boundary for non-coherent detection of unipolar 2ASK (solid circle). For comparison: decision boundary for coherent detection (dashed straight line).

at hand, bit and symbol error probability are the same. Moreover, as in Section 3.5, in the following we drop the discrete time index k as detection is done symbol by symbol independently. Since the decision regions are not congruent (and thus do not contribute equally to the average error ratio), we have to distinguish two cases.

Case A: $a = 0$ Transmitted; $\hat{a} = 2$ Detected First, we assume that the signal point $a = 0$ has been transmitted. We decide in favor of $a = 2$ – and, thus, an error occurs – if the detection variable d has magnitude larger than the threshold ρ. Since here $d = 0 + n$, we have $|d| = |n|$ and an error occurs if the zero-mean complex Gaussian variable $d = n$ with variance $\sigma_n^2 = N_0/E_g$ has magnitude greater than ρ, that is

$$\mathrm{BER}_{a=0} = \Pr\{|d| > \rho\} \,. \tag{5.38}$$

The relevant region is depicted in light gray in Figure 5.15(a). Thereby, the isolines of the Gaussian noise are given (cf. Figure A.6). The complement of this probabil-

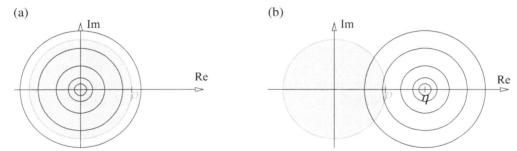

Figure 5.15 Relevant regions $|d| \le \rho$ for error ratio calculations in non-coherent OOK (gray). (a) Zero-mean Gaussian random variable. (b) Gaussian random variable with mean η. The concentric circles are the isolines of the Gaussian random variables.

ity ($\Pr\{|d| > \rho\} = 1 - \Pr\{|d| \le \rho\}$) is given by the cumulative distribution function (cdf) of the *magnitude* of a zero-mean complex Gaussian random variable n_G. As the magnitude is *Rayleigh distributed* (see Box 5.1), that is

$$\mathrm{F}_{|n_\mathrm{G}|}(\rho) = \Pr\{|n_\mathrm{G}| \le \rho\} = 1 - \mathrm{e}^{-\rho^2/\sigma_n^2} \,, \tag{5.39}$$

we have

$$\mathrm{BER}_{a=0} = \Pr\{|d| > \rho\} = 1 - \mathrm{F}_{|d|}(\rho) = \mathrm{e}^{-\rho^2/\sigma_n^2} \,. \tag{5.40}$$

Case B: $a = 2$ Transmitted; $\hat{a} = 0$ Detected Second, we assume that the signal point $a = 2$ has been transmitted. We decide in favor of $a = 0$ – and, thus, an error occurs – if the detection variable d has magnitude smaller than the threshold ρ. Since here $d = 2 + n$, we have $|d| = |2 + n|$ and an error occurs if the complex Gaussian variable $d = 2 + n$ with mean 2 and variance $\sigma_n^2 = N_0/E_g$ has magnitude smaller than ρ, that is

$$\mathrm{BER}_{a=2} = \Pr\{|d| \le \rho\} \,. \tag{5.41}$$

This probability is obtained via the cdf of the magnitude of a complex Gaussian random variable $n_{\mathrm{G},\eta}$ with mean η (the region for the integration is depicted in gray in Figure 5.15(b)). As the magnitude is *Rice distributed* (see Box 5.1), that is

$$F_{|n_{G,\eta}|}(\rho) = \Pr\{|n_{G,\eta}| \le \rho\} = 1 - Q_M\left(\frac{\eta\sqrt{2}}{\sigma_n}, \frac{\rho\sqrt{2}}{\sigma_n}\right), \quad (5.42)$$

we have

$$\text{BER}_{a=2} = \Pr\{|d| \le \rho\} = F_{|d|}(\rho) = 1 - Q_M\left(\frac{\eta\sqrt{2}}{\sigma_n}, \frac{\rho\sqrt{2}}{\sigma_n}\right). \quad (5.43)$$

Average Error Probability We are now ready to give the average error probability for non-coherent demodulation of OOK. Assuming that the two signal points used are equally likely and considering that $\sigma_n^2 = E_g/N_0$ and that for OOK $E_b = 2E_g$ and $\eta = 2$, the average error ratio is given by

$$\text{BER} = \tfrac{1}{2}\text{BER}_{a=0} + \tfrac{1}{2}\text{BER}_{a=2}$$

$$= \frac{1}{2}\left(e^{-\rho^2 E_b/(2N_0)} + 1 - Q_M\left(2\sqrt{\frac{E_b}{N_0}}, \sqrt{\frac{\rho^2 E_b}{N_0}}\right)\right). \quad (5.44)$$

Note that the error probability depends on the choice of the radius ρ of the decision boundary, which in turn depends on the signal-to-noise ratio. In other words, for each value of E_b/N_0 an optimum threshold ρ can be given for the best error ratio performance. In Figure 5.16 this optimum threshold is plotted.

As can be seen, for small SNRs (large noise power), a larger radius is preferable. For sufficiently large SNR, ρ is close to one, that is, the middle point between the signal points $a = 0$ and 2. In practice, usually radius $\rho = 1$ is chosen.

Upper Bound on Error Probability Equation (5.44) gives the exact error probability of OOK but offers no insight (due to the Marcum-Q-function) to the (asymptotic) behavior. Hence, an approximation – in particular, an *upper bound* – is of interest.

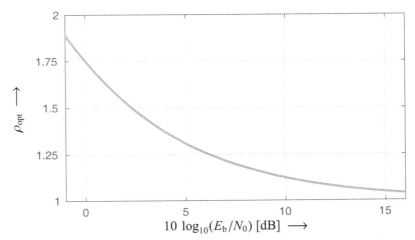

Figure 5.16 Optimum decision threshold ρ for non-coherent detection of on–off keying (plotted over the signal-to-noise ratio E_b/N_0 in dB).

5 Non-coherent Demodulation

Such an upper bound can be obtained via the following considerations. First, we set $\rho = 1$; an optimized radius gives better results, hence an upper bound results. Second, Case A leads to an analytic (and simple) expression. Case B ($a = 2$ has been transmitted), however, due to the integration over a non-centered (at the origin) Gaussian density, cannot be done analytically.

If the probability of a *correct decision* is *lower-bounded*, the *error probability* is *upper-bounded*. The probability of a correct decision is given by integrating the conditional pdf (the shifted noise pdf) over the correct decision region. This region is visualized in Figure 5.17(a). If the decision region is narrowed, and integration is done only over this smaller region, then the desired lower bound is obtained. Note that this shrinkage of the decision region is done only for the error ratio calculation – the decision region used in practice still remains the one shown in Figure 5.14.

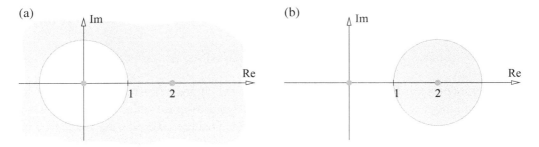

Figure 5.17 (a) Optimum decision region and (b) narrowed version for lower-bounding the probability of a correct decision ($a = 2$ transmitted).

A well-suited (for the purpose of bounding the error ratio) narrowed region is visualized in Figure 5.17(b). Using this region, the nonzero-mean Gaussian pdf is integrated over a circular region that is centered at the mean. This, however, is the same situation as integrating a zero-mean Gaussian pdf over a circular region centered at the origin. This integral has been solved in (5.23). Hence, using (5.40), we have

$$\text{BER}_{a=2} < \text{BER}_{a=0} \tag{5.45}$$

and thus

$$\text{BER} = \tfrac{1}{2}\text{BER}_{a=0} + \tfrac{1}{2}\text{BER}_{a=2}$$

$$< \text{BER}_{a=0} = e^{-E_b/(2N_0)}. \tag{5.46}$$

Non-coherent Demodulation of M-ary Unipolar ASK

Up to now, only binary transmission has been considered. The generalization to non-coherent demodulation of M-ary unipolar ASK is immediately possible.

Information is carried in the amplitude and, since the channel may rotate the signal arbitrarily, the phase of the receive signal is ignored and only the magnitude is evaluated. Hence, the decision boundaries are *concentric circles* (centered at the origin). For $M = 4$, these boundaries are visualized in Figure 5.18.

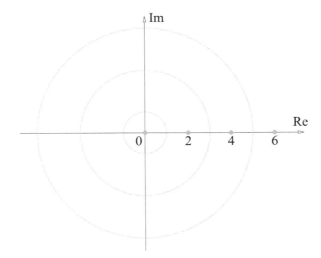

Figure 5.18 Decision boundaries for non-coherent detection of unipolar MASK ($M = 4$).

The above derived upper bound on the error ratio of OOK can be generalized straightforwardly to upper bound the (symbol) error ratio of non-coherently demodulated M-ary unipolar ASK. Here, as in Figure 5.17, the areas for correct decisions are replaced by circles of radius 1 around all signal points. We have

$$\text{SER} < e^{-E_g/N_0} = e^{-d_{\min}^2 \cdot E_b/(2N_0)}, \tag{5.47}$$

and when assuming Gray labeling (cf. Section 3.6):

$$\text{BER} \approx \frac{1}{\log_2(M)} \cdot \text{SER} = \frac{1}{\log_2(M)} e^{-d_{\min}^2 \cdot E_b/(2N_0)}. \tag{5.48}$$

Thereby, $d_{\min}^2$ is the normalized minimum squared Euclidean distance of the unipolar ASK constellation, see (3.108).

In Figure 5.19, the bit error ratio of non-coherent reception of unipolar ASK (Gray labeling) is plotted over E_b/N_0 (in dB).

The exact error ratio (solid, via numerical evaluation of the Marcum-Q-function) and the simple approximation (dashed) are given. The curves show the typical waterfall behavior, similar to the curves for coherent demodulation (cf. Figure 3.47). However, the SNR required to guarantee a desired reliability (error ratio) is significantly larger than that for coherent demodulation of bipolar ASK.

5.3.2 Comparison with Coherent Demodulation

The power/bandwidth planes in Figure 5.20 give the comparison of coherent and non-coherent detection. The roll-off factor of the $\sqrt{\text{Nyquist}}$ pulse is chosen as $\alpha = 0$, that is, the bandwidth efficiency corresponds to the rate R. As can be seen, the non-coherent schemes are up to 6 dB worse (have a poorer power efficiency) than the coherent schemes.

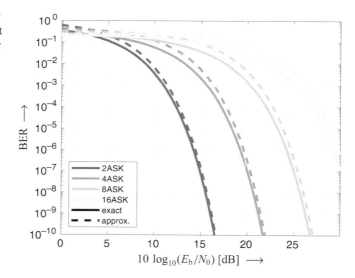

Figure 5.19 Bit error ratio of non-coherent reception of unipolar ASK over the signal-to-noise ratio (expressed as E_b/N_0 in dB).

However, one has to be careful in judging the "non-coherent" demodulation process as inefficient. Note here that *bipolar coherent* schemes are compared with *unipolar non-coherent* schemes. The question is whether the loss is caused by bipolar vs. unipolar or by coherent vs. non-coherent. To that end, we look at the symbol error probability of coherent demodulation of M-ary ASK, which was derived in (3.91) as

$$\text{SER} = \frac{2M-2}{M} Q\left(\sqrt{d_{\min}^2 \frac{E_b}{N_0}}\right). \tag{5.49}$$

Using the upper bound (3.97) on the Q-function, we can bound the error probability by

$$\text{SER} \leq \frac{M-1}{M} e^{-d_{\min}^2 E_b/(2N_0)} < e^{-d_{\min}^2 E_b/(2N_0)}. \tag{5.50}$$

A comparison with (5.47) shows that the upper bounds for coherent and non-coherent demodulation coincide. Hence, the non-coherent demodulation in and of itself does (almost) not cause a loss in power efficiency. However, a significant loss compared to coherent reception is caused because, in the case of non-coherent demodulation, only *unipolar ASK* can be used. This is the main source of the loss.

Since for unipolar ASK we have

$$\begin{aligned}
\text{E}\{|a|^2\} &= \frac{4M^2 - 6M + 2}{3} \\
&= \frac{M^2 - 1}{3} + (M-1)^2 = \sigma_a^2 + m_a^2 \\
&\approx \frac{M^2}{3} + M^2,
\end{aligned} \tag{5.51}$$

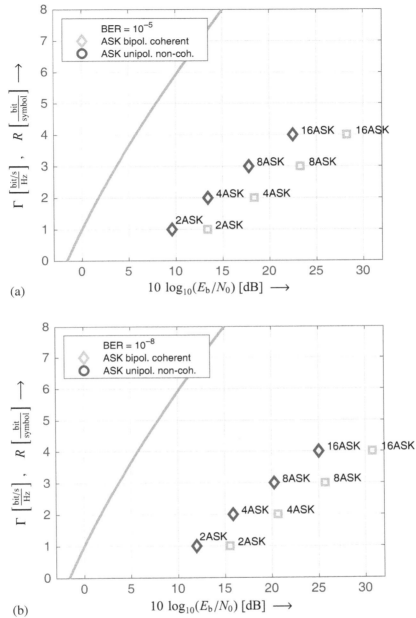

Figure 5.20 Power/bandwidth plane for ASK schemes: bipolar ASK with coherent detection and unipolar ASK with non-coherent detection. $\alpha = 0$ (i.e., power/rate plane). (a) Tolerated bit error ratio BER $\leq 10^{-5}$; (b) BER $\leq 10^{-8}$.

$m_a^2 \approx 3\sigma_a^2$ holds. In other words, three-quarters of the total transmit power is allocated to the mean; only one-quarter lies in the variance and consequently is used for representing the data. Since the factor 1/4 (in power) corresponds to 6 dB, unipolar ASK has (approximately) a 6 dB worse power efficiency compared to bipolar ASK.

5.4 Problems

5.4.1 Differential Phase-Shift Keying

Digital transmission using quaternary, differential phase-shift keying (DPSK) is considered. The mapping of the 4-ary information symbols $m[k] \in \{0, 1, 2, 3\}$ onto QPSK symbols $a[k]$ is given as follows:

$$a[k] = j^{m[k]} = e^{j\frac{\pi}{2}m[k]}, \quad \text{that is,} \quad \begin{array}{rcl} m & \to & a \\ 0 & \to & +1 \\ 1 & \to & +j \\ 2 & \to & -1 \\ 3 & \to & -j \end{array}.$$

(a) Sketch the block diagram of the modulator.

At the receiver, a matched filter followed by T-spaced sampling is used. The resulting detected variables $d[k]$ for the modulation intervals $k = -1, \ldots, 2$ are given in the following figure (complex plane).

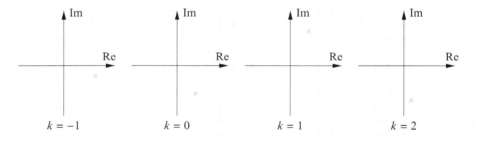

(b) Sketch the block diagram of the *differential coherent* demodulator (DCPSK).
(c) Determine the detected information symbols when DCPSK is used.
(d) Sketch the block diagram of the *differential* demodulator (DPSK).
(e) Determine the detected information symbols if DPSK is used.
(f) State the advantages and disadvantages of DPSK/DCPSK compared to coherent reception of PSK.

5.4.2 Double Differential Receiver

We consider a generalization of differential PSK (DPSK). The receiver structure is plotted in the following figure. Thereby, $d[k]$ is the detection variable at the output of the matched filter. M-ary PSK is employed.

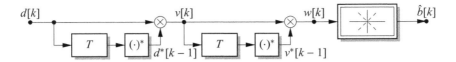

(a) Express the symbols $w[k]$ dependent on $d[k]$.
(b) Sketch the block diagram of the corresponding transmitter structure. PSK symbols $\tilde{a}[k]$, which represent the data, should enter the structure and PSK transmit symbols should be produced.
(c) Give an equation showing how $a[k]$ and $\tilde{a}[k]$ are related.

We now assume transmission over an AWGN channel with phase and frequency offset. The detection symbols are given by (noise is ignored for the moment)

$$d[k] = a[k] \cdot e^{j\phi[k]} .$$

(d) Calculate the signal $w[k]$ at the input of the PSK decision device dependent on the data symbols $\tilde{a}[k]$ for this channel.

We further assume that the phase is given by a phase ramp

$$\phi[k] = \phi_0 + \Delta k .$$

(e) Give the signal $w[k]$ for this specific assumption of the channel phase. What is the advantage of this receiver concept compared to conventional differential PSK?
(f) How does the performance qualitatively compare with that of coherent PSK and differentially detected DPSK?

5.4.3 DPSK/ASK

A data rate of 1 Mbit/s has to be transmitted by a radio-frequency signal within a frequency bandwidth of 700 kHz. Transmitter and receiver are fast-moving mobile stations and therefore sufficient carrier phase synchronization and, thus, coherent communication are not possible.

First, DPSK with differential detection is considered.

(a) Specify an appropriate DPSK scheme when a roll-off factor of the used square-root Nyquist pulses smaller than 1/3 is not tolerated in order to avoid high fluctuations of the continuous-time transmit signal. Give the rate [bit/symbol] of this scheme.

The transmit signal is attenuated by up to 140 dB (inclusive antenna gains) and at the receiver input white Gaussian noise with one-sided power spectral density 10^{-20} W/Hz is present.

(b) What average power of the transmit signal is necessary to guarantee a symbol error ratio less than 10^{-8}?

The scheme has now to be upgraded for transmission of an increased data rate of 1.5 Mbit/s while conserving as many components of the previous scheme as possible.

(c) What rate has to be chosen for the upgraded scheme?
(d) What transmit signal power is necessary for a symbol error ratio less than 10^{-8} when, for the upgraded scheme, DPSK is again applied?

An alternative design for the upgrade is to superimpose a non-coherent amplitude modulation (ASK) on the original DPSK scheme specified in part (a).

(e) Give a sketch of a signal constellation suited for such a DPSK/ASK scheme.
(f) What distances between signal amplitudes are necessary so that, in the regime of sufficient data reliability, the non-coherent ASK part does not contribute more to the symbol error ratio than the DPSK part?
Hint: Apply the analysis of the symbol error probability of the inner point of pure non-coherent ASK as a bound for the ASK part of the signal constellation of part (e).
(g) Calculate the necessary transmit signal power of the DPSK/ASK scheme for a symbol error ratio less than 10^{-8}. Compare both versions of the upgraded system with respect to power efficiency and with respect to a nonlinear behavior of the amplifier of the transmit signal.

Not only the carrier phase of the receiver input signal but also the signal attenuation fluctuates rapidly due to the fast movements of transmitter and receiver within areas of a stationary wave caused by many reflecting objects. Thus, it is not possible to fix a decision boundary for the non-coherent ASK part.

(h) Is it possible to construct a differential precoding for non-coherent ASK together with a differential detection in amplitude analogous to the phase in DPSK? If yes, sketch block diagrams for transmitter and receiver for such a differential non-coherent ASK. What effect on data reliability would be observable for such a scheme compared to the usual non-coherent ASK?

5.4.4 Sensitivity of DPSK to Carrier Frequency Offsets

The sensitivity of the symbol error ratio of M-ary DPSK with respect to a difference f_Δ of the carrier frequencies at transmitter and receiver is studied using a simple model.

In DPSK, the decision boundaries are determined by the phase of the received signal point of the preceding symbol interval. In case of a carrier frequency offset, for the current symbol to be detected, the decision boundaries are rotated by a phase offset φ_Δ.

(a) Express the phase offset φ_Δ in terms of the frequency difference f_Δ and the data rate for M-ary DPSK.

In a simple model, the loss in minimum Euclidean distance to the nearest decision boundary due to the rotation φ_Δ can be used to specify the loss in power efficiency caused by a carrier frequency offset.

(b) Calculate the Euclidean distance of the transmitted signal point (w.l.o.g., $a = 1$ can be assumed) to the decision boundary rotated by an angle φ_Δ in case of M-ary PSK.

(c) For *M*-ary PSK, derive a general formula for the phase offset of the decision boundaries that causes a loss in power efficiency of X dB. Give numerical results for $M = 4, 8$, and 16 and for $X = 0.5$ and 1 dB.

(d) Calculate the maximum frequency offset relative to the transmitted data rate for *M*-ary DPSK with $M = 4, 8$, and 16 for tolerated losses of 0.5 and 1 dB in power efficiency.

(e) Discuss the result with respect to the bandwidth efficiency of the schemes.

6 Signal Space Representation

In the last chapters, we looked at pulse-amplitude-modulated transmission; the basic principles and specific variants thereof. Only uncoded transmission and, thus, the respective symbol-by-symbol detection at the receiver have been treated. Notwithstanding that PAM is the most widely used (and simplest) version of digital transmission schemes, not all classes of modulation schemes fall under this category. For instance, digital frequency modulation (except for the MSK scheme, which is a hybrid) does not fall under the umbrella of PAM.

Consequently, in this chapter, a general view of digital modulation schemes is developed. To that end, the operations of *coding* and *modulation* are clearly separated. The key tool for the analysis and synthesis of transmission schemes is the representation of signals in a *signal space*. Based on this view, the optimum receiver for any kind of general digital modulation scheme is derived.

PREREQUISITES FOR THIS CHAPTER

- PAM transmission (Chapter 3).
- Linear algebra (orthonormal basis).

6.1 Coding and Modulation

Recall the basic block diagram of the PAM transmitter in Figure 3.1; for convenience, it is repeated in Figure 6.1. In this simplest form of digital communication schemes and assuming uncoded transmission, the source bits are grouped and m-tuples are mapped redundancy-free to a point drawn from an $M = 2^m$-ary signal constellation. The modulation step, the generation of the continuous-time transmit signal, is the filtering (pulse shaping) of the sequence of amplitude coefficients $a[k]$ with basic pulse shape $g(t)$.

At first glance, the general structure of digital modulation, sketched in Figure 6.2, looks the same. Two main building blocks are present. However, the operations of both blocks are more general.

Each digital transmitter can unambiguously be separated into the following operations.

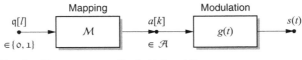

Figure 6.1 Block diagram of the PAM transmitter (without channel coding).

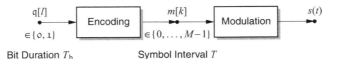

Figure 6.2 Block diagram of a general digital modulation scheme.

- *Encoding*
  In this step, *redundancy* and temporal *dependencies* (i.e., *memory*) may be introduced into the sequence $\langle m[k] \rangle$ of *signal numbers*. The signal numbers, drawn from the set $\{0, 1, \ldots, M - 1\}$, take the place of the amplitude coefficients $a[k]$ in PAM. Redundancy-free mapping is a special case of the encoding step.
- *Modulation*
  Here, *symbol by symbol* (without further redundancy or temporal dependencies), each signal number $m[k]$ is individually represented by a continuous-time *signal element* $s_m(t)$. The pulse shaping in PAM is a special form of this representation.

6.1.1 Modulation

We first look at the general version of *modulation*.

Definition 6.1 *Modulation*

The *modulation* in a digital communication scheme is defined as the representation of a sequence of signal numbers $m[k] \in \{0, 1, \ldots, M - 1\}$, $k \in \mathbb{Z}$, by a continuous-time (continuous-amplitude) signal $s(t)$ in such a way that:

- each signal number $m[k]$ is mapped onto a *signal element* per modulation interval T, independently of preceding and subsequent signal numbers

$$m[k] \quad \mapsto \quad s_{m[k]}(t - kT) \ ;$$

- the signal elements are taken from the *signal set*

$$\mathcal{S} \stackrel{\text{def}}{=} \{s_0(t), s_1(t), \ldots, s_{M-1}(t)\} \tag{6.1}$$

with cardinality $M = |\mathcal{S}|$.

Since M signal elements are available, the rate of modulation is

$$R_\mathrm{m} = \log_2(M) \ . \tag{6.2}$$

- The transmit signal $s(t)$ (ECB signal) is given by the *superposition* of all signal elements

$$s(t) = \sum_{k=-\infty}^{\infty} s_{m[k]}(t - kT) \ . \qquad (6.3)$$

Thus, in this generalized meaning, *modulation* specifies the step-by-step representation of information (a sequence of abstract symbols) by a physical, continuous-time signal.

Example 6.1 PAM and Signal Elements

Assume PAM transmission with basic pulse $g(t)$ and signal constellation

$$\mathcal{A} = \{a^{(0)}, \ a^{(1)}, \ \ldots, \ a^{(M-1)}\} \ .$$

Here, the signal elements are given by

$$s_i(t) = a^{(i)} g(t) \ , \qquad i \in \{0, 1, \ldots, M-1\} \ , \qquad (6.4)$$

that is, the signal elements are all (complex) weighted versions of the basic pulse shape. For a fixed pulse $g(t)$, the weights (amplitude coefficients) $a^{(m)}$ thus characterize the signal elements completely and are synonymous with the signal number $m[k]$. The sets $\mathcal{A}$ and

$$\mathcal{S} = \{a^{(0)} g(t), \ a^{(1)} g(t), \ \ldots, \ a^{(M-1)} g(t)\}$$

equivalently specify the modulation scheme for a given basic pulse $g(t)$.

Example 6.2 Signal Elements

An example of signal elements for binary transmission ($M = 2$) is given in Figure 6.3 Note that this is an academic example; due to their spectral properties, these signal elements are not well-suited in practice but may serve as an illustration of the basic principles. Notice, these two signal elements are not scaled versions of a single basic pulse shape – it is not a PAM scheme.

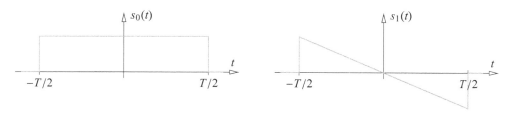

Figure 6.3 Signal elements for a theoretical binary transmission.

Example 6.3 Frequency-Shift Keying

In Figure 6.4, an example of $M = 4$-ary frequency-shift keying is given. Note that for illustration purposes, here the RF signal elements are given. It is clearly visible that using such signal elements, the information is represented in the instantaneous frequency of the signal. Again, this modulation strategy does not fall under the PAM category.

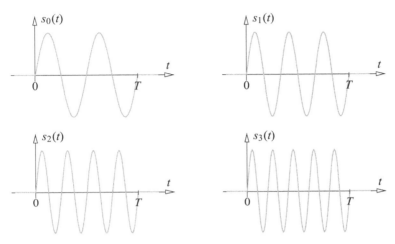

Figure 6.4 RF signal elements of $M = 4$-ary frequency-shift keying.

Example 6.4 Signal Elements

A further example of binary transmission is given in Figure 6.5. Here, the signal elements are in the RF domain and both decay smoothly for $|t| \to \infty$. It can be shown that the signal elements are orthogonal to each other. Such elements (when truncated to a finite duration and shifted for causality) are well suited in practice (see also Section 4.2.1, Figure 4.27).

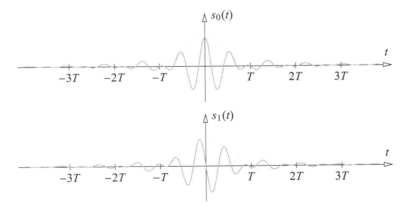

Figure 6.5 Signal elements for binary transmission.

Box 6.1 Orthogonality of Vectors and Functions

Euclidean Space $\mathbb{R}^n$
Let the vectors $a, b \in \mathbb{R}^n$ (with components a_i, b_i, respectively) be given. The vectors are *orthogonal* to each other (the angle between them is 90°) if their scalar product is zero:

$$\sum_i a_i \cdot b_i \stackrel{!}{=} 0.$$

Complex-Valued Vectors
For vectors $a, b \in \mathbb{C}^n$ the respective condition reads

$$\sum_i a_i \cdot b_i^* \stackrel{!}{=} 0.$$

Complex-Valued Functions
Similarly, given (complex-valued) functions $a(x)$ and $b(x)$, the functions are *orthogonal* to each other if

$$\int a(x) \cdot b^*(x) \, \mathrm{d}x \stackrel{!}{=} 0$$

(formally obtained for $a_i \to a(x)$, $b_i \to a(x)$, and $\sum_i \to \int \cdot \, \mathrm{d}x$).

Constraints on the Signal Elements

As stated in the above definition, the modulation is done individually, symbol by symbol. All dependencies are taken into account in the encoding step – this is the sharp division line between both operations. As a consequence, we demand that the *demodulation*, the receiver-side counterpart, can also be performed in a symbol-by-symbol fashion without any loss of information. All dependencies (memory) are taken into account in the respective subsequent *decoding* step.

In order to guarantee that the demodulation of a signal impaired by additive white Gaussian noise (AWGN) can be done individually per modulation step k, it has to be ensured that no interference between successive symbol intervals occurs. This is achieved by demanding that the signal elements are *orthogonal* to each other when shifting them by integer multiples of T. Owing to this constraint, the symbol intervals are decoupled.

Mathematically, the following *temporal orthogonality* has to be met by the signal elements $s_i(t)$:

$$\int_{-\infty}^{\infty} s_i(t + \lambda T) \cdot s_j^*(t) \, \mathrm{d}t \stackrel{!}{=} \begin{cases} E_{ij}, & \lambda = 0 \\ 0, & \lambda \in \mathbb{Z} \setminus \{0\} \end{cases}, \quad \forall i, j \in \{0, 1, \ldots, M-1\} \quad (6.5)$$

E_{ij}: cross energy between the signal elements $s_i(t)$ and $s_j(t)$
$E_{ii} \stackrel{\text{def}}{=} E_i$: energy of the signal element $s_i(t)$.

Note that signal elements which are time-limited to a single symbol interval of duration T obviously fulfill the temporal orthogonality condition as shifted (by λT) versions do not overlap.

Moreover, the constraint on temporal orthogonality of the signal elements is a generalization of the $\sqrt{\text{Nyquist}}$ property of the basic pulse $g(t)$ in PAM (cf. (3.46) in Chapter 3). For $i = j$, (6.5) specializes to

$$\int_{-\infty}^{\infty} s_i(t + \lambda T) \cdot s_i^*(t) \, dt = E_i \cdot \delta[\lambda], \qquad \forall i \in \{0, 1, \ldots, M-1\}. \tag{6.6}$$

This means that each signal element $s_i(t)$ itself has to have the $\sqrt{\text{Nyquist}}$ property.

It will be shown in Section 6.4 that using such an approach, an information-lossless transition from discrete-time sequences to continuous-time signals (modulation) and back to discrete time via suitable filtering and sampling (demodulation) is possible in general.

6.1.2 Average Power Spectral Density

Let the signal elements and their spectra (Fourier transforms) be given as

$$s_i(t) \circ\!\!-\!\!\bullet S_i(f). \tag{6.7}$$

In *uncoded transmission* the signal elements are used equiprobably and statistically independently over the time steps, that is, $\langle m[k] \rangle$ is a uniform i.i.d. sequence. As explained above, given the sequence of signal numbers, the continuous-time transmit signal (ECB domain) is generated as

$$s(t) = \sum_{k=-\infty}^{\infty} s_{m[k]}(t - kT). \tag{6.8}$$

The autocorrelation function of this cyclo-stationary process is given by

$$\mathrm{E}\{s(t+\tau)\, s^*(t)\} = \mathrm{E}\Bigg\{ \sum_{l=-\infty}^{\infty} s_{m[l]}(t+\tau - lT) \sum_{k=-\infty}^{\infty} s^*_{m[k]}(t - kT) \Bigg\}$$

$$= \sum_{l=-\infty}^{\infty} \sum_{k=-\infty}^{\infty} \mathrm{E}\{s_{m[l]}(t+\tau - lT)\, s^*_{m[k]}(t - kT)\}. \tag{6.9}$$

The *average power spectral density* of the transmit signal is given by the Fourier transform of the *average autocorrelation function*

$$\bar{\phi}_{ss}(\tau) \stackrel{\text{def}}{=} \frac{1}{T} \int_0^T \mathrm{E}\{s(t+\tau)\, s^*(t)\} \, dt, \tag{6.10}$$

which leads to

$$\bar{\Phi}_{ss}(f) \stackrel{\text{def}}{=} \mathscr{F}_\tau\{\bar{\phi}_{ss}(\tau)\} = \mathscr{F}_\tau\left\{\frac{1}{T}\int_0^T E\{s(t+\tau)\,s^*(t)\}\,dt\right\}. \quad (6.11)$$

Using the definition of the Fourier transform and taking into account that integration and expectation are both linear operators and may be exchanged, (6.11) can be written as

$$\bar{\Phi}_{ss}(f) = \int_{-\infty}^{\infty} \frac{1}{T} \int_0^T E\{s(t+\tau)\,s^*(t)\}\,dt\,e^{-j2\pi f\tau}\,d\tau$$

$$\stackrel{(6.9)}{=} \frac{1}{T}\int_{-\infty}^{\infty}\int_0^T \sum_{l=-\infty}^{\infty}\sum_{k=-\infty}^{\infty} E\{s_{m[l]}(t+\tau-lT)\,s^*_{m[k]}(t-kT)\}\,e^{-j2\pi f\tau}\,dt\,d\tau$$

$$= \frac{1}{T}\int_0^T \sum_{l=-\infty}^{\infty}\sum_{k=-\infty}^{\infty} E\left\{\int_{-\infty}^{\infty} s_{m[l]}(t+\tau-lT)\,e^{-j2\pi f\tau}\,d\tau\,s^*_{m[k]}(t-kT)\right\}\,dt$$

with $\mathscr{F}\{s_i(t)\} = S_i(f)$ and $\mathscr{F}\{s_i(t-\Delta)\} = S_i(f)\,e^{-j2\pi f\Delta}$

$$= \frac{1}{T}\int_0^T \sum_{l=-\infty}^{\infty}\sum_{k=-\infty}^{\infty} E\left\{S_{m[l]}(f)\,e^{-j2\pi f(lT-t)}\,s^*_{m[k]}(t-kT)\right\}\,dt$$

$$= \frac{1}{T}\sum_{l=-\infty}^{\infty} E\left\{S_{m[l]}(f)\int_0^T \sum_{k=-\infty}^{\infty} s^*_{m[k]}(t-kT)\,e^{+j2\pi f(t-kT)}\,dt\right\}e^{+j2\pi f(k-l)T}$$

with $\int_0^T \sum_{k=-\infty}^{\infty} x(t-kT)\,dt = \int_{-\infty}^{\infty} x(t)\,dt$

$$= \frac{1}{T}\sum_{l=-\infty}^{\infty} E\left\{S_{m[l]}(f)\int_{-\infty}^{\infty} s^*_{m[k]}(t)\,e^{+j2\pi ft}\,dt\right\}e^{+j2\pi f(k-l)T}$$

with $\mathscr{F}\{s_i^*(-t)\} = S_i^*(f)$

$$= \frac{1}{T}\sum_{l=-\infty}^{\infty} E\{S_{m[l]}(f)\,S^*_{m[k]}(f)\}\,e^{+j2\pi f(k-l)T}$$

using $l = k + \kappa$ and since $m[k]$ is a stationary sequence

$$= \frac{1}{T}\sum_{\kappa=-\infty}^{\infty} E\{S_{m[k+\kappa]}(f)\,S^*_{m[k]}(f)\}\,e^{-j2\pi f\kappa T}. \quad (6.12)$$

Since we restrict ourselves to the *uncoded case*, the expectation over the spectra of the signal elements is given by

$\kappa = 0$: $\quad \mathrm{E}\left\{S_{m[k+\kappa]}(f)\, S^*_{m[k]}(f)\right\} = \mathrm{E}\{|S_m(f)|^2\}$

$$= \frac{1}{M} \sum_{m=0}^{M-1} |S_{m[k]}(f)|^2 \stackrel{\text{def}}{=} \bar{P}(f)\,, \qquad (6.13)$$

$\kappa \neq 0$: $\quad \mathrm{E}\{S_{m[k+\kappa]}(f)\, S^*_{m[k]}(f)\} = \frac{1}{M} \sum_{m_1=0}^{M-1} \frac{1}{M} \sum_{m_2=0}^{M-1} S_{m_1}(f)\, S^*_{m_2}(f)$

$$= \left(\frac{1}{M} \sum_{m_1=0}^{M-1} S_{m_1}(f)\right)\left(\frac{1}{M} \sum_{m_2=0}^{M-1} S^*_{m_2}(f)\right)$$

$$\stackrel{\text{def}}{=} \bar{S}(f)\,\bar{S}^*(f) = |\bar{S}(f)|^2\,. \qquad (6.14)$$

Thus, we arrive at

$$\bar{\Phi}_{ss}(f) = \frac{1}{T} \sum_{\kappa=-\infty}^{\infty} \mathrm{E}\left\{S_{m[k+\kappa]}(f)\, S^*_{m[k]}(f)\right\} e^{-j2\pi f \kappa T}$$

$$= \frac{1}{T}\left(\bar{P}(f) + \sum_{\substack{\kappa=-\infty \\ \kappa \neq 0}}^{\infty} |\bar{S}(f)|^2 e^{-j2\pi f \kappa T} - |\bar{S}(f)|^2 + |\bar{S}(f)|^2 e^{-j2\pi f 0 T}\right)$$

$$= \frac{1}{T}\left(\bar{P}(f) - |\bar{S}(f)|^2\right) + \frac{1}{T}|\bar{S}(f)|^2 \sum_{\kappa=-\infty}^{\infty} e^{-j2\pi f \kappa T}\,,$$

using $\sum_{\kappa=-\infty}^{\infty} e^{-j2\pi f \kappa T} = \frac{1}{T} \sum_{\mu=-\infty}^{\infty} \delta(f - \frac{\mu}{T})$

$$= \frac{1}{T}(\bar{P}(f) - |\bar{S}(f)|^2) + \frac{1}{T^2}|\bar{S}(f)|^2 \sum_{\mu=-\infty}^{\infty} \delta\left(f - \frac{\mu}{T}\right)\,. \qquad (6.15)$$

In summary, we have

$$\bar{\Phi}_{ss}(f) = \frac{1}{T}\left(\bar{P}(f) - |\bar{S}(f)|^2\right) + \frac{1}{T^2}|\bar{S}(f)|^2 \sum_{\mu=-\infty}^{\infty} \delta\left(f - \frac{\mu}{T}\right) \qquad (6.16)$$

with

$$\bar{P}(f) \stackrel{\text{def}}{=} \frac{1}{M} \sum_{i=0}^{M-1} |S_i(f)|^2\,,$$

$$\bar{S}(f) \stackrel{\text{def}}{=} \frac{1}{M} \sum_{i=0}^{M-1} S_i(f) \quad \circ\!\!-\!\!\bullet \quad \bar{s}(t) \stackrel{\text{def}}{=} \frac{1}{M} \sum_{i=0}^{M-1} s_i(t)\,.$$

As for conventional PAM (cf. (3.15)), Dirac pulses will be present in the power spectral density if the set of signal elements has nonzero mean ($\bar{s}(t) \neq 0$, $\forall t$). Otherwise, the average psd is simply proportional to the average of the squared magnitudes of the spectra of the signal elements, that is, $\bar{\Phi}_{ss}(f) = \frac{1}{T}\bar{P}(f)$.

Of course, for conventional PAM with $s_i(t) = a^{(i)} g(t)$ and $a^{(i)} \in \mathcal{A}$, (6.16) specializes to (3.15).

6.1.3 Channel Coding

We now turn to the other block of the transmitter, the encoding operation.

> **Definition 6.2** *Encoding*
>
> In the *encoding* process, the (binary) information symbols from the source are converted into signal numbers. Thereby, *redundancy* may be introduced and *dependencies* between successive symbols may be caused.

Further, we formally decompose the encoder into a *channel encoder* and a *mapper*, shown in Figure 6.6.

- The sequence $\langle q[l] \rangle$ of source symbols is binary and i.i.d. (see Chapter 1); the clocking is given by the bit duration T_b.
- Without loss of generality, the code symbols $c[\nu]$ are assumed to be binary (drawn from $\mathbb{F}_2$); the symbol interval is T_c.
- The signal numbers $m[k]$ are taken from $\{0, 1, \ldots, M-1\}$; the symbol interval is T, the duration of the modulation interval.

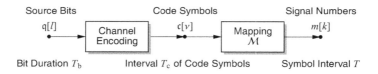

Figure 6.6 Block diagram of the encoder.

> **Definition 6.3** *Channel Coding*
>
> By *channel coding* constraints are imposed on the sequence of code symbols such that only specific sequences, the *codewords*, are available for transmission. Thus, the sequence of code symbols contains *redundancy*, that is, more sequences of code symbols would be possible than allowed by the channel code. The *code* is the set of allowed sequences/codewords.
>
> There are two basic types of channel coding in use: *block coding* and *trellis coding*. For details, see Appendix B.

By choosing the code, that is, by selecting the allowed sequences of code symbols (the codeword):

- special features may be induced on the transmit signal (called *line coding*, e.g., spectral or run-length properties) and/or
- the resilience of the signal against noise and interference may be improved (*channel coding* in a stricter sense for *error control coding*, namely error detection and correction), that is, the power efficiency of the communication scheme may be improved.

We will have a deeper look into the effect of combining coding and modulation in Section 6.6.

As communication usually takes place in *bursts*, that is, transmission of finite-length sequences, the derivation of the signal-space concept is done for a block code. The generalization to trellis codes, however, is straightforward.

Channel Encoding

The *encoding* of a block code is the bijective mapping of words of length K of binary source symbols (bits) onto words of length N of (w.l.o.g.) binary code symbols ($K \leq N$). That is, at time step μ

$$\mathbf{q}[\mu] = [q_0\ q_1\ q_2\ \ldots\ q_{K-1}] \mapsto \mathbf{c}[\mu] = [c_0\ c_1\ c_2\ \ldots\ c_{N-1}] \ .$$

The words may be cut out from the stream of symbols as

$$q_i = q[l = i + \mu K] \in \{0,\ 1\} \qquad c_j = c[\nu = j + \mu N] \in \{0,\ 1\}, \qquad \mu \in \mathbb{Z} \ .$$

The (block) *code* $\mathbf{C}$ is the set of *codewords* $\mathbf{c}$ that can be generated by the encoding process (and which are then used for transmission), that is

$$\mathbf{C} = \{\mathbf{c} \mid \mathbf{q} \mapsto \mathbf{c},\ \forall \mathbf{q}\} \ .$$

The most important *parameters* of a block code are

- codelength $\quad N$
  (i.e., number of binary symbols within one codeword)
- message (information) length $\quad K$
  (i.e., number of source bits per codeword)
- code rate $\qquad R_\mathrm{c} = \frac{1}{N} \log_2 \left(2^K\right) = \frac{K}{N}\ \left[\frac{\text{bit}}{\text{codesymbol}}\right]$
  (i.e., average information per code symbol)
  For redundancy-free sourcesymbols, the codewords are a-priori equiprobable. From continuity of information flow, $KT_\mathrm{b} = NT_\mathrm{c}$ follows, that is, $R = \frac{T_\mathrm{c}}{T_\mathrm{b}}$.

For $K < N$ **redundancy** is introduced – not all words of length N with binary code symbols are used for transmission. Thereby, error detection and/or error correction is enabled.

Mapping

The interface between channel coding and modulation is the *mapper* M. Here, the transition from abstract (finite-field) elements to real-valued integers takes place. Short blocks of U code symbols are mapped to V consecutive signal numbers. In the best case we have

$$2^U = M^V . \qquad (6.17)$$

Often, $V = 1$ is preferred, in particular when the number M of signal elements is a power of two. Then, $\log_2(M)$ binary code symbols are grouped and mapped to a signal element, similar to the mapping of blocks of binary symbols to signal points in conventional PAM (cf. Chapter 3). Mappers for very small V may be implemented by means of lookup tables.

In some situations, especially if $\log_2(M)$ is not a rational number, we have

$$2^U < M^V . \qquad (6.18)$$

Then (a small amount of) redundancy is added in the mapping step. Given M, the block lengths U and V should be chosen as small as possible and such that the additional redundancy is as small as possible. Note that the code rate is lowered by the factor $U/(V \log_2(M))$, called the *mapping efficiency*. As memory may be introduced by the mapper, it is a part of the encoder. A possible practical implementation (for large U) of such a mapper is described in Section B.3 of Appendix B.

Example 6.5 Binary Code Symbols with Ternary Modulation

When $M = 2$ or $M = 4$, a single code symbol ($U = \log_2(2) = 1$) or blocks of two code symbols ($U = \log_2(4) = 2$) are mapped to the signal element of one ($V = 1$) modulation step, respectively. Things change if we consider a ternary modulation scheme, that is, $M = 3$.

When choosing $V = 1$, a single code symbol ($U = 1$) can be mapped per mapping step ($2 < 3$). The mapping efficiency is $1/(1 \log_2(3)) = 0.631$.

When choosing $V = 2$, three code symbols ($U = 3$) can be mapped per mapping step ($8 < 9$). The mapping efficiency is $2/(3 \log_2(3)) = 0.946$.

A quite good fit is achieved for the parameters $U = 11$ and $V = 7$. Here, the set of $2^U = 2048$ binary blocks of length 11 is mapped onto the set of $3^7 = 2187$ ternary blocks of length 7, yielding a mapping efficiency of $11/(7 \log_2(3)) = 0.9915$.

Finally, the (overall) rate R of the digital communication scheme can be written as[1]

$$R = \frac{T}{T_b} = \frac{T V}{T_b V} \cdot \frac{\log_2(M)}{\log_2(M)}$$

blockwise mapping: $T_c U = T V$ and rearrangement

$$= \frac{T_c}{T_b} \frac{U}{V \log_2(M)} \cdot \log_2(M) ,$$

---
[1] Note that the following definitions are generalizations of (1.3) to redundant mappings.

or in summary

$$R = R_c \cdot R_m \quad \left[\tfrac{\text{bit}}{\text{modulation step}}\right] \text{ or } \left[\tfrac{\text{bit}}{\text{symbol}}\right], \quad (6.19)$$

with

$$R_c = \frac{T_c}{T_b} \frac{U}{V \log_2(M)}, \quad (6.20)$$

$$R_m = \log_2(M). \quad (6.21)$$

Example 6.6 Rates of Coded Modulation Schemes

We now consider some examples.

- Binary block code $K = 500$, $N = 1000$ with binary modulation (e.g., 2ASK)
$$R = R_c = 0.5 \left[\tfrac{\text{bit}}{\text{symbol}}\right].$$
- Binary block code $K = 750$, $N = 1000$ with 16-ary modulation (e.g., 16QAM)
$$R = R_c \cdot \log_2(16) = K/N \cdot 4 = 3 \left[\tfrac{\text{bit}}{\text{symbol}}\right].$$
- Binary block code $K = 700$, $N = 1100$ with 3-ary (ternary) modulation (e.g., 3ASK), mapping with $U = 11$ and $V = 7$ (i.e., codewords with 700 information bits are mapped to 700 signal elements)
$$R = R_c \cdot R_m = \frac{700}{1100} \frac{11}{7 \log_2(3)} \cdot \log_2(3) = \frac{100}{100} = 1 \left[\tfrac{\text{bit}}{\text{symbol}}\right]$$

(cf. also Chapter 1, Equation (1.3)).

6.1.4 Equivalent Coding and Modulation

The split into encoding and modulation up to now has been done on a *transmitter-based view*. However, even if the channel induces dispersive distortions and, thus, at the *receiver side* the signal elements do not fulfill the temporal orthogonality condition (6.5), an equivalent representation via *equivalent encoding* and *equivalent signal elements* can be given.

This means that not only deliberate encoding at the transmitter side has to be encountered in the encoding block, but also unintentional dispersion (temporal dependencies) in the channel. The optimum receiver then performs symbol-by-symbol demodulation with respect to the equivalent signal elements followed by decoding of the inherent code. Since the memory effects of the channel are taken into account in the decoding part, the demodulation part "only" has to take the additive channel noise into account. In other words, the demodulation is done as for the AWGN channel.

Figure 6.7
Decomposition into equivalent encoding and modulation using equivalent (temporal orthogonal) signal elements.

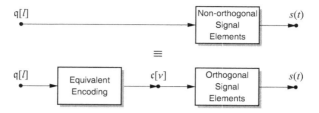

The decomposition into an equivalent encoding and the use of equivalent signal elements is depicted in Figure 6.7.

This decomposition is illustrated in the following example.

Example 6.7 Equivalent Encoding and Signal Elements

Assume that the two signal elements in Figure 6.8 are given, violating the temporal orthogonality condition (6.5).[2]

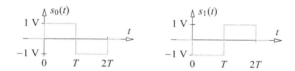

Figure 6.8 Signal elements, violating the temporal orthogonality condition.

An example of a signal may look as given in Figure 6.9.

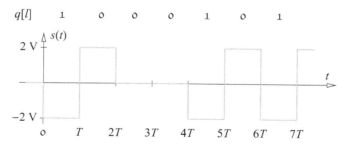

Figure 6.9 Example signal.

---

[2] These signal elements may be present at the receiver if the transmitter uses bipolar 2ASK with a rectangular pulse shape and the channel (due to reflection) has an impulse response $h_C(t) = \delta(t) - \delta(t - T)$, that is, an inverted echo delayed by T, which is equally strong as the direct (non-delayed) path is received.

This uncoded transmission using the non-orthogonal signal elements is equivalently represented using the three equivalent, temporally orthogonal signal elements together with the equivalent encoder and mapping shown in Figure 6.10. A direct inspection shows that for the example source sequence in Figure 6.9, the same transmit signal is produced.

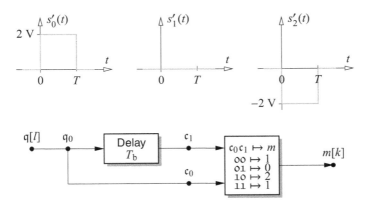

Figure 6.10 Equivalent, temporally orthogonal signal elements and equivalent encoder and mapping.

Obviously, the original binary scheme has rate $R = 1$ – of course, the equivalent interpretation also has this rate. According to (6.19), this rate factors into (equivalent) coding and modulation rates as follows:

$$R = \frac{T_b/\log_2(4)}{T_b} \frac{2}{1\,\log_2(3)} \cdot \log_2(3) = \frac{1}{\log_2(3)} \cdot \log_2(3) = R_c \cdot R_m \,. \qquad (6.22)$$

This example is continued in Example 6.17 (AMI code).

The previous example follows a simple but very general method to construct a pair of equivalent coding and modulation: define all possible signal trajectories within a modulation interval of duration T as the equivalent, time-limited signal elements. The requested orthogonality of signal elements with respect to time shifts of multiples of T holds per definition. In order to obtain valid signals, only selected sequences of signal elements are allowed. These constraints on sequences of signal elements for producing valid signals are devoted to the coding part. By this method, an optimum and straightforwardly implementable receiver structure is obtained for a very broad field of applications, for example, nonlinear dispersive signal distortion, either by the transmitter or the channel.

Example 6.8 Equivalent Encoding and Signal Elements

Any kind of linear or nonlinear intersymbol interference (ISI) over L modulation intervals due to dispersive signal distortions can be modeled by a trellis encoder with a shift-register memory of $L - 1$ delay elements. Thus, L consecutive information symbols together select a signal element.

An example of an end-to-end pulse (transmit filter and channel) that extends over three symbol intervals T is illustrated in Figure 6.11.

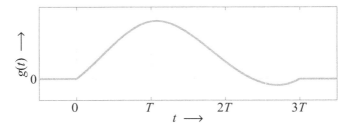

Figure 6.11 Pulse dispersed over three modulation intervals of a distorted 2ASK signal.

All possible (noise-free) receive signals are plotted on top of each other in Figure 6.12 (eye pattern) for the case when (uncoded) bipolar 2ASK signaling is employed. The interval $t \in [0, T)$ is highlighted.

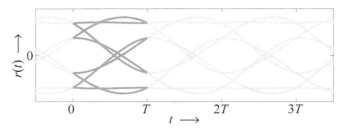

Figure 6.12 Eye pattern for the pulse of Figure 6.11 and uncoded bipolar 2ASK.

As can be seen from the pattern, eight time-limited signal elements, shown in Figure 6.13, are equivalently present. The equivalent encoder together with the mapper to the signal number are also shown. The binary 3-tuple $[c_2\ c_1\ c_0]$ that is mapped to the respective signal number is shown above the relevant figure parts.

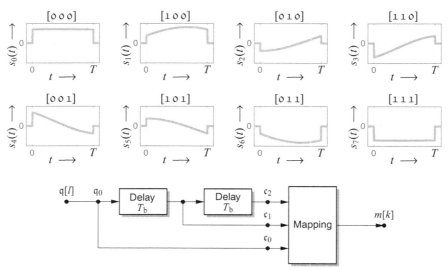

Figure 6.13 Equivalent time-limited signal elements for the pulse of Figure 6.11 and bipolar 2ASK, and equivalent encoder/mapper.

6.2 Signal Representation via Orthonormal Basis Functions

We have seen that in its general form, a digital modulation scheme is characterized by the set of signal elements

$$S = \{ s_0(t), s_1(t), \ldots, s_{M-1}(t) \} . \tag{6.23}$$

The only constraint on these signal elements is that they have to meet the temporal orthogonality condition (6.5). The main question to be answered is: What is the optimum receiver for such a modulation scheme?

To this end, for the design and analysis, it is more convenient not to operate directly on the signal elements, but to represent them in some space; it is advantageous to establish a geometrical view of the involved signals. This is first done for the signal elements and the transmit signal built thereof. Then, we consider the receive signal (Section 6.4).

6.2.1 Definition of a Signal Space

Basis Functions

Just like in a conventional (Euclidean) vector space, to represent the signal elements we need a *basis*. Thereby, a basis where the basis vectors are pairwise orthogonal and all have the same length, an *orthonormal basis*, is of particular interest. Since here functions have to be represented, basis functions are required, in particular, a set of *orthonormal basis functions*. Let D denote the dimensionality of the space, then the set of basis functions is

$$\mathcal{G} = \{ g_0(t), g_1(t), \ldots, g_{D-1}(t) \} . \tag{6.24}$$

For the basis functions we demand the following *twofold orthogonality condition*: The functions $g_j(t)$ have to be pairwise orthogonal (w.r.t. the function index j) and orthogonal with respect to temporal shifts by integer multiples of the symbol duration T. Moreover, all basis functions should have the same energy E_g. Mathematically, we demand

$$\int_{-\infty}^{\infty} g_j(t + \lambda T) \, g_l^*(t) \, dt = E_g \cdot \delta[\lambda] \cdot \delta[j - l] = \begin{cases} E_g, & \lambda = 0 \text{ and } j = l \\ 0, & \lambda \neq 0 \text{ or } j \neq l \end{cases} . \tag{6.25}$$

Note that the basis functions (as signal elements) usually represent signals in the ECB domain and thus, in general, they are complex-valued.

Example 6.9 Set of Orthonormal Basis Functions

Figure 6.14 shows a system of four orthonormal basis functions $g_j(t)$.

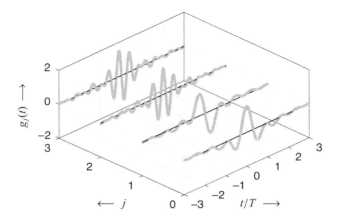

Figure 6.14 Example basis functions $g_j(t)$.

The twofold orthogonality condition (6.25) is visualized by plotting the *normalized crosscorrelations* $\varphi_{ji}(t) \stackrel{\text{def}}{=} \frac{1}{E_g}(g_j(t) * g_i^*(-t)) = \frac{1}{E_g} \int g_j(t+\tau) g_i^*(\tau) \, d\tau$. From (6.25), $\varphi_{ji}(t)$ has to be zero for all $t = \lambda T$, except for $\lambda = 0$ when $i = j$. This can be observed in the plots of Figure 6.15.

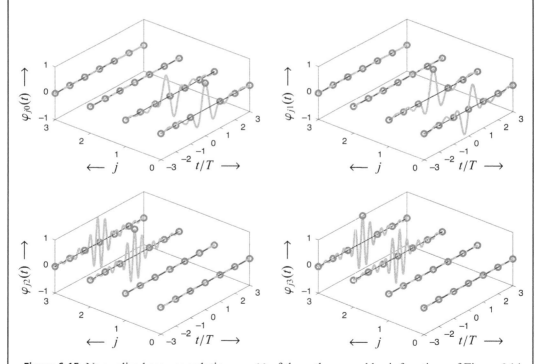

Figure 6.15 Normalized crosscorrelations $\varphi_{ji}(t)$ of the orthonormal basis functions of Figure 6.14.

Representation of the Signal Elements

In the Euclidean vector space, any point can be expressed as a linear combination of the basis vectors; the linear (scaling) factors are the coordinates with respect to the used basis. The same holds for the signal space. The signal elements are given by *linear combinations* of the basis functions, that is

$$s_i(t) = \sum_{j=0}^{D-1} s_{j,i}\, g_j(t)\,, \qquad i \in \{0,\, 1,\, \ldots,\, M-1\}\,, \tag{6.26}$$

where $s_{j,i}$ is the *linear factor* of the ith signal element with respect to the jth basis function. In summary, an *orthonormal expansion* of the signal elements is used:

- a *D-dimensional* space is spanned by the D orthonormal basis functions;
- the linear factors $s_{j,i}$ are the *coordinates* of the signal elements with respect to the basis.

This means that we have two views on the signal elements. On the one hand, they can be viewed as signals $s_i(t)$, that is, as functions over time; on the other hand, they can be viewed as a point $\mathbf{s}_i$ (column vector) in D-dimensional space:

$$s_i(t) = \sum_{j=0}^{D-1} s_{j,i}\, g_j(t) \quad \longleftrightarrow \quad \mathbf{s}_i = \begin{bmatrix} s_{0,i} \\ \vdots \\ s_{D-1,i} \end{bmatrix} \tag{6.27}$$

linear combination of basis functions $\longleftrightarrow$ *point in D-dimensional space*

Note that the choice of the set $\mathcal{G}$ of basis functions $g_j(t)$ is almost arbitrary. It only has to be guaranteed that all signal elements can be represented in this basis. Moreover, it is desired that D is as small as possible (in particular, not larger than the number M of signal elements). In the following, we always assume that D is as small as possible. Then, D is the *dimensionality of the modulation scheme per modulation step*. Note that "dimensionality" always has to be understood with respect to *complex* dimensions; the basis functions and the linear factors are allowed to be complex-valued.

A comparison of (6.26) and (6.4) reveals that digital PAM requires a single basis function $g_0(t) = g(t)$ per modulation interval. PAM has the (complex) dimension $D = 1$. The amplitude coefficients are nothing other than the coordinates, that is, $a^{(m)} = s_{0,m}$.

Representation of Signals – Sequences of Signal Elements

We now consider a block of N modulation intervals. The transmit signal within this temporal block is given by

$$s(t) = \sum_{k=0}^{N-1} s_{m[k]}(t - kT)\,. \tag{6.28}$$

It can equivalently be represented as the sequence of (temporally shifted) signal elements

$$\langle\, s_{m[0]}(t),\ s_{m[1]}(t-T),\ \ldots,\ s_{m[N-1]}(t-(N-1)T)\,\rangle \tag{6.29}$$

or, in view of (6.27), as the sequence of points $s_{m[k]}$.

Owing to the forced temporal orthogonality, the N time steps are mutually orthogonal. Each time step is characterized by D mutually orthogonal basis functions, that is, in a D-dimensional (complex-valued) space. The concatenation of the spaces per modulation interval gives the signal in an ND-dimensional space:

$$s(t) \longleftrightarrow [\,\mathbf{s}_{m[0]},\ \mathbf{s}_{m[1]},\ \ldots,\ \mathbf{s}_{m[N-1]}\,] = \begin{bmatrix} s_{0,m[0]} & \cdots & s_{0,m[N-1]} \\ \vdots & & \vdots \\ s_{D-1,m[0]} & \cdots & s_{D-1,m[N-1]} \end{bmatrix} \tag{6.30}$$

$$\hat{=} [\,s_{0,m[0]},\ \ldots,\ s_{D-1,m[0]},\ s_{0,m[1]},\ \ldots,\ s_{D-1,m[1]},\ \ldots,$$
$$s_{0,m[N-1]},\ \ldots,\ s_{D-1,m[N-1]}\,] \ .$$

We can state:

> *A signal over N modulation intervals is represented as a point in an ND-dimensional space.*

Box 6.2 Notes on the Notation of Vectors and Matrices

- Vectors can (completely equivalently) be written as either *column* or *row* vectors.
- We usually use *column vectors* when we deal with the *dimensions of the signal space* and we usually use *row vectors* when we deal with *time steps/sequences*.
- Hence, the best representation of N modulation intervals, each from a D-dimensional signal space, is a $D \times N$ matrix. However, a matrix can unambiguously be *reshaped* to a row/column vector and vice versa. We always use the notation that is more convenient in the respective situation.
- Finally, the squared Euclidean norm of a vector $\mathbf{x} = [x_0, \ldots, x_{N-1}]$ is defined as

$$||\mathbf{x}||^2 \stackrel{\text{def}}{=} \sum_{l=0}^{N-1} |x_l|^2 \ .$$

Since matrices $X = [x_{l,m}]_{\substack{l=0,\ldots,L-1 \\ m=0,\ldots,M-1}}$ can be reshaped to vectors, we have as the "squared Euclidean norm" of a matrix (called the squared *Frobenius norm*)

$$||\mathbf{x}||_{\text{F}}^2 \stackrel{\text{def}}{=} \sum_{l=0}^{L-1} \sum_{m=0}^{M-1} |x_{l,m}|^2 \ .$$

Generation of the Transmit Signal

The expansion of the signal elements in the signal space spanned by the basis functions also gives an interpretation of how the transmit signal in general digital transmission schemes can be generated. The general expression (6.3) gives the intuition of "copy and paste" – copies of the signal elements are joined to form the transmit signal.

However, the transmit signal can be written as

$$s(t) = \sum_{k=-\infty}^{\infty} s_{m[k]}(t - kT)$$

using (6.26)

$$= \sum_{k=-\infty}^{\infty} \sum_{j=0}^{D-1} s_{j,m[k]} \, g_j(t - kT)$$

exchanging the sums and defining $a_j[k] \stackrel{\text{def}}{=} s_{j,m[k]}$

$$= \sum_{j=0}^{D-1} \left(\sum_{k=-\infty}^{\infty} a_j[k] \, g_j(t - kT) \right)$$

$$= \sum_{j=0}^{D-1} s^{(j)}(t) \, . \tag{6.31}$$

This reveals that the transmit signal in general digital transmission schemes can be seen as the sum of D parallel PAM signals $s^{(j)}(t)$; the basis functions $g_j(t)$ are employed as the D different basic pulse shapes. The corresponding block diagram of the transmitter is depicted in Figure 6.16.

The different views on the generation of the transmit signal are visualized in the following examples.

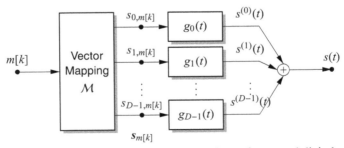

Figure 6.16 Block diagram of the transmitter of a general digital modulation scheme.

Example 6.10 Signal Elements, Basis Functions, and Transmit Signal

Let the basis functions $g_0(t)$ and $g_1(t)$ be given as plotted in Figure 6.17 and let the four signal elements $s_i(t)$ be given by

$$s_0(t) = g_0(t), \qquad s_2(t) = 0.5\, g_0(t) + 0.8\, g_1(t),$$
$$s_1(t) = -g_0(t), \qquad s_3(t) = -0.5\, g_0(t) - 0.8\, g_1(t).$$

The signal elements are characterized by points in the $D = 2$-dimensional signal space (bottom right).

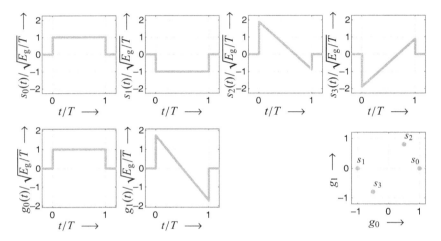

Figure 6.17 Example signal elements $s_i(t)$ and basis functions $g_j(t)$.

Assuming the sequence of signal numbers $m[k]$ given in Figure 6.18, the shown transmit signal is obtained. The decomposition of $s(t)$ into $D = 2$ PAM signals using the basis functions as basic pulse shapes is given in the bottom two plots. The sum of $s^{(0)}(t)$ and $s^{(1)}(t)$ gives $s(t)$.

The set of amplitude coefficients in the branches can be read off from the definition of the signal elements or, graphically, from the points in signal space. In the first dimension (basis function $g_0(t)$), the amplitudes (linear factors) -1, -0.5, $+0.5$, and $+1$ are used. In the second dimension (basis function $g_1(t)$), the amplitudes -0.8, 0, and $+0.8$ are active. Thus, in $s^{(0)}(t)$, the pulse $g_0(t)$ is used with four different amplitudes, and in $s^{(1)}(t)$, the pulse $g_1(t)$ is scaled by three different amplitudes.

6.2 Signal Representation via Orthonormal Basis Functions

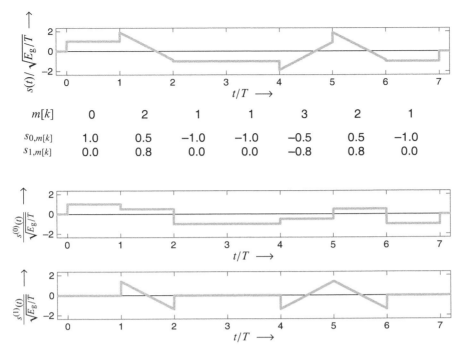

Figure 6.18 Transmit signal and decomposition into $D = 2$ PAM signals.

Calculation of the Linear Factors

Up to now, we have assumed that (given the basis functions $g_j(t)$) the signal elements are specified via selecting the linear factors. In other words, the linear factors immediately define the signal elements.

The converse procedure is also required: Given the signal elements $s_i(t)$, what are their linear factors with respect to a fixed basis?

To that end, assume that the signal elements $s_i(t)$, $i = 0, \ldots, M - 1$, and the basis functions $g_j(t)$, $j = 0, \ldots, D - 1$, are known. As in a Euclidean space, the linear factor (coordinate) $s_{j,i}$ is given by the *projection* of $s_i(t)$ onto the basis function $g_j(t)$. Since the energy of the basis functions is E_g, the projection has to be normalized to that energy (cf. Box 3.1; the coordinate of the projection of the vector b onto the vector a is given by $a^\mathsf{T} b^*/|a|$).

Mathematically, the linear factors $s_{j,i}$ of $s_i(t)$ with respect to $g_j(t)$ are calculated by

$$s_{j,i} = \frac{1}{E_g} \int_{-\infty}^{\infty} s_i(t) \, g_j^*(t) \, dt \, . \tag{6.32}$$

This can be seen from

$$
\frac{1}{E_g}\int_{-\infty}^{\infty} s_i(t)\, g_j^*(t)\, dt \stackrel{(6.26)}{=} \frac{1}{E_g}\int_{-\infty}^{\infty}\left(\sum_{l=0}^{D-1} s_{l,i}\, g_l(t)\right) g_j^*(t)\, dt
$$

$$
= \sum_{l=0}^{D-1} s_{l,i}\, \frac{1}{E_g}\int_{-\infty}^{\infty} g_l(t)\, g_j^*(t)\, dt
$$

$$
\stackrel{(6.25)}{=} \sum_{l=0}^{D-1} s_{l,i} \cdot \delta[l-j] = s_{j,i}\,. \qquad (6.33)
$$

6.2.2 Energy and Signal Space

The energy of signals can either be calculated from the time-domain signal or from the signal-space representation, that is, from the coordinates of the signal with respect to a given basis.

The *energy* of the signal elements can be calculated as follows:

$$
E_i \stackrel{\text{def}}{=} \int_{-\infty}^{\infty} |s_i(t)|^2\, dt
$$

$$
\stackrel{(6.26)}{=} \int_{-\infty}^{\infty}\left(\sum_{j=0}^{D-1} s_{j,i}\, g_j(t)\right)\left(\sum_{l=0}^{D-1} s_{l,i}^*\, g_l^*(t)\right) dt
$$

$$
= \sum_{j=0}^{D-1}\sum_{l=0}^{D-1} s_{j,i}\, s_{l,i}^* \underbrace{\int_{-\infty}^{\infty} g_j(t)\, g_l^*(t)\, dt}_{= E_g \delta[j-l]}
$$

$$
= E_g \sum_{j=0}^{D-1} |s_{j,i}|^2
$$

$$
= E_g\, \|\mathbf{s}_i\|^2\,. \qquad (6.34)
$$

Hence, we have the following simple result (general form of Parseval's theorem):

> *Energy in time domain corresponds to squared norm in signal space.*

Moreover, the *difference energy* between two signals

$$
s^{(1)}(t) = \sum_k s_{m^{(1)}[k]}(t-kT)\,,
$$
$$
s^{(2)}(t) = \sum_k s_{m^{(2)}[k]}(t-kT)\,,
$$

6.2 Signal Representation via Orthonormal Basis Functions

that is, the energy of the difference of these two signals, is given by

$$\int_{-\infty}^{\infty}\left|s^{(1)}(t)-s^{(2)}(t)\right|^2 \, dt = \int_{-\infty}^{\infty}\left|\sum_k \left(s_{m^{(1)}[k]}(t-kT) - s_{m^{(2)}[k]}(t-kT)\right)\right|^2 \, dt$$

$$\stackrel{(6.26)}{=} \int_{-\infty}^{\infty}\left|\sum_k \sum_{j=0}^{D-1} \left(s_{j,m^{(1)}[k]} - s_{j,m^{(2)}[k]}\right) g_j(t-kT)\right|^2 \, dt$$

$$= \int_{-\infty}^{\infty} \left(\sum_k \sum_{j=0}^{D-1} \left(s_{j,m^{(1)}[k]} - s_{j,m^{(2)}[k]}\right) g_j(t-kT)\right)$$

$$\cdot \left(\sum_\kappa \sum_{l=0}^{D-1} \left(s_{l,m^{(1)}[\kappa]} - s_{l,m^{(2)}[\kappa]}\right) g_l(t-\kappa T)\right)^* \, dt$$

$$= \sum_k \sum_{j=0}^{D-1} \sum_\kappa \sum_{l=0}^{D-1} \left(s_{j,m^{(1)}[k]} - s_{j,m^{(2)}[k]}\right) \left(s_{l,m^{(1)}[\kappa]} - s_{l,m^{(2)}[\kappa]}\right)^*$$

$$\cdot \underbrace{\int_{-\infty}^{\infty} g_j(t-kT) g_l^*(t-\kappa T) \, dt}_{= E_g \delta[k-\kappa]\delta[j-l]}$$

$$= E_g \sum_k \sum_{j=0}^{D-1} \left|s_{j,m^{(1)}[k]} - s_{j,m^{(2)}[k]}\right|^2$$

$$= E_g \sum_k \left|\left|s_{m^{(1)}[k]} - s_{m^{(2)}[k]}\right|\right|^2 \,. \tag{6.35}$$

Hence, we can state:

The energy of the difference of two signals is proportional to the squared Euclidean distance between the two points representing these signals in the signal space.

6.2.3 Remark on Normalization

Note that we deliberately define the energy of the basis functions to be E_g. In many textbooks, this is not done in this way. The consequences are subsequently discussed.

We assume that signals such as $s_i(t)$ and thus $s(t)$ (and $n(t)$) have the unit of the physical signal – here, as an example, we assume the unit V (volts); time t has the unit of s (seconds). As a consequence, the energy E_g (and the noise power spectral density N_0) has the unit of a (normalized) energy, specifically, $V^2/\text{Hz} = V^2 \text{s}$.

Table 6.1 collects the dimensions of the basis functions $g_j(t)$ and the consequences for the linear factors and the corresponding units of powers. In summary, when normalizing the basis functions to the energy E_g (with units $V^2 \text{s}$), all time-domain signals (including $g_j(t)$) have the unit V. The linear factors/coordinates are dimensionless. The pulse-shaping step (i.e., the transition from an abstract

Table 6.1 Normalization of the basis functions and consequences.

Definition used here	Often used in the literature
Definition of the basis functions	
$\int_{-\infty}^{\infty} g_j(t+\lambda T) \cdot g_l^*(t)\, dt$ $= \begin{cases} E_g & \lambda = 0 \text{ and } j = l \\ 0 & \text{else} \end{cases}$	$\int_{-\infty}^{\infty} g_j(t+\lambda T) \cdot g_l^*(t)\, dt$ $= \begin{cases} 1 & \lambda = 0 \text{ and } j = l \\ 0 & \text{else} \end{cases}$
$g_j(t)$ has the unit V	$g_j(t)$ has the unit $1/\sqrt{s}$
Calculation of the linear factors/projections	
$s_{j,i} = \dfrac{1}{E_g} \int_{-\infty}^{\infty} s_i(t) g_j^*(t)\, dt$	$s_{j,i} = \int_{-\infty}^{\infty} s_i(t) g_j^*(t)\, dt$
$s_{j,i}$ has no unit	$s_{j,i}$ has the unit $V\sqrt{s}$
Noise power/variance of the AWGN	
$n_j[k] = \dfrac{1}{E_g} \int_{-\infty}^{\infty} n(t) g_j^*(t-kT)\, dt$ $\sigma_n^2 = \mathrm{E}\{|n_j[k]|^2\}$	$n_j[k] = \int_{-\infty}^{\infty} n(t) g_j^*(t-kT)\, dt$ $\sigma_n^2 = \mathrm{E}\{|n_j[k]|^2\}$
σ_n^2 has no unit $\sigma_n^2 = N_0/E_g$	σ_n^2 has the unit $V^2 s$ $\sigma_n^2 = N_0$ power ↯ energy

discrete-time sequence to a physical signal) introduces the physical dimensions. Conversely, the matched filtering (due to the normalization to E_g) accepts a physical signal (unit V) and returns unitless detection samples.

In contrast, when employing the normalization given in the right column of the table (which is often done in the literature), severe confusions with respect to power (variance) versus energy/psd are induced. Here, since the functions have an unusual dimension, the power of signals is expressed in the unit of an energy, which may cause confusion. We prefer the (in our view clearer) notation in the left column of the table, although it is not as popular in the literature.

6.3 Gram–Schmidt Procedure

Given the basis functions $g_j(t)$, the signal elements $s_i(t)$ can be generated. However, typically, the situation is the other way round: the modulation scheme is specified, that is, the set of signal elements

$$\mathcal{S} \stackrel{\text{def}}{=} \{s_0(t), s_1(t), \ldots, s_{M-1}(t)\} \tag{6.36}$$

is given. The task is to find a basis, that is, a suitable set of D orthonormal basis functions $g_j(t)$, with D as small as possible, such that all signal elements can be represented as linear combinations thereof.

Note that such a basis is not unique; an infinite number of possible bases can be given. A possible and well-suited solution is to generate the orthonormal basis functions as linear combinations of the signal elements themselves. We demand

$$g_j(t) = \sum_{i=0}^{M-1} g_{i,j}\, s_i(t) \,, \tag{6.37}$$

where $g_{i,j}$ are the *orthogonalization factors*.

In a Euclidean space, this is a classical problem: the *Gram–Schmidt procedure*, well known in linear algebra, provides a solution. The main idea is to successively generate the basis vectors. Assume i vectors have already been processed and d basis vectors represent them. The next (not yet processed) vector is projected onto the d basis vectors and the projections are subtracted. In other words, all contributions that can be represented within the current basis are eliminated. If nothing remains, the vector can be represented using the current basis; no new basis vector is required. However, if the residual is nonzero, it is normalized and establishes a new dimension. The procedure is repeated until all vectors have been processed.

Example 6.11 Gram–Schmidt Procedure for Vectors in $\mathbb{R}^2$

Let the three vectors depicted on the left of Figure 6.19 be given.

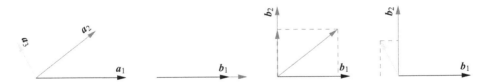

Figure 6.19 Gram–Schmidt procedure for vectors.

First, a_1 is normalized and establishes the first basis vector b_1. Next, a_2 is projected onto b_1 and this part is subtracted. The remainder (gray dashed) is normalized to establish the second basis vector b_2. When subtracting the projections of a_3 onto b_1 and b_2, nothing remains. The two basis vectors are sufficient to represent all vectors a_i.

Exactly the same method can be used in case of signal elements and basis functions. Subsequently, a pseudo-code representation of the *Gram–Schmidt procedure* for signal elements is given as follows.

Input: signal elements $s_i(t)$, $i = 0, \ldots, M - 1$
Output: basis functions $g_j(t)$, $j = 0, \ldots, D - 1$

Init:
1. first basis function $g_0(t)$ = normalized first signal element
   $D = 1$

Recursion:
2. for $i = 1, \ldots, M - 1$
3.   calculate the D projections of signal element $s_i(t)$ onto the already known basis functions
   remainder = signal element $s_i(t)$ minus sum of projections
4.   if remainder $\neq$ zero
5.   new basis function $g_D(t)$ = normalized remainder
   $D = D + 1$
       endif
   endfor

For a compact representation, a vector/matrix notation is convenient. In particular, the calculation of the signal elements from the basis functions and vice versa can be written compactly. We define:

- vector of basis functions $\quad\quad \boldsymbol{g}(t) \stackrel{\text{def}}{=} [\, g_0(t),\, g_1(t),\, \ldots,\, g_{D-1}(t) \,]$
  (row vector)

- vector of signal elements $\quad\quad \boldsymbol{s}(t) \stackrel{\text{def}}{=} [\, s_0(t),\, s_1(t),\, \ldots,\, s_{M-1}(t) \,]$
  (row vector)

- matrix of linear factors $\quad\quad \boldsymbol{S} \stackrel{\text{def}}{=} [\, s_{j,i} \,] = [\, \boldsymbol{s}_0,\, \boldsymbol{s}_1,\, \ldots,\, \boldsymbol{s}_{M-1} \,]$
  ($D \times M$ matrix)

- matrix of orthogonalization factors $\quad \boldsymbol{G} \stackrel{\text{def}}{=} [\, g_{i,j} \,]$,
  ($M \times D$ matrix)

Then, the following relations hold ($\boldsymbol{I}$ denotes the identity matrix):

- basis functions to signal elements $\quad \boldsymbol{s}(t) = \boldsymbol{g}(t) \cdot \boldsymbol{S}$
- signal elements to basis functions $\quad \boldsymbol{g}(t) = \boldsymbol{s}(t) \cdot \boldsymbol{G}$
- since $D \leq M$, we have $\quad\quad \boldsymbol{S} \cdot \boldsymbol{G} = \boldsymbol{I}_{D \times D}$.

The steps in the above pseudo-code representation of the Gram–Schmidt procedure are now given in mathematical detail using the above definitions. Thereby, the matrices $\boldsymbol{S}$ and $\boldsymbol{G}$ of the linear and orthogonalization factors are calculated as well. Note that $E_i = \int_{-\infty}^{\infty} |s_i(t)|^2 \, dt$ is the energy of the signal element $s_i(t)$ (cf. (6.5)).

6.3 Gram–Schmidt Procedure

Step 1
In the initialization, $D = 1$ is set and

$$g_0(t) = \sqrt{\frac{E_g}{E_0}} s_0(t) , \qquad (6.38)$$

that is

$$g_{00} = \sqrt{\frac{E_g}{E_0}} , \qquad s_{00} = \frac{1}{g_{00}} .$$

Step 2
In the recursion, at iteration i, the signal element $s_i(t)$ is under consideration. Up to this point, the following vectors/matrices have been built:

$$[s_0(t), \ldots, s_{i-1}(t)] = [g_0(t), \ldots, g_{D-1}(t)] \cdot \mathbf{S} , \qquad (6.39)$$

$$[g_0(t), \ldots, g_{D-1}(t)] = [s_0(t), \ldots, s_{i-1}(t)] \cdot \mathbf{G} , \qquad (6.40)$$

where $\mathbf{S}$ is a $D \times i$ matrix and $\mathbf{G}$ is an $i \times D$ matrix.

Step 3
The projections of $s_i(t)$ onto the already known basis functions $g_j(t)$ and thus the linear factors $s_{j,i}$ are calculated by

$$s_{j,i} = \frac{1}{E_g} \int_{-\infty}^{\infty} s_i(t) g_j^*(t) \, \mathrm{d}t , \qquad j = 0, 1, \ldots, D-1 . \qquad (6.41)$$

Then, the matrices $\mathbf{S}$ and $\mathbf{G}$ are updated. In $\mathbf{S}$, a new column is appended; in $\mathbf{G}$, a new row. In detail:

$$\mathbf{S} = \left[\mathbf{S} \; \middle| \; \begin{array}{c} s_{0,i} \\ \vdots \\ s_{D-1,i} \end{array} \right] , \qquad \mathbf{G} = \left[\begin{array}{c} \mathbf{G} \\ \hline 0 \; \cdots \; 0 \end{array} \right] . \qquad (6.42)$$

The part that can be expressed through the already known basis functions is subtracted. The orthogonal remainder calculates as

$$\Delta s_i(t) \stackrel{\mathrm{def}}{=} s_i(t) - \sum_{j=0}^{D-1} s_{j,i} \, g_j(t) , \qquad (6.43)$$

with energy

$$\Delta E_i = E_i - E_g \sum_{j=0}^{D-1} |s_{j,i}|^2 , \qquad (6.44)$$

since the basis functions $g_j(t)$ are orthonormal and $\Delta s_i(t)$ is orthogonal to the basis functions.

Step 4
If $\Delta E_i = 0$, the signal element $s_i(t)$ can be represented completely using the already known basis functions; no new one is required.

Step 5
If $\Delta E_i \neq 0$, the signal element $s_i(t)$ cannot be represented completely using the already known basis functions; an additional basis function is required. This is given by

$$g_D(t) \stackrel{\text{def}}{=} \sqrt{\frac{E_g}{\Delta E_i}} \Delta s_i(t) , \qquad (6.45)$$

that is

$$g_{i,D} = \sqrt{\frac{E_g}{\Delta E_i}} , \qquad s_{D,i} = \frac{1}{g_{i,D}} .$$

Then, the matrices S and G are updated. In S, a new row is appended; in G, a new column. In detail:

$$S = \left[\begin{array}{c} S \\ \hline 0 \cdots 0 \; s_{D,i} \end{array} \right] , \qquad G = \left[\begin{array}{c|c} G & \begin{array}{c} g_{0,D} \\ \vdots \\ g_{i-1,D} \\ g_{i,D} \end{array} \end{array} \right] . \qquad (6.46)$$

The elements $g_{0,D}, \ldots, g_{i-1,D}$ for the update of the matrix G can be calculated by considering $g_D(t) = \sum_{l=0}^{i} g_{l,D} \, s_l(t)$, which gives on the one hand

$$g_D(t) - g_{i,D} \, s_i(t) = \sum_{l=0}^{i-1} g_{l,D} \, s_l(t) . \qquad (6.47)$$

On the other hand

$$\begin{aligned} g_D(t) - g_{i,D} \, s_i(t) &= g_{i,D} \big(\Delta s_i(t) - s_i(t) \big) \\ &= g_{i,D} \Big(- \sum_{j=0}^{D-1} s_{j,i} \, g_j(t) \Big) \\ &= -g_{i,D} \sum_{j=0}^{D-1} s_{j,i} \sum_{l=0}^{i-1} g_{l,j} \, s_l(t) \\ &= \sum_{l=0}^{i-1} \Big(-g_{i,D} \sum_{j=0}^{D-1} s_{j,i} \, g_{l,j} \Big) s_l(t) . \end{aligned} \qquad (6.48)$$

Hence, a comparison of the coefficients results in

$$g_{l,D} = -g_{i,D} \sum_{j=0}^{D-1} s_{j,i} \, g_{l,j} \qquad l = 0, \ldots, i-1 . \qquad (6.49)$$

Finally, $D = D + 1$ is set and the algorithm continues with the next index i in Step 2 as long as $i < M$, otherwise the algorithm stops. The dimensionality of the signal space is given by the final value of D.

For completeness, we make the following remarks.

- The basis-function system $g(t)$ obtained from the Gram–Schmidt process depends on the ordering of the signal elements $s_i(t)$. A different ordering of the signal elements will yield different basis functions. For our purpose of deriving (and implementing) the optimum receiver, all bases are equivalent and in particular will finally lead to the same performance.
- Employing the Gram–Schmidt procedure, the matrices S and G are *upper triangular*.
- The energies and cross energies of the signal elements (cf. (6.5)) may be combined to the *cross-energy matrix*

$$E = [\, E_{ij}\,]_{\substack{i=0,\ldots,M-1 \\ j=0,\ldots,M-1}} = \int_{-\infty}^{+\infty} s^\mathsf{T}(t)\, s^*(t)\, \mathrm{d}t \,, \tag{6.50}$$

which can be written via the matrix of linear factors as

$$\begin{aligned}
E &= \int_{-\infty}^{+\infty} S^\mathsf{T} g^\mathsf{T}(t)\, g^*(t)\, S^*\, \mathrm{d}t \\
&= S^\mathsf{T} \int_{-\infty}^{+\infty} g^\mathsf{T}(t)\, g^*(t)\, \mathrm{d}t\, S^* \\
&= S^\mathsf{T} E_\mathrm{g}\, I\, S^* = E_\mathrm{g}\, S^\mathsf{T} S^* \,.
\end{aligned} \tag{6.51}$$

The matrix of linear factors S, which is upper triangular, can thus be obtained from the cross-energy matrix via a *Cholesky decomposition*. Moreover, as S is a $D \times M$ matrix, $D \leq M$, and E is an $M \times M$ matrix, we have

$$D = \mathrm{rank}(E) \,. \tag{6.52}$$

Example 6.12

The generation of the signal elements from the basis functions and vice versa (using the Gram–Schmidt procedure) is summarized in Figure 6.20. Starting at the bottom of the figure, two basis functions are given. They (as required) fulfill the twofold orthogonality condition; the energy is chosen to be $E_\mathrm{g} = 1\ \mathrm{V}^2 T$. Via the indicated linear combinations, three signal elements are generated (shown at the top of the figure). They are temporally orthogonal (as they are time-limited) but not all of them are orthogonal to each other. Via a Gram–Schmidt procedure (and using the given ordering of the signal elements), the basis functions can be reconstructed. The matrices S of linear factors and G of orthogonalization factors are specified. In the middle, the representation of the signals as points in the (here two-dimensional) signal space is shown.

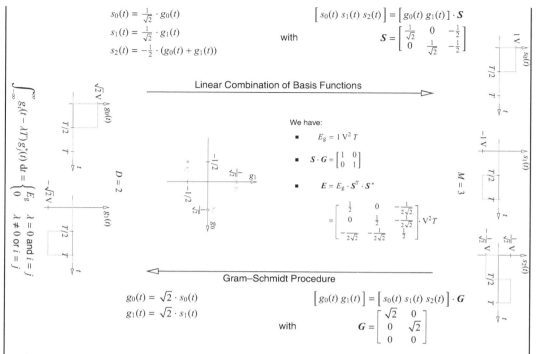

Figure 6.20 Visualization of the generation of the signal elements from the basis functions and vice versa.

6.4 Representation of the Receive Signal in Signal Space

After knowing how to represent the transmit signal in a signal space, we turn to the receive signal. To that end, we assume digital transmission over the AWGN channel, where the receive signal is given by

$$r(t) = s(t) + n(t) = \sum_k s_{m[k]}(t - kT) + n(t) . \qquad (6.53)$$

Thereby, $n(t)$ is complex Gaussian noise with noise power spectral density N_0 ($N_0/2$ per quadrature component, cf. Chapter 2).

The representation of the receive signal in the signal space follows the concepts for vectors known from linear algebra. In order to represent a vector with respect to a given basis, the vector is projected onto the basis vectors. The projection coefficients give the coordinates of the vector in the basis of interest; the tuple of coordinates specifies the vector completely.

The same concept applies to signals. The receive signal is *projected onto the basis functions*. Since the transmit signal is rasterized in symbol intervals (duration T) and the signals in each symbol interval are represented in a D-dimensional space, in each time interval k, D projections are required. This means that we have to calculate the projections of $r(t)$ onto the functions

$$\begin{matrix} g_0(t - kT) \\ \vdots \\ g_{D-1}(t - kT) \end{matrix}, \quad k \in \mathbb{Z}. \quad (6.54)$$

6.4.1 Projections and Receiver Structure

The projections, which result in the *detection variables* $d_j[k]$ (w.r.t. the basis function j at modulation interval k) calculate as

$$\begin{aligned} d_j[k] &= \frac{1}{E_g} \int_{-\infty}^{\infty} r(t) g_j^*(t - kT) \, dt \\ &= \int_{-\infty}^{\infty} r(t) \frac{1}{E_g} g_j^*(t - \tau) \, dt \Big|_{\tau=kT} \\ &= r(t) * \frac{1}{E_g} g_j^*(-t) \Big|_{t=kT}. \end{aligned} \quad (6.55)$$

Hence, $d_j[k]$ is given by filtering the receive signal $r(t)$ with a filter with impulse response $\frac{1}{E_g} g_j^*(-t)$ and sampling the output at kT. This is nothing other than the *matched filter* for the jth basis function $g_j(t)$. Since D basis functions are required to characterize the signal, the receive signal $r(t)$ has to be filtered with D matched filters (w.r.t. the basis functions $g_j(t)$, $j = 0, \ldots, D - 1$) in parallel; a *bank of matched filters*.

The detection variables of one time step can be combined into the *detection vector*

$$\boldsymbol{d}[k] \stackrel{\text{def}}{=} \begin{bmatrix} d_0[k] \\ d_1[k] \\ \vdots \\ d_{D-1}[k] \end{bmatrix}, \quad (6.56)$$

which represents the receive signal in the signal space.

The block diagram of the vector receiver, the bank of matched filters, is depicted in Figure 6.21. Note that the matched filters here are depicted as if they were non-causal (cf. the discussion in Section 3.4). The delay required for a causal implementation is not shown throughout this chapter.

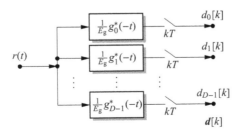

Figure 6.21 Block diagram of the vector receiver/matched-filter bank.

6.4.2 End-to-End Model

Taking the block diagram of the transmitter, as depicted in Figure 6.16, into account, the end-to-model can be established. This is shown in Figure 6.22.

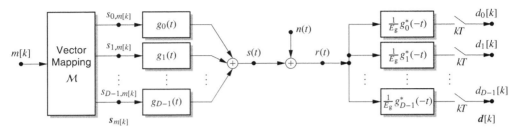

Figure 6.22 Block diagram of the end-to-end model for signal space representation.

The *useful signal* (noise-free, i.e., $r(t) = s(t)$) in the jth receiver branch at time step k calculates as

$$\tilde{d}_j[k] = \frac{1}{E_g} \int_{-\infty}^{\infty} s(t)\, g_j^*(t - kT)\, \mathrm{d}t$$

using (6.3)

$$= \frac{1}{E_g} \int_{-\infty}^{\infty} \sum_{\kappa} s_{m[\kappa]}(t - \kappa T)\, g_j^*(t - kT)\, \mathrm{d}t$$

using (6.26)

$$= \frac{1}{E_g} \int_{-\infty}^{\infty} \sum_{\kappa} \sum_{l=0}^{D-1} s_{l,m[\kappa]} g_l(t - \kappa T)\, g_j^*(t - kT)\, \mathrm{d}t$$

$$= \frac{1}{E_g} \sum_{\kappa} \sum_{l=0}^{D-1} s_{l,m[\kappa]} \underbrace{\int_{-\infty}^{\infty} g_l(t - \kappa T) g_j^*(t - kT)\, \mathrm{d}t}_{= E_g \delta[k-\kappa]\delta[l-j]}$$

taking (6.25) into account

$$= s_{j,m[k]}\,. \tag{6.57}$$

This shows that after matched filtering, the amplitude coefficient sent in branch j at time instant k is visible free of interference from other time instants or other branches. Thus:

> *The twofold orthogonality condition guarantees that in each branch and at each time instant the linear factor can be detected free of intersymbol interference.*

In order to derive the noise statistics, we first calculate the *discrete-time noise* $n_j[k]$ in the jth branch at time step k. By projection of white noise (i.e., $r(t) = n(t)$), it is given as

6.4 Representation of the Receive Signal in Signal Space

$$n_j[k] \stackrel{\text{def}}{=} \frac{1}{E_g} \int_{-\infty}^{\infty} n(t) g_j^*(t - kT) \, dt \, . \tag{6.58}$$

Note that the noise samples $n_j[k]$ do not fully capture the continuous-time white Gaussian noise $n(t)$. The useful signal is contained in the D dimensions spanned by the basis functions and thus completely picked up by the vector receiver. However, white noise spans an infinite number of dimensions. Only the dimensions in which components of the useful signal exist are of interest and relevant for making decisions. This holds by the theorem of *irrelevant data* from information theory: a variable that is statistically independent from the transmitted message is irrelevant for its estimation.

If the noise samples $n_j[k]$ are fed to the transmitter bank of basic pulse shapes in Figure 6.16, the reconstructed continuous-time noise is not $n(t)$ but "only" the *relevant part* $\check{n}(t)$, that is, the part contained in the D signal dimensions.

From the noise samples $n_j[k]$, their crosscorrelations (over time, i.e., time instant k vs. $k + \kappa$, and over the dimensions, i.e., index j vs. l) can be calculated as

$$\begin{aligned}
\phi_{n_j n_l}[\kappa] &\stackrel{\text{def}}{=} \mathrm{E}\{n_j[k + \kappa] n_l^*[k]\} \\
&= \mathrm{E}\Big\{\frac{1}{E_g} \int_{-\infty}^{\infty} n(t) g_j^*(t - (k+\kappa)T) \, dt \cdot \frac{1}{E_g} \int_{-\infty}^{\infty} n^*(\tau) g_l(\tau - kT) \, d\tau\Big\} \\
&= \frac{1}{E_g^2} \int_{-\infty}^{\infty} \int_{-\infty}^{\infty} \underbrace{\mathrm{E}\{n(t) n^*(\tau)\}}_{\stackrel{\text{WGN}}{=} N_0 \delta(t-\tau)} g_j^*(t - (k+\kappa)T) g_l(\tau - kT) \, dt \, d\tau \\
&= \frac{N_0}{E_g^2} \underbrace{\int_{-\infty}^{\infty} g_j^*(t - (k+\kappa)T) g_l(t - kT) \, dt}_{= E_g \delta[\kappa] \delta[j-l]} \\
&= \frac{N_0}{E_g} \delta[\kappa] \delta[j - l] \, .
\end{aligned} \tag{6.59}$$

Since $\phi_{n_j n_l}[\kappa] = 0$ as either $\kappa \neq 0$ or $j \neq l$, the noise samples are uncorrelated over the time steps and uncorrelated over the parallel branches, or:

The twofold orthogonality condition for the basis functions guarantees that the discrete-time noise is white with respect to time and over the dimensions.

Since $n(t)$ is Gaussian and filtering preserves the Gaussian property, the noise vector

$$\mathbf{n}[k] \stackrel{\text{def}}{=} \begin{bmatrix} n_0[k] \\ n_1[k] \\ \vdots \\ n_{D-1}[k] \end{bmatrix} \tag{6.60}$$

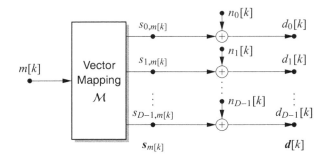

Figure 6.23 Equivalent vector channel model of general digital transmission over the AWGN channel.

is drawn from a *D-dimensional rotationally invariant Gaussian distribution* with variance N_0/E_g per (complex) dimension.

The above derived properties of the useful signal together with the statistics of the noise give rise to an *equivalent vector channel model*, depicted in Figure 6.23. This simply consists of D parallel, independent discrete-time AWGN channels; one channel for each dimension of the signal space representation.

In other words, since the detection vector is given by

$$d[k] = s_{m[k]} + n[k] \qquad (6.61)$$

digital communications can be described (per modulation interval T) as sending a *point* $s_{m[k]}$ in D-dimensional space and receiving a noisy version $d[k]$ thereof. The noise $n[k]$ is rotationally invariant Gaussian; each dimension is individually disturbed by (complex) Gaussian noise with variance $\sigma_n^2 = N_0/E_g$. In conventional PAM, only a single dimension is used. Also in this sense, any digital communication scheme may again be interpreted as a D-dimensional generalization of PAM.

6.4.3 Signal Space Representation with Reduced Dimensionality

For some modulation schemes, especially when signal elements are used that are limited to the modulation interval T, the dimensionality D of the signal space can be rather high. Thus, a large number of parallel matched filters would be required.

Often, resorting to an incomplete representation of the signal elements at the receiver side, the number of processed dimensions can be reduced while tolerating only a small loss in power efficiency. Remember that by the D dimensions only the relevant components of the receiver input signal are represented (i.e., those which are dependent on the transmit signal and not only on noise). If a system of basis functions can be found for which some dimensions contain only a small part of the energy of any signal element, these dimensions may be neglected at a small loss in power efficiency while saving a lot in complexity.[3]

Systems of basis functions with reduced dimensionality can be designed by various methods, two of them are addressed briefly here.

---

[3] Note that the same method is often applied in source coding for data compaction. Less relevant parameters at the output of a transform encoder are neglected (e.g., in standard picture and video encoding).

Sorted Gram–Schmidt Procedure

In the next step i of the Gram–Schmidt procedure (cf. page 205), not the next signal element according to its number is considered but the signal element is selected for processing for which the residual energy ΔE_i in (6.44) is maximum. Thus, the first D' dimensions represent a high portion of the energy for all signal elements, as the sequence of residual energies ΔE_i is enforced to be monotonously decreasing. To minimize the maximum unprocessed energy $\Delta E_{D'}$, this sorted Gram–Schmidt procedure (processing with pivoting) may be repeated M times, starting with each signal element. This procedure has only to be done once for the receiver design but it saves a lot of complexity at the receiver in operation.

Spectral Sampling

When all signal elements are limited to the modulation interval T, sampling may be applied in the spectral domain (cf. Fourier series). This corresponds to the use of D' basis functions

$$g'_i(t) = \sqrt{\frac{E_g}{T}} \, e^{j2\pi f_i t}, \quad 0 \le t < T, \; i = 0, 1, \ldots, D'-1, \tag{6.62}$$

with
$$f_i = \begin{cases} 0, \pm\Delta f, \pm 2\Delta f, \ldots & D' \text{ odd} \\ \pm\tfrac{1}{2}\Delta f, \pm\tfrac{3}{2}\Delta f, \ldots & D' \text{ even} \end{cases}. \tag{6.63}$$

Given a restricted dimensionality D', the spacing parameter Δf has to be optimized for minimum loss in power efficiency. Usually, non-orthogonal functions $g'_i(t)$ result. A Gram–Schmidt procedure may be applied on the set of functions in order to find an equivalent set of orthogonal basis functions $g_i(t)$. Only $D' = 2$ or $D' = 3$ dimensions are sufficient in many applications (for an example, see Section 7.4). A further important reason for the implementation of spectral sampling in practice is the existence of efficient combined synchronizers for carrier phase and symbol timing.

6.5 Maximum-Likelihood Detection and Decoding

We are now ready to derive the optimum receiver for the general form of digital communications. Thereby, two variants will be studied.

6.5.1 Vector Demodulator

As derived in the last section, using the *bank of matched filters with respect to the orthonormal basis functions* $g_j(t)$, the end-to-end discrete-time model for transmission over the continuous-time AWGN channel is given by a set of parallel AWGN channels. The detection vector calculates as

$$d[k] = s_{m[k]} + n[k]. \tag{6.64}$$

In *coded* digital transmission, the sequence of code symbols c[ν] (cf. Figure 6.6) is drawn from a code **C**, that is

$$\langle c[\nu] \rangle \in \mathbf{C}, \tag{6.65}$$

which implies (with mapping $\mathcal{M}$ from Figure 6.6)

$$\langle m[k] \rangle = \mathcal{M}\{\langle c[\nu] \rangle\} \in \mathcal{M}\{\mathbf{C}\}. \tag{6.66}$$

In coded transmission, the sequence of signal numbers $m[k]$ contains dependencies stipulated by the code. Assuming (as always) that the sequence of source bits is i.i.d. with uniformly distributed symbols $q[l]$, the codewords from the code **C** are used with equal probability. In turn, all allowed sequences $\langle m[k] \rangle$ will be a-priori equiprobable. Then, the *optimum decision rule* is given by the *maximum-likelihood (ML) strategy*. However, due to the code constraints, no symbol-by-symbol decision can be taken but entire sequences have to be detected, that is, a *sequence detection* has to be performed.

Maximum-Likelihood Sequence Detection (MLSD)
Decide in favor of the sequence of signal numbers $\langle m[k] \rangle$, allowed by the code constraints, for which the observation $\langle \mathbf{d}[k] \rangle$ has the largest probability (density):

$$\langle \hat{m}[k] \rangle = \operatorname*{argmax}_{\forall \langle m[k] \rangle \in \mathcal{M}\{\mathbf{C}\}} f_{\langle \mathbf{d} \rangle}(\langle \mathbf{d} \rangle \mid \langle \mathbf{s}_m \rangle).$$

As we have already studied in Chapter 3 for the AWGN channel, the *maximization of the pdf* is equivalent to the *minimization of the squared Euclidean distance*. Assuming a (block) code of length N, this leads to

$$\begin{aligned}
\langle \hat{m}[k] \rangle &= \operatorname*{argmax}_{\forall \langle m[k] \rangle \in \mathcal{M}\{\mathbf{C}\}} f_{\langle \mathbf{d} \rangle}(\langle \mathbf{d} \rangle \mid \langle \mathbf{s}_m \rangle) \\
&= \operatorname*{argmin}_{\forall \langle m[k] \rangle \in \mathcal{M}\{\mathbf{C}\}} \left\| [\mathbf{d}[0]\,\mathbf{d}[1]\,\ldots\,\mathbf{d}[N-1]] - [\mathbf{s}_{m[0]}\,\mathbf{s}_{m[1]}\,\ldots\,\mathbf{s}_{m[N-1]}] \right\|_F^2 \\
&= \operatorname*{argmin}_{\forall \langle m[k] \rangle \in \mathcal{M}\{\mathbf{C}\}} \sum_{k=0}^{N-1}\sum_{j=0}^{D-1} |d_j[k] - s_{j,m[k]}|^2 \\
&= \operatorname*{argmin}_{\forall \langle m[k] \rangle \in \mathcal{M}\{\mathbf{C}\}} \sum_{k=0}^{N-1} \lambda_m[k].
\end{aligned} \tag{6.67}$$

In the last step, we have defined the *metric increment* at time instant k for signal-number hypothesis m

$$\lambda_m[k] \stackrel{\text{def}}{=} \sum_{j=0}^{D-1} |d_j[k] - s_{j,m}|^2. \tag{6.68}$$

In case of uncoded transmission, this reduces to symbol-by-symbol (individual per time step k; the maximization has not to be carried out over the codewords/sequences but only over the signal number) *maximum-likelihood detection (MLD)*

$$\begin{aligned}
\hat{m}[k] &= \underset{\forall m \in \{0,1,\ldots,M-1\}}{\mathrm{argmax}} \; f_{\boldsymbol{d}}(\boldsymbol{d} \mid \boldsymbol{s}_m) \\
&= \underset{\forall m \in \{0,1,\ldots,M-1\}}{\mathrm{argmin}} \; \|\boldsymbol{d}[k] - \boldsymbol{s}_m\|^2 \\
&= \underset{\forall m \in \{0,1,\ldots,M-1\}}{\mathrm{argmin}} \; \sum_{j=0}^{D-1} |d_j[k] - s_{j,m}|^2 \\
&= \underset{\forall m \in \{0,1,\ldots,M-1\}}{\mathrm{argmin}} \; \lambda_m[k] \; .
\end{aligned} \qquad (6.69)$$

Looking at (6.67) and (6.69), the following remarks are advisable.

- *Maximum-likelihood sequence detection (MLSD)* is nothing other than *maximum-likelihood detection (MLD)* in an ND-dimensional space. The N time steps have to be treated jointly, increasing the dimensionality of the decision problem. However, the criterion (i.e., measuring squared Euclidean distances) is always the same.
- As squared Euclidean distances are measured, the primary quantity for *power efficiency* of the digital modulation scheme is the *minimum squared Euclidean distance* between the signal points in an ND-dimensional space.
- The primary task of the code in a coded transmission scheme is thus to allow only arrangements of signal points in the ND-dimensional space for which the minimum distance is maximized.

In the subsequent chapters (Chapter 7 on frequency-shift keying and Chapter 8 on transmission over dispersive channels), we will study in detail how MLSD is carried out for the particular situation.

Minimum Squared Euclidean Distance

In analogy to (6.67), where the squared Euclidean distance between receive signal and the hypothesis on the transmitted signal is evaluated, the squared Euclidean distance between signals corresponding to different sequences $\langle m[k] \rangle$ and $\langle \mu[k] \rangle$ of signal numbers is given by

$$\sum_k \sum_{j=0}^{D-1} |s_{j,m[k]} - s_{j,\mu[k]}|^2 \; .$$

Searching the minimum over all pairs of sequences gives the *minimum squared Euclidean distance*. In particular

$$d_{E,\min}^2 \stackrel{\mathrm{def}}{=} \min_{\substack{\forall \langle m[k] \rangle, \langle \mu[k] \rangle \in \mathcal{M}\{\mathbf{C}\} \\ \langle m[k] \rangle \neq \langle \mu[k] \rangle}} \sum_k \sum_{j=0}^{D-1} |s_{j,m[k]} - s_{j,\mu[k]}|^2 \; , \qquad (6.70)$$

and obeying (6.35) that the squared Euclidean distance between two points in the signal space (i.e., the measure of immunity of this pair of messages against noise) is proportional to the energy of the difference of the corresponding signals, we arrive at

$$d_{E,\min}^2 = \frac{1}{E_g} \min_{\substack{\forall s^{(m)}(t), s^{(\mu)}(t) \\ s^{(m)}(t) \neq s^{(\mu)}(t)}} \int_{-\infty}^{\infty} |s^{(m)}(t) - s^{(\mu)}(t)|^2 \, dt \,. \tag{6.71}$$

Finally, normalizing this squared Euclidean distance with respect to two times the energy per transmitted bit of information, E_b, that is

$$d_{\min}^2 \stackrel{\text{def}}{=} \frac{d_{E,\min}^2 E_g}{2 E_b} \,, \tag{6.72}$$

we arrive at the following definition.

Definition 6.4 *Normalized Minimum Squared Euclidean Distance*

The *normalized minimum squared Euclidean distance* of a digital modulation scheme is defined as

$$d_{\min}^2 = \frac{1}{2 E_b} \cdot \min_{\substack{\forall s^{(m)}(t), s^{(\mu)}(t) \\ s^{(m)}(t) \neq s^{(\mu)}(t)}} \int_{-\infty}^{\infty} |s^{(m)}(t) - s^{(\mu)}(t)|^2 \, dt \,. \tag{6.73}$$

Note that the normalization with respect to E_b is due to historical reasons – it is caused by the definition of $N_0/$ as a two-sided psd of the physical, real-valued (RF) AWGN.

Symbol Error Probability

In the same way as shown in Section 3.6, by using a *union-bound technique*, an *upper bound* on the *symbol error ratio* of a general digital transmission scheme can be given:

$$\text{SER} \leq N_{\min} \, Q\left(\sqrt{d_{\min}^2 \frac{E_b}{N_0}} \right), \tag{6.74}$$

$N_{\min}$: average number of nearest-neighbor signal points
E_b/N_0: energy per bit to noise power spectral density ratio
$d_{\min}^2$: normalized squared Euclidean distance of the scheme.

MLSD for Time-Varying Channels

In a number of applications, in particular in mobile communications, the channel is not time-invariant but the statistical properties may change over time. The simplest model for a time-variant channel is the AWGN channel model where a time-dependent scaling (fading) of the transmit signal is present. Denoting the current *fading coefficient* by $h[k] \in \mathbb{C}$, we have for the scalar ($D = 1$) case

$$d'[k] = h[k]\,a[k] + n[k]\,, \tag{6.75}$$

where $n[k]$ is Gaussian noise with (time-independent) variance σ_n^2. Instead of treating $d'[k]$, the fading may be compensated for at the receiver by scaling the received signal by $1/h[k]$ – assuming that *channel state information* (the fading coefficient) is available at the receiver. Then, a unit-gain AWGN channel with time-dependent noise variance

$$\sigma_n^2[k] \stackrel{\text{def}}{=} \sigma_n^2 \frac{1}{|h[k]|^2} \tag{6.76}$$

is present. This is shown in Figure 6.24.

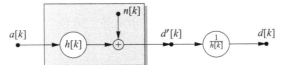

Figure 6.24 Time-variant AWGN channel with receiver-side rescaling.

In optimal decoding, this time dependency has to be taken into account. Direct calculations show that maximum-likelihood sequence detection with variable channel gain is conducted by

$$\langle \hat{m}[k] \rangle = \operatorname*{argmin}_{\forall \langle m[k] \rangle \in \mathcal{M}\{\mathbf{C}\}} \sum_k |h[k]|^2 \sum_{j=0}^{D-1} |d_j[k] - s_{j,m[k]}|^2\,. \tag{6.77}$$

The squared Euclidean distances of each time step are weighted by $|h[k]|^2$. If $|h[k]|$ is very small (extinction of the transmitted signal), the respective time step (almost) does not contribute to the sum (the decision metric). Conversely, if $|h[k]|$ is large (i.e., if good channel conditions are present and the receive signal allows a reliable decision), this contribution has to be weighted strongly. As in real life: base your decision on those, who you trust most.

6.5.2 Correlation Receiver

Up to now, we have had three different but equivalent views on *ML decoding* on the AWGN channel.

(A) Starting point is the main ML principle based on the conditional pdf (i.e., looking at probability measures)

$$\hat{s}(t) = \operatorname*{argmax}_{\forall s(t)} f_{r(t)}(r(t) \mid s(t))\,. \tag{6.78}$$

(B) On the AWGN channel, maximizing the pdf translates to minimizing the difference energy between signals, that is[4]

$$\hat{s}(t) = \underset{\forall s(t)}{\operatorname{argmin}} \int_{\mathcal{T}_s} |r(t) - s(t)|^2 \, dt \,. \tag{6.79}$$

(C) This is equivalent to minimizing the squared Euclidean distance in signal space:

$$\langle \hat{m}[k] \rangle = \underset{\forall \langle m[k] \rangle}{\operatorname{argmin}} |\boldsymbol{d} - \boldsymbol{s}_{\langle m \rangle}|^2 \,. \tag{6.80}$$

The receiver (using the bank of matched filters w.r.t. the chosen basis functions) described above utilizes strategy C. However, one can also start from variant B.

To that end, the difference energy can be written as

$$\int_{\mathcal{T}_s} |r(t) - s(t)|^2 \, dt = \int_{\mathcal{T}_s} (r(t) - s(t))(r(t) - s(t))^* \, dt$$

$$= \int_{\mathcal{T}_s} |r(t)|^2 \, dt - \int_{\mathcal{T}_s} \left(r(t) s^*(t) + r^*(t) s(t) \right) dt + \int_{\mathcal{T}_s} |s(t)|^2 \, dt$$

$$\stackrel{\text{def}}{=} E_{r(t)} - 2 \operatorname{Re} \left\{ \int_{\mathcal{T}_s} r(t) s^*(t) \, dt \right\} + E_{s(t)} \,, \tag{6.81}$$

where we have used

- energy of the receive signal

$$E_{r(t)} = \int_{\mathcal{T}_s} |r(t)|^2 \, dt \tag{6.82}$$

- energy of the hypothesis for the transmit signal

$$E_{s(t)} = \int_{\mathcal{T}_s} |s(t)|^2 \, dt = \int_{\mathcal{T}_s} \left| \sum_k s_{m[k]}(t - kT) \right|^2 dt$$

$s_m(t)$ temporal orthogonality

$$= \sum_k \int_{\mathcal{T}_s} \left| s_{m[k]}(t) \right|^2 dt$$

$$= \sum_k E_{m[k]} \,. \tag{6.83}$$

Defining

$$K_m[k] \stackrel{\text{def}}{=} \frac{1}{E_g} \int_{\mathcal{T}_s} r(t) s_m^*(t - kT) \, dt$$

$$= \left(r(t) * \frac{1}{E_g} s_m^*(-t) \right) \Big|_{t = kT} \,, \tag{6.84}$$

---

[4] Integration takes place over the support $\mathcal{T}_s$ of the transmit signal $s(t)$. Usually, a transmission burst with finite support is considered, leading to finite energies.

the correlation of the receive signal with the hypothesis for the entire transmit signal can be written as

$$\int_{\mathcal{T}_s} r(t) s^*(t)\, \mathrm{d}t = \int_{\mathcal{T}_s} r(t) \sum_k s^*_{m[k]}(t - kT)\, \mathrm{d}t$$
$$= E_{\mathrm{g}} \sum_k K_{m[k]}[k]\,. \qquad (6.85)$$

Note that $K_m[k]$ is the (normalized) correlation of the receive signal $r(t)$ with the signal element $s_m(t)$ time-shifted by kT. In other words, the quantity $K_m[k]$ is the result of *matched filtering* of the receive signal with the signal element number m and sampling at kT.

Putting everything together, we arrive at

$$\int_{\mathcal{T}_s} |r(t) - s(t)|^2\, \mathrm{d}t = E_{r(t)} - 2\,\mathrm{Re}\left\{\int_{\mathcal{T}_s} r(t) s^*(t)\, \mathrm{d}t\right\} + E_{s(t)}$$
$$= E_{r(t)} - 2\,\mathrm{Re}\left\{E_{\mathrm{g}} \sum_k K_{m[k]}[k]\right\} + \sum_k E_{m[k]}$$
$$= E_{r(t)} - E_{\mathrm{g}} \sum_k \left(2\,\mathrm{Re}\{K_{m[k]}[k]\} - E_{m[k]}/E_{\mathrm{g}}\right)$$
$$= E_{r(t)} - E_{\mathrm{g}} \sum_k \Lambda_{m[k]}[k]\,. \qquad (6.86)$$

In the last step, we have defined the *metric increment* at time instant k for signal-number hypothesis m

$$\Lambda_m[k] \stackrel{\mathrm{def}}{=} 2\,\mathrm{Re}\{K_m[k]\} - E_m/E_{\mathrm{g}}\,. \qquad (6.87)$$

The ML estimate is obtained by finding the minimizer of the difference energy (cf. (6.79)), which is equivalent to finding the minimizer of the right-hand-side expression of (6.86). As $E_{r(t)}$ is irrelevant for minimization (it does not depend on the hypothesis), ML decoding can equivalently be performed by *maximizing* the *sum metric*

$$\sum_k \Lambda_{m[k]}[k]\,.$$

For that, the *correlation vector* for step k

$$\boldsymbol{K}[k] \stackrel{\mathrm{def}}{=} \begin{bmatrix} K_0[k] \\ K_1[k] \\ \vdots \\ K_{M-1}[k] \end{bmatrix} \qquad (6.88)$$

has to be known, that is, the correlations of the receive signal with the signal elements $s_i(t), i = 0, \ldots, M-1$. In addition, the energies E_i of the signal elements are required.

As already stated, the correlations $K_i[k]$ are the result of matched filtering. Hence, the receiver consists of a *bank of matched filters* with respect to the signal elements $s_i(t)$, $i = 0, \ldots, M-1$. *Maximum-likelihood sequence detection* (coded transmission) is thus performed by

$$\langle \hat{m}[k] \rangle = \operatorname*{argmax}_{\forall \langle m[k] \rangle \in \mathcal{M}\{\mathbf{C}\}} \sum_k \left(2\operatorname{Re}\{K_{m[k]}[k]\} - E_{m[k]}/E_{\mathrm{g}} \right)$$

$$= \operatorname*{argmax}_{\forall \langle m[k] \rangle \in \mathcal{M}\{\mathbf{C}\}} \sum_k \Lambda_{m[k]}[k] . \qquad (6.89)$$

In case of uncoded transmission, this reduces to symbol-by-symbol (individual for all time steps k) *maximum-likelihood detection*

$$\hat{m}[k] = \operatorname*{argmax}_{\forall m \in \{0,1,\ldots,M-1\}} 2\operatorname{Re}\{K_m[k]\} - E_m/E_{\mathrm{g}}$$

$$= \operatorname*{argmax}_{\forall m \in \{0,1,\ldots,M-1\}} \Lambda_m[k] . \qquad (6.90)$$

The block diagram for MLD (uncoded transmission) is depicted in Figure 6.25.

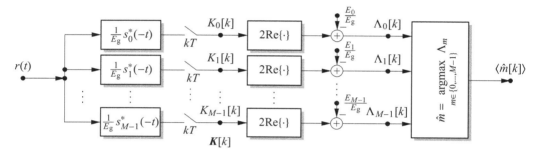

Figure 6.25 Correlation receiver or matched-filter demodulator (uncoded transmission).

This receiver concept has to be compared to the vector demodulator derived above. The one strategy uses a bank of matched filters with respect to the D orthonormal *basis functions* $g_j(t)$, the other a bank of matched filters with respect to the M signal elements $s_i(t)$. On the one hand, the *squared Euclidean distance* is *minimized*; on the other hand, the *real part of the correlation* (corrected by the energy) is *maximized*. In some sense, dual operations are present.

Since $M \geq D$, the correlation demodulator requires more matched filters compared to the vector demodulator, however, it does not require the calculation of any basis functions (which only has to be done once in system design). For $M = D$ (i.e., when all signal elements are linearly independent), both demodulation principles coincide. The calculation of squared Euclidean distances in the vector demodulator is sometimes seen as more computationally expensive than only calculating the real part of correlations in the correlation demodulator. However, the vector demodulator offers complexity savings by use of a basis with reduced dimensionality as outlined above.

It should be emphasized that both receiver concepts, the vector demodulator and the correlation receiver, implement ML(S)D in an optimum way, and thus they perform exactly the same.

6.6 Combining Coding and Modulation

After having derived the optimum receiver and having defined the parameters for assessing the power efficiency, we return to the separation of coding and modulation introduced in Section 6.1 and study the effect of deliberately combing coding and modulation in more detail.

6.6.1 Coding for Error Correction – Power Efficiency

Usually, channel codes are constructed in such a way that any two different codewords differ in a large number δ of symbols (referred to as the Hamming distance), at least in δ_min symbols, the *minimum Hamming distance*. Such codes are widespread for error control techniques, that is, *error detection* and/or *forward error correction (FEC)*, as well as for error prevention, see Appendix B. They aim to improve the *power efficiency* of the scheme.

When channel encoding is applied, redundant symbols (e.g., parity-check symbols) have to be transmitted to the receiver besides the symbols carrying information. Of course, the transmission of redundancy also consumes resources (i.e., bandwidth and transmit power). This effect is called the *rate loss of coding*.

Assume a modulation scheme is given; the uncoded operation establishes the reference. Applying channel coding, either the speed of information transmission (the data rate) has to be reduced or, when the data rate has to be kept constant, the signal bandwidth has to be increased due to faster signaling (smaller T). When, additionally, the average transmit power is fixed (i.e., the energy E_b per bit of transmitted information should remain constant) less energy E_s per modulation symbol is available. Therefore, a loss in reliability for the individually transmitted code symbols is caused by channel coding. But the exploitation of the coding constraints by the decoder usually yields a high gain in reliability, which exceeds the rate loss by far and an overall benefit with respect to power efficiency results. Note that a gain is even possible without bandwidth expansion when redundancy is transmitted by means of an expansion of the signal set of the modulation scheme.

We visualize the described effects by means of the subsequent examples.

Example 6.13 Parity-Check Code and Repetition Code

The effect of channel coding on the minimum squared Euclidean distance and the transmission rate is visualized in Figure 6.26. Thereby, binary ($R_\text{m} = 1$), bipolar transmission ($D = 1$) over $N = 3$ time intervals is assumed. The time steps establish the three dimensions depicted.

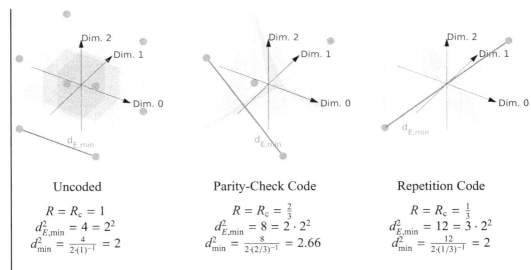

Figure 6.26 Visualization of the effect of channel coding on the minimum squared Euclidean distance. The decoding boundaries are shown in light gray.

On the left, uncoded transmission is assumed. Over the three time intervals, all combinations of "+1" and "−1" are allowed; all vertices of the cube are valid signal points. The minimum squared Euclidean distance is $d_{E,\min}^2 = 4$. Since eight combinations (over $N = 3$ time steps) exist, the rate of the scheme is $R = 1$.

In the middle, a single parity-check code is used. Here, the number of "−1" has to be even. Four valid combinations ([+ + +], [+ − −], [− + −], and [− − +]) exist. They are arranged such that the minimum squared Euclidean distance is increased to $d_{E,\min}^2 = 8$. However, the rate per three dimensions is only 2 (i.e., $R = 2/3$ per time step); a rate loss is present.

Finally, on the right, the situation for the repetition code is shown. Now, only two words ([+ + +], [− − −]) are valid. The minimum squared Euclidean distance is $d_{E,\min}^2 = 12$ but the rate per three dimensions is only 1 (i.e., $R = 1/3$).

These artificial, very short codes illustrate the trade-off between gain in distance by coding and rate loss. From $R = 1$ to $R = 2/3$, the gain in distance more than balances the rate loss as the normalized squared Euclidean distance grows. But for $R = 1/3$, the distance gain is counteracted by the increased rate loss and no coding gain remains, despite a bandwidth extension by a factor 3. (In general, repetition coding provides no gain on the AWGN channel.)

Example 6.14 Rate Loss and Power Efficiency

We consider the transmission of 1 Mbit/s (i.e., $T_b = 1$ μs) using binary, bipolar modulation (2ASK) with a $\sqrt{\text{Nyquist}}$ pulse with roll-off factor $\alpha = 0.3$. Transmission takes place over an AWGN channel with (one-sided) psd of $N_0 = 10^{-18}$ W s. The received signal power is assumed to be $S_r = 7 \times 10^{-12}$ W.

This gives, for the average received signal energy per bit of transmitted information and the signal-to-noise ratio (energy per bit to noise power spectral density ratio), respectively

$$E_b = S_r T_b = 7 \times 10^{-12} \text{ W} \cdot 1 \text{ μs} = 7 \times 10^{-18} \text{ W s}$$
$$10 \log_{10}(E_b/N_0) = 8.45 \text{ dB} .$$

- When using *uncoded transmission*, we have the following parameters:

$$T = T_b = 10^{-6} \text{ s}, \quad R = 1 \left[\tfrac{\text{bit}}{\text{symbol}}\right]$$
$$B_{\text{RF}} = 1.3 \text{ MHz}, \quad E_s = E_b .$$

The bit error ratio is approximately BER ≈ 10^{-4}.

- When *coded transmission* with a binary block code with parameters $K = 500$ and $N = 1000$ is employed, we have

$$R_c = 0.5, \quad T = T_b/2 = 5 \times 10^{-7} \text{ s}, \quad R = 0.5 \left[\tfrac{\text{bit}}{\text{symbol}}\right]$$
$$B_{\text{RF}} = 2.6 \text{ MHz}, \quad 10 \log_{10}(E_s/N_0) = 5.45 \text{ dB} .$$

The symbol error ratio for the binary symbols before channel decoding is approximately BER ≈ 5×10^{-3}.

Notice that due to the *rate loss*, the signal bandwidth is doubled and the raw error ratio is increased by a factor of 50. Via decoding, BER < 10^{-16} is achievable. Without coding, $10 \log_{10}(E_b/N_0) > 15.5$ dB would be necessary in order to achieve a comparable reliability. Thus, the channel coding scheme provides a gain in *power efficiency* of about 7 dB at the cost of decreasing the *bandwidth efficiency* by a factor of 2.

6.6.2 Coding for Bandwidth Efficiency

We now give examples to demonstrate that coding can also be used to increase the bandwidth efficiency.

Example 6.15 Trellis Code and Quaternary Modulation Scheme

In this example, we look at the interaction between the encoding and the modulation part. To that end, the set of four signal elements shown in Figure 6.27 is assumed ($R_m = 2$).

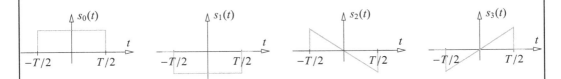

Figure 6.27 Set of signal elements.

226 6 Signal Space Representation

In case of *uncoded transmission*, since $M = 4$ signal elements are available, a rate of $R = R_\mathrm{m} = 2 \left[\frac{\text{bit}}{\text{symbol}}\right]$ can be transmitted. Assuming *natural labeling* – a pair of source bits is read as a binary representation of a natural number which selects the respective signal element – the transmit signal shown in Figure 6.28 is obtained.

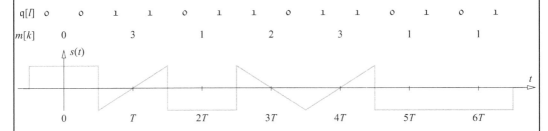

Figure 6.28 Example transmit signal. Uncoded transmission.

Now, we consider coded transmission. The trellis encoder of Figure 6.29 is used (for more details on this encoder, see Section B.2 of Appendix B). In the present case, the rate of this coded transmission system is $R = R_\mathrm{c} \cdot R_\mathrm{m} = \frac{1}{2} \cdot 2 = 1 \left[\frac{\text{bit}}{\text{symbol}}\right]$.

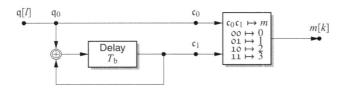

Figure 6.29 Encoding and mapping to signal numbers.

The transmit signal corresponding to an example source sequence is given in Figure 6.30.

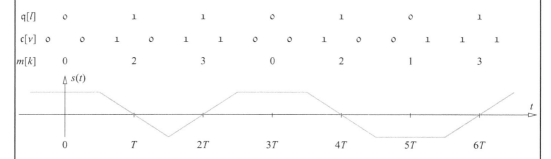

Figure 6.30 Example transmit signal. Coded transmission.

In contrast to uncoded transmission, here no discontinuities ("jumps") occur in the transmit signal. The encoding guarantees that the next signal element is chosen such that a continuous alignment is achieved. Figure 6.30 reveals that the information bit "0" can be represented by the signal elements 0 and 1, whereas the information bit "1" can be represented by the signal elements 2 and 3. Whenever the signal element 2 or 3 is used (when the source bit is 1), the sign of the signal changes (at the end of the modulation interval compared to the beginning). The memory element in the encoder counts (modulo 2) the sign changes. If the coded bit is $c_1 = 1$ (odd number of changes), the signal element $s_1(t)$ (for sending a "0") or $s_3(t)$ (for sending a "1") has to be chosen.

Because of the recursive structure of the trellis encoder in Figure 6.29, a change of a single information bit will not only change the signal of the respective modulation interval but also invert the entire subsequent transmit signal from this time instant on.

Here, the encoding for a continuous transmit signal primarily improves the spectral properties of the transmit signal despite the rate loss caused by the rate-1/2 trellis encoder. The more compact spectrum (attenuation of spectral side lobes) provides a greater benefit with respect to bandwidth efficiency than a doubling of the baud rate costs for equal data rate. Besides, now two possible transmit signals differ in at least two modulation intervals. Thus, decisions have favorably to be made over sequences of signal elements (see Appendix B) yielding a gain in power efficiency that balances the rate loss.

Example 6.16 Digital Frequency and Phase Modulation

A similar situation is discussed in the present example of a mixed frequency and phase modulation scheme that is closely related to technical practice. For an intuitive representation, RF signals for a very low carrier frequency are depicted in Figure 6.31.

Figure 6.31 Set of RF signal elements for a digital frequency and phase modulation.

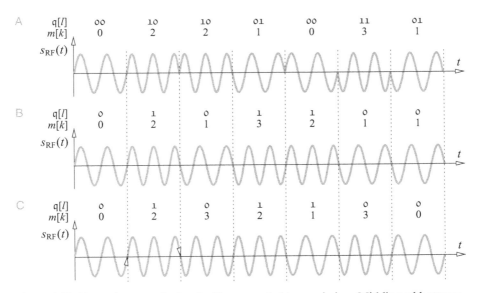

Figure 6.32 Example transmit signals. Top: uncoded transmission. Middle and bottom: coded transmission with two versions of encoding.

We have the following three examples of transmit signals, shown in Figure 6.32.

- *Signal A.* No channel coding is applied. A mixed frequency and phase modulation results, representing two bits per symbol, that is, $R = 2$ [bit/symbol]. This signal is spectrally inefficient due to discontinuities of signal slopes.
- *Signal B.* The recursive rate $R_c = 1/2$ trellis encoder of the previous example is applied here again for smoothing of the signal at the boundaries of modulation intervals. The source symbols are represented by the instantaneous frequency, whereas the redundancy serves for phase continuity of the carrier wave. A scheme with rate $R = 1$ [bit/symbol] results which, in fact, is MSK, interpreted as binary CPFSK (cf. Section 4.1.2). The recursive trellis encoder of Figures 6.29 and B.20 corresponds to the differential encoding for offset 4PSK. Here again, encoding not only improves bandwidth but also power efficiency as decisions can be made over pairs of symbols connected by carrier phase continuity. A coding gain of 3 dB balances the performance difference between 2FSK and MSK.
- *Signal C.* Now a rate $R_c = 1/2$ trellis encoder with a non-recursive FIR structure is applied (the reader may construct the non-recursive encoder as an exercise). The binary source symbols are mapped to the signal phase at the end of the modulation intervals (called phase-state mapping for CPFSK), whereas different instantaneous frequencies establish phase continuity. Here, no infinite memory is present. The power and bandwidth efficiency of signals B and C are identical, but for signal C the polarity matters (i.e., a phase reference is necessary for unambiguous data detection).

6.6.3 Line Coding

We now discuss special versions of coded PAM transmission. To that end, remember that in uncoded PAM signaling, all possible sequences of binary source symbols are mapped to all possible sequences of PAM amplitude coefficients. When all possible sequences $\langle a[k] \rangle$ are generated by the transmitter and the receiver is able to distinguish them, a so-called "transparent" PAM scheme is present.

As discussed in Section 3.2, for transparent signaling, at least the so-called *Nyquist band* – the band $[-f_N, +f_N]$, where $f_N = \frac{1}{2T}$ is the *Nyquist frequency* – has to be present in the transmit psd and has to be captured by the receiver. This fact may be illustrated by means of two extreme examples.

- The two sequences of amplitude coefficients $\langle \ldots, +1, +1, +1, +1, \ldots \rangle$ (all one) and $\langle \ldots, -1, -1, -1, -1, \ldots \rangle$ (all minus one) produce transmit signals $s(t)$ with different DC offsets. If the channel blocks the DC component (e.g., a high-pass filter), an all-zero detection sequence is obtained for both sequences of amplitude coefficients. Consequently, the DC component has to be present in the transmit signal and has to be supported by the channel.
- The alternating sequence $\langle \ldots, +1, -1, +1, -1, \ldots \rangle$ and its negative counterpart $\langle \ldots, -1, +1, -1, +1, \ldots \rangle$ produce transmit signals $s(t)$ that are periodic with frequency $\frac{1}{2T} = f_N$. If the channel blocks the Nyquist frequency f_N (e.g., by a

low-pass filter), an all-zero detection sequence is obtained for both sequences of amplitude coefficients. Consequently, the components $\pm f_N$ have to be present in the transmit signal and have to be supported by the channel.

All other sequences produce signals within these two extreme cases. Thus, the entire Nyquist band is occupied in case of transparent schemes, the channel has to pass this entire band, and the receiver has to capture the full band.

Starting from this observation, two directions are possible.

Line Coding

The aim of *line coding* is to avoid certain "unfavorable" sequences of amplitude coefficients. The most prominent example is *run-length coding*, where the number of subsequent "0" and subsequent "1" is limited to a predefined maximum. When using 2ASK signaling, no long runs of "+1" and "−1," respectively, are possible; DC sequences are avoided and the transmit psd does not contain significant components at $f = 0$. Similarly, when forcing a minimum number of equal symbols in series, fast changes are omitted and the spectral components around the Nyquist frequency are suppressed. In turn, such sequences are immune to channels with notches at $f = 0$ or $f = \frac{1}{2T}$.

In addition to this tolerance, line coding supports or even enables the receiver-side symbol frequency and phase synchronization.

Example 6.17 Alternate Mark Inversion (AMI) Code

The most simple and prominent example of a line code is the *alternate mark inversion (AMI) code*. The coding rule is that the binary source symbol "0" is represented by the amplitude coefficient $a = 0$ and the binary source symbol "1" is alternately represented by $a = +1$ and $a = -1$, respectively. The generated sequence $\langle a[k] \rangle$ of amplitude coefficients is then pulse-shaped with the basic pulse shape $g(t)$ in order to obtain the (baseband) transmit signal.

The block diagram of the transmitter and an example transmit signal (for visualization using hard keying, i.e., $g(t)$ is rectangular in the time domain) are depicted in Figures 6.33 and 6.34.

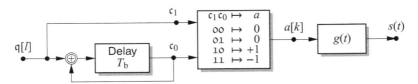

Figure 6.33 Encoding and modulation of the AMI code.

230   6 Signal Space Representation

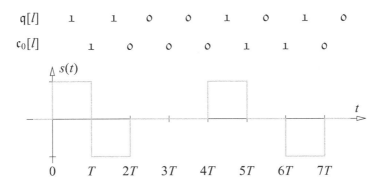

Figure 6.34 Example transmit signal when employing the AMI code.

It is straightforward to prove that the block diagram depicted in Figure 6.35 operates the same as the above-shown transmitter. However, from this block diagram the power spectral density of the sequence of amplitude coefficients $\langle a[k] \rangle$ can readily be derived (note here that $T_b = T$).

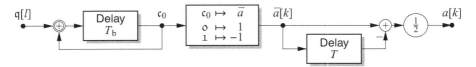

Figure 6.35 Alternative exposition of the encoding of the AMI code.

For i.i.d. redundancy-free source symbols $q[l]$, the modulo-2 accumulated symbols $c_0[l]$ are still i.i.d. and redundancy-free. Hence, after mapping, the sequence $\langle \bar{a}[k] \rangle$ is an i.i.d. redundancy-free sequence; since it is zero-mean with variance $\sigma_{\bar{a}}^2 = 1$ its psd is $\Phi_{\bar{a}\bar{a}}(e^{j2\pi fT}) = 1$. The amplitude coefficients $a[k]$ are generated from $\bar{a}[k]$ by filtering with the LTI system $H_{\mathrm{AMI}}(z) = \frac{1}{2}(1 - z^{-1})$. Thus, the power spectral density of the sequence of amplitude coefficients is given by

$$\begin{aligned}
\Phi_{aa}(e^{j2\pi fT}) &= 1 \cdot |H_{\mathrm{AMI}}(e^{j2\pi fT})|^2 \\
&= |\tfrac{1}{2}(1 - e^{-j2\pi fT})|^2 \\
&= \tfrac{1}{2}(1 - \cos(2\pi fT)) \\
&= \sin^2(\pi fT) \,,
\end{aligned} \qquad (6.91)$$

which has a spectral zero at $f = 0$; the AMI code suppresses the low-frequency components.

Note that for this feature, a high amount of redundancy is introduced by mapping binary information to a ternary signal alphabet.

Faster-than-Nyquist Signaling

In addition to or in lieu of line coding, special nonlinear signal processing methods may be employed at the receiver. In some cases it is possible to recover the data even though the channel does not pass the entire Nyquist band; the receiver seemingly does not obtain full information. However, since the amplitude coefficients are restricted to a finite set (and usually coding is applied), a detection of the transmit sequence is often still possible.

Having a bandlimited channel, the overuse of the channel might be deliberate. Such schemes are sometimes characterized as "*faster-than-Nyquist signaling.*" This approach offers interesting trade-offs between power and bandwidth efficiency for special applications, however, it is beyond the scope of this book.

6.7 Non-coherent Demodulation

The receiver concepts derived up to now are suited for *coherent* reception; the receiver has to be (perfectly) synchronized to the transmitter. However, as in Chapter 5, in some situations we are interested in *non-coherent demodulation*. The receiver-side quadrature mixer may not be synchronized; however, we assume that the *carrier frequency offset* is very small (in relation to the symbol interval, i.e., $f_\Delta T \ll 1$). For that (e.g., in mobile communications), frequency shifts due to the Doppler effect have to be compensated for. Then, a *carrier phase offset* φ_Δ remains that is constant at least over some modulation intervals T. Owing to the random nature of this offset, it is modeled as a random variable. The worst-case assumption is that the phase rotation is uniformly distributed, that is, its pdf is

$$f_{\varphi_\Delta}(\varphi_\Delta) = \begin{cases} \frac{1}{2\pi}, & -\pi \leq \varphi_\Delta < +\pi \\ 0, & \text{otherwise} \end{cases}. \tag{6.92}$$

Non-coherent demodulation is characterized by the acceptance of this uniform distribution of the carrier phase at the receiver.

As derived in (2.48), in case of up/downconversion with phase error, assuming an AWGN channel the receive signal is given by

$$r(t) = s(t) \cdot e^{j\varphi_\Delta} + n(t), \tag{6.93}$$

where φ_Δ is the (random but once drawn constant) phase error and $n(t)$ is additive white complex-valued Gaussian noise with noise power spectral density N_0 (ECB domain). The situation is also depicted in Figure 6.36.

The questions to be answered are: What is the optimum receiver for the non-coherent demodulation of a general digital modulation format and which constraints do the respective signal elements have to fulfill such that a non-coherent demodulation is possible at all?

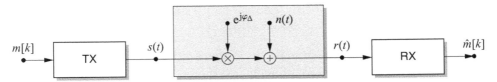

Figure 6.36 Transmission over an AWGN channel with phase error (ECB domain).

6.7.1 Optimum Non-coherent Detection

Subsequently, we consider *symbol-by-symbol non-coherent detection*. Owing to the temporal orthogonality of the signal elements (and thus of the modulation steps, cf. Section 6.1), the time intervals can be processed separately. In the following, we consequently drop the time index "k" and, without loss of generality, assume $k = 0$. Hence, a single modulation step is studied and the demodulation of this step should be done in an optimum way. If the phase offset remains constant over a number of symbol intervals, this common property over the time steps is not taken into account for improving the performance.

The phase error on the channel is modeled by a (complex) multiplication. Owing to the linearity of the pulse shaping, using (6.31) we have

$$s_m(t) \cdot e^{j\varphi_\Delta} = \left(\sum_{j=0}^{D-1} s_{j,m}\, g_j(t - kT) \right) \cdot e^{j\varphi_\Delta}$$

$$= \sum_{j=0}^{D-1} (s_{j,m}\, e^{j\varphi_\Delta})\, g_j(t - kT) \,, \qquad (6.94)$$

which means that the phase rotation on the channel can equivalently be expressed as a rotation of the coordinates in the D parallel branches.

When applying the *vector demodulator* (the bank of matched filters w.r.t. the orthonormal basis functions, cf. Figure 6.21), in case of a phase error, the equivalent vector channel model (cf. Figure 6.23) displayed in Figure 6.37 is obtained.

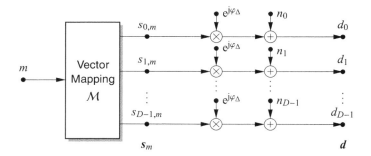

Figure 6.37 Equivalent vector channel model in case of a phase error φ_Δ.

Maximum-Likelihood Detection

The detection vector is given by

$$\boldsymbol{d} = \begin{bmatrix} d_0 \\ \vdots \\ d_{D-1} \end{bmatrix} = \boldsymbol{s}\, e^{j\varphi_\Delta} + \boldsymbol{n} \qquad (6.95)$$

$\boldsymbol{s}$: point in D-dimensional space representing the signal element (cf. (6.27))
$\boldsymbol{n}$: D-dimensional rotationally invariant Gaussian noise vector with variance $\sigma_n^2 = N_0/E_g$ per (complex) dimension.

The maximum-likelihood detection rule reads

$$\hat{m} = \operatorname*{argmax}_{\forall m \in \{0,1,\ldots,M-1\}} f_{\boldsymbol{d}}(\boldsymbol{d} \mid m)$$

$$= \operatorname*{argmax}_{\forall m \in \{0,1,\ldots,M-1\}} \int f_{\boldsymbol{d},\varphi_\Delta}(\boldsymbol{d}, \varphi_\Delta \mid m) \, \mathrm{d}\varphi_\Delta$$

$$= \operatorname*{argmax}_{\forall m \in \{0,1,\ldots,M-1\}} \log\left(\mathrm{const.} \cdot \int f_{\boldsymbol{d},\varphi_\Delta}(\boldsymbol{d}, \varphi_\Delta \mid m) \, \mathrm{d}\varphi_\Delta\right)$$

$$\stackrel{\mathrm{def}}{=} \operatorname*{argmax}_{\forall m \in \{0,1,\ldots,M-1\}} \Lambda'_m \,, \qquad (6.96)$$

where in the second-last line it has been used that any (strictly) monotonically increasing function and any (positive) scaling do not change the maximizer. In the middle line the conditional pdf $f_{\boldsymbol{d}}(\boldsymbol{d} \mid m)$ of the detection vector given the signal number is written as the marginal of the joint pdf $f_{\boldsymbol{d},\varphi_\Delta}(\boldsymbol{d}, \varphi_\Delta \mid m)$ of the detection vector and the phase error. This pdf, in turn, can be written using the chain rule[5] as

$$f_{\boldsymbol{d},\varphi_\Delta}(\boldsymbol{d}, \varphi_\Delta \mid m) = f_{\boldsymbol{d}}(\boldsymbol{d} \mid m, \varphi_\Delta) \cdot f_{\varphi_\Delta}(\varphi_\Delta \mid m) \,. \qquad (6.97)$$

As the phase error is independent of the transmitted signal (and thus of the signal number m) and as, due to the additive noise, the conditional pdf (knowing the signal and the phase error) is given by the shifted noise pdf, we arrive at

$$f_{\boldsymbol{d},\varphi_\Delta}(\boldsymbol{d}, \varphi_\Delta \mid m) = f_{\boldsymbol{n}}(\boldsymbol{d} - \mathrm{e}^{\mathrm{j}\varphi_\Delta}\boldsymbol{s}) \cdot f_{\varphi_\Delta}(\varphi_\Delta) \,. \qquad (6.98)$$

Finally, using the fact that $\boldsymbol{n}$ is D-dimensional rotationally invariant Gaussian with variance σ_n^2 per (complex) dimension and that the phase is uniformly distributed, we have

$$f_{\boldsymbol{d},\varphi_\Delta}(\boldsymbol{d}, \varphi_\Delta \mid m) = \frac{1}{(\pi\sigma_n^2)^D} \mathrm{e}^{-\|\boldsymbol{d}-\mathrm{e}^{\mathrm{j}\varphi_\Delta}\boldsymbol{s}\|^2/\sigma_n^2} \cdot \frac{1}{2\pi} \mathrm{rect}\left(\frac{\varphi_\Delta}{2\pi}\right). \qquad (6.99)$$

In order to calculate the marginal over this pdf, we interpret the squared Euclidean distance (in the exponent in (6.99)) as the difference energy (cf. (6.35)). This gives

$$\|\boldsymbol{d} - \mathrm{e}^{\mathrm{j}\varphi_\Delta}\boldsymbol{s}_m\|^2 = \frac{1}{E_\mathrm{g}} \int_{-\infty}^{\infty} \left|\check{r}(t) - \mathrm{e}^{\mathrm{j}\varphi_\Delta} s_m(t)\right|^2 \mathrm{d}t$$

$$= \frac{E_{\check{r}(t)}}{E_\mathrm{g}} + \frac{E_m}{E_\mathrm{g}} - 2\,\mathrm{Re}\!\left\{K_m \mathrm{e}^{-\mathrm{j}\varphi_\Delta}\right\}. \qquad (6.100)$$

Note that the white noise spans an infinite number of dimensions but the demodulation vector contains only D dimensions. The squared Euclidean distance thus reflects only the relevant portion, that is, $\check{r}(t) = s(t) + \check{n}(t)$, where $\check{n}(t)$ is the portion

---

[5] For pdfs, the chain rule reads $f_{a,b}(a,b) = f_a(a \mid b) f_b(b)$. In case of an additional condition to c, all pdfs are conditioned to this variable.

of the noise in the D relevant signal dimensions. Since the complex correlations perform a projection of the noise onto the D relevant dimensions (all other dimensions are ignored), we have

$$K_m = \frac{1}{E_g} \int_{-\infty}^{\infty} \check{r}(t)\, s_m^*(t)\, dt$$

$$= \frac{1}{E_g} \int_{-\infty}^{\infty} r(t)\, s_m^*(t)\, dt, \qquad m = 0, \ldots, M-1. \qquad (6.101)$$

The correlations K_m capture all information. They are sufficient to perform ML decoding; they form a *sufficient statistic*.

Putting all this together and using $\sigma_n^2 = N_0/E_g$, the joint pdf (6.99) can thus be written as

$$f_{d,\varphi_\Delta}(d, \varphi_\Delta \mid m) = \frac{1}{(\pi\sigma_n^2)^D}\, e^{-\|d - e^{j\varphi_\Delta} s\|^2/\sigma_n^2} \cdot \frac{1}{2\pi}\, \mathrm{rect}\!\left(\frac{\varphi_\Delta}{2\pi}\right)$$

$$= \frac{1}{(\pi\sigma_n^2)^D}\, e^{-(E_{\check{r}(t)}/E_g + E_m/E_g - 2\,\mathrm{Re}\{K_m e^{-j\varphi_\Delta}\})/\sigma_n^2} \cdot \frac{1}{2\pi}\, \mathrm{rect}\!\left(\frac{\varphi_\Delta}{2\pi}\right)$$

$$= \frac{1}{(\pi\sigma_n^2)^D}\, e^{-E_{\check{r}(t)}/E_g/\sigma_n^2} \cdot \frac{1}{2\pi} e^{-E_m/E_g/\sigma_n^2} \cdot e^{+2\,\mathrm{Re}\{K_m e^{-j\varphi_\Delta}\}/\sigma_n^2} \cdot \mathrm{rect}\!\left(\frac{\varphi_\Delta}{2\pi}\right)$$

$$= C_{\mathrm{irr}} \cdot \frac{1}{2\pi} e^{-E_m/N_0} \cdot e^{+2\,\mathrm{Re}\{K_m e^{-j\varphi_\Delta}\} E_g/N_0} \cdot \mathrm{rect}\!\left(\frac{\varphi_\Delta}{2\pi}\right), \qquad (6.102)$$

with the factor $C_{\mathrm{irr}} \stackrel{\text{def}}{=} \frac{1}{(\pi\sigma_n^2)^D}\, e^{-E_{\check{r}(t)}/E_g/\sigma_n^2}$ that is irrelevant for optimization (as it does not depend on the signal number m).

The metric for ML detection (6.96) then calculates as

$$\Lambda'_m \stackrel{\text{def}}{=} \log\!\left(\frac{1}{C_{\mathrm{irr}}} \cdot \int f_{d,\varphi_\Delta}(d, \varphi_\Delta \mid m)\, d\varphi_\Delta\right)$$

$$= \log\!\left(\int_{-\pi}^{\pi} \frac{1}{2\pi} e^{-E_m/N_0} \cdot e^{+2\,\mathrm{Re}\{K_m e^{-j\varphi_\Delta}\} E_g/N_0}\, d\varphi_\Delta\right)$$

$$= \log\!\left(\int_{-\pi}^{\pi} \frac{1}{2\pi} e^{+2\,\mathrm{Re}\{K_m e^{-j\varphi_\Delta}\} E_g/N_0}\, d\varphi_\Delta\right) - \frac{E_m}{N_0}$$

and with $K_m = |K_m| e^{j\varphi_K}$, $\varphi_K = \arg\{K_m\}$

$$= \log\!\left(\frac{1}{2\pi} \int_{-\pi}^{\pi} e^{+2\,E_g/N_0\, |K_m|\cdot\cos(\varphi_K - \varphi_\Delta)}\, d\varphi_\Delta\right) - \frac{E_m}{N_0}$$

$$= \log\!\left(I_0\!\left(2\frac{E_g}{N_0}|K_m|\right)\right) - \frac{E_m}{N_0}$$

$$\stackrel{\text{def}}{=} \mathrm{k_{nc}}\!\left(2\frac{E_g}{N_0}|K_m|\right) - \frac{E_m}{N_0}. \qquad (6.103)$$

Thereby, the *modified Bessel function of the first kind of order* 0

$$\mathrm{I}_0(x) \stackrel{\text{def}}{=} \frac{1}{2\pi} \int_{-\pi}^{+\pi} e^{x\cos(\varphi-\varphi_0)} \, d\varphi \qquad (6.104)$$

has been used. Note that $\mathrm{I}_0(x)$ is independent of φ_0, as the integration is carried out over one period of a periodic function.

Moreover, the *characteristic curve* for non-coherent detection

$$\mathrm{k}_{\mathrm{nc}}(x) \stackrel{\text{def}}{=} \log\left(\mathrm{I}_0(x)\right) \qquad (6.105)$$

has been defined. It can be shown that this function is monotonously increasing (see Figure 6.38) and that it is well approximated by

$$\mathrm{k}_{\mathrm{nc}}(x) \approx x - 2.4, \qquad x \gg 1.$$

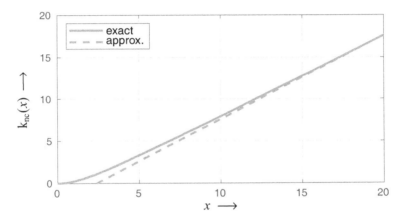

Figure 6.38 Characteristic curve for non-coherent detection.

Matched-Filter Demodulator

In summary, ML detection in case of non-coherent demodulation is performed by

$$\hat{m} = \underset{\forall m \in \{0,1,\ldots,M-1\}}{\operatorname{argmax}} \left(\mathrm{k}_{\mathrm{nc}}\!\left(2\frac{E_g}{N_0}|K_m|\right) - \frac{E_m}{N_0} \right). \qquad (6.106)$$

The corresponding block diagram of the matched-filter demodulation for non-coherent reception is depicted in Figure 6.39.

Comparison with Figure 6.36, the matched-filter demodulator for coherent reception, reveals the differences. In the coherent case, the real parts of the correlations are evaluated. In the non-coherent case, the magnitudes are processed. Compare this with the procedure in OOK (Section 5.3). In both cases, the energies of the signal elements have to be taken into account.

This general form of the receiver can be simplified if some conditions are fulfilled. In particular:

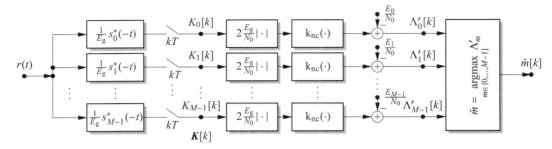

Figure 6.39 Matched-filter demodulator for non-coherent reception.

- If all signal elements have the same energy (i.e., $E_m = E_s$, $m = 0, \ldots, M-1$), the subtraction of E_m/N_0 is irrelevant and can be dropped. This, for example, is the case in digital frequency modulation (frequency-shift keying, see Chapter 7).
- If uncoded transmission using signal elements with the same energy ($E_m = E_s$, $m = 0, \ldots, M-1$) is considered, the monotonous characteristic curve can also be disregarded, which leads to

$$\hat{m} = \underset{\forall m \in \{0,1,\ldots,M-1\}}{\mathrm{argmax}} |K_m[k]| . \tag{6.107}$$

Hence, *envelope detection* is obtained. The decision is made in favor of that signal element for which the correlation has the largest envelope (magnitude of ECB signal). Envelope detection for binary transmission is depicted in Figure 6.40.

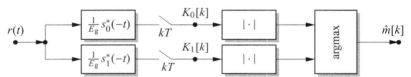

Figure 6.40 Envelope detector (uncoded binary transmission) with equal-energy signal elements.

Requirements on the Signal Elements

As already discussed in case of non-coherent detection of unipolar ASK (Section 5.3), not all signal sets are suited for non-coherent detection. The matched-filter demodulator (Figure 6.39) calculates correlations and considers only their magnitudes. In order to be able to recover data, the maximum of Λ'_m, $m = 0, \ldots, M-1$, must be unique for all hypotheses of the transmitted signal element. Using the definition of the correlations, this requires that for each μ (transmitted signal element)

$$\mathrm{k_{nc}}\left(2\frac{E_g}{N_0}\left|\int_{-\infty}^{\infty} s_\mu(t)\, s_\mu^*(t)\, \mathrm{d}t\right|\right) - \frac{E_\mu}{N_0} > \mathrm{k_{nc}}\left(2\frac{E_g}{N_0}\left|\int_{-\infty}^{\infty} s_\mu(t)\, s_m^*(t)\, \mathrm{d}t\right|\right) - \frac{E_m}{N_0} , \tag{6.108}$$

$$\forall m \in \{0, \ldots, M-1\}, \quad m \neq \mu .$$

6.7 Non-coherent Demodulation

In particular, as the phase is not exploitable for detection, this demand excludes signal elements for which $s_m(t) = e^{j\alpha} s_\mu(t)$ for any α, that is, signal elements that differ solely in the phase.

6.7.2 Multiple-Symbol Differential Detection of DPSK

We now utilize the concept of optimum non-coherent demodulation for improving the performance of differential PSK. In Section 5.2, it has been shown that differential detection of DPSK ($M \geq 4$) causes a loss of approximately 3 dB compared to the coherent reception of PSK. For convenience, the block diagram of DPSK and conventional differential detection is repeated in Figure 6.41.

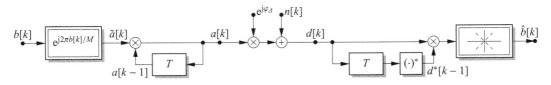

Figure 6.41 Block diagram of DPSK and conventional differential detection.

Assuming that the channel phase φ_Δ is constant over a block of symbols (not only over consecutive pairs), an improvement can be obtained by replacing the *symbol-by-symbol decisions* with detection based on *blocks of N consecutive receive symbols*. Thereby, inherently an averaging over the random phase can be obtained, leading to a more stable phase reference for differential detection (instead of a phase reference that is as noisy as the signal point itself, cf. Section 5.2). The resulting scheme is denoted as *multiple-symbol differential detection (MSDD)*.

To that end, without loss of generality, we consider the temporal block of N modulation intervals starting at $k = 0$ and ending at $k = N - 1$. That is, we look at the following transmitted symbols, collected in the transmit (row) vector:

$$\boldsymbol{a} \stackrel{\text{def}}{=} [\, a[0],\, a[1],\, a[2],\, \ldots,\, a[N-1]\,] \,. \tag{6.109}$$

Owing to the differential encoding (integration) process, and because the first transmitted symbol is used as a reference symbol, this block of N transmit symbols corresponds to a block of $N - 1$ information-carrying differential data symbols

$$\tilde{\boldsymbol{a}} \stackrel{\text{def}}{=} [\, \tilde{a}[1],\, \tilde{a}[2],\, \ldots,\, \tilde{a}[N-1]\,] \,, \tag{6.110}$$

where $a[k] = a[k-1] \cdot \tilde{a}[k]$, $k = 1, \ldots, N-1$, and $a[0]$ fixed. The demodulator will be designed to detect the transmit vector $\boldsymbol{a}$. In a final step, the differential encoding will be reversed like in DCPSK (differentiation). As in conventional DPSK, the detection processing windows overlap by one symbol such that an extra transmission of a reference symbol is not necessary.

Signal Elements

In order to apply the matched-filter demodulator from Figure 6.39 derived in the last section, we have to define signal elements $s_m(t)$ for the situation at hand. To that end, we assume an M-ary PSK signal constellation $\mathcal{A}$ and a basic pulse shape $g(t)$.

Then, there are M^N possible transmit vectors $\boldsymbol{a} \in \mathcal{A}^N$ (we enumerate them as $\boldsymbol{a}_0, \boldsymbol{a}_1, \ldots, \boldsymbol{a}_{M^N-1}$). These vectors correspond to M^N possible continuous-time transmit signals

$$s_i(t) \stackrel{\text{def}}{=} \sum_{k=0}^{N-1} a_i[k] g(t - kT), \qquad i = 0, \ldots, M^N - 1, \qquad (6.111)$$

with

$$\boldsymbol{a}_i = [\, a_i[0],\ a_i[1],\ a_i[2],\ \ldots,\ a_i[N-1]\,] \,. \qquad (6.112)$$

These M^N signals are interpreted to be the signal elements of an M^N-ary digital transmission scheme. If T is the symbol interval of the PSK scheme, NT is the corresponding symbol interval on which the detection is based.

Example 6.18 Signal Elements

For binary PSK ($M = 2$) and a block length of $N = 3$, there are $M^N = 2^3 = 8$ possible vectors $\boldsymbol{a}$ of transmit symbols. For a rectangular basic pulse shape $g(t)$, the corresponding signal elements $s_i(t)$ are sketched in Figure 6.42.

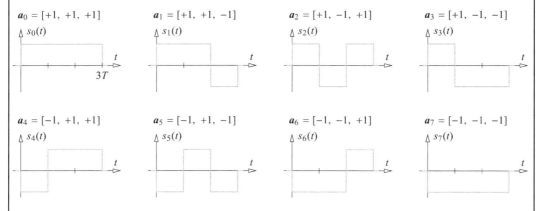

Figure 6.42 Example signal elements. $M = 2$-ary PSK, rectangular basic pulse shape, $N = 3$.

The direct application of the matched-filter demodulator (cf. Figure 6.39) would consist of a bank of M^N matched filters, each one calculating the correlation K_m with respect to one signal element $s_m(t)$ from (6.111). In detail, the correlations are calculated as

$$K_m = \frac{1}{E_g} \int_{-\infty}^{\infty} r(t)\, s_m^*(t)\, \mathrm{d}t$$

$$= \frac{1}{E_g} \int_{-\infty}^{\infty} r(t) \left(\sum_{k=0}^{N-1} a_m[k] g(t-kT) \right)^* \mathrm{d}t$$

$$= \sum_{k=0}^{N-1} (a_m[k])^* \frac{1}{E_g} \int_{-\infty}^{\infty} r(t)\, g^*(t-kT)\, \mathrm{d}t$$

$$= \sum_{k=0}^{N-1} a_m^*[k]\, d[k], \qquad (6.113)$$

where we have used the detection variable

$$d[k] = \frac{1}{E_g} \int_{-\infty}^{\infty} r(t) g^*(t-kT)\, \mathrm{d}t \qquad (6.114)$$

with respect to the basic pulse shape. Combining the detection variables for the N time steps into a vector

$$\boldsymbol{d} \stackrel{\text{def}}{=} [\, d[0],\, d[1],\, \ldots,\, d[N-1]\,], \qquad (6.115)$$

we arrive at
$$K_m = \sum_{k=0}^{N-1} a_m^*[k]\, d[k]$$

$$= \boldsymbol{d}\, \boldsymbol{a}_m^{\mathsf{H}}, \qquad (6.116)$$

where $\cdot^{\mathsf{H}}$ denotes the conjugate complex transposition (Hermitian) of a vector.

Since the signal number m and the vector $\boldsymbol{a}$ of transmit symbols are synonymous, the non-coherent ML detection rule (6.106), which here reads

$$\hat{m} = \operatorname*{argmax}_{\forall m \in \{0,1,\ldots,M^N-1\}} \left(k_{\text{nc}}\!\left(2\frac{E_g}{N_0} |K_m| \right) - \frac{E_m}{N_0} \right), \qquad (6.117)$$

can be specialized to the situation at hand into the form

$$\hat{\boldsymbol{a}} = \operatorname*{argmax}_{\forall \boldsymbol{a} \in \mathcal{A}^N} \left(k_{\text{nc}}\!\left(2\frac{E_g}{N_0} |\boldsymbol{d}\, \boldsymbol{a}^{\mathsf{H}}| \right) - \frac{E_m}{N_0} \right). \qquad (6.118)$$

We are interested in multiple-symbol differential detection of DPSK. Here, due to the PSK constellation, $|a[k]| = 1$. Hence, all M^N hypothetical signal elements $s_i(t)$ in (6.111) have the same energy $E_i = N E_g$. The correction of the energy in the decision metric is thus unessential. Moreover, as the characteristic curve $k_{\text{nc}}(x)$ is strictly monotonously increasing, it can also be neglected. Then, the scaling by $2E_g/N_0$ can also be dropped. Hence, we arrive at the following ML decision rule for MSDD of DPSK:

$$\hat{\boldsymbol{a}} = \operatorname*{argmax}_{\forall \boldsymbol{a} \in \mathcal{A}^N} \left| \boldsymbol{d}\, \boldsymbol{a}^{\mathsf{H}} \right|. \qquad (6.119)$$

The task of the receiver is thus to find that transmit vector $\boldsymbol{a}$ for which the scalar product (the correlation) with the decision vector has the largest magnitude.

Since the symbol $a[0]$ is the reference symbol and any phase offset is irrelevant – $\boldsymbol{a}$ and $e^{j\alpha}\boldsymbol{a}$ give the same metric – without loss of generality, we can assume $a_i[0] = 1$, $\forall i$. The search problem is reduced to

$$[\hat{a}[1], \hat{a}[2], \ldots, \hat{a}[N-1]] = \underset{\forall [a[1], a[2], \ldots, a[N-1]], \, a[k] \in \mathcal{A}}{\operatorname{argmax}} \left| d[0] + \sum_{k=1}^{N-1} d[k] \, a[k]^* \right|$$
$$\hat{a}[0] = 1 \,. \tag{6.120}$$

Having estimates for the transmit symbols $a[0], \ldots, a[N-1]$, the final decisions on the (differential) data symbols $\tilde{a}[k]$ are then obtained by

$$\hat{\tilde{a}}[k] = \hat{a}[k]\hat{a}^*[k-1]\,, \qquad k = 1, \ldots, N-1\,. \tag{6.121}$$

Note that, for $N = 2$, conventional differential detection results. This can be seen from

$$\hat{\tilde{a}}[1] = \hat{a}[1] \cdot \hat{a}^*[0]$$
$$= \underset{\forall a[1] \in \mathcal{A}}{\operatorname{argmax}} \left| d[0] + d[1]\,a^*[1] \right| \cdot 1 \,.$$

$d[0]$ and $d[1]$ are given complex numbers; $a[1]$ is drawn from a finite set. The magnitude of $d[0] + d[1]\,a^*[1]$ is maximized if the arguments of the complex numbers $d[0]$ and $d[1]\,a^*[1]$ are as close as possible. This leads to

$$\hat{\tilde{a}}[1] = \underset{\forall a[1] \in \mathcal{A}}{\operatorname{argmin}} \left(\arg\{d[0]\} - \arg\{d[1]\,a^*[1]\} \right)$$
$$= \underset{\forall a[1] \in \mathcal{A}}{\operatorname{argmin}} \left(\arg\{d[0]\,d^*[1]\} - \arg\{a^*[1]\} \right)$$
$$= \underset{\forall a[1] \in \mathcal{A}}{\operatorname{argmin}} \left(\arg\{d^*[0]\,d[1]\} - \arg\{a[1]\} \right) \,. \tag{6.122}$$

Hence, that PSK amplitude coefficient a has to be found whose argument is closest to that of $d^*[0]\,d[1]$. This is achieved by PSK decision on $d^*[0]\,d[1]$, which is done exactly by conventional differential detection of DPSK (cf. Figure 5.8).

We have studied a single block – the generalization to continuous transmission is straightforward by repeatedly using temporally shifted blocks. In this case, the detection blocks should overlap by one symbol – the last symbol of the preceding block is the reference of the next block. Then, only at the very beginning of the transmission does an explicit reference symbol have to be transmitted.

Employing MSDD, the power efficiency can be increased. For growing block size N, the performance approaches that of coherent transmission. However, when increasing N, the complexity increases exponentially, as in the ML rule (6.118), M^{N-1} hypotheses have to be tested instead of only M. However, efficient search algorithms exist. Finally, it has to be noted that when applying MSDD, the carrier phase has to be stable over N instead of only two symbol periods. If this cannot be guaranteed, MSDD will (significantly) degrade.

Example 6.19 Symbol Error Ratio of MSDD

Figures 6.43 and 6.44 show the symbol error ratios of DPSK. Conventional differential detection ($N = 2$, light gray solid line) and MSDD ($N > 2$) are compared. For reference, the performance of differential coherent PSK is shown (light gray dashed line). In Figure 6.43, $M = 4$-ary PSK is employed; in Figure 6.44, $M = 8$-ary PSK.

As can be seen, when increasing the block size N, the error-rate curves tend to that of coherent reception (perfect synchronization). The gap of (almost) 3 dB between differential detection of PSK and the coherent case can be bridged.

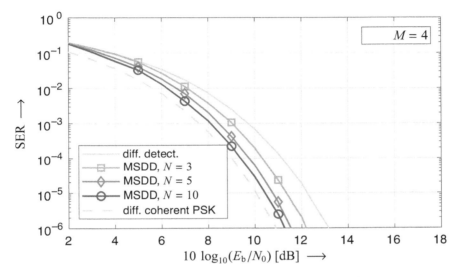

Figure 6.43 Symbol error ratios of DPSK employing MSDD. $M = 4$.

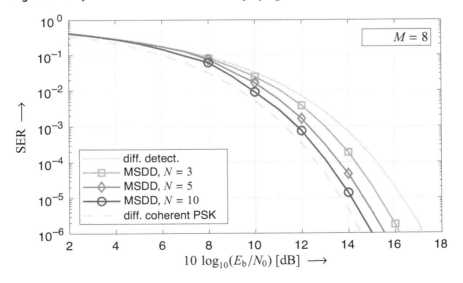

Figure 6.44 Symbol error ratios of DPSK employing MSDD. $M = 8$.

However, the gain in the error-rate curves is paid for with an increased complexity. For $M = 4$ and $N = 3, 5, 10$, the number M^{N-1} of hypotheses that have to be tested is $16, 256$, and $262\,144$, respectively. For $M = 8$, in the above example, $64, 4096$, and $134\,217\,728$ hypotheses would have to be tested, respectively, instead of eight in case of conventional differential detection. This brute-force testing is not required. In practice, efficient algorithms exist (e.g., sphere decoding and variants of sequential decoding).

Complexity Reduction

As mentioned above, the performance gain of MSDD over conventional differential detection for DPSK is due to an improved inherent phase stabilization within the decision window of $N > 2$ symbols. Usually, the subsequent detection windows are overlapping by one symbol, in other words, the detection window is always moving $N - 1$ symbols forward.

When this shift of the detection window is chosen to be less then $N - 1$ steps, say U steps, $N - U - 1$ already decided symbols of the previous block can be utilized in the current block. Now, only M^U hypotheses have to be tested while the desired phase stabilization over the entire window is maintained. This approach is called *decision-feedback MSDD (DF-MSDD)*. Because the number of hypotheses to be tested is reduced from M^{N-1} to M^U, this method may yield a large saving in complexity. Additionally, the reliability for the most recent symbol within the detection window is significantly lower than for symbols in the middle, because a neighboring symbol exists only from one side.

For $N \geq 4$ this observation leads to a scheme with high data reliability at low complexity, called *decision-feedback subset MSDD (DF-S-MSDD)*. The decision window of length N is shifted only by a single symbol, $N - 2$ symbols are fixed by the reference and the decision feedback. Thus, irrespective of the chosen window size N, only M^2 hypotheses for the two most recent symbols have to be tested. From the estimated two symbols, only the first one is accepted as a decision, whereas the last and less reliable symbol is discarded. Furthermore, by sophisticated test strategies (e.g., sphere decoding), an average number of tests quite lower than M^2 can be achieved. For an equal window size N, reliability is usually higher for DF-S-MSDD than for full MSDD because of the discarded least-reliable symbol.

6.8 Problems

6.8.1 Encoding, Mapping, and Modulation

We consider the following general structure of a coded digital PAM transmission system:

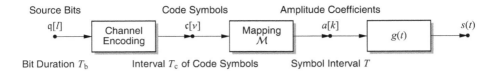

The system parameters are as follows:
- A data rate of 4 Mbit/s should be transmitted.
- A binary (7,4) Hamming code is used (cf. Appendix B). It can correct exactly one erroneous code symbol per codeword.
- A 128-ary QAM signal constellation with *Gray labeling* is employed.

(a) Give the bit duration T_b and the duration T_c of the code symbols.
(b) How many QAM symbols have to be transmitted in one second? Give the duration T of the modulation interval.
(c) How many information bits are transmitted per modulation step, that is, what is the rate of the communication scheme?

We assume that the signal-to-noise ratio is sufficiently large such that during transmission, at most errors to nearest-neighbor signal points occur.

(d) How many binary code symbols are in error when a symbol error occurs?
(e) Give the bit error ratio after the channel decoder. Give reasons for your answer.

6.8.2 Trellis Encoder

The following trellis encoder (linear time-invariant (LTI) discrete-time (T_b) system operating on binary symbols, i.e., $q[l] \in \{0, 1\}$) is given (cf. Appendix B). Thereby, the binary output sequence $c[\nu]$ is formed by parallel-to-serial conversion of the two signals $c_0[l]$ and $c_1[l]$, that is, $c[2l] = c_0[l]$ and $c[2l + 1] = c_1[l]$, or

$$\langle \ldots, c[0], c[1], c[2], c[3], \ldots \rangle = \langle \ldots, c_0[0], c_1[0], c_0[1], c_1[1], \ldots \rangle$$

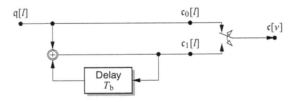

(a) Give the coded sequence that corresponds to the sequence of source symbols $q[k]$ (start with the delay element containing the value 0):

$$\langle \ 0, \ 1, \ 1, \ 0, \ 1, \ 0, \ 1, \ 1 \ \rangle$$

(b) Give the state-transition diagram of the encoder.
(c) Give the trellis diagram associated with the encoder and highlight the sequence of source symbols/coded sequence from part (a).
(d) Repeat parts (a) to (c) for the following encoder structure:

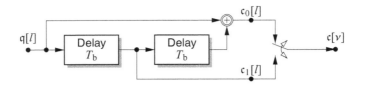

6.8.3 Equivalent Coding and Modulation

We consider binary, bipolar PAM transmission with symbol duration T and the following basic pulse shape $g(t)$:

$$g(t) = \begin{cases} 1\,\text{V}(1 - |t|/T), & \text{for } |t| \leq T \\ 0, & \text{else} \end{cases}.$$

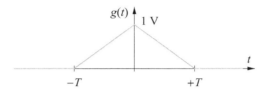

(a) Does this pulse fulfill the temporal orthogonality condition with respect to T?
(b) Sketch the transmit signal that corresponds to the following sequence of amplitude coefficients $a[k]$:

$$\langle\ -1,\ +1,\ +1,\ -1,\ -1,\ +1,\ -1\ \rangle$$

(c) Sketch all time-limited signal elements that occur within a symbol interval of duration T.
   Hint: The transients at the beginning and end of a block of finite duration are not taken into account.

We now use these time-limited signal elements to define an equivalent encoder and modulation.

(d) Give an equivalent encoder and a mapping that, using the signal elements of part (c), produce the same transmit signal as when using $g(t)$.
   To this end, assume the original mapping to be $a[k] = \begin{cases} +1, & q[k] = 0 \\ -1, & q[k] = 1 \end{cases}$.

We are now interested in an *orthonormal* basis for the equivalent signal elements.

(e) Which algorithm may be used to calculate such a basis?
(f) Give an orthonormal basis for the signal elements of part (c).
   Which dimensionality (per modulation step) has the corresponding signal space?

6.8.4 Gram–Schmidt Procedure

We consider digital transmission using a signal set with four signal elements ($M = 4$)

$$\mathcal{S} = \{s_0(t), s_1(t), s_2(t), s_3(t)\}\ .$$

The signal elements $s_i(t)$ (time-limited to the interval $[0, T]$) are sketched as follows:

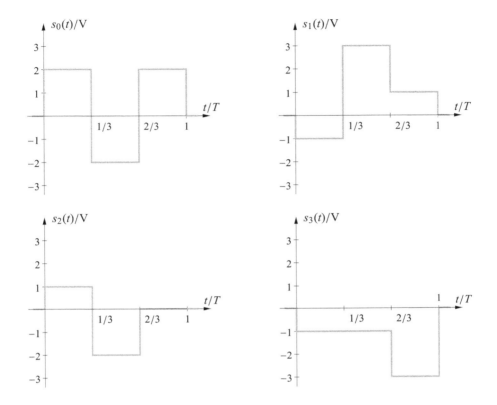

(a) Determine an orthonormal basis of the space spanned by the signal elements $s_i(t)$, $i = 0, \ldots, 3$.
Hint: Apply the Gram–Schmidt procedure to the signal elements $s_i(t)$ with the index i in ascending order. Set $E_g = 1/3 \, \text{V}^2 T$.
(b) Sketch the resulting functions of the orthonormal basis.
(c) Sketch the signal elements $s_i(t)$ as signal points in the signal space spanned by the orthonormal basis from part (a).

6.8.5 Receiver Structures and Decision Criteria

Digital PAM transmission over an AWGN channel using a 4-ary QAM constellation and a real-valued $\sqrt{\text{Nyquist}}$ pulse $g(t)$ with energy E_g is considered.
(a) Determine the signal elements $s_i(t)$ of this modulation scheme (in the ECB domain) and their respective energies E_i.
(b) Sketch the block diagram of the corresponding *correlation receiver*.
(c) Give the signal elements using an orthonormal basis. How many dimensions are required?
(d) Sketch the block diagram of the corresponding *vector receiver*. Which receiver structure would you prefer? Explain.
(e) Explain the following terms:
 • maximum-a-posteriori sequence detection (MAPSD);

- maximum-likelihood sequence detection (MLSD);
- maximum-likelihood detection (MLD).

(f) Calculate the minimum squared Euclidean distance and the normalized minimum squared Euclidean distance of this transmission scheme.

6.8.6 Alternate Mark Inversion Code

We consider the simple line code alternate mark inversion (AMI).

(a) Show that the structures in Figures 6.33 and 6.35 produced the same sequence of amplitude coefficients $a[k]$ given the same input sequence $q[l]$.
(b) Which structure would you prefer for practical implementation?

6.8.7 Multiple-Symbol Differential Detection

We consider an improved demodulation of quaternary differential phase-shift keying over the AWGN channel with an unknown (random) phase offset. The transmitter is the same as studied in Problem 5.4.1.

We assume $g(t) = 1\,\text{V} \cdot \text{rect}(t/T)$ as basic pulse shape, where T is the duration of the modulation interval. Moreover, $N = 3$ consecutive receive symbols (after matched-filtering and sampling), denoted as $d[0]$, $d[1]$, and $d[2]$ (vector $\boldsymbol{d} = [d[0], d[1], d[2]]$), are treated jointly.

(a) How does the calculation of $d[k]$ simplify for the present choice of $g(t)$?
(b) How many data-carrying symbols $\tilde{a}[k]$ are represented in the receive block of length N?
(c) Assuming $a[0] = 1$, sketch (in 3D space) the transmit signals

$$s_i(t) = \sum_{k=0}^{2} a_i[k]\, g(t - kT)$$

corresponding to all possible transmit vectors $\boldsymbol{a}_i = [a_i[0], a_i[1], a_i[2]]$ of length $N = 3$. For each signal, give the corresponding vector $\tilde{\boldsymbol{a}}$ of data-carrying symbols.

We now apply the optimum matched-filter demodulator for non-coherent reception to the signal elements (signals spanning N symbol periods) from part (c). The complex correlations thereby are calculated as

$$K_i = \boldsymbol{d}\boldsymbol{a}_i^{\mathrm{H}} .$$

(d) Show that the restriction to $a[0] = 1$ can be done without loss of generality. How can the detection problem be stated for the current situation?
(e) Perform the detection of the following vector of detection symbols:

$$\boldsymbol{d} = [\,1.17 + 1.26\text{j} \quad -1.02 + 0.02\text{j} \quad 0.09 - 0.82\text{j}\,] .$$

(f) Compare the result to that obtained from
- differential coherent demodulation
- differential demodulation.

7 Digital Frequency and Phase Modulation

In this chapter, we study *digital frequency modulation*, in particular, *frequency-shift keying (FSK)*. As already mentioned briefly in previous chapters, FSK does not fall under the umbrella of PAM schemes. Even though the frequency and phase of a carrier signal are tightly related (the instantaneous frequency is given by the derivative of the phase, cf. (2.2)), differentially encoded PSK and FSK are different families. Moreover, as we will see, in FSK, the continuity of the carrier phase plays an important role, resulting in *continuous-phase FSK (CPFSK)*. A generalization of CPFSK leads to *continuous-phase modulation (CPM)*, similar to the generalization of MSK to Gaussian MSK in Section 4.1.3. Here, only a very brief introduction to CPM will be given; an in-depth treatment of this broad field is beyond the scope of this book.

As the name suggests, in FSK the information is represented by the *instantaneous frequency*. Because the amplitude of the ECB signal and thereby the envelope of the RF signal is constant, efficient power amplification is possible. In this chapter, we assess FSK/CPM and their inherent coding. For the characterization and analysis, we have to apply the general signal space concept derived in Chapter 6.

PREREQUISITES FOR THIS CHAPTER

- PAM transmission (Chapter 3).
- Signal space representation (Chapter 6).

7.1 General Concept of Frequency-Shift Keying

We consider M-ary FSK employing *hard keying* – the signal elements are limited to the duration T of the symbol interval. In the simplest case, the signal elements are drawn uniformly and independently per modulation interval. As we will see, using coding, the power and bandwidth efficiency can be increased simultaneously. The effect that coding may lead to particular properties (e.g., continuity) of the transmit signal has been exemplarily demonstrated in the example on page 227. Here, we will treat encoding and decoding in detail.

7.1.1 Signal Elements

Bipolar Description

The signal elements of M-ary (M even) hard-keyed FSK are given in the ECB domain by

$$s_i(t) = \begin{cases} \sqrt{\frac{E_s}{T}}\, e^{j2\pi\left(i-\frac{M-1}{2}\right)h t/T}, & t \in [0, T) \\ 0, & t \notin [0, T) \end{cases}, \quad i = 0, 1, \ldots, M-1, \quad (7.1)$$

where h is the *modulation index*.

Note that, using time-limited signal elements, the temporal orthogonality condition (6.5) imposed on signal elements is obviously satisfied.

Using (2.2), the instantaneous frequency of the signal elements calculates to

$$\begin{aligned} f_{M,i}(t) &= \frac{1}{2\pi}\frac{d}{dt} 2\pi \left(i - \tfrac{M-1}{2}\right) h t/T \\ &= \left(i - \tfrac{M-1}{2}\right) h/T, \quad i = 0, 1, \ldots, M-1, \quad (7.2) \end{aligned}$$

that is, it is a multiple of h/T. The factors are $-\frac{M-1}{2}, \ldots, -\frac{1}{2}, \frac{1}{2}, \ldots, \frac{M-1}{2}$. Hence, if applicable, we call this variant FSK_b (for bipolar). The modulation index h is a design parameter of FSK and T is fixed. The quotient h/T specifies the smallest difference of instantaneous frequencies between the signal elements. This means that signal element $s_{i+1}(t)$ conducts h periods of oscillation more in one modulation interval of duration T than signal element $s_i(t)$. The modulation index h thus gives the number of oscillation cycles by which the signal elements at least differ within one modulation interval T. Note that, in the definition used, half of the signal elements have a negative instantaneous frequency, the other half a positive one. In general, in the ECB domain, signal element $s_i(t)$ fulfills $\left|\left(i - \frac{M-1}{2}\right)h\right|$ periods of oscillation in one modulation interval of duration T. Thereby, a negative frequency means an evolution in a mathematically negative direction.

Unipolar Description

Note that in the literature, an exposition of FSK also exists where the instantaneous frequencies are all positive, specifically $f_{M,i} = i h/T$, $i = 0, \ldots, M-1$. We denote this variant by FSK_u (unipolar). For FSK_u, the smallest instantaneous frequency of the RF signal is chosen as reference frequency f_0 of the ECB transform. With this reference frequency, the instantaneous frequencies of the real-valued RF signal elements are thus $f_0 + i h/T$. Hence, here *reference/transformation* frequency f_0 and *carrier* frequency (mid frequency of the RF spectrum) f_c differ. Specifically, as the mean between 0 and $M - 1$ is $\frac{M-1}{2}$, the carrier frequency is given by $f_c = f_0 + h \frac{(M-1)}{2T}$.

Properties

Subsequently, we use the description employing positive and negative instantaneous frequencies (FSK$_b$). Since in the above definition the mean over these frequencies is zero, we have $f_c = f_0$. For CPM (Section 7.4), the unipolar notation FSK$_u$ will be preferred for a concise representation.

As can be seen from (7.1), the envelopes of the signal elements are

$$|s_i(t)| = \sqrt{\frac{E_s}{T}} = \text{const.}, \qquad t \in [0, T), \quad \forall i. \qquad (7.3)$$

No amplitude modulation is present. The energy per symbol is

$$\int_0^T |s_i(t)|^2 \, dt = \int_0^T \frac{E_s}{T} \, dt = E_s. \qquad (7.4)$$

We give an example of signal elements.

Example 7.1 FSK Signal Elements

In Figure 7.1, the ECB signal elements for $M = 4$-ary FSK$_b$ with $h = 1/2$ are depicted as phasors (as a trajectory in the complex plane).

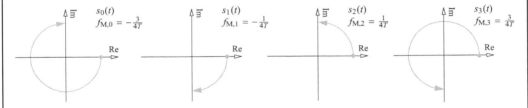

Figure 7.1 Phasors of FSK signal elements in the ECB domain. $M = 4$, $h = 1/2$.

Since $h = 1/2$, signal element $s_0(t)$ carries out $\left(0 - \frac{4-1}{2}\right)\frac{1}{2} = -\frac{3}{4}$ of an oscillation cycle, that is, 3/4 in the negative direction. Signal element $s_1(t)$ carries out a quarter of a full cycle in the negative direction, $s_2(t)$ a quarter of a full cycle in the positive direction, and $s_3(t)$ three-quarters of a full cycle in the positive direction. The constant magnitude is obvious as the phasors are segments of a circle.

Instead of the phasors, the phase can be plotted over time. This is done in Figure 7.2. As the magnitude is constant, the phase functions contain the entire information.

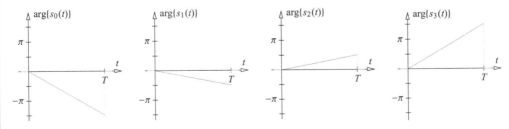

Figure 7.2 Phase of FSK signal elements in ECB domain. $M = 4$, $h = 1/2$.

The corresponding real-valued RF signal elements for $f_0 = \frac{11}{4T}$ (very small for illustration purposes) are plotted in Figure 7.3.

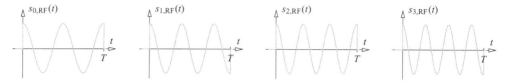

Figure 7.3 FSK signal elements in RF domain. $M = 4$, $h = 1/2$.

Since in the RF domain signal element $s_0(t)$ has instantaneous frequency $f_{RF,0} = f_0 + f_{M,0} = \frac{11}{4T} - \frac{3}{4T} = \frac{2}{T}$, two periods of the cosine wave are visible. Signal element $s_1(t)$ carries out 2.5 periods, and so on.

Similar to (6.3) for the general case, the FSK transmit signal is given by the concatenation of the signal elements, that is

$$s(t) = \sum_{k=-\infty}^{\infty} e^{j\varphi_0[k]} s_{m[k]}(t - kT) \,, \tag{7.5}$$

where $\langle m[k] \rangle$ is the sequence of signal numbers and $\varphi_0[k]$ an additional phase term.

When looking at the above example, one immediately observes that for $\varphi_0[k] =$ const. (w.l.o.g. $\varphi_0[k] = 0$), the ECB transmit signal exhibits *phase jumps* between adjacent signal elements. A *discontinuous FSK* is present.

The initial phase may be adjusted (individually for each modulation interval k) according to

$$\varphi_0[k] = \lim_{\Delta t \to 0} \arg \{ e^{j\varphi_0[k-1]} s_{m[k-1]}(T - \Delta t) \} \,, \tag{7.6}$$

that is, the starting phase at modulation interval k is chosen to be the phase at the end of modulation interval $k - 1$. Then, the phase becomes continuous (no phase jumps occur) and *continuous-phase frequency shift keying (CPFSK)* is achieved.

Thus, the starting phase has to be recursively updated, that is, the starting phase for the next symbol interval is the sum of phase increments of all previous signal elements.

As we will see in a moment, continuous-phase FSK exhibits *inherent coding*. The encoding rule (7.6) introduces dependencies between the signal elements. In FSK, such a form of coding will increase the bandwidth efficiency (a more compact spectrum of the transmit signal is present if phase jumps are avoided) and additionally will increase the power efficiency (if the receiver searches only over sequences which fulfill the continuous-phase property, i.e., decoding is performed) simultaneously.

Note that, when the instantaneous frequencies are chosen to be all positive (FSK$_u$), $f_{M,i} = i h/T$, $i = 0, \ldots, M - 1$, and the modulation index is $h = 1$ (in general, $h \in \mathbb{N}$), a CPFSK signal is always produced and no inherent coding is present (so-called *Sunde's FSK*).

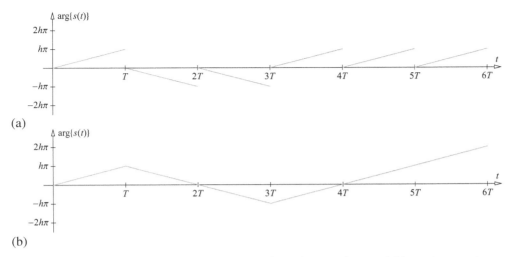

Figure 7.4 FSK signals ($M = 2$) with (a) discontinuous phase and (b) continuous phase.

Examples of FSK_b signals (phases $\arg\{s(t)\}$) with discontinuous phase and continuous phase, respectively, are plotted in Figure 7.4. To this end, we assume binary FSK ($M = 2$). The binary "0" is represented by the signal element $s_0(t)$; the binary "1" by the signal element $s_1(t)$. Both signals represent the binary source sequence ⟨1, 0, 0, 1, 1, 1⟩.

Example 7.2 Phasor Demo

As in Chapters 3 and 4, phasors are available on the companion webpage for visualization. The phasors are plotted in three-dimensional space; real and imaginary parts of the complex baseband signal $s(t)$ over time (normalized to the bit duration T_b). A number of possible signals (corresponding to different sequences of amplitude coefficients) are plotted on top of each other. One of them is plotted in red. Rotate the phasor, such that the real part (in-phase component) or the imaginary part (quadrature component) is plotted over time (eye pattern), and rotate the phasor such that the complex plane (imaginary vs. real part) is visible.

The following phasors of signals are available.

Phasor 13: Transmit signal $s(t)$ for binary ($M = 2$) discontinuous FSK with $h = 1/2$. Unipolar variant FSK_u.

Phasor 14: Transmit signal $s(t)$ for binary ($M = 2$) discontinuous FSK with $h = 1/2$. Bipolar variant FSK_b.

Phasor 15: Transmit signal $s(t)$ for binary ($M = 2$) continuous-phase FSK with $h = 1/2$. Bipolar variant FSK_b. Equal to MSK.

Phasor 16: Transmit signal $s(t)$ for binary ($M = 2$) FSK with $h = 1$. Unipolar variant FSK_u; inherently, a continuous phase is obtained.

Phasor 17: Transmit signal $s(t)$ for quaternary ($M = 4$) discontinuous FSK with $h = 1/2$. Bipolar variant FSK_b (same situation as shown in Figure 7.1).

Phasor 18: Transmit signal $s(t)$ for quaternary ($M = 4$) continuous-phase FSK with $h = 1/2$. Bipolar variant FSK_b.

7.1.2 Average Power Spectral Density of FSK

In order to calculate the average psd of FSK, we have to distinguish between the discontinuous- and continuous-phase case.

Discontinuous FSK

If the signal elements are drawn uniformly and independently over the time steps (i.e., $\varphi_0[k] = 0 \; \forall k$, no coding is used), the general psd calculation of Section 6.1 applies directly. The spectra of the FSK_b signal elements calculate as

$$s_i(t) = \sqrt{\frac{E_s}{T}} \, e^{j2\pi\left(i-\frac{M-1}{2}\right)h\,t/T} \cdot \text{rect}\left((t - T/2)/T\right)$$

$$\circ\!\!-\!\!\bullet \qquad\qquad\qquad\qquad\qquad\qquad i = 0, 1, \ldots, M - 1 \quad (7.7)$$

$$S_i(f) = \sqrt{E_s T} \, \text{si}\left(\pi(fT - (2i - M + 1)h/2)\right) \cdot e^{-j\pi fT} \cdot e^{j\pi(2i-M+1)h/2},$$

where $\text{si}(x)$ is the sinc function, see Appendix A. With the definition of the average signal element

$$\bar{s}(t) \stackrel{\text{def}}{=} \frac{1}{M} \sum_{i=0}^{M-1} s_i(t) \quad \circ\!\!-\!\!\bullet \quad \bar{S}(f) = \frac{1}{M} \sum_{i=0}^{M-1} S_i(f), \quad (7.8)$$

and using (6.16), the average psd of FSK is given by

$$\bar{\Phi}_{ss}(f) = \frac{1}{T}\left(\frac{1}{M}\sum_{i=0}^{M-1} |S_i(f)|^2 - |\bar{S}(f)|^2\right) + \frac{1}{T^2}|\bar{S}(f)|^2 \sum_{\mu=-\infty}^{\infty} \delta\left(f - \frac{\mu}{T}\right). \quad (7.9)$$

As $\bar{s}(t) \neq 0$ for all choices of M and h, discontinuous FSK will always exhibit discrete spectral lines at integer multiples of $1/T$ (unless $\bar{S}(\mu/T) = 0$).

Finally, it has to be noted that, as the signal elements are time-limited, the power spectrum cannot be bandlimited. From the above equations it can be seen that the average psd is basically the superposition of squared sinc functions. Hence, bandwidth has to be measured as $X\%$ (90% or 99%) bandwidth, as explained in Section 4.1.2.

Continuous-Phase FSK

If, according to (7.6), the initial phase of the signal element at modulation interval k is chosen to be the phase at the end of modulation interval $k - 1$, dependencies between the signal elements are introduced – an inherent coding is present.

The computation of the psds of CPFSK and CPM is rather complex. Solutions are available from general methods for the calculation of an average psd for stochastic processes that form a Markov chain. However, the result for the CPFSK case at hand (FSK$_b$, rectangularly time-limited signal elements) is short enough to be stated here. To that end, we define the following quantities:

$$c \stackrel{\text{def}}{=} \frac{\sin(M\pi h)}{M \sin(\pi h)},$$

$$\gamma_i = \pi(fT - (2i - M + 1)h/2), \qquad i = 0, 1, \ldots, M - 1, \qquad (7.10)$$

$$\alpha_{il} = \frac{\cos(\gamma_i + \gamma_l) - c \cos(\gamma_i + \gamma_l - 2\pi fT)}{1 - 2c \cos(2\pi fT) + c^2}, \qquad i, l = 0, 1, \ldots, M - 1.$$

Then, (for $h \notin \mathbb{N}$) the average psd of the ECB signal of continuous-phase FSK is given by

$$\bar{\Phi}_{ss}(f) = \frac{E_s}{M} \sum_{i=0}^{M-1} \left(\text{si}^2(\gamma_i) + \frac{2}{M} \text{si}(\gamma_i) \sum_{l=0}^{M-1} \alpha_{il} \, \text{si}(\gamma_l) \right). \qquad (7.11)$$

As can be seen from (7.11), for $h \notin \mathbb{N}$, no discrete spectral lines are present. The continuous-phase property (phase jumps are avoided) leads to a psd that only has a continuous part and (as we will see subsequently) is more compact.

Note that for the important case $Mh = 1$, the above-defined quantity c is zero. The equation for the average psd of CPFSK specializes to

$$\bar{\Phi}_{ss}(f) = \frac{E_s}{M} \sum_{i=0}^{M-1} \left(\text{si}^2(\gamma_i) + \frac{2}{M} \text{si}(\gamma_i) \sum_{l=0}^{M-1} \cos(\gamma_i + \gamma_l) \, \text{si}(\gamma_l) \right). \qquad (7.12)$$

Note that MSK with $M = 2$ and $h = 1/2$ belongs to this special case.

7.2 Binary Frequency-Shift Keying

An important special case is obtained for $M = 2$: *binary frequency-shift keying*. From (7.1), the two signal elements are here given by

$$\begin{aligned} s_0(t) &= \sqrt{\tfrac{E_s}{T}}\, e^{-j2\pi \frac{1}{2} h t/T}, \\ s_1(t) &= \sqrt{\tfrac{E_s}{T}}\, e^{+j2\pi \frac{1}{2} h t/T}, \end{aligned} \qquad t \in [0, T). \qquad (7.13)$$

Moreover, $T = T_b$, and the energy per symbol is equal to the energy per bit, $E_s = E_b$.

7.2.1 Discontinuous Binary FSK

First, we assume *discontinuous binary FSK* – the starting phase of the signal elements is always $\varphi_0[k] = 0$. Consequently, *symbol-by-symbol detection* is employed.

Normalized Minimum Squared Euclidean Distance of Discontinuous Binary FSK

As usual, we characterize the power efficiency by the normalized minimum squared Euclidean distance $d_{\min}^2$ of the modulation scheme. Note that FSK with only positive instantaneous frequencies (FSK_u) has the same power efficiency as FSK_b, which is studied here. Since $M = 2$, $d_{\min}^2$ is identical to the normalized squared Euclidean distance between the signal elements. Using (6.73), we have

$$d_{\min}^2 = \frac{1}{2E_\text{b}} \int_0^T |s_0(t) - s_1(t)|^2 \, dt$$

$$= \frac{1}{2E_\text{b}} \int_0^T (|s_0(t)|^2 - 2\text{Re}\{s_0(t)\, s_1^*(t)\} + |s_1(t)|^2)\, dt$$

$$= \frac{1}{2E_\text{b}} \left(\int_0^T |s_0(t)|^2 \, dt - \text{Re}\left\{ \int_0^T s_0(t)\, s_1^*(t)\, dt \right\} + \int_0^T |s_1(t)|^2\, dt \right)$$

with energy (7.4)

$$= \frac{E_\text{s}}{E_\text{b}} (1 - \text{Re}\{\rho\}) \, . \tag{7.14}$$

Thereby, ρ is the *correlation coefficient* between the signal elements, defined as

$$\rho \stackrel{\text{def}}{=} \frac{1}{E_\text{s}} \int_0^T s_0(t)\, s_1^*(t)\, dt$$

$$= \frac{1}{E_\text{s}} \int_0^T \sqrt{\frac{E_\text{s}}{T}} e^{-j 2\pi \frac{1}{2} h t/T} \cdot \sqrt{\frac{E_\text{s}}{T}} e^{-\left(+j 2\pi \frac{1}{2} h t/T\right)}\, dt$$

$$= \frac{1}{T} \int_0^T e^{-j 2\pi h t/T}\, dt$$

$$= \frac{1}{T} \left[\frac{1}{-j 2\pi h/T} e^{-j 2\pi h t/T} \right]_0^T$$

$$= j\frac{e^{-j 2\pi h} - e^0}{2\pi h}$$

$$= \frac{\sin(2\pi h)}{2\pi h} + j \frac{\cos(2\pi h) - 1}{2\pi h} \, . \tag{7.15}$$

Since for binary transmission $E_\text{b} = E_\text{s}$ holds, we finally arrive at

$$d_{\min}^2 = 1 - \text{Re}\{\rho\} = 1 - \frac{\sin(2\pi h)}{2\pi h} = 1 - \text{si}(2\pi h) \, . \tag{7.16}$$

The normalized minimum squared Euclidean distance $d_{\min}^2$ of binary FSK as a function of the modulation index h is plotted in Figure 7.5.

Increasing h, the distance (and thus the power efficiency) increases first. Then, however, a maximum is achieved. The maximum value of the normalized minimum squared Euclidean distance is

$$d_{\min}^2 = 1.217 \quad \text{for} \quad h = 0.715 \, . \tag{7.17}$$

For large indices, $d_{\min}^2$ converges to 1.

7.2 Binary Frequency-Shift Keying

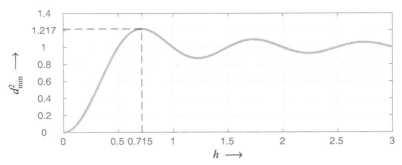

Figure 7.5 Normalized minimum squared Euclidean distance $d_{\min}^2$ of discontinuous binary FSK.

Signal Space Representation

Via the Gram–Schmidt procedure, an orthonormal basis can be calculated for the two signal elements of binary FSK. Using $E_g = E_s$, the first signal element immediately defines the first basis function, that is, $g_0(t) = s_0(t)$. Then, projecting the second signal element onto $g_0(t)$ and subtracting the projection, after normalization, the second basis function is obtained. Note that the linear factor for the second dimension is, in general, complex-valued. The signal space for binary FSK is visualized in Figure 7.6.

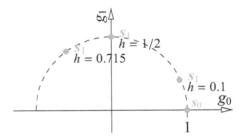

Figure 7.6 Signal space representation of binary FSK.

For $h = n/2$, $n \in \mathbb{N}$, the real part of the correlation coefficient (which is the normalized scalar product between the signal elements) is $\text{Re}\{\rho\} = 0$; the signal elements are *orthogonal* to each other (w.r.t. coherent reception). Here, $g_1(t) = s_1(t)$. The *minimum* modulation index for orthogonal signal elements is thus $h = 1/2$; this is why the respective scheme is called *minimum shift keying (MSK)* (cf. Section 4.1.2).

If h is small, the point s_1 is close to s_0; the distance is small. For $h = 0.715$, the maximum distance between the two points is obtained. Note that the energy of the signal elements is constant; the (squared) distance from the origin (cf. (6.34)) is thus constant.

Coherent Demodulation of Discontinuous Binary FSK

Applying the general concept of the matched-filter demodulator (Figure 6.25), the optimum receiver for (discontinuous) binary FSK is immediately given as depicted

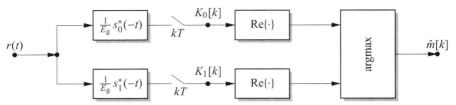

Figure 7.7 Optimum coherent matched-filter demodulator for discontinuous binary FSK.

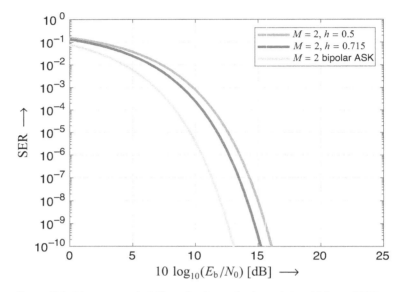

Figure 7.8 Bit error probability of coherently demodulated binary FSK.

in Figure 7.7. As $E_i = E_s = E_g$ for both signal elements, the energy correction is irrelevant.

The bit error probability (with $N_{\min} = 1$) is given by

$$\text{BER} = Q\left(\sqrt{d_{\min}^2 \frac{E_b}{N_0}}\right) = Q\left(\sqrt{(1 - \text{si}(2\pi h))\frac{E_b}{N_0}}\right). \qquad (7.18)$$

In Figure 7.8, the bit error probability is plotted over the signal-to-noise ratio for different modulation indices h. For comparison, the performance of 2ASK (identical to 4QAM) is shown. The 3 dB loss compared to ASK (which has $d_{\min}^2 = 2$) is visible; for the optimum h, a small improvement is possible over $h = 1/2$ (MSK).

Non-coherent Demodulation of Discontinuous Binary FSK

FSK can also be demodulated non-coherently. From the general concept, the optimum matched-filter demodulator (Figure 6.39) for symbol-by-symbol detection is readily specialized to the situation at hand. As $E_i = E_s = E_g$ for both signal elements, the energy correction and the nonlinear characteristic curve are irrelevant (uncoded transmission). Figure 7.9 depicts the non-coherent demodulator.

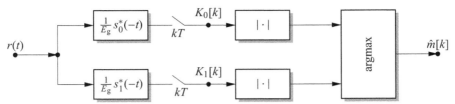

Figure 7.9 Optimum non-coherent matched-filter demodulator for discontinuous binary FSK.

In calculating the bit error probability for the non-coherent demodulation of binary FSK, we assume an AWGN channel with phase error (cf. Figure 6.36). Without loss of generality, we may assume that at modulation interval $k = 0$ the signal element $s_0(t)$ has been transmitted. The samples at the outputs of the matched filters ($t = 0$) are then given as

$$K_0 = \int (s_0(t)e^{j\varphi_\Delta} + n(t)) \frac{1}{E_g} s_0^*(t) \, dt$$

$$= e^{j\varphi_\Delta} \frac{1}{E_g} \int |s_0(t)|^2 \, dt + \int n(t) \frac{1}{E_g} s_0^*(t) \, dt$$

$$= e^{j\varphi_\Delta} + n_0 , \qquad (7.19)$$

$$K_1 = \int (s_0(t)e^{j\varphi_\Delta} + n(t)) \frac{1}{E_g} s_1^*(t) \, dt$$

$$= e^{j\varphi_\Delta} \frac{1}{E_g} \int s_0(t) s_1(t)^* \, dt + \int n(t) \frac{1}{E_g} s_1^*(t) \, dt$$

$$= e^{j\varphi_\Delta} \rho + n_1 , \qquad (7.20)$$

where the obvious definitions of the noise samples n_0 and n_1 have been used.

According to the decision rule, a bit error occurs if $|K_1| > |K_0|$ (since $s_0(t)$ has been transmitted); the bit error probability is thus given by

$$\text{BER} = \Pr\{|K_1| > |K_0|\} . \qquad (7.21)$$

In order to calculate this probability, we require the statistics of $|K_0|$ and $|K_1|$. It is straightforward to show that

$$\mathrm{E}\{|n_0|^2\} = \mathrm{E}\{|n_1|^2\} = \frac{N_0}{E_g} , \qquad (7.22)$$

$$\mathrm{E}\{n_0 n_1^*\} = \rho^* \frac{N_0}{E_g} . \qquad (7.23)$$

Owing to the Gaussianity of the channel noise $n(t)$, n_0 and n_1 are also Gaussian – $|K_0|$ and $|K_1|$ are thus *correlated Ricean-distributed random variables*.

Conducting an analysis analogous to that for DPSK in Section 5.2, we obtain after some calculations

$$\text{BER} = Q_M(x,y) - \tfrac{1}{2} e^{-(x^2+y^2)/2} I_0(xy) \qquad (7.24)$$

with

$$x \stackrel{\text{def}}{=} \sqrt{\frac{E_b}{2N_0}(1 - \sqrt{1-|\rho|^2})}, \quad y \stackrel{\text{def}}{=} \sqrt{\frac{E_b}{2N_0}(1 + \sqrt{1-|\rho|^2})}.$$

Thereby, the definitions $Q_M(x,y)$ of the *Marcum-Q-function* and $I_0(x)$ of the *zeroth-order modified Bessel function of the first kind* have been used (cf. Section 5.3).

It should be noted that for the error probability, the magnitude of the correlation coefficient ρ matters. This magnitude is plotted in Figure 7.10.

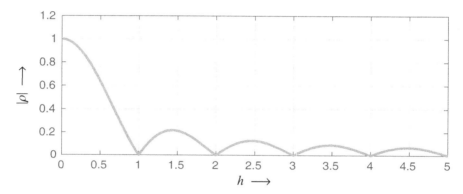

Figure 7.10 Magnitude of the correlation coefficient ρ of binary FSK.

For $h \in \mathbb{N} \setminus \{0\}$, we have $|\rho| = 0$ and *orthogonality* with respect to non-coherent demodulation is obtained. For orthogonal signal elements, we have $x = 0$, $y = \sqrt{E_b/N_0}$ and thus (using $I_0(0) = 1$ and $Q_M(0,y) = e^{-y^2/2}$), the bit error ratio expression reduces to

$$\text{BER} = \frac{1}{2} e^{-E_b/(2N_0)}. \qquad (7.25)$$

A comparison with the bit error ratio expression (5.34) of differentially detected binary DPSK reveals that for $h \in \mathbb{N} \setminus \{0\}$, 2FSK is exactly 3 dB less power-efficient than 2DPSK. The explanation of this effect is as follows. In the here considered symbol-by-symbol detection of FSK, each modulation interval of duration $T = T_b$ is treated separately. In contrast, differential demodulation of 2DPSK can be interpreted as the non-coherent demodulation with respect to the equivalent signal elements

no phase jump: $\quad s'_0(t) = g(t+T) + g(t),$
phase jump: $\quad s'_1(t) = g(t+T) - g(t),$ $\qquad (7.26)$

shown in Figure 7.11 for a rectangular basic pulse shape $g(t)$.

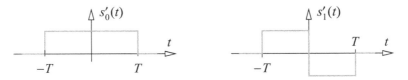

Figure 7.11 Equivalent signal elements for the non-coherent demodulation of 2DPSK.

Since

$$\int_{-\infty}^{\infty} |s'_0(t)|^2 \, dt = \int_{-\infty}^{\infty} |s'_1(t)|^2 \, dt = 2E_g , \qquad (7.27)$$

twice the impulse energy is exploited; each signal element $g(t)$ is used twice. Binary DPSK is thus 3 dB better than binary FSK with $|\rho| = 0$.

Finally, the bit error ratio of non-coherently demodulated binary FSK is displayed in Figure 7.12. Thereby, various values of the correlation coefficient are assumed: $|\rho| = 0.0, 0.1, \ldots, 0.9, 1.0$.

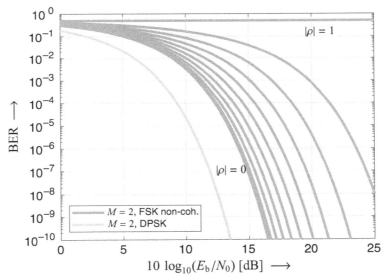

Figure 7.12 Bit error probability of non-coherently demodulated binary FSK.

For $|\rho| = 0$, that is, orthogonal signal elements (w.r.t. non-coherent demodulation), the best performance is obtained. As $|\rho| \to 1$, the performance degrades significantly. $|\rho| = 1$ is only obtained for $s_0(t) = s_1(t)$; the signals cannot be distinguished and the error ratio is BER = 1/2.

7.2.2 Binary Continuous-Phase FSK

We now turn to *binary continuous-phase FSK* – in each symbol interval, the starting phase of the signal element is chosen to guarantee a continuous phase over the entire signal. As already stated, this operation of adjusting the phase of the current signal element dependent on the previous ones introduces memory (i.e., some form of coding).

Trellis Description of Binary Continuous-Phase FSK

In order to study the effect of this inherent coding, for simplicity of exposition, we assume $Mh \in \mathbb{N} \setminus \{0\}$; this situation has already been visualized in Figure 7.4. For illustration purposes, let $h = 1/2$ (and $M = 2$ is fixed). We study the phase $\varphi(t) = \arg\{s(t)\}$ of the ECB transmit signal $s(t)$. Assume that we start (at $t = 0$) with the phase $\varphi(0) = 0$. At the end of the first modulation interval, the phase is either $\varphi(T) = -2\pi\frac{1}{2}ht/T|_{h=1/2,\ t=T} = -\pi/2$ (when sending $s_0(t)$) or $\varphi(T) = +\pi/2$ (when sending $s_1(t)$). The starting phase of the next signal element is accordingly adjusted. After the second modulation interval, the phase is now either $-\pi/2 - \pi/2 = -\pi$, $-\pi/2 + \pi/2 = 0$ or $\pi/2 - \pi/2 = 0$, $\pi/2 + \pi/2 = \pi$. Since the phase is 2π periodic ($-\pi$ and π represent the same phase), the ending phase at $t = 2T$ is either 0 or π – this procedure proceeds in the same way. The phases of all possible transmit signals (all possible binary sequences) are plotted in Figure 7.13 (cf. also Figure 4.19).[1]

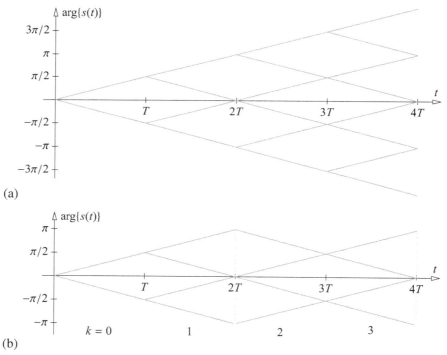

Figure 7.13 Phase trajectory of binary CPFSK. $h = 1/2$. (a) Non-reduced version; (b) 2π-reduced to the interval $(-\pi, \pi]$.

---

[1] Note that $-\pi$ and π represent the same phase. At $t = 2T \cdot z$, $z \in \mathbb{Z}$, the dotted line should indicate that these two arguments are identical. This also means that two branches (one with increasing and one with decreasing phase) leave from this starting phase. If the phase is plotted on a cylinder (cf. Figure 4.20), no discontinuities can be seen, of course.

7.2 Binary Frequency-Shift Keying

Hence, the phase at the end of modulation intervals where k is even is $\pm\pi/2$; at the end of modulation intervals where k is odd it is 0 or π. In each case, two possibilities are present that define *states*, representing the entire history. From each state, two *branches* emerge, corresponding to the two binary symbols ("0" and "1") that can be sent. The branches end in the next state. Hence, the inherent dependencies within the CPFSK signal are characterized by a *finite-state machine*, which is nothing other than a *trellis encoder* (see Section B.2 in Appendix B).

In describing the transmit signal, besides the signal elements $s_0(t)$ and $s_1(t)$ that have starting phase $\varphi_0[k] = 0$, we also need the signal elements with starting phases $-\pi/2$, $+\pi/2$, and π, respectively. Since $e^{j\phi + n\pi/2} = j^n e^{j\phi}$, $n \in \mathbb{Z}$ (i.e., a phase shift by an integer (n) multiple of $\pi/2$ corresponds to a multiplication with j^n), the signal elements are given as collected in Table 7.1. The figures show the respective phasors in the complex plane.

Table 7.1 Signal elements.

$$s_0(t) = \sqrt{\tfrac{E_s}{T}}\, e^{-j2\pi \tfrac{1}{2} h t/T},\ t \in [0, T) \qquad s_1(t) = \sqrt{\tfrac{E_s}{T}}\, e^{+j2\pi \tfrac{1}{2} h t/T},\ t \in [0, T)$$

φ_0				
$\varphi_0 = 0$	$s_0(t)$		$s_1(t)$	
$\varphi_0 = \pi/2$	$j\, s_0(t)$		$j\, s_1(t)$	
$\varphi_0 = \pi$	$(j)^2 s_0(t) = -s_0(t)$		$(j)^2 s_1(t) = -s_1(t)$	
$\varphi_0 = -\pi/2$	$(j)^3 s_0(t) = -j\, s_0(t)$		$(j)^3 s_1(t) = -j\, s_1(t)$	

Using these definitions, the phase trajectory of Figure 7.13 can be redrawn as given in Figure 7.14; the *trellis diagram* of binary CPFSK with $h = 1/2$. Five *trellis segments* – corresponding to five modulation intervals – are sketched. The states are labeled with the phase (the ending phase of merging branches is equal to the starting phase of emerging branches) and the branches are labeled with the corresponding signal element that is used. A close look reveals that the trellis diagram is time-variant and periodic – it repeats after four modulation steps. The signal elements corresponding to the branches are $\pm(-j)^k s_i(t)$.

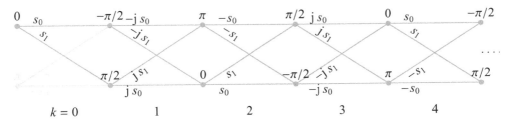

Figure 7.14 Trellis diagram of binary CPFSK with $h = 1/2$.

An alternative method to derive a time-invariant trellis is the unipolar description FSK_u, where the signal elements are either constant ($f_{M,0} = 0$) or a positive phase ramp adding π to the accumulated phase ($f_{M,1} = \frac{1}{2T}$). Note that the rotation factor $(-j)^k$ for the signal elements is equivalent to a shift of the reference frequency in the ECB transform by $\frac{1}{4T}$, that is, to a shift from bipolar to unipolar description of FSK. For FSK_u, a time-invariant trellis description results.

Each binary source sequence $\langle q[k] \rangle$ corresponds to a unique *path* through the trellis; reading off the signal elements at the respective branches gives the corresponding transmit signal $s(t)$.

Normalized Minimum Squared Euclidean Distance of Binary Continuous-Phase FSK

We are now ready to characterize the power efficiency of binary continuous-phase FSK. As per definition (6.73), the normalized minimum squared Euclidean distance of a modulation format is given by

$$d^2_{\min} = \frac{1}{2 E_\text{b}} \cdot \min_{\substack{\forall s^{(m)}(t), s^{(\mu)}(t) \\ s^{(m)}(t) \neq s^{(\mu)}(t)}} \int_{-\infty}^{\infty} |s^{(m)}(t) - s^{(\mu)}(t)|^2 \, dt \, . \tag{7.28}$$

Thereby, the minimization goes over all pairs of transmit signals. Owing to the above trellis description of binary CPFSK, this minimization is identical to searching over all pairs of paths through the trellis. Two possible paths are depicted in Figure 7.15.

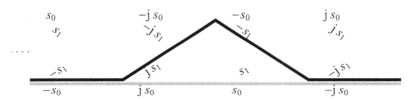

Figure 7.15 Trellis diagram of binary CPFSK ($h = 1/2$) with two possible paths (dark/middle gray).

Considering this pair of paths (they are assumed to be identical before the split and after the merge), an *upper bound* on the minimum squared Euclidean distance can be obtained; any pair of paths limits the distance. The reader is invited to show (by inspecting further pairs of paths from split to merger) that for this pair the minimum in (7.28) is indeed obtained. For path deviations over more than two symbol intervals, additional difference energy is gained and thus the squared Euclidean distance grows. So, the distance between the signal (sequence of signal elements) $\langle \cdots , j s_0, \; s_0, \; \cdots \rangle$ and $\langle \cdots , j s_1, \; -s_1, \; \cdots \rangle$ has to be calculated. Owing to the temporal orthogonality of the signal elements, it is given by

$$d_\text{E} = \int_0^T |j s_0(t) - j s_1(t)|^2 \, dt + \int_0^T |s_0(t) + s_1(t)|^2 \, dt \, . \tag{7.29}$$

Doing the same calculations as in (7.14) and considering that for $h = 1/2$ the signal elements are orthogonal, we arrive at

7.2 Binary Frequency-Shift Keying

$$d_E = 4 E_s, \quad (7.30)$$

and thus the normalized minimum squared Euclidean distance of binary CPFSK with $h = 1/2$ is ($E_b = E_s$)

$$d_{\min}^2 = \frac{1}{2 E_b} d_E = 2. \quad (7.31)$$

Remember that binary FSK with index $h = 1/2$ is *minimum-shift keying* (cf. Section 4.1.2). When interpreting MSK as an offset 4QAM scheme, $d_{\min}^2 = 2$ (cf. (4.7)); this coincides with the derived result. Only when a continuous phase is forced and this fact is utilized at the receiver is this power efficiency obtained. In the OQAM interpretation, a continuous phase is guaranteed that is (implicitly) taken into account at the receiver. In the demodulation process, symbols of duration $2T_b$ are treated, collecting twice the energy. If a continuous phase of the transmit signal is not forced (discontinuous FSK) and symbol-by-symbol (time interval T_b) detection is employed, only $d_{\min}^2 = 1$ (cf. (7.16)) can be exploited; a 3 dB loss in power efficiency is present.

Coherent Demodulation of Binary Continuous-Phase FSK

The above trellis representation leads directly to the optimum receiver for binary continuous-phase FSK with $h = 1/2$ (similar is true as long as $Mh \in \mathbb{N}$). From the general concept, the optimum matched-filter demodulator (Figure 6.39) consists of a bank of matched filters with respect to the used signal elements. Then, the real part is evaluated. As we employ $\pm(-j)^n s_i(t)$, $n = 0, \ldots, 3$ (scaled versions of $s_i(t)$, $i = 0, 1$, see Figure 7.14), we have to calculate

$$\mathrm{Re}\{(r(t) * \tfrac{1}{E_g}(\pm(-j)^n s_i^*(-t)))|_{t=kT}\} = \mathrm{Re}\{\pm(-j)^n (r(t) * \tfrac{1}{E_g} s_i^*(-t))|_{t=kT}\}$$
$$= \pm\mathrm{Re}\{(-j)^n K_i[k]\}. \quad (7.32)$$

Hence, only a bank of matched filters with respect to the signal elements $s_0(t)$ and $s_1(t)$ is required, calculating the correlations $K_0[k]$ and $K_1[k]$.

Moreover, the metric $\mathrm{Re}\{(-j)^n K_i[k]\}$ corresponding to the signal element $(-j)^n s_i(t)$ is the same as $(-1)^n \mathrm{Re}\{j^n K_i[k]\}$, which corresponds to the metric of the signal element $\pm s_i(t)$, multiplied by j^n before taking the real part. When multiplying the complex correlations $K_i[k]$ by j^k (called "de-rotation"), instead of the periodic trellis in Figure 7.14, decoding is done with respect to the regular trellis shown in Figure 7.16.

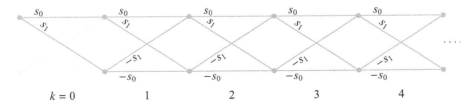

Figure 7.16 "De-rotated" trellis diagram of binary CPFSK with $h = 1/2$.

Hence, the *branch metrics* or *metric increments* (cf. (6.87)) are given by

$$\Lambda_i[k] = \pm \operatorname{Re}\{j^k K_i[k]\} , \tag{7.33}$$

where $\pm$ means "+" if the branch is labeled with $+s_i(t)$ and "−" otherwise. Since all signal elements have the same energy, the energy correction in (6.87) is not required and the factor 2 is irrelevant. Then, analogous to (6.89), a *maximum-likelihood sequence detection* is performed by

$$\langle \hat{q}[k] \rangle = \underset{\forall \langle q[k] \rangle}{\operatorname{argmax}} \sum_k \Lambda_{q[k]}[k] . \tag{7.34}$$

The search over the trellis is efficiently done employing the *Viterbi algorithm* (see Section B.2 in Appendix B).

The block diagram of the optimum receiver for binary continuous-phase FSK with $h = 1/2$ (i.e., MSK) is depicted in Figure 7.17.

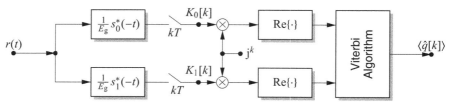

Figure 7.17 Optimum coherent receiver for binary continuous-phase FSK with $h = 1/2$ (MSK).

Note that the alternation between real and imaginary part is in line with the receiver concept for offset 4QAM (of which MSK is a special case) and the simplified receiver concept for GMSK. However, please be aware that there, the time index k refers to QAM symbols (duration $2T_b$), whereas here, it refers to FSK symbols (duration T_b).

7.2.3 Assessment of Binary FSK in the Power/Bandwidth Plane

Finally, we assess binary FSK schemes in the power/bandwidth plane. The power efficiency has just been derived via the normalized minimum squared Euclidean distance (and holds for FSK_b and FSK_u). As already noted, the spectrum of hard-keyed FSK signals has infinite bandwidth. Hence, the 90% or 99% bandwidth (see Chapter 4) has to be specified.

In Figure 7.18, the average psd of binary ($M = 2$) FSK_b with $h = 1/2$ is plotted. The medium gray curve corresponds to continuous-phase FSK_b, which for the used parameters is identical to MSK (cf. Figure 4.11), and the dark gray one to discontinuous FSK_b. The discrete spectral lines are clearly visible.[2] The 90% bandwidths of the schemes are indicated with horizontal light gray bars. For MSK it is

---

[2] Here, the continuous part and the discontinuous part of the power spectrum are plotted in the same diagram, despite the fact that they represent different physical quantities. The continuous part represents power per frequency (power density), and the weight of the discrete spectral lines represents power. For illustration purposes, the height of the lines above the continuous part is chosen to be equal to the power in a frequency band of $fT = 0.01$.

7.2 Binary Frequency-Shift Keying

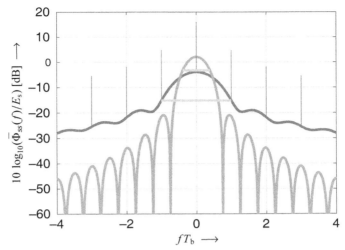

Figure 7.18 Average psd of binary FSK with $h = 1/2$. Medium gray line: continuous-phase FSK$_b$ (here MSK). Dark gray line: discontinuous FSK$_b$. Light gray horizontal lines: frequency band where 90% of the energy is contained.

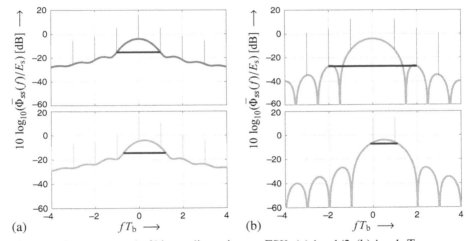

Figure 7.19 Average psd of binary discontinuous FSK. (a) $h = 1/2$. (b) $h = 1$. Top: FSK$_b$, symmetric instantaneous frequencies. Bottom: FSK$_u$, positive instantaneous frequencies.

$B_{90\%} = 0.79 \frac{1}{T_b}$ ($B_{99\%} = 1.18 \frac{1}{T_b}$, cf. Section 4.1.2); for discontinuous FSK we have $B_{90\%} = 2.00 \frac{1}{T_b}$ (and $B_{99\%} = 20.01 \frac{1}{T_b}$). Apparently, a much broader spectrum is present, caused by the phase jumps.

In Figure 7.19(a), the average psd of binary FSK with $h = 1/2$ is shown for the cases of FSK$_b$ (positive and negative instantaneous frequencies; $f_{M,0} = -\frac{1}{4T_b}$ and $f_{M,1} = +\frac{1}{4T_b}$; top) and for FSK$_u$ (instantaneous frequencies $f_{M,0} = 0$ and $f_{M,1} = \frac{1}{2T_b}$, cf. page 248; bottom). When shifting the instantaneous frequencies, the continuous part of the spectrum shifts by the respective amount. The psd (FSK$_u$) has a center frequency of $f_c T_b = h \frac{M-1}{2T_b} T_b = \frac{1}{4}$. However, the discrete spectral lines are still at integer multiples of $1/T_b$. Hence, shifting the instantaneous frequencies in discontinuous FSK does not shift the entire psd but gives (slightly) different results.[3] However, the 90% bandwidths (gray bars) are almost the same; here, $B_{90\%} = 1.94 \frac{1}{T_b}$.

In Figure 7.19(b), the average psd of binary FSK with $h = 1$ is shown. Again, the shifting of the continuous part of the spectrum is visible; here, $f_c T_b = 1/2$. However, for $\mathrm{FSK_u}$ (instantaneous frequencies $f_{M,0} = 0$ and $f_{M,1} = 1/T_b$), all spectral lines except two are suppressed as the average spectrum $\bar{S}(f)$ in (7.9) is zero at these positions. It should be noted that in this setting, a continuous phase is inherently produced (Sunde's FSK), which is not the case if positive and negative instantaneous frequencies are used. As a result, the 90% bandwidths (gray bars) differ significantly; for $\mathrm{FSK_u}$ it is $B_{90\%} = 1.23 \frac{1}{T_b}$. Keep in mind for the comparison that here approximately half of the transmit power is concentrated in the discrete lines.

Finally, in Figure 7.20, the average psds of binary discontinuous $\mathrm{FSK_u}$ (positive instantaneous frequencies) with different modulation indices are plotted. From top

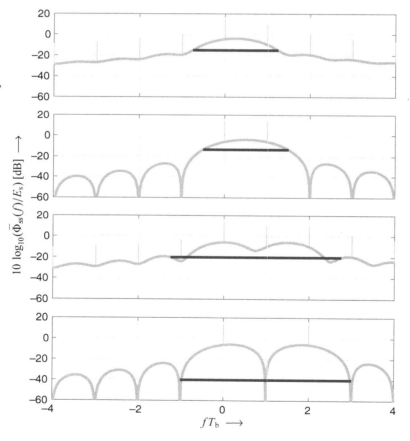

Figure 7.20
Average psd of binary discontinuous $\mathrm{FSK_u}$. Top to bottom: $h = 1/2$, 1, $3/2$, and 2.

---

[3] This effect is due to the frequency shift of the signal elements in the unipolar representation versus the bipolar representation. Thereby, different signals with different phase discontinuities are generated, leading to different psds.

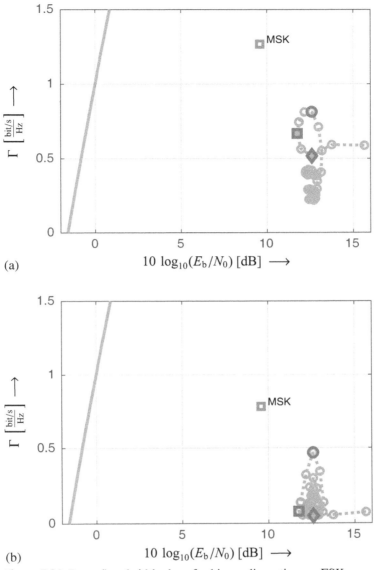

Figure 7.21 Power/bandwidth plane for binary discontinuous FSK_u schemes. The modulation index h is varied from 0.3 to 4.0 in steps of 0.1. Tolerated bit error ratio $\text{BER} \leq 10^{-5}$; spectral efficiency considering $X\%$ bandwidth. (a) 90% bandwidth; (b) 99% bandwidth.

to bottom we have $h = 1/2, 1, 3/2$, and 2. As can be seen, increasing h increases the (90%) bandwidth of the FSK scheme.

For binary FSK, the maximum of the normalized minimum squared Euclidean distance is achieved for $h = 0.715$; modulation indices $h > 0.715$ are (seemingly) useless. This is a different behavior than that of analog frequency modulation (FM). In analog FM, by increasing the frequency deviations (which there is the design parameter and the quantity corresponding to the modulation index h), increased modulation gains are enabled. This comes at the cost of an increased bandwidth (estimated via Carson's bandwidth rule). In binary FSK, when increasing the modulation index h, the $X\%$ bandwidth also increases. However, for large h, the frequency band is not utilized efficiently; "gaps" between parts with larger psd values occur (e.g., for $h = 2$, a notch at $fT_b = 1$ is visible). In order to improve the power efficiency, binary FSK has to be replaced by M-ary FSK (see the next section).

In Figure 7.21, the power/bandwidth plane is depicted for binary discontinuous FSK_u schemes (positive instantaneous frequencies). Note the different vertical scaling compared to the diagrams in Chapters 3 and 4. The modulation index h is varied from 0.3 to 4.0 in steps of 0.1; for better visibility, the points for the various h are connected by a dotted line. The rightmost point corresponds to $h = 0.3$; it has the poorest power efficiency. Increasing h increases the power efficiency (cf. Figure 7.5). The point for $h = 0.715$ is marked with a dark gray square; here the best power efficiency is obtained (leftmost point). For $h = 1$, a continuous phase is guaranteed and the best bandwidth efficiency is obtained (topmost point, dark gray circle). The point corresponding to $h = 1/2$ is also marked (dark gray diamond). For comparison, the power/bandwidth tradeoff of MSK (binary CPFSK with $h = 1/2$) is marked with a medium gray square. Contrasting both points, the jointly increased power and bandwidth efficiency by generating a continuous phase and utilizing this fact at the receiver is clearly visible.

7.3 *M*-ary Frequency-Shift Keying (*M*-ary FSK)

We now turn to M-ary discontinuous FSK ($M > 2$). Using (6.73), the normalized minimum squared Euclidean distance of the discontinuous-phase modulation scheme and *symbol-by-symbol detection* calculates as

$$\begin{aligned}
d_{\min}^2 &= \frac{1}{2E_b} \min_{\substack{\forall i,l \in \{0,1,\ldots,M-1\} \\ i \neq l}} \int_0^T |s_i(t) - s_l(t)|^2 \, dt \\
&= \frac{E_s}{E_b} \min_{\substack{\forall i,l \in \{0,1,\ldots,M-1\} \\ i \neq l}} \frac{1}{2T} \int_0^T |e^{j2\pi(i-\frac{M-1}{2})ht/T} - e^{j2\pi(l-\frac{M-1}{2})ht/T}|^2 \, dt \\
&= \frac{E_s}{E_b} \min_{\substack{\forall i,l \in \{0,1,\ldots,M-1\} \\ i \neq l}} \frac{1}{2T} \int_0^T |e^{j2\pi(i-\frac{M-1}{2})ht/T}|^2 \cdot |1 - e^{j2\pi(l-i)ht/T}|^2 \, dt \\
&= \frac{E_s}{E_b} \min_{\forall i \in \{1,\ldots,M-1\}} \frac{1}{2T} \int_0^T |1 - e^{j2\pi iht/T}|^2 \, dt \, .
\end{aligned}$$

Using the same integration as in the binary case and obeying $E_b = E_s/\log_2(M)$, we finally arrive at

$$d_{\min}^2 = \log_2(M) \min_{\forall i \in \{1,\ldots,M-1\}} (1 - \text{si}(2\pi ih)) \, . \tag{7.35}$$

Having $d_{\min}^2$, the symbol error ratio for coherent detection can be bounded. As each signal element has at most $M - 1$ (nearest) neighbors, applying union-bound

7.3 M-ary Frequency-Shift Keying (M-ary FSK)

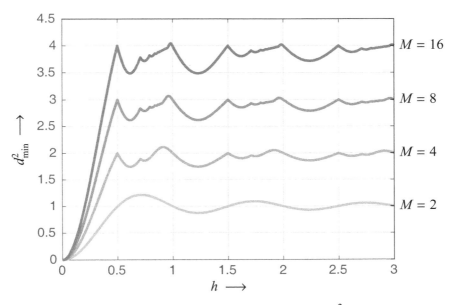

Figure 7.22 Normalized minimum squared Euclidean distance $d_{\min}^2$ of M-ary FSK.

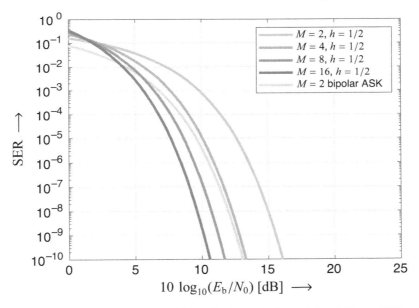

Figure 7.23 Symbol error probability of coherently demodulated M-ary FSK.

techniques (cf. Section 3.6), we have

$$\text{SER} \leq (M-1) \cdot Q\left(\sqrt{d_{\min}^2 E_b/N_0}\right). \tag{7.36}$$

The normalized minimum squared Euclidean distance $d_{\min}^2$ of M-ary FSK as a function of the modulation index h is plotted in Figure 7.22. Increasing h, the distance (and thus the power efficiency) increases but then saturates. Increasing M, the normalized minimum squared Euclidean distance can be increased. From (7.35) it can be deduced that for $h \to \infty$, the distance approaches $d_{\min}^2 \to \log_2(M)$.

Symbol error ratio curves for some non-binary FSK schemes with symbol-by-symbol coherent detection are displayed in Figure 7.23. In each case, $h = 1/2$ is chosen, that is, orthogonal signal elements are used. It can be observed that when increasing M, the power efficiency increases (as $d_{\min}^2$ increases). This is in contrast to PAM schemes, where increasing M leads to a degradation in power efficiency. However, for PAM, the bandwidth is fixed, independent of M. Instead, increasing M in FSK, the bandwidth will increase too. Thus, a classical trade-off between power and bandwidth efficiency is enabled.

7.4 Introduction to Continuous-Phase Modulation (CPM)

Continuous-phase modulation (CPM) is a straightforward generalization of continuous-phase FSK, like the advancement of MSK, interpreted as binary CPFSK with modulation index $h = 1/2$, to Gaussian MSK (GMSK). In CPM, the information is mapped to a smooth *frequency function* that represents the instantaneous frequency of a frequency-modulated (FM) transmit signal. This frequency function is a usual bipolar M-ary ASK signal with a basic pulse $g(t)$. Here, $g(t)$ is called the *frequency pulse* of the CPM scheme. The basic frequency pulse is no longer restricted to a rectangular shape over one modulation interval as for M-ary CPFSK, but it may extend over L modulation intervals.

First, a bipolar description (FSK$_b$) of a general CPM signal is introduced, like for FSK. Later on, we switch to the unipolar description FSK$_u$, which allows an easier interpretation.

7.4.1 CPM Signal and Its Interpretation

CPM Transmit Signal

In CPM, the information is first mapped to M-ary bipolar ASK coefficients (M even)

$$a[k] \in \{\pm 1, \pm 3, \ldots, \pm(M-1)\} \,. \tag{7.37}$$

The frequency function $f_M(t)$ is a usual PAM signal with basic pulse $g(t)$, the frequency pulse, and scaled with the modulation index h, that is

$$f_M(t) = h \sum_{k=-\infty}^{\infty} a[k] g(t - kT) \,. \tag{7.38}$$

The transmit signal is then expressed by its *phase function* $\varphi(t)$, which, as usual in FM, is the integral over the frequency function $f_M(t)$ and scaled by 2π. We have

$$\varphi(t) = 2\pi h \int_{-\infty}^{t} \left(\sum_{k=-\infty}^{\infty} a[k] g(t' - kT) \right) dt'$$

$$= 2\pi h \sum_{k=-\infty}^{\infty} a[k] \int_{-\infty}^{t} g(t' - kT) \, dt'$$

$$= 2\pi h \sum_{k=-\infty}^{\infty} a[k] v(t - kT), \quad (7.39)$$

and

$$s(t) = \sqrt{\frac{E_s}{T}} e^{j\varphi(t)}, \quad (7.40)$$

with

- *frequency pulse* $g(t)$
  The frequency pulse $g(t)$ is assumed to be causal and to span over L modulation intervals, that is, its support is $[0, LT)$. We always presume that this pulse is normalized to $\int_0^{LT} g(t) \, dt = \frac{1}{2}$.
- *phase pulse* $v(t)$
  The phase pulse is given as the integral over the frequency pulse, that is, $v(t) = \int_0^t g(t') \, dt'$. Owing to the normalization of $g(t)$, we have $\lim_{t \to \infty} v(t) = \frac{1}{2}$; the final contribution of a symbol $a[k]$ to the accumulated phase is thus given by $\pi h \, a[k]$.
- *modulation index* h
  The modulation index has to be a rational number, that is, $h = p/q$, $p, q \in \mathbb{N}$ and relatively prime. It is preferable that the denominator q is small.

Example 7.3 Frequency and Phase Pulses for CPM Schemes

In CPM, usually the following frequency and phase pulses are employed.

- *L-REC*: rectangular pulse over L symbol intervals

$$g(t) = \begin{cases} \frac{1}{2LT}, & 0 \le t < LT \\ 0, & \text{otherwise} \end{cases}, \quad (7.41)$$

$$v(t) = \begin{cases} 0, & t < 0 \\ \frac{1}{2LT} t, & 0 \le t < LT \\ \frac{1}{2}, & t \ge LT \end{cases}. \quad (7.42)$$

For $L > 1$, such CPM schemes are called *partial-response CPFSK*. For $L = 1$, conventional CPFSK results.

- *L-RC*: raised-cosine CPM

$$g(t) = \begin{cases} \frac{1}{LT} \sin^2\left(\frac{2\pi}{LT} t\right), & 0 \le t < LT \\ 0, & \text{otherwise} \end{cases}, \quad (7.43)$$

$$v(t) = \begin{cases} 0, & t < 0 \\ \frac{1}{2LT} t - \frac{1}{4\pi} \sin\left(\frac{2\pi}{LT} t\right), & 0 \le t < LT \\ \frac{1}{2}, & t \ge LT \end{cases}. \quad (7.44)$$

L-RC CPM may be interpreted as a direct generalization of GMSK because binary 3-RC CPM with $h = 1/2$ is very similar to GMSK.

Figure 7.24 displays these two important classes of frequency and phase pulses.

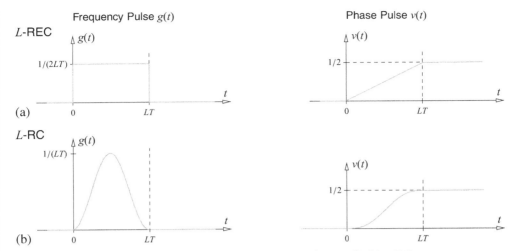

Figure 7.24 Frequency and phase pulses for CPM. (a) *L*-REC; (b) *L*-RC.

Note that DPSK with hard keying (rectangular basic pulse in the PAM notation of PSK) may also be expressed as a special CPM scheme by employing $v(t) = \frac{1}{2} \varepsilon(t)$. A frequency pulse $g(t) = \frac{1}{2} \delta(t)$ results as a hard jump in phase and corresponds to an infinite instantaneous frequency. The infinite extension of the phase pulse represents the recursive differential encoding inherent in DPSK.

Example 7.4 Phase Signals for CPM

Figure 7.25 compares the phase signals $\varphi(t)$ according to (7.39) for two kinds of pulse. In part (a), the situation for a 1-REC pulse is shown (which is identical to CPFSK); in part (b), for 3-RC. In both cases, $M = 4$ is assumed.

The smoothing of the phase function by an extended raised cosine pulse in comparison to CPFSK is clearly visible.

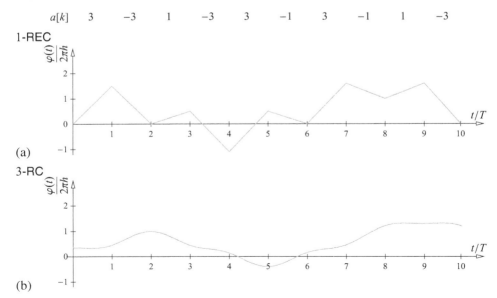

Figure 7.25 Exemplary phase signal of CPM with 4-ary amplitude coefficients.
(a) 1-REC (CPFSK); (b) 3-RC.

Discussion

As can be seen from the above exposition, *linear* intersymbol interference (ISI) over L symbols is already introduced in the ASK frequency function $f_M(f)$. Owing to the subsequent frequency modulation, *nonlinear* ISI arises from this linear ISI together with a phase continuity at the boundaries of the modulation intervals. Both phase continuity and nonlinear ISI constitute a natural kind of trellis encoding (see Appendix B), which provides a transmit signal with high bandwidth efficiency due to an attenuation of spectral side lobes and with high power efficiency due to dispersive channel encoding.

Of course, an exploitation of this coding gain requires some complexity at the receiver side for the implementation of a trellis decoder (e.g., by means of the Viterbi algorithm).

As the FM transmit signal has a strictly constant envelope, an efficient high-power amplification at the transmitter frontend is possible by low-cost devices, which may be driven at nominal power without any back-off (i.e., without any loss in efficiency). Usually, no serious signal distortion is caused by the amplification because without AM, neither AM/AM nor AM/PM conversions happen. Thus, in CPM, many advantages are combined, especially in the regime of highly power-efficient communications, at the price of some moderate receiver complexity.

Interpretation of CPM as a Trellis-Coded Signal

Following the general principle introduced in Chapter 6, the generation of a CPM signal is now separated into an inherent trellis encoding, called *continuous-phase encoder (CPE)*, and a memoryless modulation using *signal elements* that are time-limited to the modulation interval T. For that purpose, we switch from the bipolar description FSK_b to a unipolar description FSK_u, where the smallest instantaneous frequency is used as reference frequency f_0 in the ECB transform. The carrier frequency is then

$$f_c = f_0 + h \frac{M-1}{2T} . \tag{7.45}$$

It can be shown that this approach leads to a time-invariant trellis and a CPE with the minimum number of memory states.

The ECB signal with respect to the reference frequency then reads

$$s(t) = \sqrt{\frac{E_s}{T}} e^{j\psi(t)} \tag{7.46}$$

with

$$\psi(t) = \varphi(t) + 2\pi h \frac{M-1}{2T} t , \tag{7.47}$$

$$\varphi(t) = 2\pi h \sum_{k=-\infty}^{\infty} a[k] v(t - kT) . \tag{7.48}$$

Example 7.5 Bipolar and Unipolar Phase Trajectories

Figure 7.26 shows the *phase trajectories*, that is, the phases of the transmit signals, for all possible sequences of coefficients $a[k]$ plotted on top of each other (cf. Section 4.1.3). 2-RC CPM with $M = 2$ and $h = 1/2$ is assumed. In part (a), the bipolar description and in part (b), the unipolar description is given.

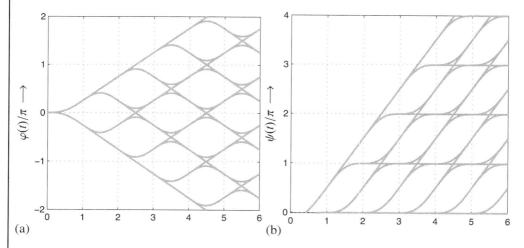

Figure 7.26 Phase trajectories for the (a) bipolar and (b) unipolar description. 2-RC CPM, $M = 2$, $h = 1/2$. Start of transmission at $t = 0$.

Without loss of generality, for the moment, we consider the newly defined phase function $\psi(t)$ within the modulation interval for $k = 0$, that is, for $0 \leq t < T$ only. This phase function is separated as follows:

$$\begin{aligned}
\psi(t) &= 2\pi h \sum_{k=-\infty}^{0} a[k] v(t-kT) + 2\pi h \frac{M-1}{2T} t, \qquad 0 \leq t < T \\
&= 2\pi h \sum_{k=-\infty}^{-L} a[k] v(t-kT) + 2\pi h \sum_{k=-L+1}^{0} a[k] v(t-kT) + 2\pi h \frac{M-1}{2T} t \\
&= 2\pi h \sum_{k=-\infty}^{-L} a[k] v(t-kT) + 2\pi h \sum_{k=-L+1}^{0} \left(a[k] v(t-kT) + \frac{M-1}{2LT} t \right) \\
&\stackrel{\text{def}}{=} \frac{2\pi}{q} \Theta_{-L} + 2\pi h \sum_{k=-L+1}^{0} b(a[k], t-kT).
\end{aligned} \qquad (7.49)$$

Hence, $\psi(t)$ is separated into the contribution Θ_{-L} of all those phase pulses that have already reached their final value $1/2$ and the superposition of the L most recent phase pulses that form the current phase trajectory.

Thereby, in order to define unipolar phase pulses $b(a,t)$, we split the phase ramp $2\pi h \frac{M-1}{2T} t$ in (7.47) (which compensates for the shift of the ECB reference frequency)

7.4 Introduction to Continuous-Phase Modulation (CPM)

into L equal parts and allocate these parts to each of the L most recent pulses. The *unipolar* phase pulse is thus defined as

$$b(a,t) \stackrel{\text{def}}{=} \begin{cases} 0, & t < 0 \\ a\, v(t) + \frac{M-1}{2LT}\, t, & 0 \leq t \leq LT \\ \frac{a+M-1}{2}, & t > LT \end{cases} \quad . \tag{7.50}$$

Owing to the 2π periodicity of the phase, the rational modulation index p/q, and the normalization of the frequency pulse, the accumulated phase Θ_{-L} is an integer from the set $\{0, 1, \ldots, q-1\}$, that is, a finite number of q different *phase states* is possible.

Via time shift by kT, the discussion can be extended to all symbol intervals $kT \leq t < (k+1)T$. The phase state relevant for modulation interval k is thus calculated by

$$\Theta_{k-L} \stackrel{\text{def}}{=} \left(p \sum_{\kappa=-\infty}^{k-L} \frac{a[\kappa] + M - 1}{2} \right) \bmod q \; . \tag{7.51}$$

Finally, by mapping the information $a[k]$ to *unipolar information symbols*

$$b[k] \stackrel{\text{def}}{=} \frac{a[k] + M - 1}{2} \in \{0, 1, \ldots, M-1\} \; , \tag{7.52}$$

the following recursion for the phase state results:

$$\Theta_{k+1} = (\Theta_k + p \cdot b[k+1]) \bmod q \; . \tag{7.53}$$

This equation describes the recursive encoder that establishes phase continuity between successive signal elements. It represents the integration of the instantaneous-frequency function to the phase function and corresponds to differential encoding because the information is mapped to the instantaneous frequency (cf. DPSK (Section 5.2)).

Looking back to (7.49), we see that the signal element in the interval $(k-1)T \leq t < kT$ is fully specified by the L most recent information symbols $b[k]$ that can be stored in a shift register with memory length $L-1$ and by the current state of the recursive phase-state memory. Thus, the current *state $S[k]$ of a Mealy machine* is given by

$$S[k] \stackrel{\text{def}}{=} \left[\Theta_{k-L},\, b[k-L+1],\, \ldots,\, b[k-1] \right] \; . \tag{7.54}$$

This Mealy machine fully specifies the memory within a CPM signal and constitutes the inherent trellis encoder for CPM, the CPE. There are $Z = q \cdot M^{L-1}$ different memory states of the CPE (M^{L-1} sequences $\langle a_{k-L+1}, \ldots, a_{k-1} \rangle$ and q phase states Θ) and $q \cdot M^L$ possible signal trajectories within one modulation interval, which

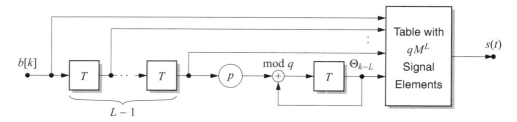

Figure 7.27 Generation of a CPM signal by an inherent trellis encoder (CPE) and a table of signal elements.

establish the *signal elements*

$$s_i(t) = \begin{cases} e^{j\psi_i(t)}, & 0 \leq t < T \\ 0, & \text{else} \end{cases}, \quad i \in \{0, 1, \ldots, qM^L - 1\}, \tag{7.55}$$

with

$$\psi_i(t) = 2\pi \left(\frac{1}{q}\Theta_i + \frac{p}{q} \sum_{\nu=-L+1}^{0} b\left(a_\nu^{(i)}, t - \nu T\right) \right). \tag{7.56}$$

Figure 7.27 shows the general structure of a CPM modulator – separated into a CPE and a table of signal elements, the modulation in the strict sense, containing the $q \cdot M^L$ signal elements. Additionally, a specific example is given subsequently.

Example 7.6 Trellis Description of CPM: $M = 2$, 2-RC, $h = 1/2$

In this example, the trellis description of CPM with $M = 2$, a 2-RC frequency pulse (thus, $L = 2$), and the modulation index $h = 1/2$ (thus, $p = 1$ and $q = 2$) is shown. The trellis encoder (the general structure from Figure 7.27 specialized to the present setting) and the corresponding trellis diagram are visualized in Figures 7.28 and 7.29. Since $M = 2$, we have $b[k] \in \{0, 1\}$ and since $q = 2$, two phase states (i.e., $\Theta_k \in \{0, 1\}$) are present. The CPE has $2 \times 2^1 = 4$ states.

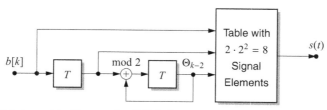

Figure 7.28 Inherent trellis encoder (CPE) for $M = 2$, 2-RC, $h = 1/2$.

7.4 Introduction to Continuous-Phase Modulation (CPM)

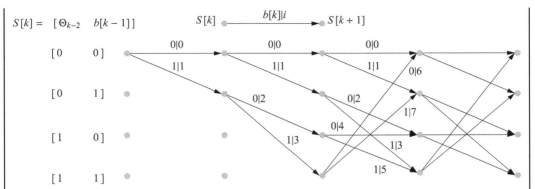

Figure 7.29 Corresponding trellis diagram.

The corresponding $2 \times 2^2 = 8$ signal elements are displayed in Figure 7.30 via their phase functions (7.56). Thereby, the signal number is given by $i = [\Theta_{k-2}, b[k-1], b[k]]_2$. These phase functions can readily be observed from Figure 7.26(b), when looking at one time interval and considering the 2π-modulo reduction of the phase.

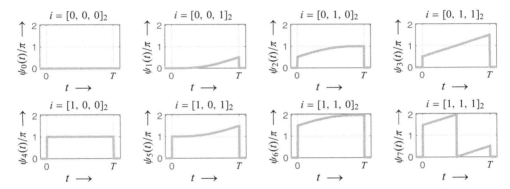

Figure 7.30 Signal elements (phase functions).

Finally, an example (phase) signal is shown in Figure 7.31; the evolution of the state is given for completeness.

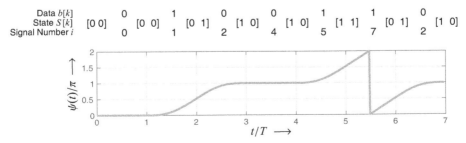

Figure 7.31 Example (phase) signal.

7.4.2 Coherent Reception and Normalized Minimum Squared Euclidean Distance

The straightforward coherent receiver for CPM consists of the downconversion of the RF signal to the ECB domain, a bank of matched filters for the (equivalent, time-limited) signal elements (which define the signal space) and sampling, followed by trellis decoding. The trellis decoding for the inherent CPE is usually done with the Viterbi algorithm (cf. Appendix B). For optimum performance, the number of $q \cdot M^{L-1}$ trellis states has to be processed. Using this optimum receiver, the subsequently derived optimum power efficiency will be achieved.

As coding is present, the difference between any two signals is spread over several modulation intervals and it is not guaranteed that the difference with smallest energy (i.e., the minimum squared Euclidean distance for a CPM scheme) is obtained by a pair of signals differing only in one information symbol. Therefore, in order to find the minimum, a search over all possible pairs of sequences of information symbols is necessary, which is equal to searching over all possible pairs of phase signals $\psi(t)$. Fortunately, it is sufficient to inspect only sequences of *difference symbols*

$$\alpha[k] \stackrel{\text{def}}{=} a^{(1)}[k] - a^{(2)}[k] \quad \in \{0, \pm 2, \ldots, \pm(2M-2)\} \tag{7.57}$$

starting, without loss of generality, at $k = 0$. By means of sophisticated tree search algorithms, the normalized minimum squared Euclidean distance

$$d_{\min}^2 = \frac{E_s/T}{2E_b} \min_{\substack{\psi^{(1)}(t), \psi^{(2)}(t) \\ \psi^{(1)}(t) \neq \psi^{(2)}(t)}} \int_{-\infty}^{\infty} \left| e^{+j\psi^{(1)}(t)} - e^{+j\psi^{(2)}(t)} \right|^2 dt \tag{7.58}$$

$$= \frac{\log_2(M)}{T} \min_{\substack{\langle \alpha[k] \rangle \\ \alpha[0] \neq 0}} \int_{-\infty}^{\infty} \left(1 - \cos\left(2\pi h \sum_{\mu=-\infty}^{\infty} \alpha[k] \, q(t - \mu T) \right) \right) dt \tag{7.59}$$

can rapidly be determined.

Figure 7.32 shows the resulting $d_{\min}^2$ over the modulation index h for 1-REC CPM (CPFSK) and 3-RC CPM for different numbers M of modulation levels. The curves for CPFSK in Figure 7.32 differ substantially from that in Figure 7.22 for FSK because now phase continuity is exploited. Substantially higher minimum squared distances can be achieved that represent the gain in power efficiency by inherent channel coding. A comparison of the results for 1-REC CPM with those for 3-RC CPM shows that the smoothing of the pulses (i.e., the introduction of nonlinear ISI) provides a remarkable additional gain in power efficiency due to an extended memory of the CPE, while the spectral efficiency is also improved significantly (see Figure 7.33).

The curves include sharp breakdowns at specific modulation indices called *weak modulation indices*. Here, memory by phase continuity is not present as different signal elements can be concatenated without or with only little dependence on neighboring signal elements and, therefore, the Euclidean distance breaks down

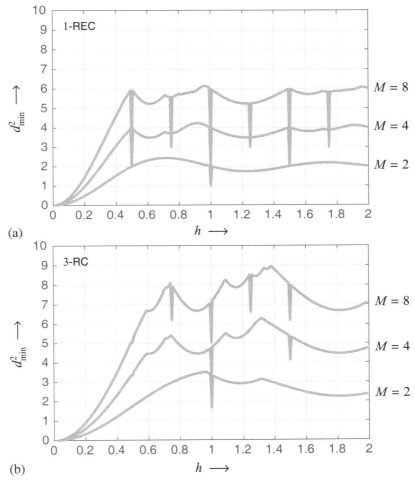

Figure 7.32 Normalized minimum squared Euclidean distance for M-ary CPM over the modulation index h. (a) 1-REC (CPFSK); (b) 3-RC.

to the value of discontinuous FSK. An obvious example is 1-REC, $M = 2$, $h = 1$ (Sunde's FSK), as both signal elements start and end at the same phase. For $M = 4$, this effect happens for pairs of signal elements when the modulation index is a multiple of $1/2$, and so on. For smoothed pulses (e.g., 3-RC CPM), the effect of weak modulation indices is less pronounced in the regime of small modulation indices as the non-recursive memory still provides good coding gains. For example, 4-ary 3-RC CPM with $h = 1/2$ is an attractive modulation scheme ($d^2_{\min} = 3.4$), although its modulation index is weak.

7.4.3 Average Power Spectral Densities of CPM

An analytic computation of the average psd of general CPM schemes is involved and needs the theory of calculation of the psd of a stochastic process, which is driven by a Markov chain. The decomposition of the signal generation into a continuous-phase encoder and a set of signal elements enables this approach.

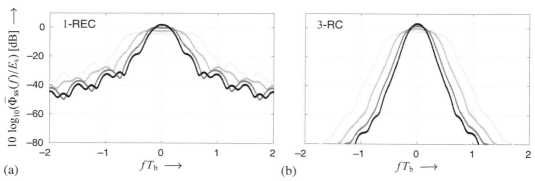

Figure 7.33 Average power spectral densities for 4-ary CPM schemes with different frequency pulses and modulation indices. (a) 1-REC (CPFSK); (b) 3-RC. Dark gray to light gray: $h = 1/4, 1/3, 1/2$, and $3/4$.

Instead, simulation together with a spectral estimation is often applied. Here, some exemplary results are presented.

In Figure 7.33, average power spectral densities of 4-ary 1-REC (CPFSK) and 3-RC CPM are depicted in logarithmic scale. Note that the frequency axis is scaled with respect to bit duration (i.e., data rate). Although the attenuation of spectral side lobes is better for CPFSK than for discontinuous FSK, a satisfactory spectral behavior will often need smoothed phase functions like for RC CPM in order to avoid adjacent channel interference or to satisfy spectral masks.

7.4.4 Power/Bandwidth Plane for CPM

From the average psds and from estimations or simulation results on the bit error ratio (BER), the power/bandwidth plane can be derived, for example, for a tolerated BER = 10^{-5} and a 99% (normalized) bandwidth, see Figure 7.34.

In these plots, the modulation index decreases along the curves from bottom left to top right. It can clearly be seen that 4-ary schemes perform much better than binary ones, but an increase to $M = 8$ offers almost no further benefits. Thus, 4-ary CPM schemes are well suited for technical practice.

The curves for 3-RC CPM show that smoothing the frequency function with the introduction of ISI clearly improves CPM with respect to all criteria except complexity. For example, the scheme 3-RC, $M = 4$, $h = 1/2$ (marked in Figure 7.34 by a square) offers a favorable compromise between performance and receiver complexity. It offers a gain in power efficiency of about 3 dB (at approximately the same bandwidth efficiency) over 3-RC, $M = 2$, $h = 1/2$ (marked by the medium gray square), and GMSK, respectively.

When the data of Figure 7.34 are compared to those for digital PAM schemes with $\sqrt{\text{Nyquist}}$ basic pulses, it should be taken into account that CPM signals have a constant envelope (i.e., no back-off is necessary), whereas for PAM, a back-off for high-power amplification (HPA) of more than 3 dB for a high roll-off factor (low spectral efficiency) and much more for a small roll-off factor is necessary in practice. Thus, only curves in a power/bandwidth plane that for PAM are shifted

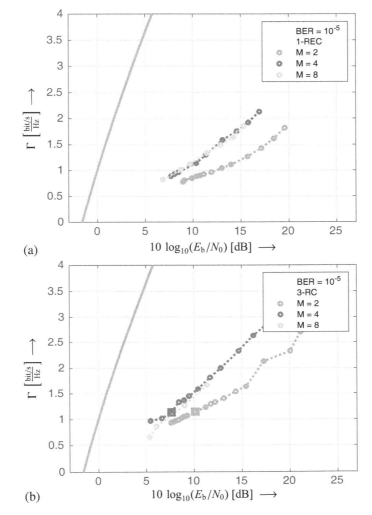

Figure 7.34
Power/bandwidth planes for CPM schemes. Spectral efficiency considering 99% bandwidth. Tolerated bit error ratio BER $\leq 10^{-5}$.
(a) 1-REC;
(b) 3-RC.

to the right by the necessary back-off are comparable to those for CPM. For such a normalization with respect to peak transmit power, CPM outperforms PAM in the low-SNR regime, even when low-complexity channel coding methods are applied to PAM (e.g., trellis-coded modulation), resulting in a receiver complexity comparable to that of CPM. In the high-SNR regime, CPM is not able to compete with M-ary QAM as only one dimension (the phase) of the complex-valued signal space is exploited by CPM.

7.4.5 Complexity Reduction of Coherent Receiver for CPM

A big saving in complexity is possible by utilizing a signal space with reduced dimensionality (cf. Section 6.4) at an almost vanishing loss in power efficiency. The method of spectral sampling with only two or three complex-valued dimensions is favorable for CPM because for this method, simple combined estimators for carrier

7 Digital Frequency and Phase Modulation

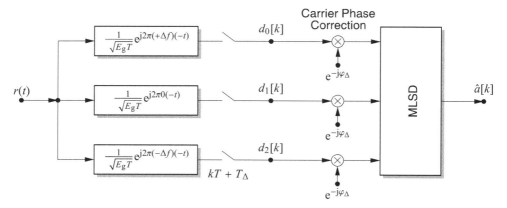

Figure 7.35 Block diagram of a CPM receiver with spectral sampling. $D = 3$.

phase offset φ_Δ and symbol-timing offset T_Δ exist. In Figure 7.35, a block diagram of such a CPM receiver is shown (without synchronization unit).

For MLSD, well-known methods of *reduced-state trellis decoding* (reduced-state sequence detection (RSSD), e.g., a combination of the Viterbi algorithm with per-state decision feedback, cf. Section 8.4) may be applied that work very well for CPM. As a result, demodulation and decoding can be done very efficiently, leading to attractive CPM schemes for technical practice.

7.5 Problems

7.5.1 Binary Frequency-Shift Keying

Binary FSK transmission with discontinuous phase and with signal elements (given in the ECB domain, T is the symbol interval)

$$
\begin{aligned}
s_0(t) &= \sqrt{\tfrac{E_g}{T}}\, e^{-j\pi h t/T}, \\
s_1(t) &= \sqrt{\tfrac{E_g}{T}}\, e^{+j\pi h t/T},
\end{aligned} \qquad t \in [0,T)
$$

and modulation index $h \in \mathbb{R}^+$ is considered.

(a) Sketch the block diagrams of the optimum
 • *coherent* demodulator and
 • *non-coherent* demodulator
 for this discontinuous FSK transmission scheme.
(b) Determine all values of the modulation index h for which the signal elements are orthogonal to each other with respect to
 • *coherent* reception and
 • *non-coherent* reception.

(c) Sketch the phasors of $s_0(t)$ and $s_1(t)$, respectively, for the smallest possible value of h, for which the requirement of part (b) for coherent detection is fulfilled.

(d) Determine the phase of the transmit signal

$$s(t) = \sum_{k=-\infty}^{\infty} s_{m[k]}(t - kT)$$

and sketch it for the following sequence of signal numbers:

$$\langle m[0], m[1], m[2], \ldots, m[4] \rangle = \langle 1, 0, 1, 1, 0 \rangle$$

for the same modulation index h as used in part (c).

(e) What disadvantage appears in part (d)? How can this disadvantage be resolved?

7.5.2 Minimum-Shift Keying I

Continuous-phase binary FSK with $h = 1/2$ is called minimum-shift keying (MSK).

(a) Show that for this setting, the average psd given in (7.12) is equal to that stated in Chapter 4.
Hint: You may use (a) $\cos(x+y)\cos(x-y) = \cos^2(x) - \sin^2(y)$; (b) $\cos(2x) = 2\cos^2(x) - 1$; (c) $2\sin(x)\cos(x) = \sin(2x)$; (d) $\sin(x \pm \pi/2) = \pm\cos(x)$

7.5.3 Minimum-Shift Keying II

We consider minimum-shift keying (MSK), that is, continuous-phase binary FSK with $h = 1/2$. The receiver according to Figure 7.17 is used. The complex correlations $K_i[k]$ over the time steps k are given as follows.

k	0	1	2	3	4	5	6
$K_0[k]$	1.57+0.67j	−0.51−0.49j	0.62+0.78j	−1.32−1.12j	0.37+0.61j	0.06+1.39j	−0.13+1.32j
$K_1[k]$	0.57−1.33j	−1.15−0.51j	1.62−0.14j	−0.69+2.12j	1.37+0.03j	0.70−0.39j	0.87−0.68j

(a) Give the decoding (branch) metrics over the time steps $k = 0, \ldots, 6$ (see the following figure).

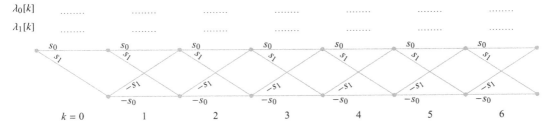

(b) Apply the Viterbi algorithm to decode the transmitted binary sequence. Assume that the sequence starts in the upper state.

7.5.4 Bipolar and Unipolar Description of CPFSK

Coherent demodulation of CPFSK can be performed by the Viterbi algorithm running on a trellis diagram, which describes the signal generation. As a simple example, 2CPFSK with modulation index $h = 1/2$ (MSK) is analyzed. First, we look at the bipolar FSK description, that is, the reference frequency f_0 of the ECB transform of the RF signal corresponds to the arithmetic mean of the instantaneous frequencies of the equally probable FSK signal elements, which is the RF carrier frequency f_c.

(a) How many signal elements are necessary for describing all signal trajectories within one symbol interval?
(b) Show that the complex correlations $K_i[k]$ of the received signal $r(t)$ with all these signal elements can efficiently be derived from two of them, that is, only two correlators are necessary.
(c) Explain why a periodic trellis diagram results for the signal description and give the period of this time-varying trellis.
(d) Give an idea how, at the receiver side, the time-variant trellis can be synchronized to the receiver input signal.

By a consecutive multiplication of the complex correlations with factors j^k (i.e., by the so-called de-rotation), a simplified, time-invariant trellis description is achieved.

(e) Can, by means of de-rotation, a synchronization of the time-variant metric calculations be avoided?
(f) Show that by means of de-rotation there is no simplification of the Viterbi decoder achieved.

We now study the unipolar description of 2CPFSK with $h = 1/2$. Here, the reference frequency f_0 of the ECB transform of the RF signal corresponds to the lowest instantaneous frequency of FSK signal elements.

(g) Sketch the phase trajectories for the unipolar FSK description. Show, by comparison with the phase trajectories for the bipolar description, that both descriptions of CPFSK indeed are equivalent.
(h) Why, in general, does this equivalence not hold for discontinuous-phase FSK?
(i) How many signal elements are now sufficient for describing all signal trajectories within one symbol interval? How can the complex correlations for all these signal elements be efficiently calculated?
(j) Deduce a trellis diagram from the phase tree of part (i). Is this trellis time-variant?
(k) Identify a minimum-distance error event in the trellis and calculate the normalized minimum squared Euclidean distance from the trellis.

7.5.5 Continuous-Phase Modulation (CPM) and Comparison with (Coded) PSK

A quaternary ($M = 4$) CPM scheme with modulation index $h = 1/2$ and a raised cosine (RC) frequency pulse over three modulation intervals ($L = 3$) offers an interesting trade-off between power and bandwidth efficiency at moderate implementation efforts.

The normalized minimum squared Euclidean distance of this 4CPM-3RC ($h = 1/2$) scheme is $d_{\min}^2 = 3.36$, the average number of nearest-neighbor error events is 1, Gray labeling is used, and the 99% signal bandwidth is $1.76/T$, where T denotes the duration of the modulation interval.

We consider transmission over an AWGN channel.

(a) Give the rate [bit/modulation step] of 4CPM-3RC.
(b) Show that the modulation index $h = 1/2$ is a weak one, that is, different amplitude coefficients may lead to the same final phase state. Why has this feature here only a small impact on power efficiency of 4CPM-3RC?
(c) Draw a power/bandwidth plane for the AWGN channel, bit error ratio 10^{-8}, spectral efficiency for 99% bandwidth. Highlight 4CPM-3RC ($h = 1/2$) in the diagram.
(d) Draw on the power/bandwidth plane the trajectory of 4PSK for $\sqrt{\text{Nyquist}}$ basic pulses with a spectral roll-off factor varying from 0 to 1. What roll-off factor results in approximately the same bandwidth efficiency as for 4CPM-3RC ($h = 1/2$)? Compare the modulation schemes.
(e) Draw the block diagram of the time-invariant continuous-phase encoder (CPE) for 4CPM-3RC, $h = 1/2$ in the ECB domain.
   Hint: Use the unipolar frequency-modulation representation with a reference frequency $f_0 = f_c - \frac{3}{4T}$ for the ECB transform (f_c is the carrier frequency of the RF signal).
(f) Specify the number of memory states of the CPE. Separate the states into those due to nonlinear ISI and those due to mapping of the information to the instantaneous frequency and phase continuity. How many different signal elements, limited to one modulation interval T, exist?
(g) How many branches have to be processed per modulation interval by a Viterbi decoder for an optimum maximum-likelihood sequence detection (MLSD)? Give also the number of processed trellis branches per decoded bit, named the *complexity number* of a trellis decoder.

Now 4QAM (equivalent to 4PSK) and 4CPM-3RC are compared with respect to the *peak power* of the transmit signal. For this purpose, a modified power/bandwidth plane is used, where the peak-to-average power ratio (PAPR) of the transmit signals is taken into account, that is, the average energy per bit is multiplied by PAPR, yielding the peak power per bit $E_{b,\text{peak}}$ (i.e., shift of values in logarithmic scale by $10\log_{10}(\text{PAPR})$ to the right). All other parameters remain the same as above.

(h) Highlight the scheme 4CPM-3RC, $h = 1/2$ in this modified power/bandwidth plane.
(i) Why is it advantageous to use offset 4QAM (4OQAM) instead of 4QAM when the peak transmit power is limited? In order to characterize 4OQAM with variable spectral roll-off factor of the basic pulse, translate the PAPR curve (offset)

of Figure 4.8 into the modified power/bandwidth plane with respect to peak power. Discuss the performance of CPM and offset-PSK under peak power limitation.

A reduction of the spectral roll-off factor for 4OQAM (cf. part (d)), together with a faster signaling (smaller T), offers the application of a redundant channel code to 4OQAM.

(j) What roll-off factor has to be used when 4OQAM is combined with a binary channel code with code rate 3/4 (coded 4OQAM) in order to achieve the same bandwidth efficiency of 4CPM-3RC (see parts (c) and (d))?

A binary convolutional code with code rate 3/4 and a trellis decoder with complexity comparable to that for 4CPM-3RC are now applied to 4OQAM with Gray labeling. This scheme *coded 4OQAM* achieves a bit error ratio 10^{-8} or less for a signal-to-noise ratio $10\log_{10}(E_b/N_0) > 6.8$ dB.

(k) Highlight this coded 4OQAM scheme in both power/bandwidth planes, for average power and for peak power.

An alternative method of bandwidth-efficient channel coding is *trellis-coded modulation* (TCM). A PSK scheme with channel coding with equal rate to 4PSK (i.e., equal baud rate and bandwidth) results for an application of a channel code with code rate 2/3 to 8PSK (TC-8PSK). Here, the redundancy of the channel code is entirely devoted to an expansion of the signal constellation and not to an increased baud rate (as discussed above for coded 4OQAM). Employing a convolutional code that needs the same decoding effort as calculated for 4CPM-3RC in part (g), a trellis-coded 8PSK (TC-8PSK) scheme with normalized minimum squared Euclidean $d_{\min}^2 = 5.76$, average number of nearest-neighbor error events equal to 4, and 32 trellis states results.

(l) Draw the trade-off curves of this TC-8PSK scheme for variable spectral roll-off factor of the basic pulse into both power/bandwidth planes, for average power and for peak power.

(m) Compare the modulation schemes (give advantages and disadvantages) 4CPM-3RC ($h = 1/2$), coded 4OQAM, and TC-8PSK with respect to (i) efficiency, (ii) demands on the high-power amplifier (HPA) at the transmitter, and (iii) receiver complexity.

8 Equalization of Dispersive Channels

Up to now, we have assumed transmission over an AWGN channel, that is, the transmit signal is corrupted by additive white Gaussian noise. Notwithstanding that the AWGN channel model is the most important one, in practice, channels often additionally cause *linear distortions* (see the discussion in Appendix D). Hence, in the present chapter, we study PAM transmission over *time-invariant linear dispersive channels*. If the channel impulse response $h_C(t)$ (cf. Figure 1.5) is not a (scaled and shifted) Dirac function, subsequently transmitted symbols smear into each other – so-called *intersymbol interference (ISI)* occurs.

In this chapter, we study receiver-side equalization strategies for linear dispersive channels. Besides the optimum procedure, which immediately follows from the general signal space concept (cf. Chapter 6), we assess low-complexity receivers, specifically *linear equalization* and *decision-feedback equalization*. In each case, we are interested in the performance that can be achieved. The loss caused by intersymbol interference will be quantified. In addition, transmitter-side techniques for pre-equalization are briefly addressed. The duality between receiver-side and transmitter-side schemes is highlighted.

PREREQUISITES FOR THIS CHAPTER

- PAM transmission (Chapter 3).
- Signal space representation (Chapter 6).
- Channel models (Chapter 1).

8.1 Situation

The situation we are dealing with is given as follows.

- *PAM Transmitter*
  We assume conventional PAM transmission (symbol interval T). The basic pulse shape used at the transmitter is now denoted by $g_T(t)$, where the subscript indicates "transmit pulse". Equivalently, its Fourier transform can be specified as

$$g_\mathrm{T}(t) \,\multimap\, G_\mathrm{T}(f)\ . \tag{8.1}$$

The PAM transmit signal is given by

$$s(t) = \sum_k a[k]\, g_\mathrm{T}(t - kT)\ , \tag{8.2}$$

where, as usual, the data symbols $a[k]$ are drawn from the signal constellation $\mathcal{A}$. We assume white, zero-mean data symbols $a[k]$, that is, $\Phi_{aa}\left(\mathrm{e}^{\mathrm{j}2\pi fT}\right) = \sigma_a^2$.

- *Channel*

  The channel is characterized by its overall *impulse response* $h_\mathrm{C}(t)$. Equivalently, its Fourier transform, the *channel transfer function*, can be specified as

$$h_\mathrm{C}(t) \,\multimap\, H_\mathrm{C}(f)\ . \tag{8.3}$$

The denominations "dispersive" and "frequency selective" designate synonymously the same phenomenon, either expressed in the time domain or in the frequency domain. They characterize the dissipation of the pulse due to channel memory.

- $h_\mathrm{C}(t) \neq \mathrm{const.} \cdot \delta(t - t_0)$, that is, the channel not only scales (const.) and delays (t_0) the transmitted signal.
- $|H_\mathrm{C}(f)| \neq \mathrm{const.}$, that is, the channel transfer function is not flat.

Of course, the channel also adds white Gaussian noise $n(t)$ (cf. Figure 1.5) with psd N_0 (ECB domain).

We always assume that the channel is (perfectly) known at the receiver – *channel state information (CSI)* is available.

8.1.1 End-to-End Pulse

As shown in Figure 8.1, transmitter-side pulse shaping and channel are in cascade. The channel itself may be the cascade of the actual channel and a noise-whitening filter (adjusted such that $\Phi_{n'n'}(f)\,|H_\mathrm{W}(f)|^2 = N_0$; cf. Section 1.3), which justifies the assumption of white noise. The problems "communication over dispersive channels" and "disturbance by colored noise" are equivalent to each other and it is sufficient to consider dispersive channels with AWGN to treat both problems.

For the calculations, it is convenient to combine transmit pulse and channel into the *end-to-end pulse* or *end-to-end pulse spectrum*

$$g(t) = g_\mathrm{T}(t) * h_\mathrm{C}(t) \quad \multimap\quad G(f) \stackrel{\mathrm{def}}{=} G_\mathrm{T}(f)\, H_\mathrm{C}(f)\ . \tag{8.4}$$

In the following, we expect the (causal) end-to-end pulse $g(t)$ to be time-limited to the interval $[0,\, (O+1)T]$, that is, it has a *limited support* of $O+1$ symbol intervals T.[1] This can be seen as a presumption on the strength of the ISI, that is, the duration

---

[1] The pulse magnitude is at least substantially greater within this time interval than outside. ISI from the residual part of $g(t)$ outside the assumed support will be disregarded in the equalization process and treated as noise. As we will see in Section 8.4, an appropriate discrete-time model can be given by a FIR system of *order O*.

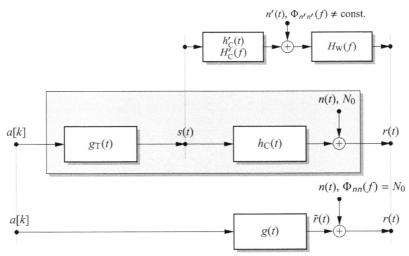

Figure 8.1 Block diagram of the end-to-end system including transmitter-side pulse shaping and channel (ECB domain). The channel $H_C(f)$ may be the cascade of the actual channel $H'_C(f)$ and a noise-whitening filter $H_W(f)$ (top).

of the (significant) dispersion. If only small distortions are caused, O will be small; if the channel impulse response spreads over numerous symbol intervals, O will be large.

This end-to-end model (bottom of the figure) shows that in sum, a PAM scheme over the AWGN channel is still present. However, the basic pulse shape $g(t)$ is active, which is composed of the deliberately used pulse $g_T(t)$ at the transmitter and the channel impulse response $h_C(t)$ that has to be accepted.

In general, this end-to-end pulse $g(t)$ will not have $\sqrt{\text{Nyquist}}$ property – it will not fulfill the temporal orthogonality condition (3.46) – which is required for intersymbol-interference-free transmission over the AWGN channel (cf. Section 3.4). Hence, ISI will occur and symbol-by-symbol detection is no longer appropriate.

The consequences are as follows.

- Each transmitted symbol disperses over a number of symbol intervals.
- Pulses of different modulation intervals smear into each other.
- Samples at the output of a matched filter cannot be detected independently.

In the literature, the following denominations are used synonymously:

- digital PAM with linear distortions;
- digital PAM with non-orthogonal pulses;
- digital PAM with ISI.

The effect of dispersion that causes intersymbol interference is visualized in Figure 8.2. In part (a), the transmit signal (assuming binary ASK with hard keying) is plotted; in part (b), the corresponding (noise-free, i.e., useful part) receive signal $\tilde{r}(t) = s(t) * h_C(t)$. As can be seen, the channel smears the pulse of duration T over

8 Equalization of Dispersive Channels

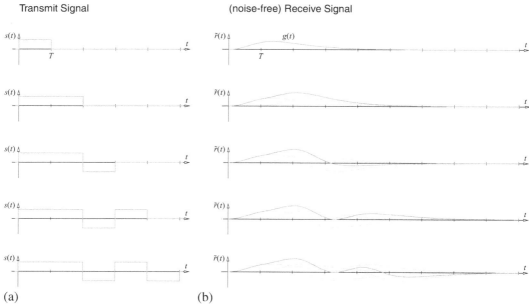

Figure 8.2 Visualization of channel dispersion leading to ISI. (a) Transmit signal; (b) receive signal.

more than five symbol intervals. In practice (e.g., DSL transmission), some hundred symbol intervals may be affected. Owing to the (assumed) linearity of the channel, the contributions of subsequently transmitted pulses are superimposed (added up) to obtain the receive signal. Looking at the last row of the figure, the transmitted ±1 sequence cannot (directly) be deduced from the signal $\tilde{r}(t)$.

8.1.2 Spectral Signal-to-Noise Ratio

For the performance analysis, it is convenient to define the following quantity.

Definition 8.1 *Spectral Signal-to-Noise Ratio*

The *spectral signal-to-noise ratio* (spectral SNR) at the receiver input is defined as

$$\mathrm{SNR}(f) \stackrel{\mathrm{def}}{=} \frac{\sigma_a^2 |G(f)|^2}{T N_0} . \qquad (8.5)$$

It is the ratio of the psd of the useful signal at the receiver input (channel output, psd of a PAM transmit signal, cf. (3.15)) and the (constant) noise psd (value N_0 in ECB domain).

Note that the spectral SNR is not affected by a receiver-input filter (e.g., a noise-whitening filter $H_\mathrm{W}(f)$), as useful signal and noise are affected the same way.

8.1.3 Receive Filter

As usual in PAM, to limit the noise bandwidth, the noisy receive signal

$$r(t) = s(t) * h_C(t) + n(t) \tag{8.6}$$

is filtered by a *receive filter* with impulse response or transfer function with

$$h_R(t) \circ\!\!-\!\!\bullet H_R(f) \tag{8.7}$$

and then *sampled* at the symbol frequency $1/T$. By this, the discrete-time sequence of receive symbols $d[k]$ is obtained.

8.1.4 Discrete-Time End-to-End Model

As already shown in Section 3.4, as we transmit discrete-time amplitude coefficients $a[k]$ and, after T-spaced sampling, obtain discrete-time detection symbols $d[k]$, end to end a discrete-time channel model can be given as shown in Figure 8.3.

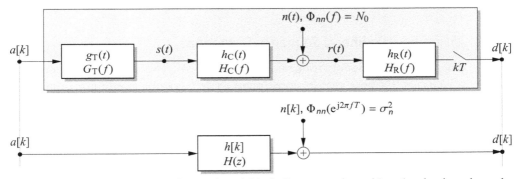

Figure 8.3 Discrete-time end-to-end model including transmitter-side pulse shaping, channel, receiver-side filtering, and sampling.

The transfer function of the cascade is given by $G_T(f) H_C(f) H_R(f) = G(f) H_R(f)$ and the filtered continuous-time noise has the psd $\Phi_{nn}(f) |H_R(f)|^2 = N_0 |H_R(f)|^2$. Since sampling corresponds to a periodic continuation of the spectrum and a weighting by $1/T$, we have for the *discrete-time end-to-end transfer function* $H(z)$ for $z = e^{j2\pi fT}$:

$$H\left(e^{j2\pi fT}\right) = \frac{1}{T} \sum_\mu G\left(f - \tfrac{\mu}{T}\right) H_R\left(f - \tfrac{\mu}{T}\right) \tag{8.8}$$

and for the *power spectral density* of the discrete-time noise sequence $n[k]$:

$$\Phi_{nn}\left(e^{j2\pi fT}\right) = \frac{N_0}{T} \sum_\mu \left|H_R\left(f - \tfrac{\mu}{T}\right)\right|^2 . \tag{8.9}$$

Note that, if $g(t)$ is a $\sqrt{\text{Nyquist}}$ pulse (with energy E_g) and $h_R(t) = \frac{1}{E_g} g^*(-t)$, that is, if the respective matched filter is applied, the discrete-time end-to-end transfer function (8.8) and the noise psd (8.9) specialize immediately to $H(e^{j2\pi fT}) = 1$ and $\Phi_{nn}(e^{j2\pi fT}) = N_0/E_g$. Hence, a *discrete-time AWGN channel* (constant signal transfer function and constant noise psd) results; no ISI is then present.

8.2 Optimum Receiver – Maximum-Likelihood Sequence Detection I

We now derive the optimum receiver for PAM transmission over a linear, dispersive channel. To that end, we assume an M-ary signal constellation $\mathcal{A}$ and expect that the end-to-end pulse shape $g(t) = g_T(t) * h_C(t)$ is known at the receiver. This is no restriction of generality, as the channel always can (and in practice will) be estimated (measured) at the receiver. Hence, channel knowledge is easy to gain.

Based on the general setting of the matched-filter receiver derived in Chapter 6, the optimum receiver for the situation at hand can be deduced. However, in order to apply the matched-filter demodulator from Figure 6.25, we have to define signal elements $s_i(t)$ for the present setting (see also the derivation in Section 6.7.2 for multiple-symbol differential detection). This can be done when considering a data sequence of finite length N; without loss of generality, we consider the modulation intervals $k = 0, \ldots, N-1$.

8.2.1 Maximum-Likelihood Sequence Detection

There are M^N possible transmit vectors $\boldsymbol{a} \in \mathcal{A}^N$ (we enumerate them as $\boldsymbol{a}_0, \boldsymbol{a}_1, \ldots, \boldsymbol{a}_{M^N-1}$). These vectors correspond to M^N possible continuous-time transmit signals

$$s_i(t) \stackrel{\text{def}}{=} \sum_{k=0}^{N-1} a_i[k] g(t - kT), \qquad i = 0, \ldots, M^N - 1 \tag{8.10}$$

with

$$\boldsymbol{a}_i = [a_i[0], a_i[1], a_i[2], \ldots, a_i[N-1]], \tag{8.11}$$

and

$$E_i = \int_{-\infty}^{\infty} |s_i(t)|^2 \, dt \tag{8.12}$$

is the energy of signal element $s_i(t)$.

These M^N signals are interpreted to be the signal elements of the digital transmission scheme. If T is the symbol interval of the PAM scheme, NT is the corresponding symbol interval on which the demodulator is based.

The direct application of the matched-filter demodulator in Figure 6.25 would consist of a bank of M^N matched filters, each one calculating the correlation $K_i = r(t) * \frac{1}{E_g} s_i^*(-t)$ with respect to one signal element $s_i(t)$. Thereof (cf. (6.87)), the decision metric is obtained as

$$\Lambda_i = 2\operatorname{Re}\left\{\left(r(t) * \frac{1}{E_g} s_i^*(-t)\right)\Big|_{t=0}\right\} - \frac{E_i}{E_g}, \tag{8.13}$$

where E_g is the energy of the end-to-end pulse $g(t)$.

8.2 Optimum Receiver – Maximum-Likelihood Sequence Detection I

Hence, maximum-likelihood detection of the substitute signal elements is given by

$$\hat{m} = \underset{i \in \{0,1,\ldots,M^N-1\}}{\operatorname{argmax}} \Lambda_i . \quad (8.14)$$

As the number m of the vector $\boldsymbol{a}_m$ is equivalent to the (finite-length) sequence $\langle a_m[0], a_m[1], a_m[2], \ldots, a_m[N-1]\rangle$ (cf. (8.11)), the ML detection (8.14) is hence synonymous with *maximum-likelihood sequence detection (MLSD)* for the (finite-length) sequence. The block diagram of the optimum receiver is depicted in Figure 8.4.

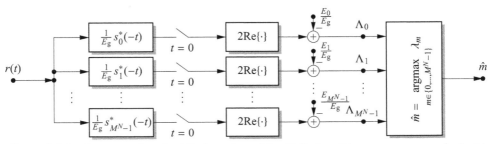

Figure 8.4 Optimum receiver for ML detection of the (finite-length) sequences $\boldsymbol{a}_i$, equivalent to the signal elements $s_i(t)$.

It is obvious that the receiver should not be implemented in this direct way. Typically, the number of matched filters would be extremely high, leading to enormous numerical complexity. Fortunately, the decision metric can be simplified, leading to a procedure applicable in practical systems.

Using the definition (8.10) of the signal elements, the above decision metric can be written as

$$\Lambda_i = 2\operatorname{Re}\left\{\left(r(t) * \frac{1}{E_g}s_i^*(-t)\right)\bigg|_{t=0}\right\} - \frac{E_i}{E_g}$$

$$= 2\operatorname{Re}\left\{\int_{-\infty}^{\infty} r(\tau) \cdot \frac{1}{E_g}\sum_{k=0}^{N-1} a_i^*[k]g^*(\tau-kT)\,\mathrm{d}\tau\right\} - \frac{1}{E_g}\int_{-\infty}^{\infty}\left|\sum_{k=0}^{N-1} a_i[k]g(t-kT)\right|^2 \mathrm{d}t$$

$$= 2\operatorname{Re}\left\{\sum_{k=0}^{N-1} a_i^*[k]\frac{1}{E_g}\int_{-\infty}^{\infty} r(\tau)g^*(\tau-kT)\,\mathrm{d}\tau\right\}$$

$$\qquad - \sum_{k=0}^{N-1}\sum_{j=0}^{N-1} a_i[j]a_i^*[k]\frac{1}{E_g}\int_{-\infty}^{\infty} g(t-jT)g^*(t-kT)\,\mathrm{d}t$$

$$= \sum_{k=0}^{N-1} 2\operatorname{Re}\left\{a_i^*[k]\,d_{\mathrm{M}}[k]\right\} - \sum_{k=0}^{N-1}\sum_{j=0}^{N-1} a_i[j]a_i^*[k]\psi_{gg}[k-j], \quad (8.15)$$

where we have used the samples at the output of the *matched filter* with respect to the end-to-end basic pulse $g(t)$:

$$d_{\mathrm{M}}[k] \stackrel{\text{def}}{=} \frac{1}{E_g} \int_{-\infty}^{\infty} r(t)\, g^*(t - kT)\, \mathrm{d}t \tag{8.16}$$

and the *normalized sampled autocorrelation* of the end-to-end pulse $g(t)$:

$$\psi_{gg}[k] \stackrel{\text{def}}{=} \frac{1}{E_g} \int_{-\infty}^{\infty} g(t + kT)\, g^*(t)\, \mathrm{d}t = \frac{1}{E_g} \varphi_{gg}(kT)$$

$$\updownarrow \tag{8.17}$$

$$\Psi_{gg}\left(\mathrm{e}^{\mathrm{j}2\pi fT}\right) \stackrel{\text{def}}{=} \frac{1}{E_g T} \sum_{\mu} \left| G\left(f - \tfrac{\mu}{T}\right) \right|^2 .$$

Note that above, we have assumed that the support of $g(t)$ is limited to a duration of $(O+1)T$. This immediately gives

$$\psi_{gg}[k] = 0, \qquad \text{for} \quad |k| > O. \tag{8.18}$$

Equation (8.15) shows that the decision metric Λ_i can be calculated completely from the samples $d_{\mathrm{M}}[k]$ at the output of the matched filter with respect to the end-to-end response $g(t)$. In other words:

The matched filter provides sufficient statistics for the optimum receiver.

Hence, not a bank of M^N filters has to be implemented, but only a single matched filter for the end-to-end pulse $g(t)$. At the output of this filter, T-spaced samples are taken. All further processing is done on these samples $d_{\mathrm{M}}[k]$.

The matched-filter frontend is visualized in Figure 8.5.

Up to now, we have derived the decision metric for a block of N symbols. If we require the decision metric for a block of $N+1$ symbols, it can be calculated as follows. First we note that

$$\Lambda\big|_{a[0],\ldots,a[N-1],a[N]} \stackrel{(8.15)}{=} \sum_{k=0}^{N} 2\mathrm{Re}\left\{a^*[k]\, d_{\mathrm{M}}[k]\right\} - \sum_{k=0}^{N} \sum_{j=0}^{N} a[j] a^*[k]\, \psi_{gg}[k-j]. \tag{8.19}$$

Compared to (8.15), the upper limits of the sums are now N instead of $N-1$. The first sum can be partitioned as

$$\sum_{k=0}^{N} 2\mathrm{Re}\left\{a^*[k]\, d_{\mathrm{M}}[k]\right\} = 2 \sum_{k=0}^{N-1} \mathrm{Re}\left\{a^*[k]\, d_{\mathrm{M}}[k]\right\} + 2\mathrm{Re}\left\{a^*[N]\, d_{\mathrm{M}}[N]\right\} \tag{8.20}$$

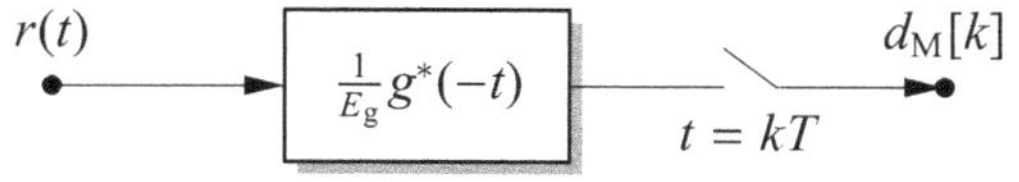

Figure 8.5 Optimum receiver frontend for MLSD.

8.2 Optimum Receiver – Maximum-Likelihood Sequence Detection I

and the double sum as

$$\sum_{k=0}^{N}\sum_{j=0}^{N} a[j]a^*[k]\psi_{gg}[k-j] = \sum_{k=0}^{N-1}\sum_{j=0}^{N-1} a[j]a^*[k]\psi_{gg}[k-j] + a[N]a^*[N]\psi_{gg}[0]$$

$$+ a^*[N]\sum_{j=0}^{N-1} a[j]\psi_{gg}[N-j] + a[N]\sum_{k=0}^{N-1} a^*[k]\psi_{gg}[k-N]$$

$$= \sum_{k=0}^{N-1}\sum_{j=0}^{N-1} a[j]a^*[k]\psi_{gg}[k-j] + a[N]a^*[N]\psi_{gg}[0]$$

$$+ \mathrm{Re}\left\{2a^*[N]\sum_{k=0}^{N-1} a[k]\psi_{gg}[k-N]\right\}. \tag{8.21}$$

Combining both parts and considering (8.15), the decision metric (8.19) for the one-sample-longer sequence can be written as

$$\Lambda|_{a[0],\ldots,a[N-1],a[N]} = \lambda|_{a[0],\ldots,a[N-1]}$$

$$+ 2\mathrm{Re}\left\{a^*[N]\,d_M[N]\right\}$$

$$- a[N]a^*[N]\psi_{gg}[0] - \mathrm{Re}\left\{2a^*[N]\sum_{k=0}^{N-1} a[k]\psi_{gg}[k-N]\right\}$$

$$= \lambda|_{a[0],\ldots,a[N-1]} \tag{8.22}$$

$$+ \mathrm{Re}\left\{a^*[N]\left(2d_M[N] - a[N]\psi_{gg}[0] - 2\sum_{l=1}^{O} a[N-l]\psi_{gg}[l]\right)\right\}.$$

In the last step, we have used that the normalized sampled autocorrelation $\psi_{gg}[k]$ of the end-to-end pulse $g(t)$ is zero for $|k| > O$ (cf. (8.18)).

This equation shows that the decision metric can be *calculated recursively* – that is, that for a block of length $N + 1$ can be calculated from that for a block of length N (considering the symbols $a[0], \ldots, a[N-1]$) plus a *metric increment* (which considers $a[N]$). This recursive calculation of the metric is key for complexity-efficient implementations of MLSD.

In summary, MLSD for the intersymbol-interference channel is done by finding that sequence $\langle a[k]\rangle$ which has the largest accumulated metric. The accumulated metric for a sequence of finite length $N+1$ can be calculated using that of a sequence of finite length N – thus, the metric for any length of a sequence can be calculated recursively, time step by time step adding a metric increment. However, the metric increment depends (via the sum) on the "history" of the sequence, the last O time steps. This history defines a *state* at time step $k-1$ (the superscript enumerates the states)

$$S[k-1] = S^{(a)} \stackrel{\text{def}}{=} \left[a^{(a)}[k-1],\, a^{(a)}[k-2],\, \ldots,\, a^{(a)}[k-O]\right]. \tag{8.23}$$

Having calculated the matched-filter output $d_\mathrm{M}[k]$ ($\psi_{gg}[k]$ is precalculated and known), the *branch metric* (metric increment) at time step k depends on the previous state $S^{(a)}$ and the current symbol $a[k]$. Hence, we have

$$\Lambda^{(a)\to(b)}[k] = \mathrm{Re}\left\{a^*[k]\left(2d_\mathrm{M}[k] - a[k]\psi_{gg}[0] - 2\sum_{l=1}^{O} a^{(a)}[k-l]\psi_{gg}[l]\right)\right\} \quad (8.24)$$

$$= \mathrm{fct}_\mathrm{B}\left(S[k-1],\,a[k]\right).$$

Given the state $S[k-1]$ and the current symbol $a[k]$, in the next time step a new state $S[k]$ is reached:

$$S[k] = S^{(b)} \stackrel{\mathrm{def}}{=} \left[a[k],\,a^{(a)}[k-1],\,\ldots,\,a^{(a)}[k-O+1]\right] \quad (8.25)$$

$$= \mathrm{fct}_\mathrm{S}\left(S[k-1],\,a[k]\right).$$

The two functions $\mathrm{fct}_\mathrm{S}(\cdot,\cdot)$ and $\mathrm{fct}_\mathrm{B}(\cdot,\cdot)$ for the state transition and the branch metric define a *finite-state machine* (for details, see Appendix B). Here, the memory of the finite-state machine has a pure shift-register structure. The possible state transitions over time are visualized by a *trellis diagram*. Since the state comprises the last O amplitude coefficients that are drawn from an M-ary constellation, M^O states are present. Each path through the trellis uniquely corresponds to a data sequence $\langle a[k]\rangle$.

Thus, MLSD turns into searching the *path through the trellis* that has *maximum accumulated metric*

$$\Lambda_\mathrm{acc} = \sum_k \Lambda^{(a)\to(b)}[k], \quad (8.26)$$

where the *branch metric* (state-dependent, branch from state $S^{(a)}$ to state $S^{(b)}$) is given by (8.24). This task is solved in an optimal way by the *Viterbi algorithm* (see Section B.2 in Appendix B).

The main idea of the Viterbi algorithm is that due to the additive metric update out of all paths that merge into a certain state at time step k only that with the currently best (here: highest) path metric can be part of the final best path. Hence, all other merging paths can be discarded from the further search. In each time step, M^O paths (one for each state) have to be traced in parallel for MLSD.

Example 8.1 Trellis Diagram for MLSD

Figure 8.6 shows a trellis diagram assuming $a[k] \in \{+1,-1\}$ ($M = 2$) and $O = 2$. Three segments corresponding to three modulation intervals are depicted. Here, $M^O = 2^2 = 4$ states are present.

8.2 Optimum Receiver – Maximum-Likelihood Sequence Detection I

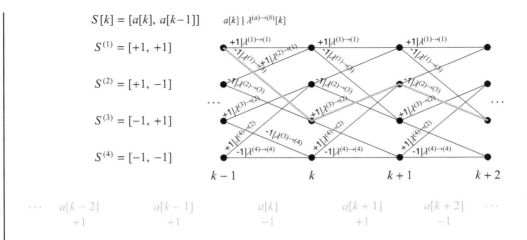

Figure 8.6 Trellis diagram (three segments/modulation intervals). $M = 2$ and $O = 2$.

The path through the trellis corresponding to the particular given sequence of amplitude coefficients $a[k]$ is highlighted.

8.2.2 Power Efficiency of MLSD

As usual in digital communications, the symbol error ratio is well approximated by

$$\text{SER} \approx N_{\min} \, Q\left(\sqrt{d_{\min}^2 \frac{E_b}{N_0}}\right), \tag{8.27}$$

where $N_{\min}$ here is the average number of nearest-neighbor sequences.

As derived in Chapter 6, the *normalized minimum squared Euclidean distance* is thereby given as

$$d_{\min}^2 = \frac{1}{2 E_b} \cdot \min_{\substack{\forall \tilde{r}^{(m)}(t), \tilde{r}^{(\mu)}(t) \\ \tilde{r}^{(m)}(t) \neq \tilde{r}^{(\mu)}(t)}} \int_{-\infty}^{\infty} |\tilde{r}^{(m)}(t) - \tilde{r}^{(\mu)}(t)|^2 \, dt , \tag{8.28}$$

where the minimum has to be taken over all pairs of different *noise-free receive signals* $\tilde{r}(t)$. It is not the distance between the transmitted sequences that determines the performance, but the distance of the sequences present at the receiver input, which takes the effect of dispersion into account. It will be shown below that dispersive channels always cause a loss in $d_{\min}^2$ compared to the AWGN channel.

Since, as discussed above, the noise-free receive signals are PAM signals, we have

$$\Delta = \int_{-\infty}^{\infty} |\tilde{r}^{(m)}(t) - \tilde{r}^{(\mu)}(t)|^2 \, dt$$

$$= \int_{-\infty}^{\infty} \left| \sum_{k=-\infty}^{\infty} a^{(m)}[k] g(t - kT) - \sum_{k=-\infty}^{\infty} a^{(\mu)}[k] g(t - kT) \right|^2 dt$$

$$= \int_{-\infty}^{\infty} \left| \sum_{k=-\infty}^{\infty} \Delta_a[k] g(t - kT) \right|^2 dt , \tag{8.29}$$

where we have introduced the *difference symbols* $\Delta_a[k] \stackrel{\text{def}}{=} a^{(m)}[k] - a^{(\mu)}[k]$.

The normalized minimum squared Euclidean distance of a transmission scheme over a dispersive channel (transmit pulse and channel impulse response are combined into $g(t)$) is thus obtained by searching over all *difference sequences* $\langle \Delta_a[k] \rangle$ not equal to the all-zero sequence, that is

$$d_{\min}^2 = \frac{1}{2 E_b} \cdot \min_{\forall \langle \Delta_a[k] \rangle \neq \langle 0 \rangle} \int_{-\infty}^{\infty} \left| \sum_{k=-\infty}^{\infty} \Delta_a[k] g(t - kT) \right|^2 dt . \qquad (8.30)$$

Moreover, since (we use $|x|^2 = x \cdot x^*$)

$$\int_{-\infty}^{\infty} \left| \sum_{k=-\infty}^{\infty} \Delta_a[k] g(t - kT) \right|^2 dt = \int_{-\infty}^{\infty} \sum_k \Delta_a[k] g(t - kT) \sum_l \Delta_a^*[l] g^*(t - lT) \, dt$$

$$= \sum_k \sum_l \Delta_a[k] \Delta_a^*[l] \int_{-\infty}^{\infty} g(t - kT) g^*(t - lT) \, dt$$

$$= \sum_k \sum_l \Delta_a[k] \Delta_a^*[l] E_g \psi_{gg}[l - k] , \qquad (8.31)$$

with $\psi_{gg}[k]$ as defined in (8.17), we can also write

$$d_{\min}^2 = \frac{E_g}{2 E_b} \cdot \min_{\forall \langle \Delta_a[k] \rangle \neq \langle 0 \rangle} \sum_k \sum_l \Delta_a[k] \Delta_a^*[l] \psi_{gg}[l - k] . \qquad (8.32)$$

The minimization in (8.32) over the difference sequences is equivalent to searching for two paths through the trellis for which the minimum difference energy occurs.

We now show how $d_{\min}^2$ can be calculated efficiently. Given the pulse $g(t)$, which was assumed to be time-limited, the sampled normalized autocorrelation $\psi_{gg}[k]$ can be calculated. It is zero for $|k| > O$. Moreover, we may restrict to causal difference sequences $\langle \Delta_a[k] \rangle$ of length L, that is, $\Delta_a[k] = 0$, $k < 0$ and $k \geq L$. Then, the double sum in (8.32) reduces to finite sums and can be written as a *quadratic form* ($\cdot^{\text{H}}$ denotes the conjugate complex transposition (Hermitian) of a vector)

$$\sum_{k=0}^{L-1} \sum_{l=0}^{L-1} \Delta_a[k] \Delta_a^*[l] \psi_{gg}[l - k] = \boldsymbol{\Delta}_a \boldsymbol{\Psi} \boldsymbol{\Delta}_a^{\text{H}} \qquad (8.33)$$

with the definitions

$$\boldsymbol{\Delta}_a \stackrel{\text{def}}{=} \begin{bmatrix} \Delta_a[0] & \Delta_a[1] & \cdots & \Delta_a[L-1] \end{bmatrix} , \qquad (8.34)$$

$$\boldsymbol{\Psi} \stackrel{\text{def}}{=} \begin{bmatrix} \psi_{gg}[0] & \cdots & \psi_{gg}[O] & 0 & \cdots & 0 \\ \vdots & \ddots & & \ddots & \ddots & \vdots \\ \psi_{gg}[-O] & & \ddots & & \ddots & 0 \\ 0 & \ddots & & \ddots & & \psi_{gg}[O] \\ \vdots & \ddots & \ddots & & \ddots & \vdots \\ 0 & \cdots & 0 & \psi_{gg}[-O] & \cdots & \psi_{gg}[0] \end{bmatrix} . \qquad (8.35)$$

Performing a Cholesky factorization of the positive definite Hermitian Toeplitz matrix $\boldsymbol{\Psi}$ (see Appendix A) yields

$$\boldsymbol{\Psi} \stackrel{\text{def}}{=} \boldsymbol{BB}^{\mathsf{H}}, \tag{8.36}$$

where the matrix $\boldsymbol{B} = [b_{i,j}]_{\substack{i=1,\ldots,L \\ j=1,\ldots,L}}$ is *upper triangular*. Further, we can write

$$\boldsymbol{\Delta}_a \boldsymbol{\Psi} \boldsymbol{\Delta}_a^{\mathsf{H}} = \boldsymbol{\Delta}_a \boldsymbol{BB}^{\mathsf{H}} \boldsymbol{\Delta}_a^{\mathsf{H}} = \|\boldsymbol{\Delta}_a \boldsymbol{B}\|^2, \tag{8.37}$$

where $\|\cdot\|$ denotes the Euclidean norm.

Hence, determining the normalized minimum squared Euclidean distance amounts to the task

$$d_{\min}^2 = \frac{E_{\text{g}}}{2 E_{\text{b}}} \cdot \min_{\forall \boldsymbol{\Delta}_a \in \mathcal{D}} \|\boldsymbol{\Delta}_a \boldsymbol{B}\|^2, \tag{8.38}$$

which is a (restricted) *shortest vector problem*, that is, we have to find the shortest vector in the *lattice* defined by the generator matrix $\boldsymbol{B}$. Thereby, $\mathcal{D}$ is the set of vectors of dimension L with components drawn from the set of difference symbols and not equal to the all-zero vector. For ASK signal constellations, this means

$$\mathcal{D} = \left\{ \boldsymbol{\Delta}_a \mid \Delta_a[k] \in \{-2(M-1),\ldots,-2,0,2,\ldots,2(M-1)\},\ \begin{array}{l} k=0,\ldots,L-1 \\ \boldsymbol{\Delta}_a \neq \boldsymbol{0} \end{array} \right\}. \tag{8.39}$$

For QAM signaling, a respective definition is easily possible.

Finally, having performed the Cholesky factorization, a simple lower bound can be stated (see Section A.3 in Appendix A):

$$d_{\min}^2 \geq \frac{E_{\text{g}}}{E_{\text{b}}} 2 |b_{1,1}|^2. \tag{8.40}$$

Example 8.2 Minimum Distance

This example shows that for different pulses, other difference sequences $\langle \Delta_a[k] \rangle$ lead to the minimum difference energy. ASK signaling is presumed.

In Figure 8.7(a), an end-to-end pulse $g_1(t)$ of short duration is assumed; in part (b), a pulse $g_2(t)$ of longer duration. In the top row, the situation for the difference sequence $\langle \ldots, 0, 2, 0, \ldots \rangle$ is plotted; in the bottom row, the situation for the difference sequence $\langle \ldots, 0, 2, -2, 0, \ldots \rangle$. The integral over the squared signal determines the difference energy and thus the minimum distance.

For the short pulse, the minimum is obtained for the top difference sequence. For the broader one, the bottom difference sequence leads to the minimum difference energy.

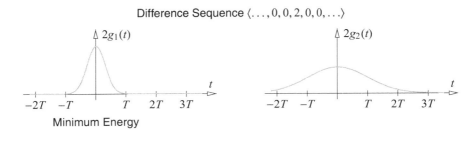

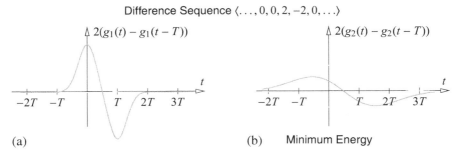

(a)   (b)

Figure 8.7 Difference signal for pulse of (a) short and (b) longer duration. Two difference sequences (the light gray curves are the addends).

Example 8.3 Loss in Minimum Distance Compared to AWGN Channel

We assess the loss

$$\vartheta^2 \stackrel{\text{def}}{=} \frac{d_{\min}^2}{d_{\min,\text{AWGN}}^2}$$

in minimum distance compared to an AWGN channel. To that end, we assume a coaxial cable (cf. Appendix D) with white noise and rectangular transmitter pulses (hard keying). The squared magnitude of the end-to-end transfer function is thus given by

$$|G_{\text{T}}(f) \cdot H_{\text{C}}(f)|^2 = T \, \text{si}^2(\pi f T) \cdot 10^{-\alpha_{\text{c}} \sqrt{2|f|T}/10} \, , \quad (8.41)$$

where α_{c} is the characteristic attenuation at the Nyquist frequency $f_{\text{N}} = \frac{1}{2T}$.

The difference sequences $\langle \Delta_a[k] \rangle$ leading to the minimum distance for the given interval of α_{c} (in dB) are summarized in Table 8.1.

Table 8.1 Difference sequences leading to the minimum distance.

$M = 2$ ASK		$M = 4$ ASK	
α_{c} [dB]	$\langle \Delta_a[k] \rangle$	α_{c} [dB]	$\langle \Delta_a[k] \rangle$
0...12	$\langle 2 \rangle$	0...12	$\langle 2 \rangle$
12...68	$\langle 2,-2 \rangle$	12...42	$\langle 2,-2 \rangle$
68...	$\langle 2,-2,-2,2 \rangle$	42...70	$\langle 2,-4,2 \rangle$
		70...	$\langle 2,-6,6,-2 \rangle$

Figure 8.8 shows that the loss increases as the characteristic attenuation at the Nyquist frequency of the coaxial cable increases. As this number increases, the differences in the magnitude of the channel transfer function within the transmission band increase. Hence, the dispersion of the channel increases. As for an $M = 4$-ary constellation more difference sequences exist, the minimum distance can be smaller and, thus, the loss may be higher. In the present example, this can be seen for attenuations larger than approximately 42 dB. The dashed line shows the simple upper bound on the loss obtained from the lower bound (8.40) on $d_{\min}^2$.

The light gray curve corresponds to a transmission of a single pulse at $k = 0$, that is, the artificial situation of no ISI and therefore the best achievable SNR at the output of the matched filter with respect to the end-to-end pulse. It is denoted as *matched-filter bound* and is subsequently derived and explained in detail in Section 8.3.1.

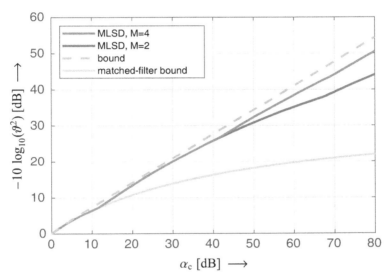

Figure 8.8 Loss in minimum distance compared to AWGN channel over the characteristic attenuation at the Nyquist frequency of the coaxial cable.

8.3 Linear and Nonlinear Equalization

As we have seen, the optimum receiver performs maximum-likelihood sequence detection via the Viterbi algorithm. However, it has to operate on a trellis with M^O states. If "large" ($M \gg 2$) constellations are employed or if the channel exhibits significant intersymbol interference (O large), a huge complexity may be present. Note that, even for $M = 16$-ary QAM transmission and $O = 5$, the number of states would be more than one million. Hence, *reduced-complexity receivers* are of practical interest.

In the context of MLSD it has been shown that the matched filter (matched to transmitter pulse and channel) is the optimum way to go from the continuous-time receive signal to the discrete-time detection variables. This principle holds in general. For a correct sampling phase, an optimum detection is possible based on T-spaced samples at the output of the *matched filter* for $g(t)$. In principle, using the matched filter, no loss of information with respect to the transmitted message is caused by this transition from continuous to discrete time.[2] Hence, the derivation of any equalization technique for practice may start from this discrete-time signal (cf. also Problem 8.6.3).

This means that the receiver can always be decomposed into

matched filter — T-spaced sampling — discrete-time processing

Since discrete-time filters have a frequency-periodic transfer function, the receive filter can hence always be written as

$$H_R(f) = \frac{1}{E_g} G^*(f) F\left(e^{j2\pi fT}\right), \qquad (8.42)$$

where $F(z)$ denotes the transfer function (z-transform) of the discrete-time (feed-forward) part. The decomposition of the receive filter is depicted in Figure 8.9.

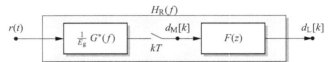

Figure 8.9 Decomposition of the receive filter: matched filter with respect to $g(t)$, sampling at kT, and discrete-time filtering by $F(z)$.

Since the matched filter plays an important role and is the starting point for other receiver concepts, it is advisable to state the discrete-time end-to-end model including the matched filter. The block diagram of the discrete-time end-to-end model employing the matched filter is depicted in Figure 8.10.

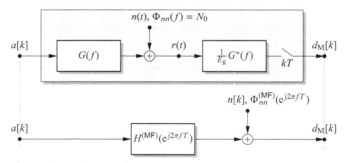

Figure 8.10 Discrete-time end-to-end model for digital PAM employing the matched filter.

---

[2] It should be noted that the T-spaced samples at the output of the matched filter are sufficient for the optimum detection of the data sequence $\langle a[k] \rangle$. As the sampling theorem is not satisfied, they do not fully represent the continuous-time receiver input signal $r(t)$ but only the relevant part (cf. Section 6.7), which, however, is sufficient. Moreover, T-spaced sampling is only information-lossless if the sampling phase is correct; as we ignore the delay that would be required for a causal implementation of the filter, the sampling instants are kT.

Using (8.8) and (8.9), the respective transfer function and the noise psd, respectively, are given by (the denomination ".(MF)" is used to tag the quantity that results when the matched filter is used)

$$h^{(\mathrm{MF})}[k] = \psi_{gg}[k] \quad \circ\!\!-\!\!\bullet \quad H^{(\mathrm{MF})}\left(e^{j2\pi fT}\right) = \Psi_{gg}\left(e^{j2\pi fT}\right),$$

$$\phi_{nn}^{(\mathrm{MF})}[k] = \frac{N_0}{E_g}\psi_{gg}[k] \quad \circ\!\!-\!\!\bullet \quad \Phi_{nn}^{(\mathrm{MF})}\left(e^{j2\pi fT}\right) = \frac{N_0}{E_g}\Psi_{gg}\left(e^{j2\pi fT}\right).$$

(8.43)

Thereby, the normalized sampled autocorrelation $\psi_{gg}[k]$ of the end-to-end pulse $g(t)$ and its Fourier transform (cf. (8.17)) have been used.

Note that in the case of an MF receiver we have:

$$\Phi_{nn}^{(\mathrm{MF})}\left(e^{j2\pi fT}\right) = \frac{N_0}{E_g}H^{(\mathrm{MF})}\left(e^{j2\pi fT}\right).$$

8.3.1 Matched-Filter Bound

We are now ready to derive an upper bound on the performance of any receive strategy for linearly distorting channels. To that end, we assume that:

- the matched filter constitutes the receiver frontend;
- only a single pulse (w.l.o.g. data symbol $a[0] = 1$) has been transmitted.

This means that ISI, which degrades performance, is ignored but the energy of the single pulse is perfectly captured (see the SNR maximization in Section 3.4).

The SNR at the detection instant is given by

$$\mathrm{SNR}_d^{(\mathrm{MF})} \stackrel{\text{def}}{=} \frac{\sigma_a^2 \left|h^{(\mathrm{MF})}[0]\right|^2}{\sigma_n^2}.$$

(8.44)

Since

$$h^{(\mathrm{MF})}[0] = T\int_{-\frac{1}{2T}}^{\frac{1}{2T}} H^{(\mathrm{MF})}\left(e^{j2\pi fT}\right) e^{+j2\pi 0 t}\, df = \int_{-\frac{1}{2T}}^{\frac{1}{2T}} \Psi_{gg}\left(e^{j2\pi fT}\right) df$$

and

$$\sigma_n^2 = T\int_{-\frac{1}{2T}}^{\frac{1}{2T}} \Phi_{nn}^{(\mathrm{MF})}\left(e^{j2\pi fT}\right) df = T\int_{-\frac{1}{2T}}^{\frac{1}{2T}} \frac{N_0}{E_g}\Psi_{gg}\left(e^{j2\pi fT}\right) df$$

we have

$$\mathrm{SNR}_d^{(\mathrm{MF})} = \frac{\sigma_a^2 \left|T\int_{-\frac{1}{2T}}^{\frac{1}{2T}} \Psi_{gg}\left(e^{j2\pi fT}\right) df\right|^2}{T\int_{-\frac{1}{2T}}^{\frac{1}{2T}} \frac{N_0}{E_g}\Psi_{gg}\left(e^{j2\pi fT}\right) df} = T\int_{-\frac{1}{2T}}^{\frac{1}{2T}} \frac{\sigma_a^2 E_g}{N_0}\Psi_{gg}\left(e^{j2\pi fT}\right) df.$$

(8.45)

Furthermore, considering the definition (8.5) of the spectral SNR at the receiver input

$$\mathrm{SNR}(f) = \frac{\sigma_a^2 |G(f)|^2}{T N_0}, \qquad (8.46)$$

and using this definition, (8.43) and (8.17), we have the important relation

$$\frac{\sigma_a^2 E_g}{N_0} \Psi_{gg}\left(\mathrm{e}^{\mathrm{j}2\pi fT}\right) = \frac{\sigma_a^2 E_g}{N_0} \frac{1}{E_g T} \sum_\mu |G\left(f - \tfrac{\mu}{T}\right)|^2 = \sum_\mu \mathrm{SNR}\left(f - \tfrac{\mu}{T}\right). \qquad (8.47)$$

Finally, defining the *folded spectral SNR* as

$$\widetilde{\mathrm{SNR}}\left(\mathrm{e}^{\mathrm{j}2\pi fT}\right) \stackrel{\mathrm{def}}{=} \sum_\mu \mathrm{SNR}\left(f - \tfrac{\mu}{T}\right), \qquad (8.48)$$

and plugging this into (8.45) gives the *matched-filter bound* as

$$\mathrm{SNR}^{(\mathrm{MF})} = T \int_{-\frac{1}{2T}}^{\frac{1}{2T}} \widetilde{\mathrm{SNR}}\left(\mathrm{e}^{\mathrm{j}2\pi fT}\right) \mathrm{d}f \qquad (8.49)$$

arithmetic mean over the folded spectral SNR.

Knowing the spectral SNR at the receiver input, the folded version can be calculated. The arithmetic mean (see Box 8.1) over this function simply gives the matched-filter bound.

Example 8.4 Folded Signal-to-Noise Ratio

Assume that a transmission scheme is given, where the spectral signal-to-noise ratio $\mathrm{SNR}(f)$ has the shape as depicted in Figure 8.11(a). As the spectral SNR is a multiple of $|G(f)|^2$, the spectral shape of the transmit pulse $G_\mathrm{T}(f)$ and the channel transfer function $H_\mathrm{C}(f)$ (which may include a whitening filter, see Figure 8.1) determine the spectral SNR.

The corresponding folded spectral signal-to-noise ratio $\widetilde{\mathrm{SNR}}\left(\mathrm{e}^{\mathrm{j}2\pi fT}\right)$ is shown in Figure 8.11(b). It is the aliased version of $\mathrm{SNR}(f)$ and, thus, frequency-periodic.

If $\mathrm{SNR}(f)$ satisfies the Nyquist criterion in the frequency domain (cf. Section 3.4), the folded SNR is a constant. This is exactly the case when there is no ISI and the simple AWGN channel model applies.

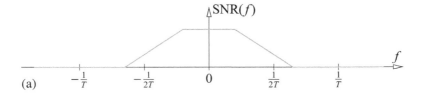

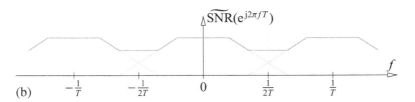

Figure 8.11 (a) Spectral signal-to-noise ratio and (b) corresponding folded spectral signal-to-noise ratio.

Box 8.1 Arithmetic, Geometric, and Harmonic Mean

Let the following be given:
(a) N real numbers $x_1, x_2, \ldots x_N$; (b) real-valued function $f(x)$, $x \in [0, I]$

In mathematics, the following means are common:

- arithmetic mean

$$A \stackrel{\text{def}}{=} \frac{1}{N} \sum_{i=1}^{N} x_i \qquad A \stackrel{\text{def}}{=} \frac{1}{I} \int_0^I f(x)\, dx \,;$$

- geometric mean

$$G \stackrel{\text{def}}{=} \sqrt[N]{\prod_{i=1}^{N} x_i} \qquad G \stackrel{\text{def}}{=} \exp\left\{\frac{1}{I} \int_0^I \log(f(x))\, dx\right\};$$

- harmonic mean

$$H \stackrel{\text{def}}{=} \frac{1}{\frac{1}{N} \sum_{i=1}^{N} \frac{1}{x_i}} \qquad H \stackrel{\text{def}}{=} \frac{1}{\frac{1}{I} \int_0^I \frac{1}{f(x)}\, dx}\,.$$

It can be shown that for $x_i > 0$, or $f(x) > 0$, respectively, the following relation holds:

$$H \leq G \leq A \,.$$

The means are equal if and only if

$$x_i = c, \quad \forall i \qquad f(x) = c, \quad x \in [0, I]\,.$$

Note that the geometric mean ($x_i \geq 0$) can be written as

$$\sqrt[N]{\prod_{i=1}^{N} x_i} = \left(\prod_{i=1}^{N} x_i\right)^{1/N} = \exp\left\{\frac{1}{N} \log\left(\prod_{i=1}^{N} x_i\right)\right\} = \exp\left\{\frac{1}{N} \sum_{i=1}^{N} \log(x_i)\right\}\,.$$

This shows the connection to the formula in case of functions.

8.3.2 Discrete-Time Part of the Receive Filter

In view of Figure 8.9, employing purely the matched filter as receiver frontend corresponds to the choice $F(z) = 1$. This leads to a discrete-time end-to-end model in Figure 8.10 where the useful transfer function (see (8.43)) is $\Psi_{gg}\left(e^{j2\pi fT}\right)$ and the noise psd is $\frac{N_0}{E_g}\Psi_{gg}\left(e^{j2\pi fT}\right)$. Neither a flat transfer function (no ISI) nor a flat noise psd (white noise) are present.

Hence, in the next step, we discuss how and why to choose the discrete-time part $F(z)$ to achieve these extreme cases of either a flat transfer function or a flat noise psd. As already discussed in Chapter 1, intersymbol interference and non-white noise are two sides of the same coin – they can be transformed into each other. In the context of equalization, this transformation leads to different receive strategies. Figure 8.12 depicts the three mentioned strategies. To that end, the sampling after the matched filter is shifted to the output of the entire receive filter; the discrete-time part is contained as a frequency-periodic transfer function.

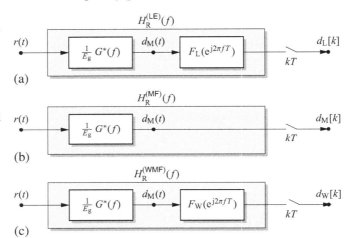

Figure 8.12 Versions of the receive filter by choosing $F(z)$ suitably. (a) Linear equalization; (b) matched-filter receiver; (c) whitened-matched filter.

8.3.3 Linear Equalization

The aim of *linear equalization (LE)* is to adjust the receive filter in such a way that *symbol-by-symbol decisions free of intersymbol interference are enabled*. Hence, the linearly distorting channel is linearly equalized.

Receive Filter and Discrete-Time End-to-End Model

As given above (cf. Figure 8.10), the end-to-end impulse response/transfer function including the matched filter is given by (cf. (8.43))

$$h^{(\mathrm{MF})}[k] = \psi_{gg}[k] \quad \circ\!\!-\!\!\bullet \quad H^{(\mathrm{MF})}\left(e^{j2\pi fT}\right) = \Psi_{gg}\left(e^{j2\pi fT}\right). \qquad (8.50)$$

We require that in the linearly equalized detection symbols $d_\mathrm{L}[k]$ (see Figure 8.12(a)) no (residual) intersymbol interference is present. This is achieved if the discrete-time end-to-end transfer function equals one. Hence, in order to linearly equalize the channel, we choose

$$F_L(z) = \frac{1}{\Psi_{gg}(z)} \qquad (8.51)$$

where

$$\Psi_{gg}(z) \stackrel{\text{def}}{=} \sum_k \psi_{gg}[k] z^{-k} \qquad (8.52)$$

is the z-transform of the impulse response $\psi_{gg}[k]$ corresponding to the spectrum

$$\Psi_{gg}\left(e^{j2\pi fT}\right) = \frac{1}{E_gT} \sum_\mu \left|G\left(f - \tfrac{\mu}{T}\right)\right|^2 . \qquad (8.53)$$

$\Psi_{gg}(z)$ is also the so-called *analytic continuation* of $\Psi_{gg}\left(e^{j2\pi fT}\right)$, that is, the extension of the transfer function from the unit circle (argument $e^{j2\pi fT}$) to the entire z-plane.

Considering that $G(f) = G_T(f)H_C(f)$, the entire receive filter for linear equalization is thus given by

$$H_R^{(LE)}(f) = \frac{\frac{1}{E_g} G^*(f)}{\frac{1}{E_gT} \sum_\mu \left|G\left(f - \tfrac{\mu}{T}\right)\right|^2} = \frac{T\, G_T^*(f)\, H_C^*(f)}{\sum_\mu \left|G_T\left(f - \tfrac{\mu}{T}\right) H_C\left(f - \tfrac{\mu}{T}\right)\right|^2} . \qquad (8.54)$$

As the entire intersymbol interference is *forced to zero*, this strategy is also called *zero-forcing linear equalization (ZF-LE)*.

At this point it is advisable to emphasize that the receive filter does not totally equalize the *continuous-time* cascade $G_T(f)H_C(f)$ (e.g., via $H_R(f) = \frac{1}{G_T(f)H_C(f)}$). Only the signal at the *sampling instants* has to be equalized. The continuous-time end-to-end impulse response has to be a *Nyquist function* – it has to have equidistant zeros at integer multiples of T. Hence, the receive filter that enforces this Nyquist property in an optimum way, is often called the *optimum Nyquist filter (ONF)*.

Using (8.8) and (8.9) it is straightforward to show that when employing the linear receive filter according to (8.54), the discrete-time end-to-end impulse response/signal transfer function including the receive filter and sampling are given by

$$h^{(LE)}[k] = \delta[k] \quad \circ\!\!-\!\!\bullet \quad H^{(LE)}\left(e^{j2\pi fT}\right) = 1 . \qquad (8.55)$$

The psd of the noise $n[k]$ at the output of the receive filter $H_R^{(LE)}(f)$ reads

$$\begin{aligned}
\Phi_{nn}^{(LE)}\left(e^{j2\pi fT}\right) &= \Phi_{nn}^{(MF)}\left(e^{j2\pi fT}\right) \left|F_L\left(e^{j2\pi fT}\right)\right|^2 \\
&= \frac{N_0}{E_g} \Psi_{gg}\left(e^{j2\pi fT}\right) \frac{1}{\left|\Psi_{gg}\left(e^{j2\pi fT}\right)\right|^2} \\
&= \frac{N_0}{E_g} \frac{1}{\Psi_{gg}\left(e^{j2\pi fT}\right)} . \qquad (8.56)
\end{aligned}$$

Hence, the psd of the noise is in general not constant – *the noise is colored* and its power is usually *enhanced*.

To summarize, when employing the receive filter $H_{\text{R}}^{(\text{LE})}(f)$ for linear equalization, the useful signal is equalized, no intersymbol interference is present, but the additive (Gaussian) noise is colored. The discrete-time end-to-end model is given by

$$d_{\text{L}}[k] = a[k] + n[k] \ . \tag{8.57}$$

Based on the detection symbols $d_{\text{L}}[k]$, symbol-by-symbol threshold decisions (uncoded transmission) on the amplitude coefficients $a[k]$ can be taken.

The entire transmission model is depicted in Figure 8.13.

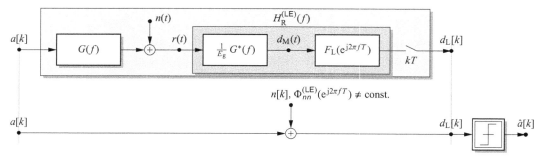

Figure 8.13 Entire transmission model when employing the receive filter $H_{\text{R}}^{(\text{LE})}(f)$.

Signal-to-Noise Ratio

The signal-to-noise ratio when applying $H_{\text{R}}^{(\text{LE})}(f)$ for linear equalization can immediately be given. Since the power of the (zero-mean) useful data is σ_a^2 and the noise power, active at the input of the threshold device, is given by the integral over the psd, the SNR when performing zero-forcing linear equalization calculates as

$$\text{SNR}^{(\text{LE})} = \frac{\sigma_a^2}{T \int_{-\frac{1}{2T}}^{\frac{1}{2T}} \Phi_{nn}^{(\text{LE})}\left(e^{j2\pi fT}\right) df}$$

using (8.56)

$$= \frac{1}{T \int_{-\frac{1}{2T}}^{\frac{1}{2T}} \frac{N_0}{\sigma_a^2 E_g} \frac{1}{\Psi_{gg}\left(e^{j2\pi fT}\right)} df} \ . \tag{8.58}$$

Finally, using (8.47), we arrive at

$$\text{SNR}^{(\text{LE})} = \frac{1}{T \int_{-\frac{1}{2T}}^{\frac{1}{2T}} \frac{1}{\widetilde{\text{SNR}}\left(e^{j2\pi fT}\right)} df} \tag{8.59}$$

harmonic mean over the folded spectral SNR.

Knowing the spectral SNR at the receiver input, the folded version can be calculated. The harmonic mean (see Box 8.1) over this function simply gives the SNR for linear equalization. Note that the harmonic mean (over non-negative quantities) is always smaller than (or equal to) the arithmetic mean. Hence, the linear equalization strategy has a poorer performance than the limit given by the matched-filter bound.

Finally, note that besides the *zero-forcing* version of linear equalization, an optimization of the receive filter also exists according to the *minimum mean-squared error (MMSE)* criterion.[3] Thereby, the ISI is not completely eliminated as in ZF linear equalization but a compromise between ISI suppression and noise enhancement is found. Some (small) gains in performance can be achieved. MMSE versions are beyond the scope of this book.

Example 8.5 Loss in Minimum Distance Compared to AWGN Channel

This example continues Example 8.3. We assess the loss ϑ^2 in minimum distance (i.e., performance) compared to an AWGN channel. A coaxial cable with white noise and rectangular transmitter pulses (hard keying) is again assumed.

In Figure 8.14, the loss over the characteristic attenuation (in dB) of the coaxial cable at the Nyquist frequency is plotted. As can be seen, in all cases it increases significantly with the attenuation. Linear equalization has a much higher loss and, thus, poorer performance than MLSD.

However, this strategy requires much less effort; only linear filtering is carried out, followed by a symbol-by-symbol threshold decision. Compared to MLSD, which has to run a Viterbi algorithm, LE is a very-low-complexity scheme.

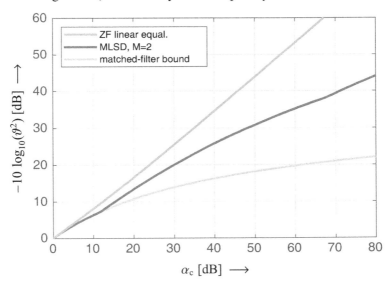

Figure 8.14 Loss in minimum distance compared to AWGN channel over the characteristic attenuation of the coaxial cable at the Nyquist frequency.

---

[3] The terms "zero forcing" and "minimum mean-squared error" denote *criteria* under which the system is optimized. They do not – even though erroneously used in the literature – denote receive strategies.

8.3.4 Whitened-Matched Filter and Decision-Feedback Equalization

We now turn to the other extreme case discussed above and shown in Figure 8.12(c): the discrete-time part $F_W(z)$ should be adjusted such that white noise is present after sampling the filtered received signal. Since the entire continuous-time receive filter (i) whitens the noise and (ii) is matched to the channel, it is called the *whitened-matched filter (WMF)*.

Receive Filter and Discrete-Time End-to-End Model

In order to obtain white noise, the discrete-time part $F_W(z)$ has to be chosen such that the colored noise after the matched filter is white with some variance $\sigma_{n,(\mathrm{WMF})}^2$, that is

$$\Phi_{nn}^{(\mathrm{WMF})}\left(\mathrm{e}^{\mathrm{j}2\pi fT}\right) = \Phi_{nn}^{(\mathrm{MF})}\left(\mathrm{e}^{\mathrm{j}2\pi fT}\right) \cdot \left|F_W\left(\mathrm{e}^{\mathrm{j}2\pi fT}\right)\right|^2 \stackrel{!}{=} \mathrm{const.} = \sigma_{n,(\mathrm{WMF})}^2 \ . \quad (8.60)$$

In order to solve this equation, we first write

$$F_W(z) = F_L(z) \, H(z) \ , \quad (8.61)$$

that is, $F_W(z)$ is conceptually split into $F_L(z)$ for linear equalization cascaded with an additional filter $H(z)$. Using (8.43) and (8.56), this gives

$$\begin{aligned}
\Phi_{nn}^{(\mathrm{WMF})}\left(\mathrm{e}^{\mathrm{j}2\pi fT}\right) &= \Phi_{nn}^{(\mathrm{MF})}\left(\mathrm{e}^{\mathrm{j}2\pi fT}\right) \cdot \left|F_L\left(\mathrm{e}^{\mathrm{j}2\pi fT}\right)\right|^2 \cdot \left|H\left(\mathrm{e}^{\mathrm{j}2\pi fT}\right)\right|^2 \\
&= \Phi_{nn}^{(\mathrm{LE})}\left(\mathrm{e}^{\mathrm{j}2\pi fT}\right) \cdot \left|H\left(\mathrm{e}^{\mathrm{j}2\pi fT}\right)\right|^2 \\
&= \frac{N_0}{E_g} \frac{1}{\Psi_{gg}\left(\mathrm{e}^{\mathrm{j}2\pi fT}\right)} \cdot \left|H\left(\mathrm{e}^{\mathrm{j}2\pi fT}\right)\right|^2 \\
&\stackrel{!}{=} \sigma_{n,(\mathrm{WMF})}^2 \ .
\end{aligned} \quad (8.62)$$

At this point we remark that in the last equation only the (squared) *magnitude* of $H(z)$ is constrained by (8.62). Thus, the *phase* of $H(z)$ can be chosen to meet additional properties. Moreover, any scaling of $H(z)$ is possible since it affects useful signal and noise in the same way.

To find a solution for the transfer function $H(z)$, we seek a factorization of $\Psi_{gg}(z)$. Because $\psi_{gg}[k]$ is a (normalized) autocorrelation, it can be written as $\psi_{gg}[k] = b[k] * b^*[-k]$. Equivalently, $\Psi_{gg}(z)$ can be split as follows[4] $\left(z^{-*} \stackrel{\mathrm{def}}{=} (z^*)^{-1}\right)$:

$$\Psi_{gg}(z) = B(z) \cdot B^*(z^{-*}) \ , \quad (8.63)$$

where, as in (8.51), $\Psi_{gg}(z)$ is the analytic continuation of $\Psi_{gg}\left(\mathrm{e}^{\mathrm{j}2\pi fT}\right)$. Thereby, the transfer function $B(z)$ should correspond to a *causal* impulse response $b[k]$. Moreover, since $\psi_{gg}[k]$ is assumed to be time-limited, this also holds for $b[k]$. In

---

[4] Note that for $z = \mathrm{e}^{\mathrm{j}2\pi fT}$, (8.63) results in $\Psi_{gg}\left(\mathrm{e}^{\mathrm{j}2\pi fT}\right) = B\left(\mathrm{e}^{\mathrm{j}2\pi fT}\right) \cdot B^*\left(\mathrm{e}^{\mathrm{j}2\pi fT}\right) = \left|B\left(\mathrm{e}^{\mathrm{j}2\pi fT}\right)\right|^2$. Hence, (8.63) is a generalization from the unit circle to the entire z-plane.

summary, $B(z)$ may be chosen to be causal and with order O; it can be written as (cf. Section A.3 of Appendix A)

$$B(z) = b[0] + b[1]\,z^{-1} + b[2]\,z^{-2} + \cdots + b[O]\,z^{-O} \, . \tag{8.64}$$

We may now choose

$$H(z) = \frac{1}{b[0]}\, B(z) \, , \tag{8.65}$$

thus, $H(z) = 1 + h[1]\,z^{-1} + h[2]\,z^{-2} + \cdots + h[O]\,z^{-O}$.

Since $F_W(z) = F_L(z)\, H(z)$ and, according to (8.51), $F_L(z) = \frac{1}{\Psi_{gg}(z)}$, the discrete-time end-to-end model is given by the following transfer function and power spectral density:

$$\begin{aligned} H^{(\mathrm{WMF})}\left(e^{j2\pi fT}\right) &= H^{(\mathrm{MF})}\left(e^{j2\pi fT}\right) F_W\left(e^{j2\pi fT}\right) \\ &= \Psi_{gg}\left(e^{j2\pi fT}\right) \frac{1}{\Psi_{gg}\left(e^{j2\pi fT}\right)}\, H\left(e^{j2\pi fT}\right) \\ &= H\left(e^{j2\pi fT}\right) \, , \end{aligned} \tag{8.66}$$

$$\begin{aligned} \Phi_{nn}^{(\mathrm{WMF})}\left(e^{j2\pi fT}\right) &= \frac{N_0}{E_g}\, \frac{1}{\Psi_{gg}\left(e^{j2\pi fT}\right)} \cdot \left|H\left(e^{j2\pi fT}\right)\right|^2 \\ &= \frac{N_0}{E_g}\, \frac{1}{\Psi_{gg}\left(e^{j2\pi fT}\right)}\, \frac{1}{b[0]} B\left(e^{j2\pi fT}\right) \cdot \frac{1}{b^*[0]} B^*\left(e^{j2\pi fT}\right) \\ &= \frac{N_0}{E_g}\, \frac{1}{|b[0]|^2} = \sigma_{n,(\mathrm{WMF})}^2 \, . \end{aligned} \tag{8.67}$$

As can be seen from (8.66), the scaling of $B(z)$ in (8.65) guarantees that the end-to-end impulse response $h[k] \circ\!\!-\!\!\bullet H(z)$ is causal and *monic*, that is, the first tap equals one. Moreover, (8.67) shows that in order to have the lowest noise power, $|b[0]|$ should be as large as possible. It can be shown that among all causal transfer functions $B(z)$ that fulfill (8.63), the *minimum-phase* transfer function has the largest first tap (for the properties of minimum-phase systems, see Section A.3.2 of Appendix A). In Section A.3.2 of Appendix A, it is proven that for a causal and minimum-phase polynomial $B(z)$, the following relation holds:

$$|b[0]|^2 = \exp\left\{T \int_{-\frac{1}{2T}}^{\frac{1}{2T}} \log\left(\Psi_{gg}\left(e^{j2\pi fT}\right)\right)\, df\right\} \, . \tag{8.68}$$

In summary, due to the choice (8.65), $H(z)$ is:

- causal and of order O;
- minimum-phase;
- monic (the first tap, $h[0]$, is one).

Thus, end-to-end, a causal, minimum-phase, and monic signal transfer function and white noise are present. The discrete-time end-to-end model is given by

$$d_W[k] = a[k] * h[k] + n[k] \, . \tag{8.69}$$

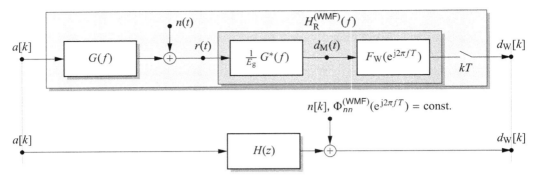

Figure 8.15 Entire transmission model when employing the receive filter $H_R^{(\text{WMF})}(f)$.

This is visualized in Figure 8.15.

All parts of the receive filter can be combined into the entire filter, resulting in the *whitened-matched filter* (cf. (8.54))

$$H_R^{(\text{WMF})}(f) \stackrel{\text{def}}{=} \frac{\frac{1}{E_g} G^*(f)}{\frac{1}{E_g T} \sum_\mu \left| G\left(f - \frac{\mu}{T}\right)\right|^2} H\left(e^{j2\pi fT}\right)$$

$$= \frac{1}{E_g} \frac{G_T^*(f) H_C^*(f)}{\Psi_{gg}\left(e^{j2\pi fT}\right)} \frac{1}{b[0]} B\left(e^{j2\pi fT}\right)$$

$$= \frac{1}{E_g} \frac{G_T^*(f) H_C^*(f)}{B\left(e^{j2\pi fT}\right) B^*\left(e^{j2\pi fT}\right)} \frac{1}{b[0]} B\left(e^{j2\pi fT}\right)$$

$$= \frac{1}{E_g} G_T^*(f) H_C^*(f) \cdot \frac{1}{b[0] B^*\left(e^{j2\pi fT}\right)} \ . \tag{8.70}$$

Decision-Feedback Equalization

In contrast to linear equalization where no ISI remains but the noise is colored, here ISI as defined by $H(z)$ is present but white noise is active. Owing to the interference, symbol-by-symbol decisions cannot be immediately generated based on $d_W[k]$. However, a "trick" can be used.

Assume that the amplitude coefficient $a[k-1]$ at time step $k-1$ (the previous one) has been correctly detected. Since we know the channel impulse response $h[k]$ and the transfer function $H(z)$ at the receiver side, we know the intersymbol interference this amplitude coefficient causes. This known interference can simply be subtracted (via the filter $H(z) - 1$ that has the required impulse response $h[k] - \delta[k]$, the end-to-end impulse response without the leading tap $h[0] = 1$) from the discrete-time receive signal $d_W[k]$. If we have done this up to time step $k-1$, since $h[k]$ is causal, the current time step k is free of interference. Since $h[0] = 1$, we "see" $a[k] + n[k]$, where $n[k]$ is the white Gaussian noise. Hence, the current symbol is free of interference with additive white Gaussian noise; via threshold decision an estimate $\hat{a}[k]$ can be generated. This procedure is repeated over the time steps. Since available decisions

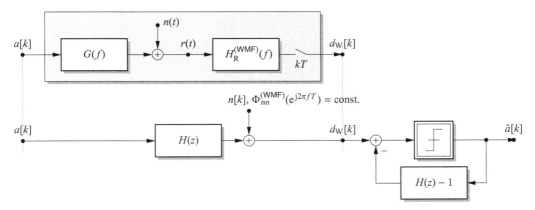

Figure 8.16 Decision-feedback equalization.

are fed back to subtract the interference, this strategy is called *decision-feedback equalization (DFE)*.

The respective transmission system using the discrete-time end-to-end model from Figure 8.15 and the DFE receiver structure is depicted in Figure 8.16.

The mode of operation of DFE may be explained from the receiver structure in this figure. For correct decisions, the useful signal (symbols $a[k]$) passes the nonlinear decision device unaltered, but the noise is blocked. For the useful signal, the decision device is a shortcut. If the decision device were replaced by a shortcut, the feedback loop would implement the transfer function $1/H(z)$. Linear equalization is present for the useful signal but no noise enhancement occurs.

It should be mentioned that DFE requires immediate decisions for equalization. As a consequence, the combination with channel coding, where decoding is done over an entire receive block, is not straightforward. In addition, in case of decision errors, error propagation in the feedback loop will occur. However, this effect is typically moderate. Both problems can be avoided if DFE is replaced by an equivalent transmitter-side structure. Such strategies are called *precoding* (see Section 8.5).

Example 8.6 DFE

In order to visualize the operation of DFE, we assume that the amplitude coefficients are drawn from $a[k] \in \{-1, +1\}$. The channel is expected to have the transfer function $H(z) = 1 + 0.7\,z^{-1} + 0.1\,z^{-2}$, that is, the taps after the leading one are $h[1] = 0.7$ and $h[2] = 0.1$.

Table 8.2 gives the transmitted symbols together with their caused post-cursors. Transmission is expected to start at $k = 0$. The detection symbols $d_W[k]$ (assuming $n[k] = 0$) are given by summing up all post-cursors of the transmitted symbols.

Successively, the coefficient $a[k]$ is detected and the interference (its post-cursors) caused by it is subtracted.

Table 8.2 Example operation of decision-feedback equalization.

k	0	1	2	3	4	5	6	7
$a[0]=-1$	-1	-0.7	-0.1					
$a[1]=1$		1	0.7	0.1				
$a[2]=-1$			-1	-0.7	-0.1			
$a[3]=1$				1	0.7	0.1		
$a[4]=1$					1	0.7	0.1	
$a[5]=1$						1	0.7	0.1
...								
$d_W[k]$		0.3	-0.4	0.4	1.6	1.8	...	...
$a[0]=-1$ can be detected and its interference (post-cursors)		-0.7	-0.1					
can be subtracted, leading to			-0.3	0.4	1.6	1.8		
$a[1]=1$ can be detected and its interference (post-cursors)			0.7	0.1				
can be subtracted, leading to				0.3	1.6	1.8		
$a[2]=-1$ can be detected and its interference (post-cursors)				-0.7	-0.1			
can be subtracted, leading to					1.7	1.8		
$a[3]=1$ can be detected and its interference (post-cursors)					0.7	0.1		
can be subtracted, leading to						1.7		
$a[4]=1$ can be detected ...								
...								

Signal-to-Noise Ratio

The signal-to-noise ratio when applying $H_R^{(\text{WMF})}(f)$ for decision-feedback equalization can immediately be given when assuming that the past O decisions are correct, that is, no error propagation exists. Thus, an upper bound on the SNR is derived, but this bound is quite tight in the region of interest (low error probabilities).

Since the power of the (zero-mean) useful data is σ_a^2 and the noise power, active at the input of the threshold device, is given by $\sigma_{n,(\text{WMF})}^2$, the SNR when performing zero-forcing decision-feedback equalization calculates as

$$\text{SNR}^{(\text{DFE})} = \frac{\sigma_a^2}{\sigma_{n,(\text{WMF})}^2}$$

using (8.67)

$$= \frac{\sigma_a^2 E_g}{N_0} |b[0]|^2$$

using (8.68)

$$= \frac{\sigma_a^2 E_g}{N_0} \exp\left\{ T \int_{-\frac{1}{2T}}^{\frac{1}{2T}} \log\left(\Psi_{gg}\left(e^{j2\pi fT}\right)\right) df \right\}$$

$$= \exp\left\{ T \int_{-\frac{1}{2T}}^{\frac{1}{2T}} \log\left(\frac{\sigma_a^2 E_g}{N_0} \Psi_{gg}\left(e^{j2\pi fT}\right)\right) df \right\}. \quad (8.71)$$

Finally, considering (8.47), we arrive at

$$\boxed{\text{SNR}^{(\text{DFE})} = \exp\left\{ T \int_{-\frac{1}{2T}}^{\frac{1}{2T}} \log\left(\widetilde{\text{SNR}}\left(e^{j2\pi fT}\right)\right) df \right\} \quad (8.72)}$$

geometric mean over the folded spectral SNR.

Again, knowing the spectral SNR at the receiver input, the folded version can be calculated. The geometric mean (see Box 8.1) over this function simply gives the SNR for decision-feedback equalization with optimum receiver frontend. Note that the geometric mean (over non-negative quantities) is always smaller than (or equal to) the arithmetic mean but larger than (or equal to) the harmonic mean. Hence, decision-feedback equalization has a poorer performance than the limit given by the matched-filter bound but a better performance than linear equalization.

Finally, note that besides the *zero-forcing* version of decision-feedback equalization, an optimization according to the *minimum mean-squared error* criterion also exists. Some (small) gains in performance can be achieved. MMSE versions are beyond the scope of this book.

Example 8.7 Loss in Minimum Distance Compared to AWGN Channel

This example continues Examples 8.3 and 8.5. We again assess the loss ϑ^2 in minimum distance (i.e., performance) compared to an AWGN channel. A coaxial cable with white noise and rectangular transmitter pulses (hard keying) are assumed.

In Figure 8.17, the loss over the characteristic attenuation (in dB) of the coaxial cable at the Nyquist frequency is plotted. As can be seen, decision-feedback equalization has only a somewhat higher loss than MLSD; for small α_c almost no loss occurs. The loss is much smaller (better performance) than that of linear equalization.

DFE also requires only little effort; only linear filtering is carried out, followed by symbol-by-symbol threshold decision and feedback of the decisions. Hence, DFE is a low-complexity but well-performing scheme.

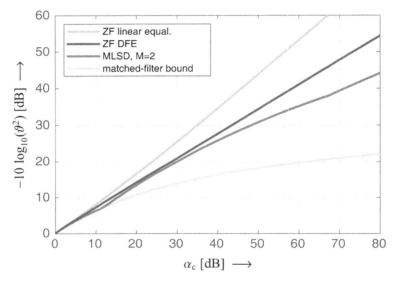

Figure 8.17 Loss in minimum distance compared to AWGN channel over the characteristic attenuation of the coaxial cable at the Nyquist frequency.

DFE and Channel Capacity

Finally, an important property of DFE should briefly be stated. To that end, we assume that the channel $H_C(f)$ has low-pass characteristics and that the transmit filter $G_T(f)$ (pulse shaping) is rectangular in the frequency domain (width $1/T$, cf. (3.62)). Thus, we have

$$G(f) = \sqrt{E_g T} \operatorname{rect}(fT) H_C(f) . \tag{8.73}$$

As no aliasing takes place, the folded SNR reads, for $f \in \left[-\frac{1}{2T}, \frac{1}{2T}\right]$

$$\widetilde{\mathrm{SNR}}\left(e^{j2\pi fT}\right) = \frac{\sigma_a^2}{TN_0} |G(f)|^2 = E_g T \frac{\sigma_a^2}{TN_0} |H_C(f)|^2 = \frac{S_s T}{N_0} |H_C(f)|^2 , \tag{8.74}$$

where we have used the definition (3.16) of the transmit power $S_s = \sigma_a^2 E_g/T$ for zero-mean data symbols.

The SNR for (zero-forcing) DFE then reads

$$\text{SNR}^{\text{(DFE)}} = \exp\left\{T \int_{-\frac{1}{2T}}^{\frac{1}{2T}} \log\left(\frac{S_s T}{N_0} |H_C(f)|^2\right) df\right\}$$

or

$$\log_2\left(\text{SNR}^{\text{(DFE)}}\right) = T \int_{-\frac{1}{2T}}^{\frac{1}{2T}} \log_2\left(\frac{S_s T}{N_0} |H_C(f)|^2\right) df . \tag{8.75}$$

We now interpret this equation.

Owing to the subtraction of the interference, DFE results in an AWGN channel with signal-to-noise ratio $\text{SNR}^{\text{(DFE)}}$. The *capacity in bit per channel use* of this created *discrete-time* AWGN channel (assuming i.i.d. Gaussian transmit symbols; Shannon's capacity formula) is well approximated for large SNRs by

$$C = \log_2\left(1 + \text{SNR}^{\text{(DFE)}}\right) \approx \log_2\left(\text{SNR}^{\text{(DFE)}}\right) . \tag{8.76}$$

The left-hand expression of (8.75) is the capacity of the discrete-time end-to-end channel.

The underlying channel has spectral SNR $\frac{S_s T}{N_0}|H_C(f)|^2$ within the used bandwidth. Hence, the *capacity in bit per seconds* of this *continuous-time* non-flat channel with AWGN is well approximated for large SNRs by

$$C_T = \int_{-\frac{1}{2T}}^{\frac{1}{2T}} \log_2\left(1 + \frac{ST}{N_0} |H_C(f)|^2\right) df \approx \int_{-\frac{1}{2T}}^{\frac{1}{2T}} \log_2\left(\frac{ST}{N_0} |H_C(f)|^2\right) df .$$
$$\tag{8.77}$$

The channel is used every T seconds; hence, the capacity per use is $C_T T$. This is the right-hand expression of (8.75).

Consequently, (8.75) reveals that the discrete-time AWGN channel created by DFE has (for large SNR) the same capacity as the underlying continuous-time channel. No capacity is lost; DFE fully utilizes the capacity provided by the channel. In this sense, DFE is an optimal equalization method with respect to information theory.

8.3.5 Noise Prediction

The main problem of linear equalization is *noise enhancement*; the power of the noise at the output of the linear equalizer (optimum Nyquist filter) is increased and the noise samples are correlated. As given in (8.56), psd and thus the autocorrelation sequence are given by

$$\Phi_{nn}^{(\text{LE})}\left(e^{j2\pi fT}\right) = \frac{N_0}{E_g} \frac{1}{\Psi_{gg}\left(e^{j2\pi fT}\right)}$$

$$\updownarrow \qquad (8.78)$$

$$\phi_{nn}^{(\text{LE})}[\kappa] = \frac{N_0}{E_g} \mathscr{F}^{-1}\left\{\frac{1}{\Psi_{gg}\left(e^{j2\pi fT}\right)}\right\}.$$

Since $\phi_{nn}^{(\text{LE})}[\kappa] \neq \sigma_n^2 \delta[\kappa]$, subsequent samples are correlated; they exhibit dependencies. These correlations may be exploited to obtain an SNR gain. This is possible by *noise prediction*; the respective structure is shown in Figure 8.18.

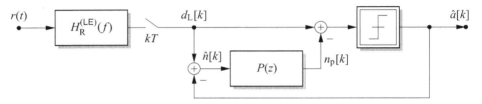

Figure 8.18 Linear equalization followed by noise prediction.

Employing the optimum linear equalizer, the decision variable is given by $d_\text{L}[k] = a[k] + n[k]$. Hence, assuming correct decisions $\hat{a}[k]$, estimates of the noise samples may be generated via

$$\hat{n}[k] = d_\text{L}[k] - \hat{a}[k]. \qquad (8.79)$$

Having estimates on the past noise samples (since these samples are correlated), a *prediction* $n_\text{p}[k]$ of the current noise sample can be generated. This is done via the linear *prediction filter* $P(z)$. When subtracting the prediction from the decision variable, only the *prediction error* $n_\epsilon[k] \stackrel{\text{def}}{=} n[k] - n_\text{p}[k]$ is active as noise at the input of the threshold device. Hence, by adjusting the prediction filter suitably, a *prediction gain* can be achieved.

Since the noise psd/autocorrelation sequence of the noise is known (see (8.78)), the optimum predictor can be given analytically. Let

$$P(z) = \sum_{k=1}^{\rho} p[k] z^{-k} \qquad (8.80)$$

be a strictly causal ($p[0] = 0$) FIR filter of order ρ, by which the prediction value

$$n_\text{p}[k] = \sum_{\kappa=1}^{\rho} p[\kappa] \hat{n}[k - \kappa] \qquad (8.81)$$

is calculated based on the ρ last noise estimates.

The desired coefficients $p[k]$, $k = 1, \ldots, \rho$, for smallest variance of the residual noise $n_\epsilon[k]$, are calculated by solving the *Yule–Walker equations* (for details on linear prediction, see Section A.5 of Appendix A):

$$\begin{bmatrix} \phi_{nn}^{(LE)}[0] & \phi_{nn}^{(LE)}[-1] & \cdots & \phi_{nn}^{(LE)}[-\rho+1] \\ \phi_{nn}^{(LE)}[1] & \ddots & \ddots & \vdots \\ \vdots & \ddots & \ddots & \phi_{nn}^{(LE)}[-1] \\ \phi_{nn}^{(LE)}[\rho-1] & \cdots & \phi_{nn}^{(LE)}[1] & \phi_{nn}^{(LE)}[0] \end{bmatrix} \cdot \begin{bmatrix} p[1] \\ p[2] \\ \vdots \\ p[\rho] \end{bmatrix} = \begin{bmatrix} \phi_{nn}^{(LE)}[1] \\ \phi_{nn}^{(LE)}[2] \\ \vdots \\ \phi_{nn}^{(LE)}[\rho] \end{bmatrix}. \qquad (8.82)$$

The block diagram in Figure 8.18 can be transformed into a fully equivalent structure. Since $P(z)$ is a linear filter, it can be implemented twice and instead of doing the subtraction at the input, it can be moved to the output. The resulting alternative structure is shown in Figure 8.19.

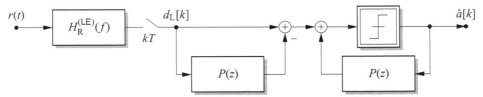

Figure 8.19 Equivalent DFE-like structure of linear equalization followed by noise prediction.

A comparison of this figure with Figure 8.16 immediately reveals that it is the same structure as in DFE. Here the feedforward discrete-time part

$$H_P(z) = 1 - P(z) \qquad (8.83)$$

is present. Note that $1-P(z)$ is the prediction-error filter and, if the prediction order ρ is sufficiently large (here $\rho \geq O$), the prediction error is white (otherwise not all correlations have been exploited). Hence, in the optimum case $H_P(z) = H(z)$, as derived for DFE. The noise is whitened but intersymbol interference is introduced; these interferences are readily canceled by the feedback structure.

The noise predictor structure offers the opportunity to employ a predictor filter of a smaller order $\rho < O$. An exchange between order ρ of the filter $H_P(z)$ and the variance of the residual noise is enabled. For $\rho = 0$, pure linear equalization with $\sigma_{n_\epsilon}^2 = \sigma_{n,(LE)}^2$ is present. For $\rho \geq O$, the whitened-matched filter occurs and $\sigma_{n_\epsilon}^2 = \sigma_{n,(WMF)}^2$. Especially when sophisticated sequence-detection techniques are applied (cf. Sections 8.2 and 8.4), a free choice of the prediction order ρ, that is, the pulse length $\rho + 1$, allows for an optimum trade-off between complexity and noise variance by incomplete noise whitening.

8.4 Maximum-Likelihood Sequence Detection II

In Section 8.2, we derived the optimum receiver for PAM transmission over a linear, dispersive channel. Maximum-likelihood sequence detection is performed based on the samples at the output of the *matched filter* for the entire pulse $g(t)$ seen at the receiver (transmitter pulse plus channel). A suitable branch metric has been defined and the search over all possible transmit sequences may be conducted by the Viterbi algorithm, delivering the sequence with the *largest* sum metric. This concept followed directly from the general *matched-filter* or *correlation receiver* displayed in Figure 6.39.

However, in Chapter 6, the strategy based on the *vector demodulator* of Figure 6.21 was also presented. There, a decoding metric based on the squared Euclidean distance is utilized and the sequence with the *smallest* accumulated metric is searched. Both strategies have the same performance.

This also applies to maximum-likelihood sequence detection for intersymbol-interference channels. The suitable receiver is obtained by employing the *whitened-matched filter* (8.70), that is, the optimum receive filter for decision-feedback equalization. However, in contrast to DFE, the interference is not successively canceled (looking only at the best symbol for the current step), but a search over all possible sequences is conducted.

8.4.1 End-to-End Model

As derived in Section 8.3.4, the discrete-time end-to-end model when applying the whitened-matched filter is given by (see (8.69))

$$d_{\mathrm{W}}[k] = a[k] * h[k] + n[k] = u[k] + n[k], \tag{8.84}$$

where we have introduced the noise-free channel output symbols

$$u[k] \stackrel{\text{def}}{=} a[k] * h[k]. \tag{8.85}$$

The data symbols are drawn from an M-ary constellation; $a[k] \in \mathcal{A}$. The end-to-end impulse response $h[k]$ is that of a causal, minimum-phase, monic finite-impulse response (FIR) system of order O, that is

$$H(z) = 1 + \sum_{k=1}^{O} h[k] z^{-k}, \tag{8.86}$$

which gives

$$u[k] = a[k] + \sum_{l=1}^{O} a[k-l] h[l]. \tag{8.87}$$

The additive Gaussian noise $n[k]$ is white with variance $\sigma_n^2 = \sigma_{n,\text{(WMF)}}^2$. For convenience, the block diagram is repeated in Figure 8.20, showing the tapped-delay-line representation.

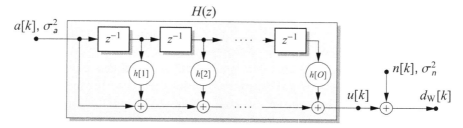

Figure 8.20 End-to-end model when applying the whitened-matched filter.

8.4.2 Maximum–Likelihood Approach

Owing to the presence of intersymbol interference, symbol-by-symbol detection is not possible. Consequently, the search has to be done over the entire transmit sequence $\langle a[k]\rangle$; maximum-likelihood sequence detection has to be performed. Since the noise is white and Gaussian, the ML decision rule can be written as[5]

$$\widehat{\langle a[k]\rangle} = \operatorname*{argmax}_{\langle a[k]\rangle} f_{\langle d_\mathrm{W}[k]\rangle}(\langle d_\mathrm{W}[k]\rangle \mid \langle a[k]\rangle)$$

and for white Gaussian noise

$$= \operatorname*{argmax}_{\langle a[k]\rangle} \left(\mathrm{const.}\ \exp\left\{-\left\|\langle d_\mathrm{W}[k]\rangle - \langle a[k] * h[k]\rangle\right\|^2 / \sigma_n^2\right\}\right)$$

$$= \operatorname*{argmin}_{\langle a[k]\rangle} \left\|\langle d_\mathrm{W}[k]\rangle - \langle a[k] * h[k]\rangle\right\|^2$$

$$= \operatorname*{argmin}_{\langle a[k]\rangle} \sum_k \left| d_\mathrm{W}[k] - a[k] - \sum_{l=1}^{O} a[k-l]\, h[l]\right|^2 . \qquad (8.88)$$

Hence, we have the result that MLSD is performed by searching for the data sequence $\langle a[k]\rangle$, for which the noise-free channel output sequence $\langle u[k]\rangle = \langle a[k] * h[k]\rangle$ has *minimum squared Euclidean distance* to the observed sequence $\langle d_\mathrm{W}[k]\rangle$ at the output of the whitened-matched filter (cf. (6.67)).

The key to conducting MLSD in practice is again the description of the dependencies introduced by the channel as *finite-state machine*; the tool for carrying out MLSD is then again the *Viterbi algorithm*.

Here the history, the last O amplitude coefficients (which are the content of the delay line of the FIR channel model, cf. Figure 8.20), define the state at time step $k-1$:

$$S[k-1] \stackrel{\text{def}}{=} S^{(a)} = \left[a^{(a)}[k-1],\, a^{(a)}[k-2],\, \ldots,\, a^{(a)}[k-O]\right]. \qquad (8.89)$$

Given the current state $S[k-1]$ and the current symbol $a[k]$, in the next time step a *new state $S[k]$* is reached:

$$S[k] = S^{(b)} \stackrel{\text{def}}{=} \left[a[k],\, a^{(a)}[k-1],\, \ldots,\, a^{(a)}[k-O+1]\right] \qquad (8.90)$$
$$= \mathrm{fct}_S\left(S[k-1],\, a[k]\right).$$

Having calculated the WMF output $d_\mathrm{W}[k]$, the *branch metric* at time step k depends on the previous state and the current symbols $a[k]$, specifically

$$\lambda^{(a)\to(b)}[k] = \left|d_\mathrm{W}[k] - a[k] - \sum_{l=1}^{O} a[k-l]\, h[l]\right|^2 \qquad (8.91)$$
$$= \mathrm{fct}_B\left(S[k-1],\, a[k]\right).$$

---

[5] The squared Euclidean norm of a sequence has to be understood in the same way as the squared Euclidean norm of a vector.

As above, the possible state transitions over time are visualized via a *trellis diagram*. Since the state comprises the last O amplitude coefficients that are drawn from an M-ary constellation, M^O states are present. Each path through the trellis uniquely corresponds to a data sequence $\langle a[k] \rangle$.

Thus, in this variant, MLSD is carried out by searching the *path through the trellis* that has *minimal accumulated metric*

$$\lambda_{\text{acc}} = \sum_k \lambda^{(a) \to (b)}[k], \qquad (8.92)$$

where the *branch metric* (state-dependent, branch from state $S^{(a)}$ to state $S^{(b)}$) is given by (8.91). This task is solved in an optimal way by the *Viterbi algorithm* (see Section B.2 in Appendix B).

Note that both variants of MLSD for linearly distorting channels

- using the matched filter (introduced by G. Ungerböck in 1974)
- using the whitened-matched filter (introduced by D. Forney in 1972)

perform exactly the same. They require a Viterbi algorithm that operates on the same number of states (M^O). The only difference is that the matched-filter version ("Ungerböck metric" (8.24)) requires only the calculation of real parts; the other ("Forney metric" (8.91)) has to calculate quadratic distances. This is the same difference as between the correlation receiver and the vector demodulator in Chapter 6.

Example 8.8 Finite-State Machine and Trellis Diagram

We continue Example 8.6, where the end-to-end channel was expected to have the transfer function $H(z) = 1 + 0.7 z^{-1} + 0.1 z^{-2}$ of order $O = 2$ and binary ($M = 2$) transmission $a[k] = \{+1, -1\}$ was used. Figure 8.21 shows the FIR channel model.

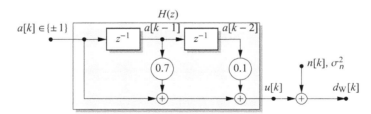

Figure 8.21 FIR channel model with $O = 2$ and $M = 2$.

Four trellis segments corresponding to four modulation intervals are depicted in Figure 8.22. In the present case, $2^2 = 4$ states are present. The branches are labeled with the channel input symbol $a[k]$ causing this transition and the noise-free channel output symbol $u[k]$ that is present for the assumed state (history, content of the delay elements) and the assumed input.

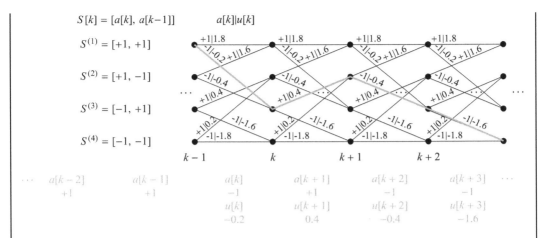

Figure 8.22 Trellis diagram corresponding to the above channel. $O = 2$ and $M = 2$.

The path through the trellis corresponding to the particular given sequence of amplitude coefficients $a[k]$ is highlighted and the corresponding sequence of noise-free channel output symbols $u[k]$ is noted.

8.4.3 Reduced-State Sequence Detection

Finally, we note that MLSD and DFE are two extreme cases. MLSD searches over all possible sequences – characterized by a trellis diagram with M^O states. In contrast, DFE tracks only a single path through the trellis. In each step a decision is taken and the interference is eliminated.

Decision-feedback sequence detection (DFSD), as a particular representative of so-called *reduced-state sequence detection* schemes, enables a trade-off between both extrema – a trade-off between complexity and performance is possible. The interpretation follows directly from the FIR channel model, depicted in Figure 8.23.

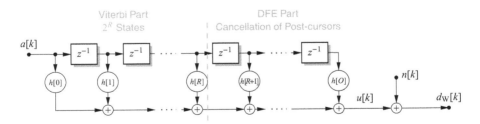

Figure 8.23 FIR end-to-end model when applying the whitened-matched filter.

The history of the transmitted sequence – graphically the delay line – is split into two parts. The post-cursors of the previous R symbols are taken into account in the sequence search, that is, in the Viterbi algorithm. The post-cursors of earlier symbols are canceled, that is, subtracted via the DFE principle. For each trellis state in the reduced trellis an individual DFE is applied. Hence, we have the following possibilities.

- *MLSD:* $R = O$, that is, Viterbi algorithm with M^O states.
- *DFSE:* $0 < R < O$, that is, Viterbi algorithm with M^R states for $R + 1$ taps; the ISI of the remaining taps is canceled by DFE (per state).
- *DFE:* $R = 0$, that is, only a single state is present; the ISI of all post-cursors is canceled.

Note that for most applications, a small number R of taps in the sequence-detection part is necessary to achieve the performance of MLSD very closely. This is because the channel impulse response is minimum-phase, which means that the energy is concentrated in the first few taps. Thus, DFSE is a highly efficient complexity-reduction method.

8.5 Transmitter-Side Channel Pre-equalization

Up to now, the linear dispersive channel has been equalized at the receiver side. The channel produces intersymbol interference; it is handled once it has occurred. To that end, we have always assumed that the receiver knows the channel characteristics (channel transfer function, psd of the noise), that is, the receiver has *channel state information (CSI)*. Using known training sequences, the acquisition of CSI at the receiver by *channel estimation techniques* is easily possible and daily practice.

In some situations it is favorable to have CSI also at the transmitter. Thereby, the transmit signal may be adopted by the actual channel, that is, the channel may be *pre-equalized*. In bidirectional point-to-point communication links, CSI that is acquired at the receiver can be forwarded to the transmitter using control channels that are present anyway to establish/monitor the data link.

The exploitation of CSI at the transmitter may allow for high gains in power efficiency, especially when highly efficient channel coding with long codewords cannot be applied due to restrictions on the latency and/or on the implementation complexity[6] (cf. also Chapter 9).

8.5.1 Transmitter-Side Preprocessing

Remember that the generation of the PAM transmit signal is given by the filtering (pulse shaping) of the discrete-time sequence $\langle a[k] \rangle$ of amplitude coefficients (cf. Chapter 3). As a consequence, the adaptation of the transmitter can be done by

---

[6] Even if no actual CSI is available at the transmitter, some average CSI (e.g., derived from physical channel properties) may be utilized. In such situations, an approximate adjustment of the transmit signal combined with a residual (linear, adaptive) equalization at the receiver is a suitable strategy.

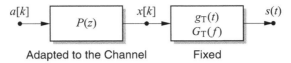

Figure 8.24 Transmitter-side preprocessing (here by a linear filter $P(z)$).

adjusting the transmit filter $G_T(f)$ suitably and/or by preprocessing the discrete-time sequence data sequence, that is, by generating a suitable *channel input sequence* $\langle x[k] \rangle$ from the given sequence $\langle a[k] \rangle$ (cf. Figure 8.24).

In practice, the transmit filter $G_T(f)$ is preferably kept fixed, even though its adaptation offers more degrees of freedom (the transfer function can be adapted for all frequency points f; in case of discrete-time processing, only a frequency-periodic filter can be adapted). Thus, dual to what is done at the receiver, the pre-equalization is performed in the discrete-time domain by means of digital signal processing followed by the fixed transition from discrete to continuous time.

Spectral Signal-to-Noise Ratio

We have seen that to assess the performance of equalization schemes, the spectral signal-to-noise ratio and its folded version play an important role. For a zero-mean i.i.d. sequence $\langle a[k] \rangle$ of amplitude coefficients with variance σ_a^2, we have

$$\text{SNR}(f) = \frac{\sigma_a^2 |G(f)|^2}{T N_0} = \frac{\sigma_a^2 |G_T(f) H_C(f)|^2}{T N_0}, \tag{8.93}$$

$$\widetilde{\text{SNR}}\left(e^{j2\pi fT}\right) = \sum_\mu \text{SNR}\left(f - \tfrac{\mu}{T}\right). \tag{8.94}$$

When changing the statistical properties of the discrete-time sequence or when changing the transmit filter, these quantities will change. However, in the following we will express the results in terms of the folded spectral SNR, that is (8.93) and (8.94), which is present if no adaptation of the transmitter were done.

Transmit Power

Applying a preprocessing filter $P(z)$, the average psd of the continuous-time transmit signal is given by

$$\bar{\Phi}_{ss}(f) = \Phi_{xx}\left(e^{j2\pi fT}\right) \frac{|G_T(f)|^2}{T} \tag{8.95}$$

with

$$\Phi_{xx}\left(e^{j2\pi fT}\right) = \sigma_a^2 \left|P\left(e^{j2\pi fT}\right)\right|^2, \tag{8.96}$$

for a zero-mean i.i.d. sequence $\langle a[k]\rangle$ with variance σ_a^2. Thus, the average transmit power of the continuous-time transmit signal $s(t)$ amounts to

$$S_s = \int_{-\infty}^{\infty} \bar{\Phi}_{ss}(f)\, df = \frac{1}{T}\int_{-\infty}^{\infty} \Phi_{xx}\left(e^{j2\pi fT}\right) |G_T(f)|^2\, df$$

$$= \frac{1}{T}\sum_{\mu} \int_{-\frac{1}{2T}}^{\frac{1}{2T}} \Phi_{xx}\left(e^{j2\pi(f-\mu/T)T}\right) \left|G_T\left(f - \tfrac{\mu}{T}\right)\right|^2\, df$$

$$= \frac{1}{T}\int_{-\frac{1}{2T}}^{\frac{1}{2T}} \Phi_{xx}\left(e^{j2\pi fT}\right) \sum_{\mu}\left|G_T\left(f - \tfrac{\mu}{T}\right)\right|^2\, df .$$

If $g_T(t) \circ\!\!-\!\!\bullet\, G_T(t)$ is a $\sqrt{\text{Nyquist}}$ pulse we have $\sum_\mu \left|G_T\left(f - \tfrac{\mu}{T}\right)\right|^2 = E_T T$, where $E_T = \int |g_T(t)|^2\, dt$ is the pulse energy. This gives

$$S_s = \frac{E_T}{T} T \int_{-\frac{1}{2T}}^{\frac{1}{2T}} \Phi_{xx}\left(e^{j2\pi fT}\right)\, df = \frac{E_T}{T} \sigma_x^2 . \tag{8.97}$$

Hence, expecting a $\sqrt{\text{Nyquist}}$ transmit pulse $g_T(t)$, the transmit power remains the same as without preprocessing if $\sigma_x^2 = \sigma_a^2$ is guaranteed.

End-to-End Model

As the T-spaced discrete-time samples $d_W[k]$ at the output of the whitened-matched filter are the optimum starting point for decision-feedback equalization and for maximum-likelihood sequence detection, we again assume the whitened-matched filter as the receiver frontend. Hence, we have (see (8.69))

$$d_W[k] = a[k] * h[k] + n[k] . \tag{8.98}$$

Thereby, the end-to-end impulse response $h[k]$ (comprising transmit filter, channel, and receive filter) is that of a causal, minimum-phase, monic finite-impulse response (FIR) system of order O, that is

$$H(z) = 1 + \sum_{k=1}^{O} h[k]\, z^{-k} . \tag{8.99}$$

The additive Gaussian noise $n[k]$ is white with variance $\sigma_n^2 = \sigma_{n,(\text{WMF})}^2$. For the block diagram, see Figure 8.20.

8.5.2 Linear Pre-equalization

Similar to receiver-side equalization, the basic version of transmitter-side preprocessing is *linear pre-equalization (LPE)*. As was derived in Section 8.3, the discrete-time feedforward parts of the receive filters $H_R^{(\mathrm{LE})}(f)$ and $H_R^{(\mathrm{WMF})}(f)$ differ in the causal, minimum-phase, monic transfer function $H(z)$, that is, $H_R^{(\mathrm{WMF})}(f) = H_R^{(\mathrm{LE})}(f)\, H\!\left(e^{j2\pi fT}\right)$. Hence, when applying the whitened-matched filter as the receiver frontend, the end-to-end transfer function $H(z)$ at hand may be equalized via $1/H(z)$. The same situation as when directly utilizing the optimum Nyquist filter (ONF) arises; in $d_L[k]$ no intersymbol interference is active and symbol-by-symbol decisions can be taken. Since $H(z)$ is minimum-phase, its inverse is stable. The situation is shown in the top part of Figure 8.25.

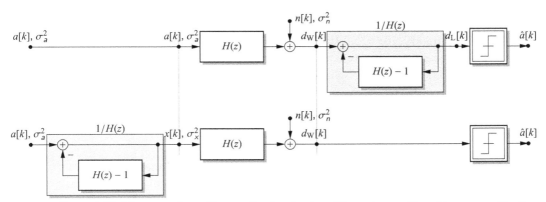

Figure 8.25 Linear receiver-side equalization (top) and linear transmitter-side pre-equalization (bottom).

The idea of linear pre-equalization is to move the linear filter $1/H(z)$ from the receiver to the transmitter. The situation is depicted in the bottom part of Figure 8.25. The data sequence $\langle a[k]\rangle$ is filtered by $P(z) = 1/H(z)$ in order to generate the channel input sequence $\langle x[k]\rangle$. This sequence is then fed to the transmit filter and transmitted over the channel. At the receiver, as the systems with transfer functions $1/H(z)$ and $H(z)$ cancel each other, intersymbol-interference-free samples with additive white Gaussian noise are obtained. A threshold device can immediately generate decisions $\hat{a}[k]$ on the amplitude coefficients.

Signal-to-Noise Ratio

The main disadvantage of receiver-side linear equalization is the noise enhancement due to the filter $1/H(z)$; the noise samples at the input of the decision device have noise power

$$\sigma^2_{n,(\mathrm{LE})} = \sigma^2_{n,(\mathrm{WMF})} \cdot T \int_{-\frac{1}{2T}}^{\frac{1}{2T}} \frac{1}{\left|H\!\left(e^{j2\pi fT}\right)\right|^2}\, df = \sigma^2_{n,(\mathrm{WMF})} \cdot \gamma^2_{\mathrm{enh}} . \qquad (8.100)$$

Compared to the variance of the noise samples at the output of the whitened-matched filter, a *noise enhancement* by the factor γ^2_{enh} is present.

A noise enhancement is not active in case of linear pre-equalization. However, the *transmit power* is increased. Assuming a white data sequence, the power of the channel input symbols calculates as

$$\sigma_x^2 = \sigma_a^2 \cdot T \int_{-\frac{1}{2T}}^{\frac{1}{2T}} \frac{1}{\left|H\left(e^{j2\pi fT}\right)\right|^2} \, df = \sigma_a^2 \cdot \gamma_{\text{enh}}^2 \, . \tag{8.101}$$

Since the same system is active in both cases, the same power enhancement occurs. For a fair comparison, the transmit power has to be fixed to a desired level; here we fix it to the same level σ_a^2 as in case of no preprocessing. Consequently, the samples have to be scaled and $\frac{1}{\gamma_{\text{enh}}} x[k]$ has to be actually sent. At the receiver, this scaling has to be compensated for by multiplying $d_W[k]$ by γ_{enh}. This, however, also scales the noise samples and the same noise power $\sigma_{n,(\text{LE})}^2 = \sigma_{n,(\text{WMF})}^2 \cdot \gamma_{\text{enh}}^2$ as in the case of linear receiver-side equalization results.

In summary, for equal transmit power, linear receiver-side equalization and linear transmitter-side pre-equalization perform exactly the same. The SNR$^{(\text{LPE})}$ is given by (8.59), that is, by the harmonic mean over the folded spectral SNR (8.93), (8.94).

8.5.3 Tomlinson–Harashima Precoding (THP)

In Section 8.3, it was shown that decision-feedback equalization is optimum in principle to treat linear dispersive channels. In particular, the capacity of the underlying channel can be exploited. However, in order to approach capacity and, thus, achieve high power efficiency, the application of channel coding is required. Thereby, the main principle is to perform decisions over long codewords. Unfortunately, DFE requires immediate decisions in order to be able to cancel the post-cursors of the channel impulse response. Additionally, in DFE error propagation occurs, especially at low SNRs, the regime where channel coding is profitable. Unfortunately, no immediate combination of DFE with channel coding schemes is possible.

These problems can be overcome by a transmitter-side equivalent to the DFE feedback loop, that is, by a *nonlinear version of pre-equalization*, also called *precoding*. To that end, in the top part of Figure 8.26 the transmission scheme employing DFE is repeated. Compared to linear equalization (top part of Figure 8.25), the nonlinear decision device is incorporated into the feedback loop.

Basic Operation

An elegant precoding method, called *Tomlinson–Harashima precoding (THP)*, was independently proposed by Tomlinson and Harashima/Miyakawa. The block diagram is depicted in the bottom part of Figure 8.26. The main idea is to include a *modulo device* in the feedback loop (depicted as a sawtooth device). If replaced by a shortcut, linear pre-equalization via $1/H(z)$ would be present, as it would be if the decision device in the DFE loop was replaced by a shortcut. Like the decision device, the modulo device has to be adjusted to the signal constellation – the decision device has to match the spacing of the signal points, the modulo device has to match the boundary of the signal constellation.

8.5 Transmitter-Side Channel Pre-equalization

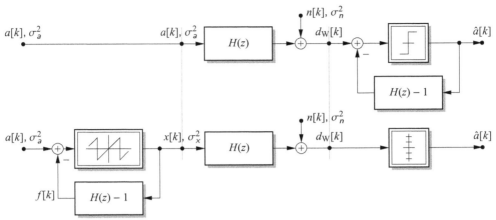

Figure 8.26 Decision-feedback equalization (top) and Tomlinson–Harashima precoding (bottom).

For brevity, we assume an M-ary ASK constellation $\mathcal{A} = \{\pm 1, \pm 3, \ldots, \pm(M-1)\}$. Here, the (symmetric) modulo operation is defined as

$$x = \mathrm{mods}_{2M}(w) \quad \in [-M, M) \quad (8.102)$$
$$= w + 2M\xi, \quad \text{with} \quad \xi \in \mathbb{Z}, \quad (8.103)$$

that is, a unique integer multiple of $2M$ is added or subtracted from the input symbol w, such that x lies in the interval $[-M, M)$.

Let $f[k]$ denote the output of the filter $H(z) - 1$ in the feedback. Then, the channel symbol is given by

$$\begin{aligned} x[k] &= \mathrm{mods}_{2M}(a[k] - f[k]) = a[k] - f[k] + 2M\,\xi[k] \\ &= (a[k] + 2M\,\xi[k]) - f[k] \stackrel{\text{def}}{=} v[k] - f[k]. \end{aligned} \quad (8.104)$$

When employing linear pre-equalization, the transmit signal is given by $a[k] - f[k]$. Hence, in THP not the data symbols themselves are linearly pre-equalized but the *effective data symbols* $v[k]$. The effective data symbols are given by the data symbol $a[k]$ plus a suitably chosen integer multiple of $2M$. This means that each data symbol is *represented multiply* – out of the effective data symbols $a[k] + 2M\,\xi[k]$ that represent $a[k]$ and are *modulo congruent* to $a[k]$, the one is chosen such that the channel symbol falls in the interval $[-M, M)$. An example of a periodically extended signal constellation is shown in Figure 8.27.

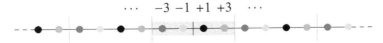

Figure 8.27 4-ary ASK signal constellation (the points within the shaded range) and periodically extended set. Points with the same shading represent the same data.

Signal-to-Noise Ratio

This periodic extension/modulo reduction has two important consequences. First, as the effective data sequence $\langle v[k] \rangle$ is pre-equalized, at the decision device the signal $v[k] + n[k]$ is present. The decision device has to take the modulo congruence into account: by simply performing the same modulo reduction as applied at the transmitter (this is visualized in Figure 8.26 by the multilevel threshold device). The spacing of the signal points is still the same as for a conventional ASK constellation and the noise is not affected. Hence, the reliability is the same. Only the average number of nearest neighbors is increased slightly to 2.

Second, if the order O of the impulse response is not too small and when the coefficients $h[k]$ are not all integers, the channel symbols $x[k]$ are almost uniformly distributed over the interval $[-M, M)$. Thus, for an M-ary ASK constellation, the variance of the channel symbols is given by

$$\sigma_x^2 = \int_{-M}^{M} x^2 \frac{1}{2M} \, dx = \frac{M^2}{3} = \sigma_a^2 \cdot \gamma_{\text{prec}}^2 \, . \tag{8.105}$$

Since the variance of an M-ary ASK constellation is $\sigma_a^2 = \frac{M^2-1}{3}$, only a very small increase in average transmit power by the factor

$$\gamma_{\text{prec}}^2 = \frac{M^2}{M^2 - 1} \tag{8.106}$$

occurs when employing THP. This factor is $10 \log_{10}\left(\gamma_{\text{prec}}^2\right) = 1.25$ dB for $M = 2$ and negligible for $M > 2$ as $10 \log_{10}\left(\gamma_{\text{prec}}^2\right) = 0.28$ dB for $M = 4$ and $10 \log_{10}\left(\gamma_{\text{prec}}^2\right) = 0.07$ dB for $M = 8$.

Hence, in summary, THP has almost the same transmit power as conventional transmission. The same signal-point spacing and noise power as in DFE are active. THP achieves the same SNR as DFE; given by (8.72), that is, by the geometric mean over the folded spectral SNR (8.93), (8.94).

The main advantages of THP over DFE are that no error propagation occurs in the feedback loop, and, even more important, no immediate decisions are required at the receiver – THP can straightforwardly be combined with channel coding schemes. THP enables a clear separation between equalization and channel coding. Coding schemes designed for the AWGN channel can be applied and a performance very close to limits from information theory is achievable.

We note finally that for square QAM constellations THP is done in the same way as for ASK; now in-phase and quadrature components are individually modulo-reduced. The initial QAM constellation is periodically extended to the entire complex plane. Details and generalizations can be found in the literature.

8.5.4 Combined Transmitter- and Receiver-Side Linear Equalization

Up to now, equalization has been done either completely at the receiver or at the transmitter. It is natural to split the equalization task between both sides. This balancing may provide a gain in power efficiency.

As described at the beginning of this subsection, we assume that the transmit filter is fixed and the spectral SNR and its folded version are given by (8.93) and (8.94), respectively. Now, it is more convenient to start the derivations assuming that the optimum Nyquist filter $H_R^{(LE)}(f)$ is applied as receiver frontend, or, equivalently, the whitened-matched filter $H_R^{(WMF)}(f)$ followed by linear equalization via $1/H(z)$. Thus, the model of Figure 8.13 shows the setting – the signal transfer function is one (the signal is equalized) but the additive noise is colored with psd (cf. (8.56) and (8.47)):

$$\Phi_{nn}^{(LE)}\left(e^{j2\pi fT}\right) = \frac{\sigma_a^2}{\widetilde{SNR}\left(e^{j2\pi fT}\right)} . \tag{8.107}$$

Procedure

From analog transmission, especially from frequency modulation, it is well known that in case of colored noise the SNR can be improved by a pre-filter at the transmitter, called *pre-emphasis*, together with its counterpart at the receiver side, called *de-emphasis*. Both filters together are called the *pre-emphasis/de-emphasis method (PD)*. The main idea is to enhance the signal in bands with high noise power in order to allow for an attenuation of the noise by the receiver-side filter without distorting the useful signal; this balances signal and noise over the entire frequency band.

Such a procedure can also be applied to the setting at hand – the respective block diagram is depicted in Figure 8.28. A linear preprocessing filter $P(z)$ is employed at the transmitter and its inverse $D(z) = 1/P(z)$ is used at the receiver.

The question is how to choose $P(z)$ (and thus $D(z)$) optimally. First, for a fair comparison, the transmit power should be fixed. As shown above (see (8.97)), $\sigma_x^2 = \sigma_a^2$ has to hold. Second, the performance is maximized if the power of the filtered noise samples at the output of $D(z)$ is minimized. This establishes an optimization problem.

Optimization

The optimum filters can be determined by means of calculus of variations. For this, the following abbreviations are defined:

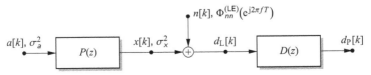

Figure 8.28 Transmission model with pre-emphasis and de-emphasis.

$$F \stackrel{\text{def}}{=} fT, \tag{8.108}$$

$$N(F) \stackrel{\text{def}}{=} \Phi_{nn}^{(\text{LE})}\left(e^{j2\pi fT}\right), \tag{8.109}$$

$$A(F) \stackrel{\text{def}}{=} \left|D\left(e^{j2\pi fT}\right)\right|^2. \tag{8.110}$$

The noise power to be minimized and the constraint of a fixed average transmit power are then written as

$$\sigma_{n,D}^2 = T \int_{-\frac{1}{2T}}^{\frac{1}{2T}} \Phi_{nn}^{(\text{LE})}\left(e^{j2\pi fT}\right) \left|D\left(e^{j2\pi fT}\right)\right|^2 df$$

$$= \int_{-\frac{1}{2}}^{\frac{1}{2}} N(F)\, A(F)\, dF, \tag{8.111}$$

$$\sigma_x^2 = T \int_{-\frac{1}{2T}}^{\frac{1}{2T}} \frac{\sigma_a^2}{\left|D\left(e^{j2\pi fT}\right)\right|^2} df \stackrel{!}{=} \sigma_a^2$$

$$1 \stackrel{!}{=} \int_{-\frac{1}{2}}^{\frac{1}{2}} \frac{1}{A(F)}\, dF. \tag{8.112}$$

Objective and constraint are combined to a *Lagrange function*, where λ^2 is a *Lagrange multiplier*. Let $B(F)$ describe an arbitrary deviation from the optimum $A(F)$ and ϵ a weight factor. The Lagrange function is then given by

$$L(\epsilon) = \int_{-\frac{1}{2}}^{\frac{1}{2}} N(F)\,(A(F) + \epsilon B(F))\, dF + \lambda^2 \left(\int_{-\frac{1}{2}}^{\frac{1}{2}} \frac{1}{A(F) + \epsilon B(F)}\, dF - 1\right). \tag{8.113}$$

In the optimum, $L(\epsilon)$ has to exhibit a local minimum at $\epsilon = 0$, independent of the choice of $B(F)$, that is

$$\left.\frac{\partial}{\partial \epsilon} L(\epsilon)\right|_{\epsilon=0} = \int_{-\frac{1}{2}}^{\frac{1}{2}} \left(N(F) B(F) - \lambda^2 \frac{B(F)}{A^2(F)}\right) dF \stackrel{!}{=} 0. \tag{8.114}$$

This equation holds for any $B(F)$ if and only if

$$N(F) = \frac{\lambda^2}{A^2(F)}. \tag{8.115}$$

This yields the desired solution for the squared magnitude of the receiver-side filter:

$$\left|D\left(e^{j2\pi fT}\right)\right|^2 = \sqrt{\frac{\lambda^2}{\Phi_{nn}^{(\text{LE})}(e^{j2\pi fT})}} \stackrel{(8.107)}{=} \frac{\lambda}{\sigma_a} \sqrt{\widetilde{\text{SNR}}\left(e^{j2\pi fT}\right)}. \tag{8.116}$$

The phase function is arbitrary as it is reversed by $P(z)$. This result shows that the receiver-side filter $D(z)$ removes the *square root* of the coloring of the noise. More noise decorrelation would cause too much signal enhancement at the transmitter side by $P(z)$, which would finally result in an SNR loss.

Interpretation

An interpretation of the result follows when the end-to-end model applying the optimum Nyquist filter $H_{\text{R}}^{(\text{LE})}(f)$ is not assumed as the starting point (cf. Figure 8.13), but that applying the whitened-matched filter $H_{\text{R}}^{(\text{WMF})}(f)$ (cf. Figure 8.15). As has been seen in Section 8.3, the end-to-end signal transfer function is $H(z)$ and white noise is present. Moreover, (8.62) and (8.47) show that (c_{h} is a constant)

$$\left| H\left(e^{j2\pi fT}\right) \right|^2 = c_{\text{h}} \widetilde{\text{SNR}}\left(e^{j2\pi fT}\right) . \tag{8.117}$$

Consequently, when applying the whitened-matched filter, we have for the magnitudes of the transmitter-side pre-filter and receiver-side post-filter

$$\left| P\left(e^{j2\pi fT}\right) \right| \sim \frac{1}{\sqrt{\left| H\left(e^{j2\pi fT}\right) \right|}} , \tag{8.118}$$

$$\left| \frac{D\left(e^{j2\pi fT}\right)}{H\left(e^{j2\pi fT}\right)} \right| \sim \frac{1}{\sqrt{\left| H\left(e^{j2\pi fT}\right) \right|}} . \tag{8.119}$$

This means that the equalization of the ISI-producing filter $H(z)$ is split into two parts, a pre-equalization $P(z)$ at the transmitter and a receive filter $D(z)/H(z)$ at the receiver, both equalizing the *square-root* of the channel. In contrast to pure linear (transmitter- or receiver-side) equalization via $1/H(z)$, here the equalization task is geometrically split between both sides. Hence, this method may also be called *root–root equalization (RRE)*. We use this denomination in the following.

Signal-to-Noise Ratio

Using (8.112), the Lagrange multiplier can be determined as

$$\int_{-\frac{1}{2}}^{\frac{1}{2}} \frac{1}{A(F)}\, dF = T \int_{-\frac{1}{2T}}^{\frac{1}{2T}} \frac{\sigma_a}{\lambda} \frac{1}{\sqrt{\widetilde{\text{SNR}}\left(e^{j2\pi fT}\right)}}\, df \stackrel{!}{=} 1 ,$$

$$\frac{\lambda}{\sigma_a} = T \int_{-\frac{1}{2T}}^{\frac{1}{2T}} \frac{1}{\sqrt{\widetilde{\text{SNR}}\left(e^{j2\pi fT}\right)}}\, df . \tag{8.120}$$

The noise power at the output of $D(z)$ calculates as (cf. (8.111))

$$\begin{aligned}
\sigma_{n,(\text{RRE})}^2 &= T \int_{-\frac{1}{2T}}^{\frac{1}{2T}} \Phi_{nn}^{(\text{LE})}\left(e^{j2\pi fT}\right) \left| D\left(e^{j2\pi fT}\right) \right|^2 df \\
&= T \int_{-\frac{1}{2T}}^{\frac{1}{2T}} \frac{\sigma_a^2}{\widetilde{\text{SNR}}\left(e^{j2\pi fT}\right)} \frac{\lambda}{\sigma_a} \sqrt{\widetilde{\text{SNR}}\left(e^{j2\pi fT}\right)}\, df \\
&= \sigma_a^2 \frac{\lambda}{\sigma_a} T \int_{-\frac{1}{2T}}^{\frac{1}{2T}} \frac{1}{\sqrt{\widetilde{\text{SNR}}\left(e^{j2\pi fT}\right)}}\, df \\
&\stackrel{(8.120)}{=} \sigma_a^2 \left(T \int_{-\frac{1}{2T}}^{\frac{1}{2T}} \frac{1}{\sqrt{\widetilde{\text{SNR}}\left(e^{j2\pi fT}\right)}}\, df \right)^2 .
\end{aligned} \tag{8.121}$$

The signal-to-noise ratio is thus given by

$$\mathrm{SNR}^{(\mathrm{RRE})} = \frac{\sigma_a^2}{\sigma_{n,(\mathrm{RRE})}^2}$$

$$= \left(T \int_{-\frac{1}{2T}}^{\frac{1}{2T}} \left(\widetilde{\mathrm{SNR}} \left(\mathrm{e}^{\mathrm{j}2\pi fT} \right) \right)^{-1/2} \mathrm{d}f \right)^{-2} . \quad (8.122)$$

This SNR is a *Hölder mean* for the Hölder power $-1/2$ (see Box 8.2) over the folded spectral SNR valid for the initial situation. Analogous to the popular denomination *root mean square (RMS)* for the Hölder mean power 2, the Hölder mean power $-1/2$ may be denoted as *squared harmonic mean root (SHR)*.

As stated at the beginning of this subsection, all results are expressed in terms of the spectral SNR that is active if the transmitter-side pre-filter is not (yet) present. Of course, the same results are obtained when $P(z)$ is assumed to be already known and the optimum linear equalizer (see Section 8.3) for this new situation including $P(z)$ is designed.

First, by application of the pre-emphasis filter $P(z)$, the psd of the transmit signal $x[k]$ is changed, which results in a changed folded spectral SNR with respect to the transmit signal $x[k]$. We have

$$\widetilde{\mathrm{SNR}}^{(\mathrm{P})} = \sum_\mu \frac{\sigma_a^2 \left| G_\mathrm{T}\left(f - \frac{\mu}{T}\right) \right|^2 \left| P\left(\mathrm{e}^{\mathrm{j}2\pi(f-\mu/T)T}\right) \right|^2}{T N_0}$$

$$= \widetilde{\mathrm{SNR}} \left(\mathrm{e}^{\mathrm{j}2\pi fT} \right) \left| P\left(\mathrm{e}^{\mathrm{j}2\pi(f-\mu/T)T}\right) \right|^2$$

$$= \widetilde{\mathrm{SNR}} \left(\mathrm{e}^{\mathrm{j}2\pi fT} \right) \frac{1}{\left| D\left(\mathrm{e}^{\mathrm{j}2\pi(f-\mu/T)T}\right) \right|^2}$$

$$= \frac{\sigma_a}{\lambda} \sqrt{\widetilde{\mathrm{SNR}} \left(\mathrm{e}^{\mathrm{j}2\pi fT} \right)} . \quad (8.123)$$

Second, the SNR when applying the optimum linear equalizer is given by the harmonic mean over the folded spectral SNR (cf. (8.59)). Thus, we arrive at

$$\begin{aligned}
\mathrm{SNR}^{(\mathrm{LE}),(\mathrm{P})} &= \left(T \int_{-\frac{1}{2T}}^{\frac{1}{2T}} \left(\widetilde{\mathrm{SNR}}^{(\mathrm{P})} \left(\mathrm{e}^{\mathrm{j}2\pi fT} \right) \right)^{-1} \mathrm{d}f \right)^{-1} \\
&= \frac{\sigma_a}{\lambda} \left(T \int_{-\frac{1}{2T}}^{\frac{1}{2T}} \left(\widetilde{\mathrm{SNR}} \left(\mathrm{e}^{\mathrm{j}2\pi fT} \right) \right)^{-1/2} \mathrm{d}f \right)^{-1} \\
&\stackrel{(8.120)}{=} \left(T \int_{-\frac{1}{2T}}^{\frac{1}{2T}} \left(\widetilde{\mathrm{SNR}} \left(\mathrm{e}^{\mathrm{j}2\pi fT} \right) \right)^{-1/2} \mathrm{d}f \right)^{-2} \\
&= \mathrm{SNR}^{(\mathrm{RRE})} . \quad (8.124)
\end{aligned}$$

Box 8.2 *Generalized Means*

Besides the well-known arithmetic, geometric, and harmonic mean, a more general family of means, called *generalized mean, power mean*, or *Hölder mean* is defined in mathematics.

Let N positive real numbers x_i, $i = 1, \ldots, N$, be given. The Hölder mean of power $\rho \in \mathbb{R}$ of these numbers is defined as

$$M_\rho \stackrel{\text{def}}{=} \left(\frac{1}{N} \sum_{i=1}^{N} x_i^\rho \right)^{\frac{1}{\rho}} . \qquad (8.125)$$

Choosing ρ suitably, the well-known means are special cases thereof. Obviously, for $\rho = 1$ the *arithmetic mean (AM)* and for $\rho = -1$ the *harmonic mean (HM)* are obtained. As is shown subsequently, for $\rho = 0$ the *geometric mean (GM)* turns up. The *root mean square (RMS)* is the Hölder mean for $\rho = 2$ and for $\rho = -1/2$ the *squared harmonic mean root (SHR)* appears. Finally, for $\rho \to -\infty$ the *minimum operator* and for $\rho \to \infty$ the *maximum operator* result. Table 8.3 collects these special cases.

Table 8.3 Hölder means.

ρ	Mean
$-\infty$	Minimum
-1	Harmonic Mean
$-\frac{1}{2}$	Squared Harmonic Mean Root
0	Geometric Mean
1	Arithmetic Mean
2	Root Mean Square
∞	Maximum

For $\rho < p$, the Hölder means are related to each other by

$$M_\rho \leq M_p ,$$

where equality only holds if $x_1 = \cdots = x_N$. Hence, the following relation (as a generalization of the arithmetic–geometric mean inequality) holds:

$$M_{-\infty} \leq M_{-1} \leq M_{-\frac{1}{2}} \leq M_0 \leq M_1 \leq M_2 \leq M_\infty .$$

The Hölder mean over N positive real numbers can be generalized to the Hölder mean over a positive real function $f(x)$, $x \in [0, I]$:

$$M_\rho \stackrel{\text{def}}{=} \left(\frac{1}{I} \int_0^I f^\rho(x) \, \mathrm{d}x \right)^{\frac{1}{\rho}} . \qquad (8.126)$$

It has the same properties and relations as the mean over numbers.

Figure 8.29 visualizes symbolically the relations between the discussed means.

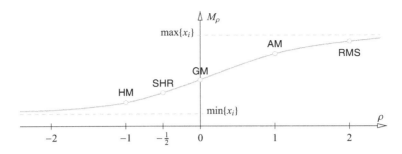

Figure 8.29 Relations between the various means (symbolic representation).

Some remarks follow:
- The Hölder mean can be written as

$$M_\rho = \left(\frac{1}{N}\sum_{i=1}^{N} x_i^\rho\right)^{\frac{1}{\rho}} = \exp\left\{\log\left(\left(\frac{1}{N}\sum_{i=1}^{N} x_i^\rho\right)^{\frac{1}{\rho}}\right)\right\} = \exp\left\{\frac{1}{\rho}\log\left(\frac{1}{N}\sum_{i=1}^{N} x_i^\rho\right)\right\}.$$

For $\rho \to 0$ the argument of the exponential function approaches

$$\lim_{\rho \to 0} \frac{\log\left(\frac{1}{N}\sum_{i=1}^{N} e^{\rho \log(x_i)}\right)}{\rho} \stackrel{\text{L'Hôpital}}{=} \lim_{\rho \to 0} \frac{\frac{1}{N}\sum_{i=1}^{N} \log(x_i) e^{\rho \log(x_i)}}{\frac{1}{N}\sum_{i=1}^{N} e^{\rho \log(x_i)}} = \frac{\frac{1}{N}\sum_{i=1}^{N} \log(x_i)}{\frac{1}{N}\sum_{i=1}^{N} 1}$$

$$= \frac{1}{N}\sum_{i=1}^{N} \log(x_i),$$

thus we have

$$M_0 = \exp\left\{\frac{1}{N}\sum_{i=1}^{N} \log(x_i)\right\} = \sqrt[N]{\prod_{i=1}^{N} x_i},$$

which is the geometric mean.
- Moreover, for $\rho \to \infty$, the Hölder mean gives $\left(x_{\max} \stackrel{\text{def}}{=} \max\{x_1,\ldots,x_N\}\right)$

$$\lim_{\rho \to \infty} \left(\frac{1}{N}\sum_{i=1}^{N} x_i^\rho\right)^{\frac{1}{\rho}} = x_{\max} \lim_{\rho \to \infty} \left(\frac{1}{N}\sum_{i=1}^{N}\left(\frac{x_i}{x_{\max}}\right)^\rho\right)^{\frac{1}{\rho}} = x_{\max}.$$

Similarly, for $\rho \to -\infty$, the Hölder mean gives $\left(x_{\min} \stackrel{\text{def}}{=} \min\{x_1,\ldots,x_N\}\right)$

$$\lim_{\rho \to -\infty}\left(\frac{1}{N}\sum_{i=1}^{N} x_i^\rho\right)^{\frac{1}{\rho}} = x_{\min}\lim_{\rho\to-\infty}\left(\frac{1}{N}\sum_{i=1}^{N}\left(\frac{x_i}{x_{\min}}\right)^\rho\right)^{\frac{1}{\rho}} = x_{\min}\lim_{p\to\infty}\frac{1}{\left(\frac{1}{N}\sum_{i=1}^{N}\left(\frac{x_{\min}}{x_i}\right)^p\right)^{\frac{1}{p}}} = x_{\min}.$$

- To prove that for any $\rho < p$ the relation $M_\rho \leq M_p$ holds, we define the function $f(x) = x^{\frac{p}{\rho}}$, $x > 0$. We restrict ourselves to the case $\rho, p > 0$; the other cases ($\rho, p < 0$ and $\rho < 0, p > 0$) are proven in the same way. It is straightforward to show that for $\rho, p > 0$ the function $f(x)$ is monotonously increasing and convex on its domain. By Jensen's inequality, we have

thus
$$f\left(\frac{1}{N}\sum_{i=1}^{N} x_i^\rho\right) \le \frac{1}{N}\sum_{i=1}^{N} f(x_i^\rho),$$

$$\left(\frac{1}{N}\sum_{i=1}^{N} x_i^\rho\right)^{\frac{p}{\rho}} \le \frac{1}{N}\sum_{i=1}^{N} x_i^p$$

and finally ($p > 0$)

$$\left(\frac{1}{N}\sum_{i=1}^{N} x_i^\rho\right)^{\frac{1}{\rho}} \le \left(\frac{1}{N}\sum_{i=1}^{N} x_i^p\right)^{\frac{1}{p}}.$$

(For ρ, $p < 0$ the function $f(x)$ is monotonously increasing and concave; for $\rho < 0$, $p > 0$ it is monotonously decreasing and convex.)

Example 8.9 Loss in Minimum Distance Compared to AWGN Channel

This example continues Examples 8.3, 8.5, and 8.7. We assess the loss ϑ^2 in minimum distance (i.e., performance) compared to an AWGN channel. Transmission of a rectangular transmitter pulse (hard keying) over a coaxial cable and disturbance by white noise is assumed.

In Figure 8.30, the loss over the characteristic attenuation (in dB) of the coaxial cable at the Nyquist frequency is plotted. It can be seen that combined transmitter- and receiver-side linear equalization (root–root equalization) slightly outperforms linear equalization completely done at one side. However, decision-feedback equalization or Tomlinson–Harashima precoding are still preferable.

Note that the curves are plotted over α_c, the characteristic attenuation at the Nyquist frequency $f_N = \frac{1}{2T}$. For comparison with the results in the next chapter (performance of OFDM in very-high-bit-rate digital subscriber lines, VDSL), we now assume a Nyquist frequency $f_N = 8.499$ MHz (which corresponds to the VDSL-2 standard G.993.2, profile 8d). Comparing (8.41) and (D.3), one can deduce that α_c and the cable length of a standard coaxial cable with attenuation $a_{\mathrm{dB}} = 2.46 \left[\frac{\mathrm{dB}}{\mathrm{km}}\right]$ are related by

$$-\alpha_c \sqrt{2 \cdot \tfrac{1}{2}}/20 \stackrel{!}{=} -a_{\mathrm{dB}}\, \ell_{\mathrm{coax}} \sqrt{\tfrac{8.499\ \mathrm{MHz}}{1\ \mathrm{MHz}}}/20, \qquad (8.127)$$

which leads to

$$\ell_{\mathrm{coax}} = \alpha_c \cdot \frac{1}{\sqrt{8.499}} \frac{1}{2.46} \left[\frac{\mathrm{km}}{\mathrm{dB}}\right]. \qquad (8.128)$$

The second x-axis in Figure 8.30 shows the corresponding length of the coax cable.

Finally, as shown in Appendix D, in the frequency band of interest, the cable length ℓ_{coax} of a standard coaxial cable and that, $\ell_{\mathrm{twistpair}}$, of a twisted-pair line with 0.4 mm diameter of the conductor are approximately related by

$$\ell_{\mathrm{twistpair}} \approx 0.086 \cdot \ell_{\mathrm{coax}}. \qquad (8.129)$$

The third x-axis in Figure 8.30 shows the corresponding length of the twisted-pair line.

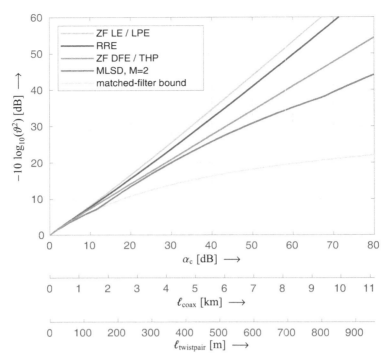

Figure 8.30 Loss in minimum distance compared to AWGN channel over the characteristic attenuation of the coaxial cable at the Nyquist frequency.

8.6 Problems

8.6.1 End-to-End Model for Transmission over ISI Channels

Digital transmission over a linearly distorting channel is considered. The block diagram (ECB domain) of the PAM transmission scheme including the discrete-time end-to-end model is shown in Figure 8.3.

Determine:
(a) the transfer function $H(z)$ of the discrete-time end-to-end model;
(b) the respective impulse response $h[k]$;
(c) the power spectral density $\Phi_{nn}\left(e^{j2\pi fT}\right)$ of the discrete-time noise $n[k]$;
(d) the respective autocorrelation function $\phi_{nn}[\kappa]$.

8.6.2 End-to-End Model for √Nyquist Pulses

We assume that $g(t)$ is a $\sqrt{\text{Nyquist}}$ pulse (with energy E_g) and the respective matched filter is applied at the receiver.

(a) Show that the discrete-time end-to-end transfer function (8.8) and the noise psd (8.9) specialize to $H\left(e^{j2\pi fT}\right) = 1$ and $\Phi_{nn}\left(e^{j2\pi fT}\right) = N_0/E_g$.
(b) What type of end-to-end channel model is present in this case?

8.6.3 Optimum Receive Filter

We want to prove that the optimum receive filter for PAM transmission over linearly distorting channels is always given as a cascade of the matched filter $\frac{1}{E_g}G^*(f)$, sampling with frequency $1/T$, and a discrete-time filter $F(z)$.

To this end, we first assume that an *arbitrary* receive filter with transfer function $H_R^{(a)}(f)$ is given.

(a) Give the discrete-time end-to-end transfer function and the power spectral density of the resulting discrete-time noise.

We now replace the arbitrary receive filter by the cascade of *matched* filter, sampling, and discrete-time filter.

(b) Sketch this cascade.
(c) Determine the transfer function of the discrete-time filter $F(z)$ such that the same end-to-end transfer function as in part (a) is obtained.
(d) Give the power spectral density of the resulting discrete-time noise $n[k]$ if the discrete-time filter $F(z)$ from part (c) is used.
(e) Compare the power spectral densities of part (a) and (d). What can be deduced from this comparison?

8.6.4 Maximum-Likelihood Sequence Detection I

We consider binary ASK transmission ($a[k] \in \{+1, -1\}$) over an intersymbol-interference channel. The end-to-end impulse response when using the *matched filter* for $g(t) \stackrel{\text{def}}{=} g_T(t) * h_C(t)$ reads

$$\psi_{gg}[k] = \begin{cases} 1.0, & k = 0 \\ 0.6, & |k| = 1 \\ 0.2, & |k| = 2 \\ 0, & \text{else} \end{cases}.$$

For $k < 0$ a sufficiently long sequence of "+1" has been transmitted.

(a) How long has this "training" sequence to be, in order to guarantee that a well-defined state in the trellis diagram characterizing the internal memory of the channel is reached?
(b) How many states has the trellis, on which maximum-likelihood sequence detection (MLSD) may be performed?

The sequence of samples $d_M[k]$ at the output of the matched filter is given by

$$\langle\, d_M[0],\, d_M[1],\, d_M[2],\, d_M[3],\, d_M[4]\,\rangle = \langle\, 2.49,\, 1.56,\, 1.18,\, 0.48,\, -1.49\,\rangle.$$

(c) Perform maximum-likelihood sequence detection based on the Viterbi algorithm. Give the estimates $\hat{a}[k]$ for $k = 0, \ldots, 4$.
(d) Calculate the normalized minimum squared Euclidean distance that is present when transmitting over this channel.

8.6.5 Maximum-Likelihood Sequence Detection II
Again, binary ($a[k] \in \{+1, -1\}$) transmission over an intersymbol-interference channel is considered. The end-to-end discrete-time model when applying the *whitened-matched filter* reads, in the z-domain

$$H(z) = 1 + 0.5\, z^{-1} + 0.2\, z^{-2} .$$

(a) Sketch the end-to-end transmission model using a linear discrete-time system with delay elements. What characteristics does the additive noise sequence have?
(b) Construct the trellis diagram corresponding to this channel.

For $k < 0$ a sufficiently long sequence of "+1" has been transmitted. For $k \geq 0$ the sequence of samples $d_W[k]$ at the output of the whitened-matched filter is given by

$$\langle\, d_W[0],\, d_W[1],\, d_W[2],\, d_W[3],\, d_W[4],\, d_W[5]\, \rangle = \langle\, -0.4,\, 0.6,\, 1.5,\, 1.5,\, -0.4,\, -1.2\, \rangle .$$

(c) Perform maximum-likelihood sequence detection based on the Viterbi algorithm. Give the estimates $\hat{a}[k]$ for $k = 0, \ldots, 5$.

8.6.6 Spectral Signal-to-Noise Ratio
Digital PAM (symbol duration T) with a transmit basic pulse

$$g_T(t) = \sqrt{\frac{E_g}{T}}\, \text{si}\!\left(\frac{\pi t}{T}\right)$$

is assumed. Digital baseband transmission over a coaxial cable is studied. The magnitude of the transfer function for this cable is given as

$$|H_C(f)| = e^{-\frac{\alpha}{2}\sqrt{|2fT|}} ,$$

where $\alpha \in \mathbb{R}^+$ is assumed to be given and fixed. At the receiver side, additive white Gaussian noise with one-sided power spectral density N_0 is superimposed.

(a) Calculate and sketch the spectral signal-to-noise ratio (SNR) and its folded version.
(b) Calculate the matched-filter bound, that is, the upper bound on the SNR.
   Hint: For $x \geq 0$ the following equation holds:

$$\int e^{\alpha \sqrt{x}}\, dx = \frac{2}{\alpha^2} e^{\alpha \sqrt{x}} (\alpha \sqrt{x} - 1) .$$

(c) What SNR can be achieved when using (zero-forcing) decision-feedback equalization (DFE)?
(d) Compare your results from parts (b) and (c).

8.6.7 Comparison of Equalization Schemes
We analyze digital binary, bipolar ASK baseband transmission with signal points ± 1 over a linear, dispersive channel and disturbance by white Gaussian noise with one-sided power spectral density N_0. At the receiver input the basic PAM pulse $g(t)$ is a triangle over two symbol intervals T (cf. the following figure).

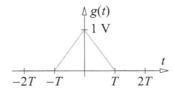

(a) Calculate the energy E_g of the basic pulse. Determine the average power of the (noise-free) receive signal and the energy E_b per bit. Give also the spectrum of the basic pulse $g(t)$.

(b) Calculate the autocorrelation $\varphi_{gg}(\tau)$ of $g(t)$ for shifts $\tau \in \lambda T$, $\lambda \in \mathbb{Z}$. Show that orthogonality with respect to time shifts by multiples of T does not hold, that is, ISI is present. Determine the order O of the ISI.

(c) Show by inspecting very short sequences of difference symbols that here ISI causes no SNR loss with respect to the matched-filter bound. Give the minimum normalized squared Euclidean distance of the transmission scheme when applying maximum-likelihood sequence detection (MLSD).

(d) Sketch the (noise-free) receiver input signal for the sequence of data symbols

$$\langle 1,\ 1,\ -1,\ 1,\ -1,\ -1,\ 1 \rangle$$

and identify all the different signal elements within one symbol interval T.

(e) Represent these signal elements by means of points in a suitably defined signal space. Give the orthonormal basis functions and draw a diagram of the signal points in the signal space.

Recalculate the average energy per bit using this signal-space representation.

(f) Derive an equivalent encoder and mapper for generation of the (noise-free) receiver input signal using the signal elements identified in part (e).

(g) Sketch the trellis diagram for MLSD with the Viterbi algorithm for the encoder of part (f) over six symbol intervals and characterize the branches by the corresponding signal elements.

Identify a minimum distance error event from split to merge from a correct path.

Calculate again the minimum normalized squared Euclidean distance from a corresponding minimum distance pair of paths in the trellis.

A continuous-time matched filter, matched to $g(t)$, is applied to the receiver input signal before T-spaced sampling for an optimum transition from continuous-time to discrete-time signal processing.

(h) Determine the impulse response of this matched filter.

Specify the normalized discrete-time end-to-end impulse that is identical to the normalized sampled autocorrelation $\psi_{gg}[k]$ and give its z-transform $\Psi(z)$ and its discrete-time Fourier transform $\Psi\left(e^{j2\pi fT}\right)$.

Hint: Use the results from part (b).

(i) Calculate the transfer function of the (non-causal) receiver-input filter for optimum *zero-forcing linear equalization (LE)* and sketch its magnitude over the frequency range $-2/T < f < +2/T$.
(j) Calculate the spectral signal-to-noise ratio at the receiver input and its folded version. Sketch both quantities over the range $-2/T < f < +2/T$.
(k) Determine the $\text{SNR}^{(\text{LE})}$ at the optimum sampling instances for optimum zero-forcing linear equalization and the loss of LE with respect to MLSD.
Hint: The following relation holds: $\int_0^{1/2} \frac{1}{1+\frac{1}{2}\cos(2\pi x)} \, dx = \frac{1}{\sqrt{3}}$.

For the analysis of MLSD employing the whitened-matched filter (WMF) and for the analysis of decision-feedback equalization (DFE), a spectral factorization of $\Psi(z)$ from part (h) has to be performed.

(l) Write $\Psi(z)$ in the form

$$\Psi(z) = c\left(z - z_0\right)\left(1 - z_0^{-*} z^{-1}\right)$$

with $|z_0| > 1$ and give a causal, minimum-phase $B(z)$ such that

$$\Psi(z) = B(z)\, B^*(z^{-*}) \, .$$

Determine the monic, causal, and minimum-phase end-to-end discrete-time impulse and the variance of the i.i.d. Gaussian noise samples at the output of the WMF.
(m) Derive the equivalent encoder for the (noise-free) discrete-time output signal of the WMF and sketch the corresponding trellis diagram. Identify a minimum distance error event and calculate again the minimum normalized squared Euclidean distance for MLSD.
(n) Determine $\text{SNR}^{(\text{DFE})}$ of a DFE receiver using the output signal of the WMF calculated in part (l).
Give the SNR loss of DFE with respect to MLSD and the matched-filter bound.
(o) Sketch a block diagram of a transmission scheme applying Tomlinson–Harashima precoding (THP) corresponding to DFE with WMF.
Within what interval and how are the transmit symbols distributed? What SNR loss arises for THP with respect to DFE?
(p) Determine the (non-causal) continuous-time impulse response of the WMF and sketch it. Explain why – at least in theory – an infinite delay would be necessary for an exact causal implementation of the WMF.
Sketch the continuous-time end-to-end pulse including transmit pulse, channel, and whitened-matched filter. Mark the nonzero samples at $t = kT$.
Hint: Apply the formula of geometric series $\sum_{i=0}^{\infty} q^i = \frac{1}{1-q}$ for $|q| < 1$ to the remaining discrete-time equalization term $1/B^*(z^{-*})$ in the WMF.
(q) Give the transfer functions of the optimum pre-emphasis and de-emphasis filters for linear equalization split into transmitter- and receiver-side filters

(root–root equalization, RRE) and calculate the corresponding $\text{SNR}^{(\text{RRE})}$ at the input of the slicer.

Hint: The following holds: $\int_0^{1/2} \frac{1}{\sqrt{1+\frac{1}{2}\cos(2\pi x)}} \, \mathrm{d}x = 0.5273$.

(r) Sort the equalization strategies MLSD, LE, DFE, and RRE according to their performance for the present scenario.

9 Orthogonal Frequency-Division Multiplexing

In the last chapter, we were concerned with transmission over linearly distorting channels. We showed how to optimally treat intersymbol interference (ISI) and how to build low-complexity receivers for this task. ISI occurs if the channel transfer function is not flat within the transmission band. Hence, an obvious strategy to avoid ISI is to divide the transmission band into a large number of subbands which are used individually and in parallel. If these bands are small enough such that fluctuations of the channel transfer function can be ignored, no linear distortions will occur that would have to be equalized.

In this chapter, we study this main idea in the particular form of *orthogonal frequency-division multiplexing (OFDM)*. It is shown that even starting from the frequency-division multiplexing idea, the key principle behind OFDM is blockwise transmission and the use of suitable transformations at transmitter and receiver. We assess OFDM and show how the resulting parallel data transmission can be optimized. OFDM is compared with the equalization schemes that were discussed in the previous chapter.

PREREQUISITES FOR THIS CHAPTER

- PAM transmission (Chapter 3).
- Signal space representation (Chapter 6).
- Equalization of dispersive channels (Chapter 8).

9.1 Basics of Multi-Carrier Transmission

We are still concerned with the situation described in Section 8.1, that is, transmission over a linear, dispersive channel with impulse response/transfer function

$$h_C(t) \circ\!\!-\!\!\bullet\ H_C(f) \qquad (9.1)$$

and (if applicable, due to a whitening filter) additive white Gaussian noise $n(t)$ (psd N_0 in the ECB domain, cf. Figure 8.1) is considered.

We assume that the channel (or the part we are allowed to use) has bandwidth B_{RF} – hence the transmit signal is allowed to have a bandwidth of at most B_{RF}.

9.1.1 Single-Carrier Transmission

In conventional PAM transmission – also called *single-carrier transmission* – the single scheme uses the entire bandwidth. As shown in Chapter 3, the smallest symbol duration (fastest transmission) of a PAM scheme over this channel, that is, of a PAM scheme that does not exceed the bandwidth, is given by (Nyquist bandwidth)

$$T \stackrel{\text{def}}{=} \frac{1}{B_{\text{RF}}} . \tag{9.2}$$

This smallest value will be achieved if the basic pulse shape is rectangular in the frequency domain (cf. (3.62)).

9.1.2 Multi-Carrier Transmission – Frequency Multiplexing

As already stated, the obvious idea of *multi-carrier transmission* is to divide the transmission band with bandwidth B_{RF} into, say D, *subbands*, *(sub)carriers*, or *subchannels* of width

$$B_{\text{S}} \stackrel{\text{def}}{=} \frac{B_{\text{RF}}}{D} . \tag{9.3}$$

The smallest symbol duration for the PAM transmission schemes within the subbands (D individual, identical PAM schemes are assumed) is then

$$T_{\text{s}} \stackrel{\text{def}}{=} \frac{1}{B_{\text{S}}} = \frac{D}{B_{\text{RF}}} = DT . \tag{9.4}$$

In the context of multi-carrier transmission it is usual to assume that the channel transmission band (in the ECB domain) is located at $f \in [-B_{\text{S}}/2, B_{\text{RF}} - B_{\text{S}}/2]$ (mainly on the positive part of the frequency axis). The reference frequency f_0 in the ECB transformation and the center frequency differ. Then, the carrier frequency in the dth subband is given by

$$f_{\text{c},d} = d \cdot B_{\text{S}} = \frac{d}{T_{\text{s}}}, \quad d = 0, 1, \ldots, D-1 . \tag{9.5}$$

The philosophy of splitting the channel bandwidth into a (large) number of subchannels is visualized in Figure 9.1 and the consequences on the bandwidth of the channels and the symbol duration are sketched in Figure 9.2.

In single-carrier transmission the single scheme uses the entire bandwidth, thus the symbol duration T is short. In contrast, in multi-carrier transmission when splitting the initial transmission band into D subbands (which are used individually in parallel), the respective symbol duration has to be prolonged by the factor D. The gray-shaded "area" in the time/frequency plane is the same for both approaches.

9.1.3 Denominations and Applications

As we will see subsequently, some adaptations of the main idea presented have to be done in order to obtain practical schemes. The resulting schemes are known under the denominations

9 Orthogonal Frequency-Division Multiplexing

- orthogonal frequency-division multiplexing (OFDM)
- discrete multitone (DMT)
- multi-carrier modulation (MCM).

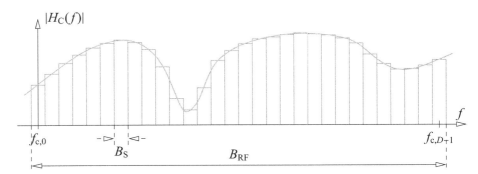

Figure 9.1 Splitting of a frequency-selective channel with bandwidth B_{RF} into a large number D of subchannels (bandwidth B_S) that are used individually in parallel.

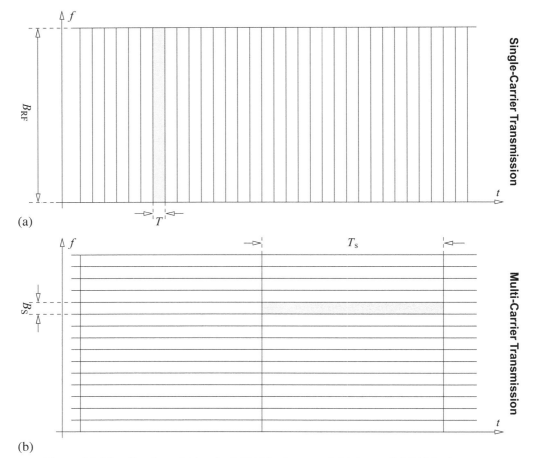

Figure 9.2 Visualization of (a) a single PAM scheme and (b) D parallel PAM schemes over time and frequency.

Nowadays, OFDM is the most popular modulation format; nearly all newly designed digital transmission systems use OFDM/DMT or a variant thereof. OFDM is standardized, among others, in the following transmission systems:

- digital audio broadcasting (DAB);
- digital video broadcasting (DVB), in particular DVB-T2 and DVB-C2;
- asymmetric digital subscriber lines (ADSL) and very-high-bit-rate digital subscriber lines (VDSL);
- wireless local area network (WLAN), in particular IEEE 802.11 a/g/n/ac/ax;
- 4G cellphone and following standards.

9.2 Basic Principles of OFDM

9.2.1 OFDM Transmitter

The representation of using D independent PAM schemes in parallel as shown in Figure 9.2 has to be understood as the main idea but it does not directly give a practical implementation. D independent PAM transmitters/receivers and modulators/demodulators would be required, leading to an infeasible implementation effort. Instead, the following specifications lead to OFDM as it is used in practice.

Assumption 1

Instead of pulse shaping (symbol interval T_s) followed by modulation (frequency shift), each modulator uses its *individual pulse shape* $g_d(t)$, $d = 0, \ldots, D-1$, whereby the spectrum $G_d(f) \circ\!\!-\!\!\bullet\, g_d(t)$ is concentrated at the respective carrier frequency $f_{c,d}$ according to (9.5). Thus, the pulses $g_d(t)$ are *orthogonal*.

This leads to the transmitter structure shown in Figure 9.3. Note that this picture coincides with the signal-space representation of general digital communication schemes discussed in Chapter 6; in particular, see Figure 6.16.

Up to now, the transmit signal (ECB domain) is given by

$$s(t) = \sum_{j=0}^{D-1} s_j(t) , \qquad (9.6)$$

where

$$s_j(t) = \sum_{k_s} a_j[k_s] \cdot g_j(t - k_s T_s) \qquad (9.7)$$

is the PAM transmit signal of the jth branch (modulator), k_s is the respective discrete time index counting the symbols, and $a_j[k_s]$ is the amplitude coefficient in branch j at symbol interval k_s.

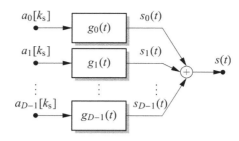

Figure 9.3
Transmitter structure using D individual pulse shapes.

Assumption 2

The pulse shapes $g_j(t)$ are expected to be composed of D *chips* $g_T(t)$. The chips have duration $T = T_s/D$.

Denoting the *chip weights* by $w_{l,j}$, we have

$$g_j(t) = \sum_{l=0}^{D-1} w_{l,j}\, g_T(t - lT)\,, \qquad (9.8)$$

that is, the pulses $g_j(t)$ are themselves PAM signals using the chips $g_T(t)$ as basic pulse shape.

Combining (9.6), (9.7), and (9.8), we arrive at

$$s(t) = \sum_{j=0}^{D-1} \sum_{k_s} a_j[k_s] \sum_{l=0}^{D-1} w_{l,j}\, g_T(t - lT - k_s T_s)$$

$$= \sum_{k_s} \sum_{l=0}^{D-1} \underbrace{\sum_{j=0}^{D-1} w_{l,j}\, a_j[k_s]}_{\stackrel{\text{def}}{=} x_l[k_s],\ \text{see (9.11)}}\, g_T(t - (l + Dk_s)T)$$

$$= \sum_{k_s} \sum_{l=0}^{D-1} x_l[k_s]\, g_T(t - (l + Dk_s)T)\,. \qquad (9.9)$$

We introduce the definitions $k \stackrel{\text{def}}{=} l + Dk_s$ (from which $l = k \bmod D$ and $k_s = \lfloor k/D \rfloor$ are obtained) and $x[k] \stackrel{\text{def}}{=} x_{k \bmod D}[\lfloor k/D \rfloor]$ to obtain a compact representation of the transmit signal in the form

$$s(t) = \sum_k x[k]\, g_T(t - kT)\,. \qquad (9.10)$$

Hence, the OFDM transmit signal is a PAM signal with amplitude coefficients $x_l[k_s]$ and basic pulse shape $g_T(t)$. The generation of the transmit signal is done blockwise. In (9.9), the outer sum with index k_s runs over the symbols with duration T_s – we call them *blocks*. The inner sum in (9.9) runs over the chips of duration T.

The condensed expression (9.10) shows that OFDM corresponds to classical PAM with the parameters modulation interval T, basic pulse $g_T(t)$, and *chip amplitude coefficients* $x[k]$. Given the D data symbols $a_j[k_s]$, $j = 0,\ldots,D-1$, the D chips/channel symbols are calculated block by block (fixed k_s) according to

9.2 Basic Principles of OFDM

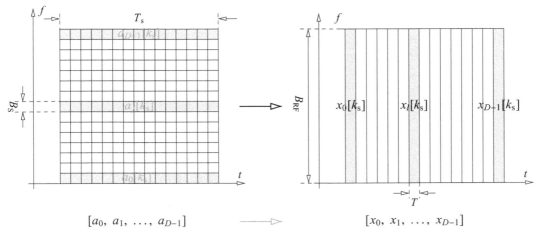

Figure 9.4 Conversion of the D parallel data symbols $a_j[k_s]$ considering the chip structure of the pulses $g_j(t)$ into the PAM data symbols $x_l[k_s]$ weighting the basic pulse shape $g_T(t)$.

$$x_l[k_s] \stackrel{\text{def}}{=} \sum_{j=0}^{D-1} w_{l,j}\, a_j[k_s] \,. \tag{9.11}$$

The essential difference from classical PAM is the blockwise mapping of information symbols from a signal constellation to continuously distributed (complex-valued) amplitude coefficients. This conversion is depicted in Figure 9.4.

9.2.2 Block Transmission

In summary, a *block transmission over the ISI channel* is present. The channel still causes *intersymbol interference*; the PAM symbols $x_l[k_s]$ interfere within the block. However, we want to avoid *interblock interference*. These interferences are visualized in Figure 9.5. As an example, the channel distorts the transmitted pulse $g_T(t)$ (whatever it looks like) to the depicted shape. Its decaying part ("tail") smears into the subsequent symbols.

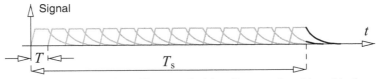

Figure 9.5 Visualization of intersymbol (medium gray) and interblock (dark gray) interference.

Before we discuss how the interblock interference can be avoided, we establish a discrete-time end-to-end model. As shown in Figure 9.6, the chip $g_T(t)$, which is the transmit pulse shape, the channel with impulse response $h_C(t)$, and the receive filter plus T-spaced sampling are combined into the model. For the moment, we use the *whitened-matched filter* (8.70), as derived in the last chapter, since it is matched to the channel and whitens the noise at its output. Hence:

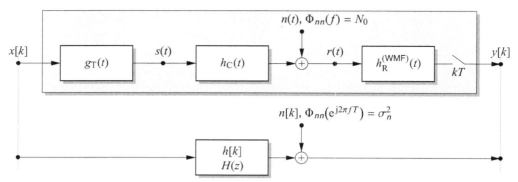

Figure 9.6 Discrete-time end-to-end model combining the chip pulse shape $g_T(t)$, the channel with impulse response $h_C(t)$, and the receive filter plus T-spaced sampling.

- the end-to-end impulse response $h[k]$/transfer function $H(z)$ is monic, minimum-phase, and of order O;
- additive white Gaussian noise $n[k]$ with variance σ_n^2 is present.

Based on this discrete-time end-to-end model, we can study the interference. As the channel is of order O (see the calculation in Section 8.3), each transmitted symbol $x_l[k_s]$ causes O post-cursors leading to intersymbol interference within the block. As we will see later, this *intrablock ISI* is skillfully processed by OFDM. In addition, at the end of each block, O symbols would smear into the next block. This *interblock interference* can be avoided by introducing a *guard space* or *guard interval* of at least O chips between adjacent blocks. The following strategies for utilizing this guard interval are common.

- *Optimum Block Transmission*
  After each block of D channel symbols (at least) O zeros are transmitted; consecutive blocks are separated by O zeros. Since the channel is modeled by an FIR system of order O (cf. Figure 8.20), this clears the channel memory.

  Using this strategy, no transmit power is wasted. At the transmitter, the D channel symbols are calculated from the D data symbols. However, it can be shown (cf. Problem 9.6.1) that for an optimum adaptation to the channel, here pre- and postprocessing (the above derived conversion from $a_j[k_s]$ to $x_l[k_s]$ and back) become channel-dependent, that is, they have to be adapted to the actual channel impulse response $h[k]$.

- *Unique-Word OFDM*
  A different preprocessing that is not channel-dependent exists as well. Instead of an all-zero word, any other fixed word can be inserted between the blocks, which, for example, may be utilized for synchronization. This version is known as *unique-word OFDM* (UW-OFDM).

  Here, the transmitter-side processing (encoding) is more involved, because here the prefix (i.e., the unique word) has to be placed within the DFT block. A sophisticated postprocessing allows for some improvements of data reliability as the guard interval can be exploited as a sort of coding redundancy. A short introduction to UW-OFDM is given in Appendix C.

- *Cyclic-Prefix OFDM*

    The most common strategy is the *partial cyclic repetition of the transmit block*; a *cyclic prefix*[1] is appended. Thereby, a small increase in transmit power (due to transmitting the extra symbols) is caused. However, as we will subsequently show, pre- and postprocessing will be channel-independent; in particular, they can be implemented with only little computational complexity.

    Cyclic-prefix OFDM is the most popular variant and most often used in practice. It is commonly known under the abbreviation OFDM. Subsequently, we will only consider this version.

9.2.3 Cyclic Repetition

The use of a partial cyclic repetition is motivated by Figure 9.7. In part (a), a block of (here $D = 8$) symbols is depicted.

Linear Convolution

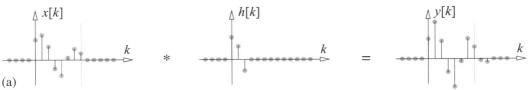

(a)

Cyclic Convolution

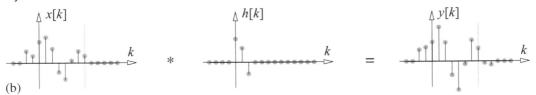

(b)

Figure 9.7 (a) Linear convolution of a block and (b) linear convolution of a cyclically prefixed block (i.e., cyclic convolution within the block).

Assuming a channel of order $O = 2$ (its impulse response is plotted in the middle column), the (noise-free) receive signal as shown on the right-hand side is produced. As the channel is modeled as an LTI system, the output $y[k]$ is given by the *linear convolution*

$$y[k] = x[k] * h[k] = \sum_{l=0}^{O} x[k-l]\,h[l] \qquad (9.12)$$

of the input $x[k]$ with the system's impulse response $h[k]$.

If, instead of a single block, a periodic signal were present, the channel output would also be a periodic signal (with the same period D). Note that one period of the output is connected to one period of the input by the *cyclic convolution*

---

[1] The viewpoints of inserting a number of symbols *after* each block or *before* each block are synonymous as a sequence of blocks is transmitted.

$$y[k] = x[k] \circledast h[k] \overset{\text{def}}{=} \sum_{l=0}^{O} x[(k-l)_{\text{mod } D}] h[l], \qquad (9.13)$$

where $(\kappa)_{\text{mod } D}$ denotes modulo reduction of the integer κ to the set $0, 1, \ldots, D-1$.

In order to have this connection for the block starting at $k = 0$, a periodic signal is not necessarily required. Owing to the FIR property of the channel, O symbols cyclically prefixing the block are sufficient (here $O = 2$). At the channel output, the O symbols preceding the block can be ignored; the same holds for the O symbols succeeding the block.

Hence, in summary, we have:

> *When prefixing a block of D symbols with (at least) O symbols obtained by cyclically repeating the block, the action of a channel with impulse response of order O – the linear convolution of the channel input with the impulse response – is converted into a cyclic convolution.*

Based on this fundamental principle, in OFDM the block processing pictured in Figure 9.8 is employed.

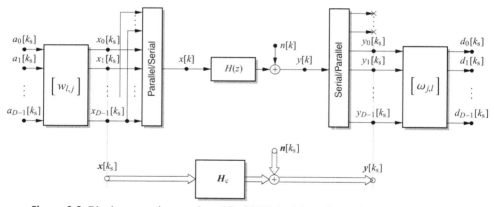

Figure 9.8 Block processing employed in OFDM with cyclic prefix and corresponding vector/matrix model.

- Given the D data symbols $a_j[k_s]$, $j = 0, \ldots, D-1$, of a block at time index k_s, the D channel symbols $x_l[k_s]$, $l = 0, \ldots, D-1$, are calculated according to (9.11).
- A cyclic prefix of $D_g \geq O$ symbols is introduced; the last D_g symbols are prefixed to the block of D symbols.
- The entire block of $D + D_g$ symbols is (serially) transmitted over the channel (LTI system with impulse response $h[k]$ of order O).
- At the receiver, the guard interval is discarded – a block of again D symbols $y_l[k_s]$, $l = 0, \ldots, D-1$, is extracted.

- From the block of D channel output symbols, D *detection symbols* $d_j[k_\mathrm{s}]$, $j = 0, \ldots, D-1$, are calculated using weights[2] $\omega_{j,l}$.

This processing is done for each block (index k_s) separately. Concatenating the blocks of $D + D_\mathrm{g}$ channel symbols, a continuous transmission of blocks is obtained. This sequential transmission of blocks is depicted in Figure 9.9.

Owing to this construction, the blocks (D symbols) are separated by the guard intervals (D_g symbols). The duration of a block is thus $T_\mathrm{s} = (D + D_\mathrm{g}) \cdot T$. The transients between the blocks (darker gray parts in Figure 9.9) are ignored; they are absorbed in the guard period and do not appear in the extracted blocks.

For a compact notation, we resort to a vector/matrix notation (see also the bottom of Figure 9.8). The block of the D transmit symbols $x_l[k_\mathrm{s}]$ is given by the (row) vector

$$\boldsymbol{x}[k_\mathrm{s}] \stackrel{\text{def}}{=} [\, x_0[k_\mathrm{s}], \ldots, x_{D-1}[k_\mathrm{s}] \,] \,. \tag{9.14}$$

The block of receive symbols is then given by

$$\boldsymbol{y}[k_\mathrm{s}] \stackrel{\text{def}}{=} [\, y_0[k_\mathrm{s}], \ldots, y_{D-1}[k_\mathrm{s}] \,]$$
$$= \boldsymbol{x}[k_\mathrm{s}] \cdot \boldsymbol{H}_\mathrm{c} + \boldsymbol{n}_\mathrm{c}[k_\mathrm{s}] \,, \tag{9.15}$$

where $\boldsymbol{n}_\mathrm{c}[k_\mathrm{s}]$ is the vector of D white Gaussian noise samples.[3] It is straightforward to show that due to the cyclic convolution, the end-to-end channel matrix is given by the $D \times D$ *circulant matrix* (its rows/columns are cyclically shifted versions of each other)

$$\boldsymbol{H}_\mathrm{c} = \begin{bmatrix} h[0] & h[1] & \cdots & h[O] & & 0 \\ & \ddots & \ddots & & \ddots & \\ & & \ddots & \ddots & & h[O] \\ h[O] & & 0 & \ddots & \ddots & \vdots \\ \vdots & \ddots & & & \ddots & h[1] \\ h[1] & \cdots & h[O] & & & h[0] \end{bmatrix}. \tag{9.16}$$

In other words, $\boldsymbol{y}[k_\mathrm{s}] = \boldsymbol{x}[k_\mathrm{s}] \cdot \boldsymbol{H}_\mathrm{c}$ is the vector/matrix equivalent of (9.13). From the blocks $\boldsymbol{x}[k_\mathrm{s}]$ at the transmitter to the blocks $\boldsymbol{y}[k_\mathrm{s}]$ at the receiver, we thus have the model depicted in Figure 9.10 (gray box).

Owing to the non-diagonal structure of $\boldsymbol{H}_\mathrm{c}$, the transmit symbols $x_l[k_\mathrm{s}]$ still interfere at the receiver side. Note that, if $H(z) = 1$, that is, $h[0] = 1$ and $h[k] = 0$, $k \neq 0$, the channel matrix $\boldsymbol{H}_\mathrm{c}$ would be the identity matrix $\boldsymbol{I}$ and parallel, decoupled transmission would be present. The question now is, how to perform the pre- and postprocessing such that even if $H(z) \neq \text{const.}$, parallel, independent channels (scaling λ_j, noise $n_j[k_\mathrm{s}]$) are present, meaning that

---

[2] Note that these weights $\omega_{j,l}$ at the receiver side differ from the weights $w_{l,j}$ at the transmitter side. The weights are calculated subsequently.

[3] When employing, as assumed, the whitened-matched filter, the noise samples are white. When other receiver frontends, where noise whitening is not performed perfectly, are present, the elements of $\boldsymbol{n}_\mathrm{c}[k_\mathrm{s}]$ will exhibit correlations. The situation of correlated noise samples is addressed in Appendix C.

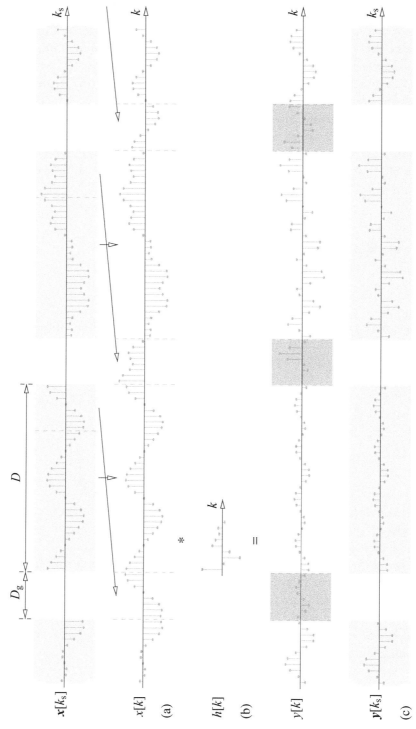

Figure 9.9 Sequential transmission of blocks. (a) Insertion (partial cyclic repetition) of symbols at the transmitter; (b) transmission over the channel (convolution with channel impulse response); (c) extraction of blocks at the receiver. Real-valued example.

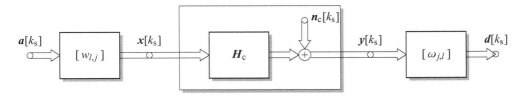

Figure 9.10 Vector/matrix channel model induced by the block processing of Figure 9.8.

$$d_j[k_s] = \lambda_j\, a_j[k_s] + n_j[k_s]\,, \quad j = 0, \ldots, D-1\,. \quad (9.17)$$

From linear algebra it is known that the *modal matrix* – the *matrix of the eigenvectors* – associated with a given matrix can be used to transform that matrix to a diagonal structure (see Box 9.1). Fortunately, circulant matrices have the property that independent of their entries (the impulse response $\langle h[k]\rangle$), the modal matrix is given by the (normalized) *discrete Fourier transform (DFT) matrix*

$$\boldsymbol{W}_D \stackrel{\text{def}}{=} \frac{1}{\sqrt{D}}\begin{bmatrix} 1 & 1 & 1 & \cdots & 1 \\ 1 & e^{-j\frac{2\pi}{D}} & e^{-j\frac{2\pi}{D}2} & \cdots & e^{-j\frac{2\pi}{D}(D-1)} \\ 1 & e^{-j\frac{2\pi}{D}2} & & & \vdots \\ \vdots & \vdots & & & \vdots \\ 1 & e^{-j\frac{2\pi}{D}(D-1)} & \cdots & \cdots & e^{-j\frac{2\pi}{D}(D-1)^2} \end{bmatrix}. \quad (9.18)$$

Thereby, the normalization is done such that $\boldsymbol{W}_D$ is a unitary matrix, that is, $\boldsymbol{W}_D^{-1} = \boldsymbol{W}_D^{\mathsf{H}}$.

Consequently, parallel, independent channels from $\boldsymbol{a}[k_s]$ to $\boldsymbol{d}[k_s]$ in Figure 9.10 are obtained if for the transmitter-side processing the *inverse DFT matrix* (which is the Hermitian DFT matrix, $\boldsymbol{W}_D^{\mathsf{H}}$) and for the receiver-side processing the *DFT matrix* are employed, that is

$$w_{l,j} = \frac{1}{\sqrt{D}} e^{j\frac{2\pi}{D}lj}\,, \qquad \omega_{j,l} = \frac{1}{\sqrt{D}} e^{-j\frac{2\pi}{D}jl}\,. \quad (9.19)$$

In total we have

$$\boldsymbol{W}_D^{\mathsf{H}} \cdot \boldsymbol{H}_{\mathrm{c}} \cdot \boldsymbol{W}_D = \mathrm{diag}(\lambda_0, \ldots, \lambda_{D-1})\,, \quad (9.20)$$

with

$$\lambda_j = \sum_{k=0}^{D-1} h[k] \cdot e^{-j\frac{2\pi}{D}jk} = \sum_{k=0}^{Q} h[k] \cdot z^{-k}\bigg|_{z=e^{j\frac{2\pi}{D}j}}$$

$$= H\left(e^{j2\pi j/D}\right)\,, \quad j = 0, \ldots, D-1\,. \quad (9.21)$$

This equation states nothing other than the well-known fact that a cyclic convolution in the time domain corresponds to an element-wise multiplication in the frequency domain:

$$y[k] = x[k] \circledast h[k] \quad \circ\!\!-\!\!\bullet \quad Y_j = X_j \cdot H_j\,, \quad j = 0, \ldots, D-1\,, \quad (9.22)$$

where X_j, H_j, and Y_j are the elements of the DFT of $x[k]$, $h[k]$, and $y[k]$, respectively.

> **Box 9.1** *Modal Matrix*
>
> Let an $N \times N$ (square) matrix A be given. To each matrix of order N, N (right) *eigenvectors* v_j and corresponding *eigenvalues* λ_j are associated, which fulfill the equation (for simplicity we assume that the eigenvectors are all distinct)
>
> $$A v_j = \lambda_j v_j , \quad j = 0, \ldots, N - 1 .$$
>
> This set of equations can compactly be written as
>
> $$A [v_0, \ldots, v_{N-1}] = [v_0, \ldots, v_{N-1}] \begin{bmatrix} \lambda_0 & & 0 \\ & \ddots & \\ 0 & & \lambda_{N-1} \end{bmatrix} .$$
>
> Defining the *modal matrix*, the matrix of (right) eigenvectors of A, as
>
> $$M \stackrel{\text{def}}{=} [v_0, \ldots, v_{N-1}]$$
>
> and multiplying the above equation by M^{-1} from the left, we arrive at
>
> $$M^{-1} A M = \begin{bmatrix} \lambda_0 & & 0 \\ & \ddots & \\ 0 & & \lambda_{N-1} \end{bmatrix} = \operatorname{diag}(\lambda_0, \ldots, \lambda_{N-1}) ,$$
>
> that is, via the modal matrix, the matrix A is transformed into the diagonal matrix of its eigenvalues.

In summary, in OFDM, the following procedure is carried out:

> - The data symbols $a_d[k_s]$ are interpreted to be the frequency-domain symbols.
> - The inverse DFT of $a[k_s] = [a_0[k_s], \ldots, a_{D-1}[k_s]]$ is calculated to obtain the time-domain transmit symbols
>
> $$x[k_s] = a[k_s] W_D^H ,$$
>
> that is
>
> $$x_l[k_s] = \frac{1}{\sqrt{D}} \cdot \sum_{j=0}^{D-1} a_j[k_s] \cdot e^{+j \frac{2\pi}{D} j l} , \quad l = 0, 1, \ldots, D - 1 .$$
>
> - The cyclic prefix is inserted and the block is serially transmitted over the channel. At the receiver, the samples in the guard interval are ignored and the receive block is extracted.
> - The DFT of $y[k_s]$ is calculated to obtain
>
> $$d[k_s] = y[k_s] W_D ,$$

that is

$$d_j[k_s] = \frac{1}{\sqrt{D}} \cdot \sum_{l=0}^{D-1} y_l[k_s] \cdot e^{-j\frac{2\pi}{D}jl} , \quad j = 0, 1, \ldots, D-1 .$$

- The detection symbols $d_j[k_s]$ can again be interpreted to be in the frequency domain.

9.2.4 End-to-End Model of OFDM

The above derivations have already revealed the end-to-end model of OFDM transmission. The ISI channel is transformed into D parallel, independent channels with individual scaling. The model is depicted in Figure 9.11.

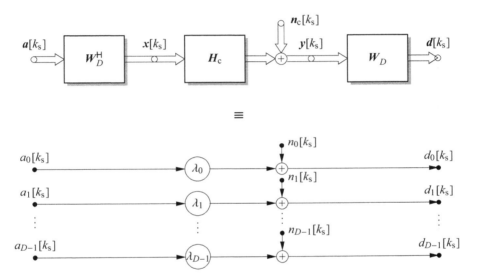

Figure 9.11 End-to-end model of OFDM.

Signal Scaling

As derived above, the scaling factor in the jth subchannel is given by

$$\lambda_j = \sum_{k=0}^{O} h[k] \cdot e^{-j\frac{2\pi}{D}jk} = H\left(e^{j2\pi j/D}\right) , \quad j = 0, \ldots, D-1 . \quad (9.23)$$

In mobile communications, the channel taps $h[k]$ are often modeled as (independent) rotationally invariant complex-valued Gaussian random variables. Then, the gains λ_j are also complex-valued rotationally invariant Gaussian random variables. Hence, $|\lambda_j|$ here is *Rayleigh distributed*.

Noise Variance

It can be shown (see Appendix C) that the noise variance in the jth subchannel is given by

$$\sigma_{n,j}^2 = \sum_{\kappa=-D+1}^{D-1} \left(1 - \frac{|\kappa|}{D}\right) \phi_{nn}[\kappa] \cdot \mathrm{e}^{-\mathrm{j}\frac{2\pi}{D}j\kappa}, \qquad j = 0,\ldots,D-1. \qquad (9.24)$$

As long as for the autocorrelation sequence of the discrete-time noise $\phi_{nn}[\kappa] = 0$, $|\kappa| \geq D_{\mathrm{g}}$, holds, the variance is well approximated by

$$\sigma_{n,j}^2 \approx \sum_{\kappa=-\infty}^{\infty} \phi_{nn}[\kappa] \cdot \mathrm{e}^{-\mathrm{j}\frac{2\pi}{D}j\kappa} = \Phi_{nn}\left(\mathrm{e}^{\mathrm{j}2\pi j/D}\right), \qquad j = 0,\ldots,D-1. \qquad (9.25)$$

In particular, if $n[k]$ is white with variance σ_n^2 (i.e., $\Phi_{nn}\left(\mathrm{e}^{\mathrm{j}2\pi j/D}\right) = \sigma_n^2$), then $\sigma_{n,j}^2 = \sigma_n^2$, $\forall j$ ($\boldsymbol{W}_D$ is unitary; any unitary transformation of white noise results in white noise).

Hence, in summary, employing the whitened-matched filter and the block processing with IDFT at the transmitter and DFT at the receiver D parallel, independent AWGN subchannels with gains λ_j, $j = 0,\ldots,D-1$, and noise variance σ_n^2 are obtained.

9.2.5 Discussion

Note the following properties of OFDM.

- The signals $\boldsymbol{a}[k_{\mathrm{s}}]$ and $\boldsymbol{x}[k_{\mathrm{s}}]$ are related via the (inverse) DFT; $\boldsymbol{a}[k_{\mathrm{s}}]$ is the spectrum of $\boldsymbol{x}[k_{\mathrm{s}}]$, the transmit signal is specified in the frequency domain. Because of that, the subchannels are also called *(sub)carriers*.
- As $\boldsymbol{W}_D^{\mathrm{H}}$ is (defined to be) unitary, the variance of the transmit symbols $x[k]$ is given by the average over the variances $\sigma_{a,j}^2 = \mathrm{E}\{a_j[k_{\mathrm{s}}]\}$ of the data symbols $a_j[k_{\mathrm{s}}]$:

$$\sigma_x^2 = \frac{1}{D} \sum_{j=0}^{D-1} \sigma_{a,j}^2 . \qquad (9.26)$$

- As usual in digital communications, the transmit symbols $x_l[k_{\mathrm{s}}]$ are complex-valued (ECB signal). In some situations (baseband transmission, e.g., ADSL), however, a real-valued transmit signal is required. This is achieved by enforcing the *Hermitian symmetry*

$$a_j[k_{\mathrm{s}}] = a_{D-j}^*[k_{\mathrm{s}}], \qquad j = 1,\ldots,D/2-1 \qquad (9.27)$$

(which means that $a_0[k_{\mathrm{s}}]$ and $a_{D/2}[k_{\mathrm{s}}]$ have to be real-valued). Performing the DFT, directly a real-valued transmit signal results. However, in this case, only one half of the spectrum can be chosen arbitrarily.

- OFDM is a rather low-complexity scheme as the DFT is usually carried out in the form of the *fast Fourier transform (FFT)* since D is typically a power of 2.

  If D is sufficiently large, the loss in power and bandwidth efficiency due to the guard interval can be neglected. Nevertheless, if D is large, the latency in the system is large as a block processing is performed.

- Owing to the superposition within the IDFT, for i.i.d. data symbols in all carriers the OFDM transmit signal has approximately the same statistics as discrete-time *white Gaussian noise* (cf. the central limit theorem). The continuous-time transmit signal has a very high peak-to-average power ratio that grows with D. As a consequence (cf. Chapter 4), OFDM is sensitive to nonlinear power amplification. For more details and possible countermeasures, see Section 9.5.

9.2.6 Average Power Spectral Density of OFDM Signals

It can be shown (see Appendix C) that when using a DFT of length D and a guard interval of length D_g ($T = T_s/(D + D_g)$, $D_s = D + D_g$), the average power spectral density of the continuous-time transmit signal $s(t)$ can be calculated as

$$\bar{\Phi}_{ss}(f) = \frac{|G_T(f)|^2}{T} \cdot \frac{D_s}{D} \sum_{j=0}^{D-1} \sigma_{a,j}^2 \sum_{\mu=-\infty}^{+\infty} \text{si}^2\left(\pi(fT_s - jT_s/T_u - \mu D_s)\right) \quad (9.28)$$

$$= \frac{|G_T(f)|^2}{T} \cdot \frac{1}{D_s D} \sum_{j=0}^{D-1} \sigma_{a,j}^2 \frac{\sin^2\left(\pi(fT_s - jT_s/T_u)\right)}{\sin^2\left(\pi(fT_s - jT_s/T_u)/D_s\right)}. \quad (9.29)$$

The interpretation is as follows.

- The blocks are (rectangularly) limited to the block duration T_s. This gives rise to the $\text{si}(x) = \sin(x)/x$ shape (in the psd si^2) of the individual carriers (parallel data symbols).
- The power of the carriers is proportional to the variance $\sigma_{a,j}^2$ of the data symbols.
- The signal generated by the IDFT is a discrete-time signal. Hence, its (power) spectrum is frequency-periodic. This can be seen from the fact that the carriers repeat periodically (sum over the index μ in (9.28)). Alternatively, the sum over shifted si^2 functions can be written as given in (9.29) (cf. Appendix C).
- As in any PAM scheme, pulse shaping finally limits the bandwidth as it suppresses the spectral replicas. This gives rise to the factor $|G_T(f)|^2$ in (9.28).

  However, in contrast to conventional PAM, here the spectrum can also be influenced by the choice of the variance $\sigma_{a,j}^2$ of the data symbols. For instance, it is common practice to exclude a block of consecutive carriers (index set $\mathcal{J}$) at the band edges from transmission, that is, $\sigma_{a,j}^2 = 0$ for $j \in \mathcal{J}$. This may relax the pulse shaping significantly; only a moderate low-pass filtering will be required, which might already be inherently present in the used transmitter components of the antenna feed.

Example 9.1 Power Spectral Density of OFDM Transmit Signals

Figures 9.12 to 9.15 show average psds of OFDM transmit signals. In Figures 9.12 and 9.13, $D = 32$ is assumed. 24 carriers are active ($\sigma_{a,j}^2 = $ const., $j = 0,\ldots,23$); the others are inactive ($\sigma_{a,j}^2 = 0$, $j = 24,\ldots,31$). In Figure 9.12 no guard interval is used; in Figure 9.13 a guard interval of $D_g = 4$ symbols is expected.

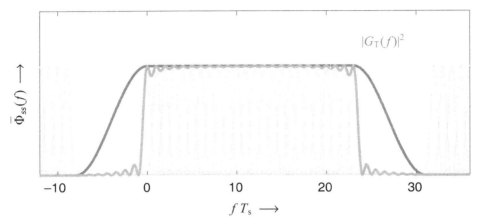

Figure 9.12 Average psd of OFDM transmit signal. $D = 32$, $D_g = 0$; 24 carriers active.

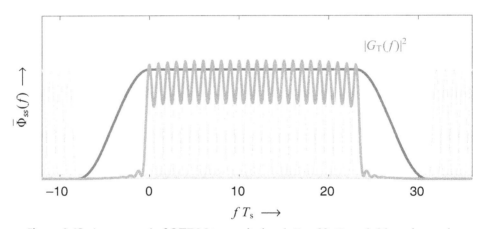

Figure 9.13 Average psd of OFDM transmit signal. $D = 32$, $D_g = 4$; 24 carriers active.

The si^2-shape of the subcarriers is clearly visible (drawn in gray). Also the periodicity of the carriers over the frequency can be recognized. The transmit filter $G_T(f)$ can have wide slopes, as only the spectral replicas have to be suppressed. The resulting psd (drawn in medium gray) has almost a brick-wall shape.

When a cyclic prefix is introduced the block duration increases, leading to smaller si^2-functions. They no longer add up to a constant; some fluctuations in the psd remain.

Additionally, there are no longer zeros of the si-functions at the nominal frequencies of the adjacent subcarriers. Note that these zeros in Figure 9.12 are often taken for illustration of the orthogonality of the subcarrier signals. At the transmitter side, after the insertion of the guard interval no orthogonality is present. After the removal of the guard interval at the receiver, orthogonality is restored.

In Figures 9.14 and 9.15, $D = 128$ is used. 96 carriers are active ($\sigma_{a,j}^2 = \text{const.}$, $j = 0,\ldots,95$), the others are inactive ($\sigma_{a,j}^2 = 0$, $j = 96,\ldots,127$). A cyclic prefix of $D_g = 8$ symbols is expected. Both figures show the same psd but in linear (Figure 9.14) and logarithmic (Figure 9.15) scale.

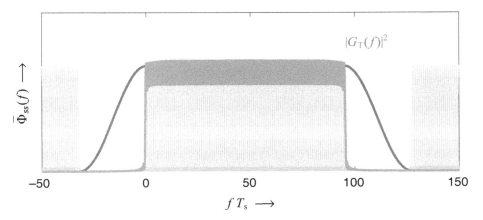

Figure 9.14 Average psd of OFDM transmit signal (linear scale). $D = 128$, $D_g = 8$; 96 carriers active.

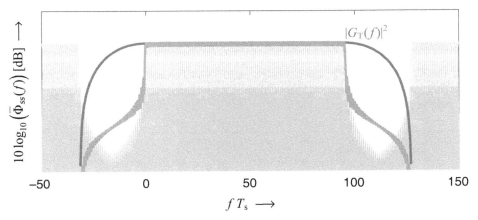

Figure 9.15 Average psd of OFDM transmit signal (logarithmic scale). $D = 128$, $D_g = 8$; 96 carriers active.

Again the almost brick-wall shape is visible. In the logarithmic scale, some side lobes can be recognized. Again, the transmit filter can have a wide transition band and, thus, it is easily implementable.

9.2.7 Signal-to-Noise Ratio

As derived in the last section, OFDM transforms the ISI channel into a set of independent, parallel channels with channel gains λ_j and white noise with variance $\sigma_{n,j}^2$. In mobile communications, where the taps of the channel impulse response often can be modeled as zero-mean rotationally invariant complex Gaussian, the magnitude of the channel gains λ_j is Rayleigh distributed – when going across the carriers, a *flat-fading channel* is present.

Assuming that uncoded transmission (employing the same signal constellation $\mathcal{A}$ with variance σ_a^2) is used over the active subchannels ($a_j[k_s] \in \mathcal{A}, \forall j$), the average symbol error ratio is dominated by the worst-case subchannel and, thus, well approximated by

$$\text{SER} = \max_{j=0,\ldots,D-1} \text{SER}_j \;,$$

where SER_j is the symbol error ratio in subchannel j. The respective effective SNR of OFDM is thus given by

$$\text{SNR}^{(\text{OFDM})} = \min_{j=0,\ldots,D-1} \text{SNR}_j = \min_{j=0,\ldots,D-1} \frac{\sigma_a^2 |\lambda_j|^2}{\sigma_{n,j}^2} \;.$$

Using (9.23) and (9.25), this leads to

$$\text{SNR}^{(\text{OFDM})} = \min_{j=0,\ldots,D-1} \frac{\sigma_a^2 \left|H\left(\mathrm{e}^{\mathrm{j}2\pi j/D}\right)\right|^2}{\Phi_{nn}\left(\mathrm{e}^{\mathrm{j}2\pi j/D}\right)} \;.$$

When applying the whitened-matched filter (see Section 8.3), we have from (8.67) $\Phi_{nn}^{(\text{WMF})}\left(\mathrm{e}^{\mathrm{j}2\pi f T}\right) = \sigma_{n,(\text{WMF})}^2$ and from (8.62) $\left|H\left(\mathrm{e}^{\mathrm{j}2\pi f T}\right)\right|^2 = \sigma_{n,(\text{WMF})}^2 \frac{E_g}{N_0} \Psi_{gg}\left(\mathrm{e}^{\mathrm{j}2\pi f T}\right)$. This gives

$$\text{SNR}^{(\text{OFDM})} = \min_{j=0,\ldots,D-1} \sigma_a^2 \frac{E_g}{N_0} \Psi_{gg}\left(\mathrm{e}^{\mathrm{j}2\pi j/D}\right) = \min_{j=0,\ldots,D-1} \widetilde{\text{SNR}}\left(\mathrm{e}^{\mathrm{j}2\pi j/D}\right) \;,$$

where in the last step (8.47) has been used. Hence, for large D, the SNR of uncoded OFDM is given by

$$\text{SNR}^{(\text{OFDM})} = \min_{f \in [-\frac{1}{2T}, \frac{1}{2T}]} \widetilde{\text{SNR}}\left(\mathrm{e}^{\mathrm{j}2\pi f T}\right) \tag{9.30}$$

minimum over the folded spectral SNR.

Again, the SNR is given by a *Hölder mean* over the folded spectral SNR; in case of uncoded OFDM it is the Hölder mean of power $-\infty$.

Owing to this poor performance, OFDM will never be used without further coding or optimization techniques. Thereby, we have to distinguish between two application scenarios.

- *Broadcast Scenarios*
  In broadcast scenarios (e.g., DAB, DVB) a single transmitter serves a (large) number of users. Here, all (active) carriers use the same signal constellation and same power. Channel coding (typically across the carriers) is used – *coded OFDM (COFDM)* is applied – to compensate for the poor carriers.
- *Point-to-Point Scenarios*
  In point-to-point scenarios (e.g., DSL or WLAN) a specific transmitter/receiver pair is present. Here, knowing the channel transfer function at the transmitter, the modulation in each carrier can be adapted to the current situation. This means that the total transmission rate and the total transmit power are distributed in an optimized way over the carriers. This is the topic of the next section.

9.3 OFDM with Channel Knowledge at the Transmitter

We now assume that a point-to-point communication scenario is present and that channel state information is available at the transmitter. Since uncoded OFDM transmission performs very poorly when all subchannels are utilized the same way (same signal constellation/rate, same transmit power), a straightforward idea to improve performance is to distribute rate and power in an optimized way. Basically, "good" subchannels carry more information, "bad" ones may be switched off. This is the task of so-called *loading algorithms*.

9.3.1 Situation

The starting point for the following optimization is the end-to-end model of OFDM depicted in Figure 9.11; it is repeated in Figure 9.16 for convenience.

It is beneficial to abbreviate the transmit powers in the parallel channels by $S_j = \sigma_{a,j}^2$. The rate (in $\frac{\text{bit}}{\text{symbol}}$) in channel j is denoted by R_j.

Then, the constraints are as follows.

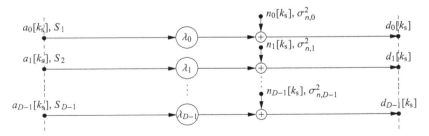

Figure 9.16 Parallel, independent channels for optimal distribution of rate and power.

- The *total transmit power* should be fixed, that is, we demand

$$S_T = \sum_{j=0}^{D-1} S_j . \qquad (9.31)$$

- The *total rate* should be fixed, that is, we demand

$$R_T = \sum_{j=0}^{D-1} R_j . \qquad (9.32)$$

In the course of the development of OFDM-based DSL techniques in the 1990s, a number of loading algorithms (i.e., algorithms for rate and power assignment to the parallel channels) have been invented. The most prominent ones are those of Chow, Cioffi, and Bingham (1995), which uses the channel capacity as the criterion of optimality, and of Hughes-Hartogs (1987/89), which successively assigns the bits to the subchannels.

Here, we will address the problem of finding a rate and power distribution that directly minimizes the (uncoded) error probability. Such an algorithm has been proposed by Fischer and Huber (1996).

9.3.2 Optimum Loading

In each of the parallel subchannels, PAM transmission with QAM constellations of adapted cardinality is used. The situation for one subchannel (number j) is depicted in Figure 9.17.

The R_j bits are mapped (mapping $\mathcal{M}$) to a QAM constellation with 2^{R_j} signal points and with a spacing of the signal points of 2, that is, the distance of a signal point to the nearest decision boundary is 1 (as introduced in Section 3.3). In order to adjust the transmit power S_j, the amplitude coefficient a_j is multiplied by the (real-valued) scaling factor V_j. The scaled point is transmitted over the channel that scales the signal by λ_j and adds Gaussian noise with variance $\sigma_{n,j}^2$. At the receiver side, a threshold decision is performed to obtain an estimate of a_j. Via the inverse mapping, the source bits are recovered.

The symbol error ratio for this setting is given by (cf. Chapter 3, $Q(x)$ denotes the complementary Gaussian integral function)

$$\text{SER}_j = N_{\min,j} \cdot Q\left(\sqrt{\text{SNR}_{0,j}}\right), \quad \text{with} \quad \text{SNR}_{0,j} = \frac{|\lambda_j|^2 V_j^2}{\sigma_{n,j}^2/2} , \qquad (9.33)$$

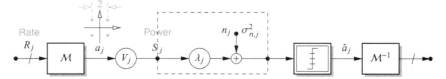

Figure 9.17 Situation for one subchannel.

where $N_{\min,j}$ is the number of nearest-neighbor signal points for the constellation used in this channel. Note that SNR_0 is the relation of the spacing of the signal points at the receiver to the noise power per *real* dimension (hence, one half of the total noise power). The usual SNR (signal power in relation to the noise power) is related to this quantity by $\text{SNR} = \frac{\sigma_a^2}{2}\text{SNR}_0$.

It is intuitive that the average error ratio (over the parallel channels) is minimized if all channels exhibit the same SER_j; otherwise, the worst-case channel will dominate. Since $N_{\min,j}$ is close to 4 for large QAM constellations and this prefactor is of less importance, the demand of same error ratios is equivalent to the demand

$$\text{SNR}_{0,j} = \frac{V_j^2}{Q_j/2} \stackrel{!}{=} \text{SNR}_0 \,, \quad \forall j \,, \tag{9.34}$$

where we have used the abbreviation $Q_j \stackrel{\text{def}}{=} \sigma_{n,j}^2/|\lambda_j|^2$. Hence, the optimization task is

$$\text{SNR}_0 \longrightarrow \max \tag{9.35}$$

under the constraints (9.31) and (9.32).

The problem can be solved analytically if we relax it. Instead of treating only integer rates R and, hence, constellations with cardinalities being a power of two, we allow real-valued rates and, hence, hypothetical QAM signal constellations with $M = 2^R \notin \mathbb{N}$ points. The variance of such constellations is approximated by $\sigma_a^2 = \frac{2(M-1)}{3} \approx \frac{2}{3}2^R$.

The average transmit power in subchannel j then calculates as

$$S_j = V_j^2 \cdot \frac{2}{3}2^{R_j}$$

with (9.34) solved for V_j^2

$$= \text{SNR}_0 \frac{Q_j}{2} \frac{2}{3} 2^{R_j} \,. \tag{9.36}$$

This gives, for the power constraint:

$$S_\text{T} = \sum_{j=0}^{D-1} S_j = \frac{\text{SNR}_0}{3} \sum_{j=0}^{D-1} Q_j \, 2^{R_j} \tag{9.37}$$

and thus

$$\text{SNR}_0 = \frac{3 S_\text{T}}{\sum_{j=0}^{D-1} Q_j \, 2^{R_j}} \,. \tag{9.38}$$

Hence, the maximization of SNR_0 is identical to a minimization of the denominator under the additional constraint of a fixed total rate. This final problem can be solved via *Lagrangian optimization*. The respective Lagrange function combining the target function and the additional constraint is thus given by

$$L = \sum_{j=0}^{D-1} Q_j \, 2^{R_j} - \mu \left(\sum_{j=0}^{D-1} R_j - R_\text{T} \right) \,. \tag{9.39}$$

In the optimum, all partial derivatives of L with respect to the rates R_j have to be zero, that is

$$\frac{\partial L}{\partial R_j} = \log(2)\, Q_j\, 2^{R_j} - \mu \stackrel{!}{=} 0, \qquad j = 0, 1, \ldots, D-1. \tag{9.40}$$

This set of equations is solved for

$$Q_j\, 2^{R_j} \stackrel{!}{=} \text{const.}, \qquad j = 0, 1, \ldots, D-1. \tag{9.41}$$

The constant on the right-hand side is determined if we take the product over all terms in (9.41), leading to

$$\prod_{l=0}^{D-1} Q_l\, 2^{R_l} = \prod_{l=0}^{D-1} \text{const.},$$

$$2^{\sum_{l=0}^{D-1} R_l} \prod_{l=0}^{D-1} Q_l = (\text{const.})^D,$$

the constant holds for any j

$$2^{R_T} \prod_{l=0}^{D-1} Q_l = \left(Q_j\, 2^{R_j} \right)^D,$$

$$2^{R_T/D} \left(\prod_{l=0}^{D-1} Q_l \right)^{1/D} = Q_j\, 2^{R_j}, \tag{9.42}$$

$$\log\left(2^{R_T/D} \left(\prod_{l=0}^{D-1} Q_l \right)^{1/D} \Big/ Q_j \right) = R_j. \tag{9.43}$$

With (9.41), that is, $Q_j\, 2^{R_j} = \text{const.}$, and (9.34), that is, $\text{SNR}_0 = \frac{V_j^2}{Q_j/2} = \text{const.}$, one immediately obtains $V_j^2\, 2^{R_j} = \text{const.}$, which means that the transmit powers in the parallel channels are all the same.

In summary, resubstituting $Q_j = \sigma_{n,j}^2/|\lambda_j|^2$, the optimal rate and power distribution are given by

$$\boxed{\begin{aligned} R_j &= \frac{R_T}{D} + \frac{1}{D} \log_2\left(\frac{\prod_{l=0}^{D-1} \sigma_{n,l}^2/|\lambda_l|^2}{(\sigma_{n,j}^2/|\lambda_j|^2)^D} \right), \\ S_j &= \frac{S_T}{D}. \end{aligned}} \tag{9.44}$$

Having the individual rates, the achieved performance (in terms of the signal-to-noise ratio SNR_0) can be calculated by (9.38). For comparison, we assume a *non-loaded scheme*, that is, D parallel PAM schemes, each with average rate R_T/D and signal point spacing 2. Here, the sum transmit power would be

$$S_\mathrm{T} = D \cdot \frac{2}{3} 2^{R_\mathrm{T}/D} \ . \tag{9.45}$$

The SNR of *optimally loaded OFDM* with the same total rate and same total power thus calculates as (cf. (9.38))

$$\mathrm{SNR}_0 = \frac{3 S_\mathrm{T}}{\sum_{j=0}^{D-1} Q_j \, 2^{R_j}}$$

using (9.42)

$$= \frac{3 S_\mathrm{T}}{\sum_{j=0}^{D-1} \sqrt[D]{2^{R_\mathrm{T}} \cdot \prod_{l=0}^{D-1} Q_l}}$$

using (9.45)

$$= \frac{3 D \cdot \frac{2}{3} 2^{R_\mathrm{T}/D}}{D 2^{R_\mathrm{T}/D} \sqrt[D]{\prod_{l=0}^{D-1} Q_l}}$$

using $Q_j = \sigma_{n,j}^2/|\lambda_j|^2$

$$= \frac{1}{\sqrt[D]{\prod_{l=0}^{D-1} \sigma_{n,l}^2/(2|\lambda_l|^2)}} \ . \tag{9.46}$$

In summary, knowing the OFDM end-to-end model – the signal gains λ_j and the noise variances $\sigma_{n,l}^2$ – the optimal rate distribution and the final performance can be calculated. Note that a study of (9.44) reveals that if the ratio Q_j of noise power and squared channel gain in subchannel j is larger than the *geometric mean* of this ratio over all channels, that is, $\sqrt[D]{\prod_{l=0}^{D-1} \sigma_{n,l}^2/|\lambda_l|^2}$, the rate R_j will be smaller than the average rate R_T/D. Otherwise, if the ratio is smaller (better channel conditions), the rate will be larger.

The rates obtained from (9.44) are in general non-integer and may even be negative. Hence, in practice, these rates are quantized to non-negative integers (thereby not changing the sum rate). Channels with (virtual) negative R_j are excluded from transmission. Moreover, the maximum rate (maximum size of the constellation) has to be limited. For example, in DSL transmission rates up to $R = 15$ (i.e., constellations with cardinalities up to $2^{15} = 32\,768$ points) are standardized. In a final step, the power may be adjusted according to the quantized rate to fine-tune performance.

Example 9.2 Digital Subscriber Line Upstream Transmission

We show the effect of loading for a VDSL-2 system (downstream transmission) according to the ITU standard G.993.2; profile 8d is selected. Here, the symbol rate is $1/T = 17.664$ MHz, a DFT of length $D = 4096$ is employed. Since in DSL baseband transmission is used, at most half of the carriers can be used. In the VDSL-2 system, the number of the highest supported downstream carrier is 1971. In contrast to the actual standard, for simplicity, we do not take into account that blocks of carriers are omitted since they are reserved for the upstream (frequency-division multiplexing). A total rate of 50 Mbit/s is desired. This means that each OFDM frame contains 12 500 bits. Transmission takes place over a standard telephone cable (twisted-pair line) with 0.4 mm diameter and length $\ell_{\text{twistpair}} = 700$ m (cf. Appendix D). Since downstream transmission is studied, white noise is expected.

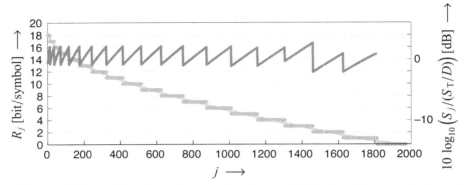

Figure 9.18 Rate (left axis, medium gray) and power distribution (right axis, dark gray) using optimal loading. No limitation of the maximum constellation size.

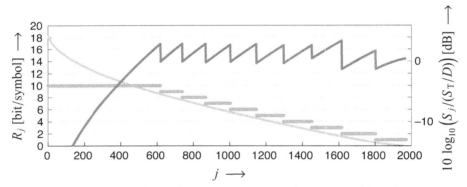

Figure 9.19 Rate (left axis, medium gray) and power distribution (right axis, dark gray) using optimal loading. Limitation of the maximum constellation size to 2^{10}.

Figure 9.18 shows the rate (left axis, medium gray) and power distribution (right axis, dark gray) for no limitation of the maximum constellation size. Figure 9.19 shows the situation for a limitation of the maximum constellation to 1024-ary QAM, that is, a maximum rate of 10. The light gray curve is the non-quantized rate distribution according to (9.44).

Telephone cables exhibit strong low-pass characteristics – the attenuation and the noise level grow over the frequency. Hence, the carriers at lower frequencies (lower index j) have better conditions and are used to carry larger rates. A number of carriers with large index are switched off ($R_j = 0$). The power distribution compensates for the quantization of the rates to integers; when decreasing the rate by one bit, the power can be decreased by approximately 3 dB. It increases again as the channel conditions get worse over the carriers. The carriers that could support larger rates but are limited to $R_j = 10$ need much less power to achieve the same error ratio.

Figure 9.20 shows the loss in minimum distance compared to the AWGN channel over the cable length ℓ (cf. Examples 8.3, 8.5, 8.7, and 8.9 in Chapter 8). As the length increases, dispersion and attenuation increase too, leading to a larger loss. Naturally, if the maximum rate per carrier is limited, the loss increases. A limitation to $R = 15$, as done in the VDSL-2 standard, causes almost no loss.

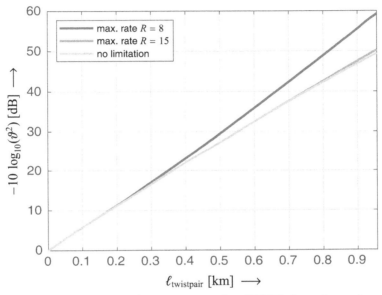

Figure 9.20 Loss in minimum distance compared to AWGN channel over the cable length $\ell_{\text{twistpair}}$. Different limitations of the maximum rate per carrier.

Example 9.3 Bit Loading in a VDSL Modem

Figure 9.21 shows the view from the diagnostic menu of a VDSL modem (FritzBox®) in operation. The estimated SNR over the carriers and the rate distribution are shown. In part (a) a data rate of 25 Mbit/s is active; in part (b) a data rate of 50 Mbit/s over the same telephone line.

In ADSL/VDSL separated blocks of carriers (i.e., separated subbands) are allocated for down- and upstream transmission in order to avoid interference by near-end crosstalk (NEXT) in multi-pair cables.

It can be seen that for the double data rate, all carriers are loaded with approximately the doubled number of bits.

370 | **9 Orthogonal Frequency-Division Multiplexing**

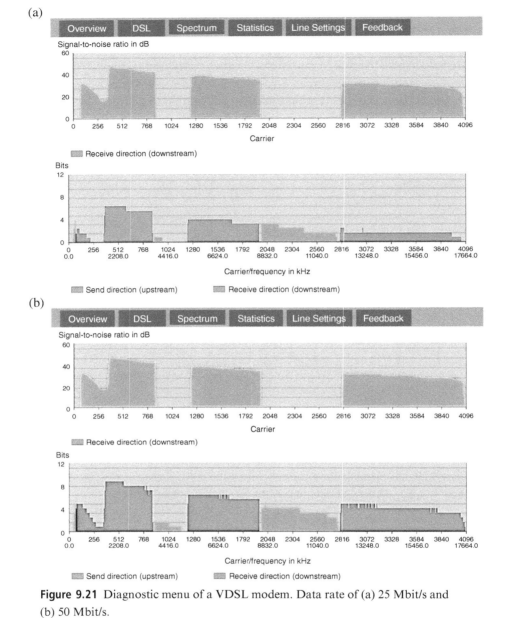

Figure 9.21 Diagnostic menu of a VDSL modem. Data rate of (a) 25 Mbit/s and (b) 50 Mbit/s.

9.3.3 Signal-to-Noise Ratio

In order to calculate the SNR of loaded OFDM, we first look at the influence of the receiver frontend filter used for the transition from continuous-time to discrete-time signals.

To that end, we resort to the general concept presented in Section 8.3: the receiver can always be optimally decomposed into the matched filter (w.r.t. the end-to-end pulse $g(t)$ comprising the transmit filter and the channel), T-spaced sampling, and discrete-time processing with the transfer function $F(z)$ (cf. (8.42)). Hence, the end-to-end discrete-time transfer function (cf. Figure 9.6) is given by

$$H^{(\text{OFDM})}\left(e^{j2\pi fT}\right) = \Psi_{gg}\left(e^{j2\pi fT}\right) \cdot F\left(e^{j2\pi fT}\right), \tag{9.47}$$

and the psd of the discrete-time noise by

$$\Phi_{nn}^{(\text{OFDM})}\left(e^{j2\pi fT}\right) = \frac{N_0}{E_g}\Psi_{gg}\left(e^{j2\pi fT}\right) \cdot \left|F\left(e^{j2\pi fT}\right)\right|^2. \tag{9.48}$$

Note that $F(z)$ may be chosen for linear equalization, for noise whitening (whitened-matched filter), or according to any other criterion. In any case, the scaling in the parallel channels is then given by (9.23)

$$\lambda_j = H^{(\text{OFDM})}\left(e^{j2\pi j/D}\right) \tag{9.49}$$

and the noise powers are given by (9.24)

$$\sigma_{n,j}^2 = \Phi_{nn}^{(\text{OFDM})}\left(e^{j2\pi j/D}\right). \tag{9.50}$$

The quotient Q_j used above then calculates as

$$\begin{aligned}
Q_j &= \frac{\sigma_{n,j}^2}{|\lambda_j|^2} = \frac{\Phi_{nn}^{(\text{OFDM})}\left(e^{j2\pi j/D}\right)}{\left|H^{(\text{OFDM})}\left(e^{j2\pi j/D}\right)\right|^2} \\
&= \frac{\frac{N_0}{E_g}\Psi_{gg}\left(e^{j2\pi j/D}\right) \cdot \left|F\left(e^{j2\pi j/D}\right)\right|^2}{\left|\Psi_{gg}\left(e^{j2\pi j/D}\right) \cdot F\left(e^{j2\pi j/D}\right)\right|^2} \\
&= \frac{N_0}{E_g}\frac{1}{\Psi_{gg}\left(e^{j2\pi j/D}\right)}.
\end{aligned} \tag{9.51}$$

Note that these factors Q_j are samples of the noise psd (8.56) at the output of the optimum Nyquist filter (ONF, optimal linear equalization). This can be explained as follows: using OFDM, the channel distortions are transformed to the frequency domain – samples of the channel transfer function give the signal scaling and samples of the noise psd give the noise power in the parallel channels. When scaling the individual, parallel channels by $1/\lambda_j$, that is, when compensating for the scaling factors at the receiver (applying a gain control), unit-gain AWGN channels with noise variance $Q_j = \sigma_{n,j}^2/|\lambda_j|^2$ are present. However, since unit-gain AWGN channels are forced over the entire transmission band, this is nothing other than implementing linear equalization via component-wise scaling in the frequency domain. Consequently, the same noise power distribution over the carriers as in linear equalization is present.

Furthermore, this fact shows that the discrete-time processing $F(z)$ is of less importance in OFDM; its action can be counteracted by scaling the carriers. However, it is of importance that $F(z)$ is chosen in such a way that the order O of the discrete-time end-to-end impulse response $h[k]$ is shorter than the length D_g of the guard interval. Otherwise, interblock interference would occur. This can be accomplished by calculating $F(z)$ based on the principle for noise prediction (cf. Figure 8.19).

Having the parallel channel model, we are now ready to calculate the SNR of OFDM with loading. To that end, we assume point-to-point transmission and that rate and power are distributed in an optimal way over the carriers. Employing the above derived loading algorithm, and taking the relation between SNR and SNR_0 into account, we have for the SNR

$$\text{SNR}^{(\text{OFDM,load})} = \frac{\sigma_a^2}{2} \text{SNR}_0^{(\text{OFDM,load})}$$

using (9.46)

$$= \frac{\sigma_a^2}{2} \frac{1}{\sqrt[D]{\prod_{j=0}^{D-1} \sigma_{n,j}^2/(2|\lambda_j|^2)}}$$

using (9.51)

$$= \frac{\sigma_a^2}{\sqrt[D]{\prod_{j=0}^{D-1} \frac{N_0}{E_g} \frac{1}{\Psi_{gg}\left(e^{j2\pi j/D}\right)}}}$$

$$= \sqrt[D]{\prod_{j=0}^{D-1} \frac{\sigma_a^2 E_g}{N_0} \Psi_{gg}\left(e^{j2\pi j/D}\right)}$$

using (8.47)

$$= \sqrt[D]{\prod_{j=0}^{D-1} \widetilde{\text{SNR}}\left(e^{j2\pi j/D}\right)}$$

$$= \exp\left\{\frac{1}{D} \log\left(\prod_{j=0}^{D-1} \widetilde{\text{SNR}}\left(e^{j2\pi j/D}\right)\right)\right\}$$

$$= \exp\left\{\frac{1}{D} \sum_{j=0}^{D-1} \log\left(\widetilde{\text{SNR}}\left(e^{j2\pi j/D}\right)\right)\right\} . \quad (9.52)$$

Asymptotically, for a large number D of carriers, we thus have

$$\text{SNR}^{(\text{OFDM,load})} \xrightarrow{D\to\infty} \exp\left\{\int_0^1 \log\left(\widetilde{\text{SNR}}\left(e^{j2\pi\nu}\right)\right) d\nu\right\}$$

$$\stackrel{\nu=fT}{=} \exp\left\{T \int_{-\frac{1}{2T}}^{\frac{1}{2T}} \log\left(\widetilde{\text{SNR}}\left(e^{j2\pi fT}\right)\right) df\right\} . \quad (9.53)$$

Hence, in summary, we have for the SNR of optimally loaded OFDM

$$\text{SNR}^{(\text{OFDM,load})} = \exp\left\{T \int_{-\frac{1}{2T}}^{\frac{1}{2T}} \log\left(\widetilde{\text{SNR}}\left(e^{j2\pi fT}\right)\right) df\right\} \quad (9.54)$$

geometric mean over the folded spectral SNR.

9.4 Comparison of Single- and Multi-Carrier Modulation

Since the invention of OFDM in the 1960s there has been controversy over which strategy – single-carrier or multi-carrier modulation – is superior. A close look at both competing approaches shows that they are dual with respect to some aspects.

- *Single-Carrier Transmission*
  Here, the operation is *time-invariant*; all symbols over time are treated the same way. Moreover, when considering PAM with decision-feedback equalization, the operation at the receiver is *nonlinear*.
- *Multi-Carrier Transmission*
  The operation of OFDM is *time-variant*; symbols are grouped in blocks, only some are repeated during the guard interval. Yet, all operations (in particular the DFT) are *linear*.

Subsequently, we will answer the question of which scheme (potentially) has the better performance. Thereby, we resort to the achievable signal-to-noise ratio $\text{SNR}^{(\cdots)}$ as the criterion of optimality, which, via the Q-function, means that we compare the error ratio performance. The channel is completely characterized by the *folded spectral signal-to-noise ratio* $\widetilde{\text{SNR}}\left(e^{j2\pi fT}\right)$ according to (8.46) and (8.48).

We also distinguish between schemes that do not require channel state information at the transmitter and those that utilize such knowledge. In each case, perfect channel knowledge is expected at the receiver side.

9.4.1 No Channel State Information at the Transmitter

Single-Carrier Transmission

First, we review the performance of single-carrier transmission. In Chapter 8 we derived the achievable SNRs for PAM over a linear, dispersive channel.

- *Linear Equalization*
  Using (zero-forcing) linear equalization by employing the optimal Nyquist filter (8.54) as receiver frontend, the SNR is given by the *harmonic mean* over the folded spectral SNR

$$\text{SNR}^{(\text{LE})} = \frac{1}{T \int_{-\frac{1}{2T}}^{\frac{1}{2T}} \frac{1}{\widetilde{\text{SNR}}\left(e^{j2\pi fT}\right)} \, df}$$
$$= M_{-1}\left(\widetilde{\text{SNR}}\left(e^{j2\pi fT}\right)\right) . \tag{9.55}$$

  Note that the harmonic mean corresponds to the Hölder mean (power mean) $M_\rho(\cdot)$ for power $\rho = -1$ (cf. Box 8.2).
- *Decision-Feedback Equalization*
  In contrast, when employing the whitened-matched filter (8.70) as receiver frontend and performing the detection via (zero-forcing) decision-feedback equalization, the SNR is given by the *geometric mean* (which is the Hölder mean for $\rho = 0$)

over the folded spectral SNR

$$\begin{aligned}\text{SNR}^{(\text{DFE})} &= \exp\left\{T\int_{-\frac{1}{2T}}^{\frac{1}{2T}} \log\left(\widetilde{\text{SNR}}\left(\mathrm{e}^{\mathrm{j}2\pi fT}\right)\right)\,\mathrm{d}f\right\} \\ &= M_0\left(\widetilde{\text{SNR}}\left(\mathrm{e}^{\mathrm{j}2\pi fT}\right)\right)\,.\end{aligned} \qquad (9.56)$$

Multi-Carrier Transmission

We now turn to multi-carrier transmission, in particular in the form of OFDM. Here, if no adaptation of the transmitter is done, the SNR is given by the *minimum* (which is the Hölder mean for $\rho = -\infty$) over the folded spectral SNR

$$\begin{aligned}\text{SNR}^{(\text{OFDM})} &= \min_{f\in[-\frac{1}{2T},\,\frac{1}{2T}]} \widetilde{\text{SNR}}\left(\mathrm{e}^{\mathrm{j}2\pi fT}\right) \\ &= M_{-\infty}\left(\widetilde{\text{SNR}}\left(\mathrm{e}^{\mathrm{j}2\pi fT}\right)\right)\,.\end{aligned} \qquad (9.57)$$

Discussion

Considering the relations of the Hölder means, in case of no channel state information at the transmitter, single-carrier modulation, in particular DFE, clearly outperforms OFDM.

In single-carrier modulation, the equalization process inherently averages over the frequency-selective folded spectral signal-to-noise ratio. In contrast, in OFDM no form of averaging is done as a division into independent subchannels is the opposite to averaging. Hence, such an averaging for achieving high power efficiency has to be done solely by the channel coding (COFDM). In the limit of infinite codelength, channel capacity is indeed achievable using OFDM. However, for very strict restrictions on latency of the data stream, the lack of averaging over the folded spectral SNR is a disadvantage of OFDM compared to single-carrier modulation.

For example, if extreme low latency is desired and no CSI at the transmitter is usable (e.g., broadcasting), single-carrier PAM combined with convolutional codes and Viterbi decoding employing individual DFEs per encoder state for equalization (RSSE for the product trellis of encoder and ISI) yields substantially better results than coded OFDM. (Note that in the regime of extreme low latency convolutional codes outperform any sort of block coding because, despite the short path registers for low latency, no truncation w.r.t. past symbols occurs.) For UW-OFDM this disadvantage is less pronounced due to a possible exploitation of the guard interval for coding at the price of an increased complexity.

9.4.2 Channel State Information at the Transmitter

Single-Carrier Transmission

Again, we review the performance of single-carrier transmission (cf. Chapter 8).

- *Linear Pre-equalization*
  If CSI is available at the transmitter, ZF linear pre-equalization can be utilized. As shown above, this strategy has the same SNR as its receiver-side counterpart:
  $$\text{SNR}^{(\text{LPE})} = \text{SNR}^{(\text{LE})} = M_{-1}\left(\widetilde{\text{SNR}}\left(e^{j2\pi fT}\right)\right), \tag{9.58}$$
  where $\widetilde{\text{SNR}}\left(e^{j2\pi fT}\right)$ is the folded spectral SNR for the initial (flat transmit psd) situation.

- *Tomlinson–Harashima Precoding*
  Alternatively, Tomlinson–Harashima precoding, as the receiver-side counterpart of DFE, can be employed. As shown above, ignoring the small precoding loss, this strategy has the same SNR as DFE:
  $$\text{SNR}^{(\text{THP})} = \text{SNR}^{(\text{DFE})} = M_0\left(\widetilde{\text{SNR}}\left(e^{j2\pi fT}\right)\right). \tag{9.59}$$
  The main advantage of THP over DFE is that any highly efficient channel coding scheme designed for the AWGN channel can be applied to a frequency-selective channel in an optimum way.

- *Combined Transmitter- and Receiver-Side Linear Equalization*
  Finally, the equalization task can be split between receiver and transmitter. As derived above, this pre-emphasis/de-emphasis method achieves an SNR that is given by the square harmonic mean root, which is the Hölder mean for $\rho = -1/2$, over the folded spectral SNR
  $$\begin{aligned}\text{SNR}^{(\text{RRE})} &= \left(T\int_{-\frac{1}{2T}}^{\frac{1}{2T}}\left(\widetilde{\text{SNR}}\left(e^{j2\pi fT}\right)\right)^{-1/2}df\right)^{-2} \\ &= M_{-1/2}\left(\widetilde{\text{SNR}}\left(e^{j2\pi fT}\right)\right).\end{aligned} \tag{9.60}$$

Multi-Carrier Transmission

If optimum rate and power loading are done based on CSI at the transmitter, OFDM achieves an SNR that is given by the geometric mean (Hölder mean for $\rho = 0$) over the (initial) folded spectral SNR
$$\text{SNR}^{(\text{OFDM,load})} = M_0\left(\widetilde{\text{SNR}}\left(e^{j2\pi fT}\right)\right). \tag{9.61}$$

Discussion

A comparison immediately reveals that
$$\text{SNR}^{(\text{OFDM,load})} = \text{SNR}^{(\text{DFE})} = \text{SNR}^{(\text{THP})}. \tag{9.62}$$

OFDM and DFE/THP potentially have the same performance!

This equivalence is valid under the following assumptions.

- OFDM uses a sufficiently large number of carriers and the optimum rate and power distribution is applied. The power loss due to the guard interval is ignored.
- The error propagation in the feedback loop of DFE is ignored.
- For THP the very small increase of transmit power is ignored; here, no error propagation exists.

Hence, the choice between single-carrier and multi-carrier schemes has to be based on other criteria. In particular, this includes implementation aspects, the question of how coding should be applied, and the fact whether latency due to block processing in OFDM can be tolerated.

9.4.3 Summary

We have studied the equalization of linear dispersive channels (characterized by their folded spectral signal-to-noise ratio) in Chapters 8 and 9. It has been shown that the power efficiency of the discussed basic schemes can be expressed by Hölder means over the folded spectral SNR. The different schemes thereby correspond to different Hölder powers ρ.

This result makes the theory on equalization of data signals rather compact and quite aesthetical. In Figure 9.22, the results are schematically summarized over the Hölder power ρ. Remember that the Hölder mean over a real and positive function increases for an increasing parameter ρ. Thus, the Hölder power characterizes an ordering with respect to performance of different equalization techniques.

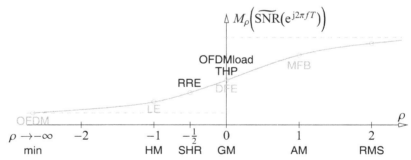

Figure 9.22 Illustration of achievable SNRs as Hölder means over the folded spectral signal-to-noise ratio. Medium gray: receiver-side techniques. Dark gray: with channel state information at the transmitter. (HM: harmonic mean, SHR: square harmonic mean root, GM: geometric mean, AM: arithmetic mean, RMS: root mean square).

Example 9.4 Digital Subscriber Line Upstream Transmission

This example continues Example 9.2 on a (simplified) VDSL-2 system.

Figure 9.23 compares the performance (expressed as loss compared to the AWGN channel) for the various schemes over the cable length $\ell_{\text{twistpair}}$. In the single-carrier case, linear equalization, decision-feedback equalization, and root–root equalization are studied. OFDM employs the loading algorithm from Section 9.4. A maximum number of 15 bits is loaded to the carriers.

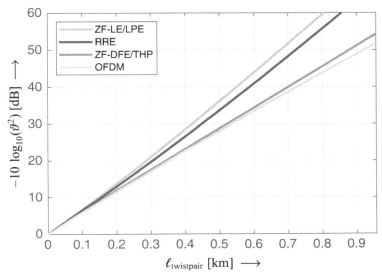

Figure 9.23 Loss in minimum distance compared to AWGN channel over the cable length $\ell_\text{twistpair}$. Single-carrier vs. multi-carrier schemes.

For moderate cable lengths, single-carrier schemes using DFE and the optimally loaded multi-carrier scheme show (almost) the same performance. Linear equalization has much poorer performance and should not be used. Linear root–root equalization leads to a result only somewhat better than ZF-LE. For longer cables OFDM has some slight advantage. This is due to the fact that loading inherently optimizes the bandwidth of the transmit signal (by setting carriers to zero). Thus, OFDM enables a fine-grained adaptation to the situation at hand, whereas in the single-carrier scheme such an adaptation is not easily possible. Only by increasing the size of the constellation and increasing the symbol duration T accordingly, the bandwidth can be decreased and, thereby, frequency bands of high attenuation can be avoided.

9.5 Peak-Power Reduction

The time-domain transmit signal of an OFDM scheme has a very high dynamic range – the crest factor ζ or the peak-to-average power ratio (**PAPR**) (cf. (3.72) and (3.73)), respectively, are extraordinarily large.

This effect is due to the inverse DFT, where each amplitude coefficient $x_l[k_\text{s}]$ in the time domain consists of a weighted superposition of all D data symbols $a_j[k_\text{s}]$ within one block. By pulse shaping with $g_\text{T}(t)$ for the generation of the continuous-time transmit signal $s(t)$, the dynamic range is even more increased (cf. Section 3.4).[4]

Driving the high-power amplifier (HPA) with a too small back-off, due to non-linear effects, *out-of-band radiation* (spectral side lobes) will be created, by which a given spectral mask for the average power spectral density (PSD) of the transmit signal will be violated. Hence, prevention against an overload of the HPA is essential. However, the effect of signal clipping on data reliability is of minor importance as in OFDM the error is distributed by the DFT at the receiver side over all subcarriers and a channel coding scheme will handle this small amount of additional noise.

When discussing methods to mitigate the problems due to the high dynamic range of the transmit signal, the *reduction of the peak power* should be addressed rather than the reduction of the PAPR, as is often done in the literature. Note that the PAPR may also be reduced by an enlargement of the average power instead of reducing the peak power, which, of course, would be counterproductive for the power efficiency of communication schemes.

9.5.1 Gaussian Signal Model and Principal Coding Approach

In Section 9.2 we have seen that a block of D time-domain transmit symbols x_l is generated from the block of data-carrying frequency-domain symbols a_j according to (the block index k_s is dropped)

$$[x_0, x_1, \ldots, x_{D-1}] = [a_0, a_1, \ldots, a_{D-1}] \, W_D^H, \qquad (9.63)$$

where W_D^H, the inverse of the normalized DFT matrix, is unitary. Each transmit symbol x_l is thus given by the superposition of D independent and zero-mean coefficients a_j. Owing to the central limit theorem of probability theory, a good approximate model for the time-domain symbols x_l is that they are independent and that their probability density function (pdf) is very close to a zero-mean rotationally invariant complex Gaussian distribution with variance σ_x^2. Subsequently, we stick to this Gaussian, independent setting.

As a simple model, the amount of generated out-of-band power is analyzed by the relative frequency of clipping events, that is, by the probability that the squared magnitude of signal samples (instantaneous power) $\gamma \stackrel{\text{def}}{=} |x_l|^2$ exceeds a given threshold γ_{cl}. This probability corresponds to the complementary distribution function of the instantaneous signal power and is denoted the *clipping probability curve*. Thus, we have to look at the distribution of the instantaneous power γ of random signals.

The squared magnitude γ of a zero-mean, rotationally invariant complex Gaussian variable with variance σ_x^2 is exponentially distributed (cf. Box 5.1, (5.26)).

---

[4] The problem of a high PAPR of the continuous-time transmit signal also arises for single-carrier PAM schemes when a basic pulse with very low roll-off factor is chosen for high bandwidth efficiency (cf. Figure 3.32). The methods discussed here for OFDM can also be applied to such single-carrier schemes.

Hence, the probability that the instantaneous power of a sample exceeds a threshold $\gamma_{\rm cl}$ is given by

$$\Pr\{\gamma > \gamma_{\rm cl}\} = 1 - \left(1 - e^{-\gamma_{\rm cl}/\sigma_x^2}\right) = e^{-\gamma_{\rm cl}/\sigma_x^2}. \tag{9.64}$$

The event that the squared magnitude of at least one sample (the sample with maximum magnitude) within a block of D samples exceeds the threshold $\gamma_{\rm cl}$ is complementary to the event that none of the independent samples exceeds this threshold. Thus, the probability for at least one clipping within a block of D values for a clipping threshold $\gamma_{\rm cl}$ (i.e., the clipping probability curve) reads

$$\mathrm{clpr}_{\rm OFDM}(\gamma_{\rm cl}) \stackrel{\text{def}}{=} \Pr\{\max |x_l|^2 > \gamma_{\rm cl}\} = 1 - \left(1 - e^{-\gamma_{\rm cl}/\sigma_x^2}\right)^D. \tag{9.65}$$

In Figure 9.24 the clipping probability curve is shown over $10 \log_{10}\left(\gamma_{\rm cl}/\sigma_x^2\right)$ for blocks of $D = 256$ complex-valued samples. If, for instance, a clipping probability less than 10^{-5} is necessary in order to avoid unlicensed out-off-band power radiation, a back-off for high-power amplification of about 12.3 dB has to be applied, which corresponds to a rather serious loss in power efficiency.

For continuous-time signals, an analysis is not as simple as for T-spaced samples, because the argument of statistical independence no longer holds due to the bandwidth limitation of the signal. However, simulations reveal that the respective clipping probability curve essentially follows the curve in Figure 9.24 with only a small shift to the right.

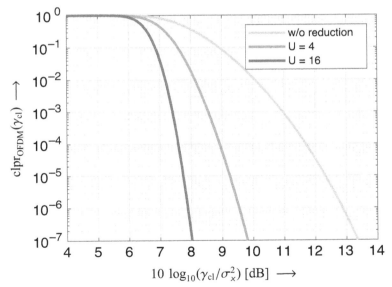

Figure 9.24 Clipping probability curve for $D = 256$. Without reduction scheme and with selected mapping (see below) with $U = 4$ and $U = 16$ candidates.

The principle idea to reduce the peak power of the transmit signal is as follows. Data vectors $\boldsymbol{a} = [a_0, a_1, \ldots, a_{D-1}]$ for which at least one sample of the corresponding transmit vector $\boldsymbol{x} = [x_0, x_1, \ldots, x_{D-1}]$ exceeds a given threshold are excluded from transmission. Since, out of the 2RD possible transmit sequences in a scheme with rate R $\left[\frac{\text{bit}}{\text{symbol}}\right]$ only $2^{(R-r_\text{P})D}$ are allowed, some sort of (channel) coding with an (average) redundancy of r_P $\left[\frac{\text{bit}}{\text{symbol}}\right]$ is present.

Considering this theoretical coding scheme, a simple non-existence bound for peak-power reduction can be derived. Using (9.65) and the law of large numbers (i.e., the relative frequency converges to the probability for an increasing number of realizations of a random experiment), there exist about $2^{RD}\left(1 - e^{-\gamma_\text{cl}/\sigma_x^2}\right)^D$ sequences that do not exceed the threshold γ_cl. If, by a perfect coding scheme, exactly those sequences are selected for communication, the following equation holds for the threshold γ_cl that need not be exceeded if an average redundancy r_P per sample is spent for peak-power reduction, although the samples are Gaussian and thus initially unlimited:

$$2^{(R-r_\text{P})D} \overset{!}{\leq} 2^{RD}\left(1 - e^{-\gamma_\text{cl}/\sigma_x^2}\right)^D \tag{9.66}$$

$$\Rightarrow \quad \gamma_\text{cl}/\sigma_x^2 \geq -\log_e\left(1 - 2^{-r_\text{P}}\right). \tag{9.67}$$

Figure 9.25 (light gray curve) shows that it might be possible to reduce the peak power very significantly by spending a rather small amount of redundancy r_P.

When restricting to sequences with limited peak power, the average power is decreased as well, because for this subset the pdf approximately follows a truncated exponential one. Thus, a smaller variance σ_P^2 of such limited Gaussian variables results. After some calculations, one obtains (cf. Figure 9.25 (middle gray curve))

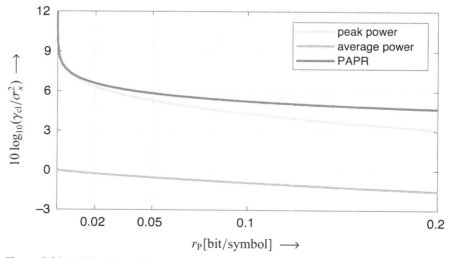

Figure 9.25 Achievable minimum peak power by means of redundancy for Gaussian random variables.

$$\sigma_P^2/\sigma_x^2 = 1 - (1 - 2^{r_P}) \log_e (1 - 2^{-r_P}) . \qquad (9.68)$$

From this equation, a lower bound on the PAPR for the Gaussian signal model with perfect coding for peak-power reduction follows (cf. Figure 9.25 (dark gray curve)):

$$\mathrm{PAPR} = \sigma_{cl}^2/\sigma_P^2 \geq \frac{-\log_e (1 - 2^{-r_P})}{1 - (1 - 2^{r_P}) \log_e (1 - 2^{-r_P})} . \qquad (9.69)$$

As these considerations for the Gaussian signal model indicate that in principle there may exist methods to overcome a high PAPR problem to a great extent, many practical approaches have been published in the literature. Here, only a small selection is briefly reviewed.

9.5.2 Peak-to-Average Power Reduction Schemes

Selected Mapping

An efficient way to reduce the peak power of blockwise organized transmit signals with high dynamic range is *selected mapping (SLM)*. This procedure follows the above considerations on coding, that is, looking for suitable signals out of the set of possible sequences by a simple scheme in practice. For this purpose, a small number U of different mappings of the given block of payload data a to tentative transmit signals $x^{(u)}$ are performed in parallel. Thereby, in OFDM the signal generation (sig. gen.) is done via the IDFT. From the candidates, the signal for which the maximum magnitude of the samples is minimum is selected for transmission (cf. Figure 9.26).

In order to correctly recover the data at the receiver, the index u^* of the used mapping has to be communicated as side information. Hence, a very small amount of $\log_2(U)$ bits of redundancy has to be spent. More generally speaking: a small amount of redundancy creates the freedom to select the signal that is beneficial with respect to a desired criterion out of a set of signals representing the same information. Note that, using SLM, any desired signal property may be controlled, not only the peak power.

For the usual choices of mappings, the different tentative transmit signals look like representing different quasi-independent data sequences. For each mapping, the clipping probability curve of (9.65) holds. When SLM is active, the threshold

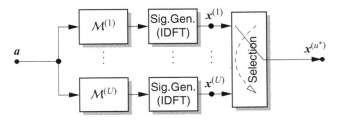

Figure 9.26 Selected mapping.

γ_{cl} for the peak power is only overshot if this happens to all U tentative signals. For the Gaussian model, this gives:

$$\text{clpr}_{\text{SLM}}(\gamma_{\text{cl}}) = (\text{clpr}_{\text{OFDM}}(\gamma_{\text{cl}}))^U = \left(1 - \left(1 - e^{-\gamma_{\text{cl}}/\sigma_x^2}\right)^D\right)^U. \tag{9.70}$$

In Figure 9.24, the clipping probability curves (9.70) for SLM for a selection out of $U = 4$ and $U = 16$ tentative signals are shown. Even a small number of different mappings (e.g., 2 or 4 bits of redundancy; average redundancy $r_P = 0.0078$ bit/sample or $r_P = 0.0156$ bit/sample, respectively) provide a rather significant reduction of the peak power. For example, only 16 different mappings permit the back-off to be reduced by about 5 dB. This corresponds to a gain of 5 dB in power efficiency for the digital communication scheme, if a clipping probability of less than 10^{-5} has to be guaranteed for a sufficiently low out-of-band power radiation.

Meanwhile, a huge number of SLM variants have been proposed. Among them, scrambling SLM (SLM-SRC), which uses binary scramblers on the source sequence to implement the U different mappings, is of particular interest. If the same scrambler but with different initializations of the memory cells (flip flops) is used, no explicit side information at all has to be communicated. At the receiver side, the corresponding FIR descrambler is applied and the different transients at the beginning of the blocks are ignored.

A further popular proposal is *partial transmit sequences (PTS)*, where sub-blocks within the vector *a* are rotated by different phase factors for different mappings such that modified signal points still emanate from the applied constellation.

Repeated Clipping and Filtering

In order to avoid uncontrolled overloading of the HPA, it is beneficial to clip the discrete-time transmit signal. Thereby, usually new undesirable spectral side lobes are generated and a filter has to be applied to suppress these new out-of-band signal components. However, this filtering typically restores some signal peaks beyond the clipping threshold.

Experience shows that by *repeated clipping and filtering* both goals (i.e., peak-power reduction and sufficient attenuation of out-of-band power) can be achieved simultaneously to some extent. Favorably, the clipping threshold is successively lowered during the iterative process. As usually the average signal power grows slightly, the gain in PAPR reduction of this method is more impressive than a peak-power reduction, but the latter criterion is more relevant in technical practice.

Tone Reservation

In *tone reservation* a few subcarriers within an OFDM block remain unloaded from payload data. By means of a sophisticated optimization process, coefficients (not restricted to a constellation) for these reserved subcarriers (tones) are generated such that after IDFT the peak power within the block is minimized. Intuitively, the time-domain signal corresponding to the reserved tones exhibits peaks, inverse to those of the signal corresponding to the payload data. Of course, due to the additional carriers, the average signal power is increased.

Constellation Expansion

Without a loss in data reliability, the outer signal points of a QAM signal constellation may be shifted in directions where the distance to all decision boundaries is not decreased. Thus, for all outer signal points, areas of a possible enlargement (i.e., for *constellation expansion*) exist. Thus, at subcarriers to which outer signal points are assigned, the coefficients may be modified (enlarged within their allowed regions) such that the peak power in the discrete-time transmit signal is reduced. A sophisticated optimization process has to be used to find suitable constellation expansions at these subcarriers.

By constellation expansion, a significant increase in average transmit power occurs. The impressive results of PAPR reduction are largely obtained by this increased average power, rather than by a decreased peak power.

9.6 Problems

9.6.1 Block Transmission over ISI Channels I

We consider a discrete-time signal $x[k]$ which is transmitted over a discrete-time, dispersive channel with a causal, finite-length impulse response $h[k]$ (FIR system).

(a) Determine the (noise-free) received signal $y[k]$ depending on the transmit signal $x[k]$ and the channel impulse response $h[k]$.

Subsequently, blockwise transmission is considered.

(b) Give in vector/matrix notation the calculation of the receive signal from the transmit signal
  - for a linear convolution with the channel impulse response and
  - for a cyclic convolution with the channel impulse response.

(c) Which methods can be used to avoid interferences between successively transmitted blocks?

9.6.2 DFT and Eigenvalue Decomposition

The idea of *multi-carrier transmission schemes* is to decompose the dispersive channel into a (large) number of parallel, intersymbol-interference-free subchannels. However, in practice, a different approach is taken to reach this aim.

(a) Give the vector/matrix notation for the discrete Fourier transform (DFT) of length D.

(b) Show that the rows/columns of the DFT matrix are the eigenvectors of a *circulant matrix*.

(c) Determine the corresponding eigenvalues.

(d) Give the *eigenvalue decomposition* of a circulant channel matrix and show that the result is a decomposition of the dispersive channel into a set of parallel, independent subchannels.

9.6.3 Block Transmission over ISI Channels II

We consider blockwise (block length D) transmission over a discrete-time, dispersive channel with causal, finite-length impulse response $h[k]$ (FIR system of order O). The blocks are separated by inserting $D_0 \geq O$ zeros between successive blocks.

(a) State the input/output relation (vector/matrix notation) between blocks of length D at the channel input and channel output blocks of length $D + D_0$.

The dispersive channel should be transformed into a set of parallel, non-frequency-selective subchannels by suitable preprocessing (at the transmitter) and postprocessing (at the receiver) of the blocks. At the transmitter, the block of transmit symbols is obtained from the block of data symbols by a multiplication with the matrix $\mathbf{A}$. At the receiver, the block of length $D + D_0$ is multiplied with the matrix $\mathbf{R}$ to obtain the decision variables.

(b) Sketch the respective block diagram of the entire transmission system.

Now, we want to determine the optimum pre-/postprocessing matrices $\mathbf{A}$ and $\mathbf{R}$, respectively.

(c) What optimization criterion (in view of digital communications) is reasonable?
(d) What additional constraints are required in order to avoid singular solutions and to guarantee parallel, decoupled channels?
(e) Give the Lagrangian corresponding to the optimization problem and solve this problem.
(f) Discuss the similarities and differences of the present scheme to conventional OFDM.

9.6.4 Loading in OFDM

We consider the distribution of power over the carriers in an OFDM scheme. To that end, the discrete-time channel transfer function is assumed to be piecewise constant, specifically

$$H\left(e^{j2\pi fT}\right) = \begin{cases} \alpha, & 0 \leq fT < 1/2 \\ \beta, & 1/2 \leq fT < 1 \end{cases}.$$

First, we expect that the average power spectral density of the transmit signal is constant over the entire frequency range $0 \leq fT < 1$.

(a) Give a relation between α and β such that the average receive power is equal to the average transmit power.

Now, we presume that $\alpha \gg \beta$ and that the relation derived in part (a) holds.

(b) How should the rate qualitatively be distributed over the carriers?
(c) Characterize the rate distribution for the limiting case $\beta \to 0$. How should the power be distributed over the carriers?
(d) Calculate the average receive power for this case.
(e) What effect of loading becomes obvious here?

9.6.5 Distribution of the Peak Power in OFDM

We consider OFDM and assume that the number of carriers, D, is large.

(a) Why is it justified to assume that the time-domain samples x_l after the inverse DFT have a complex Gaussian distribution?

Let x_l be a random variable, complex Gaussian distributed with zero mean and variance σ_x^2.

(b) Give the probability that $|x_l|^2$ does not exceed a given threshold γ_{cl}.

Assume we have D independent random variables x_l, $l = 0, \ldots, D-1$, all with the distribution of part (b).

(c) Give the probability that no $|x_l|^2$ exceeds a given threshold γ_{cl}.
(d) Give the probability that at least one $|x_l|^2$ exceeds a given threshold γ_{cl}.

Assume we have U sets of D independent random variables $x_l^{(u)}$, $l = 0, \ldots, D-1$, $u = 1, \ldots, U$, all with the above distribution.

(e) Give the probability that in all sets at least one $|x_l^{(u)}|^2$ exceeds a given threshold γ_{cl}.
(f) Explain why the result in part (e) is a reasonable estimate for the clipping probability curve of OFDM with selected mapping (SLM) for peak-power reduction.

APPENDIX A

Signals, Systems, and Processes

This appendix collects some basics on complex numbers, signals and systems, and stochastic processes. In particular, the notation used throughout the book is introduced.

A.1 Recapitulation of Signals and Systems

A.1.1 Complex Numbers

In communications, as always when one has to deal with oscillation phenomena, we heavily resort to *complex numbers*. A complex number is of the form

$$z = a + \mathrm{j}b, \qquad \text{with} \quad a, b \in \mathbb{R}, \tag{A.1}$$

where a is the *real part*, b is the *imaginary part*, and the *imaginary unit* j is defined as $\mathrm{j}^2 = -1$. Obtaining the real and imaginary part from a complex number z is also written as

$$a = \operatorname{Re}\{z\}, \qquad b = \operatorname{Im}\{z\}. \tag{A.2}$$

The *conjugate complex* number to $z = a + \mathrm{j}b$ is given by

$$z^* = a - \mathrm{j}b. \tag{A.3}$$

Instead of specifying the complex number z by its *Cartesian coordinates* a and b, it can also be given in *polar form*:

$$z = |z|\,\mathrm{e}^{\mathrm{j}\varphi}, \tag{A.4}$$

where

$$|z| = \sqrt{a^2 + b^2} \qquad \text{is the } \textit{absolute value} \text{ or } \textit{magnitude}, \tag{A.5}$$

$$\varphi = \arg(z) \qquad \text{is the } \textit{argument} \text{ or } \textit{phase}. \tag{A.6}$$

Note that, φ is not uniquely determined. Since the complex exponential function $\mathrm{e}^{\mathrm{j}x}$ is 2π periodic, the phase is also 2π periodic – all values $\varphi + 2\pi n$, $n \in \mathbb{Z}$, are equivalent. The arg-function has to be understood as returning an infinite set of values. Using the inverse to the cosine function, $y = \arccos(x)$, with domain $x \in [-1, 1]$ and range $y \in [0, 2\pi]$, we can write

$$\varphi = \arg(z) = \begin{cases} \arccos\left(\frac{a}{|z|}\right), & b \geq 0 \\ -\arccos\left(\frac{a}{|z|}\right), & b < 0 \end{cases} + 2\pi n, \quad n \in \mathbb{Z}. \tag{A.7}$$

For $n = 0$, the *principle value* in the interval $(-\pi, \pi]$ is obtained. The principle value of the argument is denoted by $\operatorname{Arg}(z)$. In some situations it is more convenient to define the principle value as the value in the interval $[0, 2\pi)$ that is obtained for $n = 0$ if $b \geq 0$ and $n = 1$ if $b < 0$.

A.1.2 Continuous-Time Signals and Systems

Signal and Spectrum

Time-domain *signals* are denoted by lowercase letters; the continuous time variable is t. The signal can be real- or complex-valued:

$$x(t) \in \mathbb{C}, \quad t \in \mathbb{R}. \tag{A.8}$$

The corresponding *Fourier transform* or *spectrum* is denoted by the corresponding uppercase letter and is defined as[1]

$$X(f) \stackrel{\text{def}}{=} \mathscr{F}\{x(t)\} = \int_{-\infty}^{\infty} x(t) \, e^{-j2\pi f t} \, dt \tag{A.9}$$

with the corresponding inverse transform

$$x(t) = \mathscr{F}^{-1}\{X(f)\} = \int_{-\infty}^{\infty} X(f) \, e^{j2\pi f t} \, df. \tag{A.10}$$

The correspondence between the time-domain signal $x(t)$ and its spectrum $X(f)$ is written in short as

$$\text{time domain} \quad x(t) \circ\!\!-\!\!\bullet X(f) \quad \text{frequency domain.}$$

Basic Signals

Some basic signals that are frequently used are defined next.

- Dirac delta function: $\quad\quad\quad \delta(t)$.

  The Dirac delta "function" (which is a generalized function or distribution) is implicitly defined via $\int_{-\infty}^{\infty} \delta(t)\, \psi(t) \, dt = \psi(0)$ for any test function $\psi(t)$.

- Step function: $\quad\quad\quad \varepsilon(t) \stackrel{\text{def}}{=} \begin{cases} 1, & t > 0 \\ 1/2, & t = 0 \\ 0, & t < 0 \end{cases}.$

---

[1] Without further notice, we assume the convergence of the integrals.

- **Signum function:** $\operatorname{sgn}(t) \stackrel{\text{def}}{=} \begin{cases} 1, & t > 0 \\ 0, & t = 0 \\ -1, & t < 0 \end{cases}$.

- **Rectangular pulse:** $\operatorname{rect}(t) \stackrel{\text{def}}{=} \begin{cases} 1, & |t| < 1/2 \\ 1/2, & |t| = 1/2 \\ 0, & |t| > 1/2 \end{cases}$.

- **Triangular pulse:** $\operatorname{tri}(t) \stackrel{\text{def}}{=} \operatorname{rect}(t) * \operatorname{rect}(t) = \begin{cases} 1+t, & -1 < t \leq 0 \\ 1-t, & 0 < t < 1 \\ 0, & |t| > 1 \end{cases}$.

- **"Sin(x)/x" function:** $\operatorname{si}(t) \stackrel{\text{def}}{=} \begin{cases} 1, & t = 0 \\ \frac{\sin(t)}{t}, & t \neq 0 \end{cases}$.

Energy, Power, and Autocorrelation of a Signal

The *energy* or *power* of signals are compared at equal impedance level R. Therefore, it is usual to specify energy and power by normalized values with respect to the reference impedance $1\,\Omega$. Throughout the book, only such normalized values are used. Consequently, the *instantaneous power* $p(t)$ of a signal $x(t)$ is defined as

$$p(t) = x(t) \cdot x^*(t) = |x(t)|^2 . \qquad (A.11)$$

The *energy* E_x of the signal $x(t)$ is defined as

$$E_x \stackrel{\text{def}}{=} \int_{-\infty}^{\infty} |x(t)|^2 \, dt \qquad (A.12)$$

if the integral is finite. In this case, the signal is called *energy-limited*. Typical examples are basic pulse shapes.

If the energy tends to infinity (i.e., the integral does not exist), a signal $x(t)$ can often be characterized by its *average power*

$$S_x \stackrel{\text{def}}{=} \lim_{\tau \to \infty} \frac{1}{2\tau} \int_{-\tau}^{\tau} |x(t)|^2 \, dt . \qquad (A.13)$$

If this integral is finite, the signal is called *power-limited*. Notice that the average power of energy-limited signals is zero by definition. Some models of signals may even not be power-limited (e.g., white noise). Of course, such signals cannot exist in reality but are often useful to obtain a simple exposition.

According to *Parseval's theorem*, the energy can also be calculated in the frequency domain

$$E_x = \int_{-\infty}^{\infty} |x(t)|^2 \, dt = \int_{-\infty}^{\infty} |X(f)|^2 \, df \, . \tag{A.14}$$

Finally, the *autocorrelation*[2] of an (energy-limited) signal $x(t)$ is defined as

$$\varphi_{xx}(\tau) \stackrel{\text{def}}{=} x(\tau) * x^*(-\tau) = \int_{-\infty}^{\infty} x(t + \tau) x^*(t) \, dt$$
$$\circ\!\!-\!\!\bullet$$
$$|X(f)|^2 = X(f) \cdot X^*(f) \, . \tag{A.15}$$

Note that $\varphi_{xx}(\tau = 0) = E_x$.

System

A *system* $\mathcal{S}$ performs a transformation of the input signal, denoted as $x(t)$, to the output signal, denoted as $y(t)$, that is

$$y(t) = \mathcal{S}\{x(t)\} \, , \tag{A.16}$$

and is visualized by the block diagram of Figure A.1.

The most important special class of systems are *linear time-invariant (LTI)* systems. For these, we demand the following.

- *Linearity*:
  for each $y_1(t) = \mathcal{S}\{x_1(t)\}$ and $y_2(t) = \mathcal{S}\{x_2(t)\}$ and $c_1, c_2 \in \mathbb{C}$, we have
  $$\mathcal{S}\{c_1 x_1(t) + c_2 x_2(t)\} = c_1 \mathcal{S}\{x_1(t)\} + c_2 \mathcal{S}\{x_2(t)\} \, .$$

- *Time Invariance*:
  for each $y(t) = \mathcal{S}\{x(t)\}$ and $t_0 \in \mathbb{R}$, we have
  $$y(t - t_0) = \mathcal{S}\{x(t - t_0)\} \, .$$

Each LTI system can be fully characterized by its *impulse response*, that is, the output signal corresponding to $x(t) = \delta(t)$:

$$h(t) \stackrel{\text{def}}{=} \mathcal{S}\{\delta(t)\} \, . \tag{A.17}$$

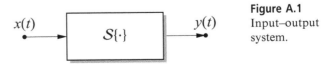

Figure A.1
Input–output system.

---

[2] Note the difference between the *autocorrelation* of an energy-limited signal and the *autocorrelation function* (acf) of a (wide-sense) stationary stochastic process.

The input–output relation is given by

$$y(t) = x(t) * h(t) = h(t) * x(t) , \qquad (A.18)$$

where "$*$" denotes the convolution

$$y(t) = \int_{-\infty}^{\infty} h(\tau)x(t-\tau)\,\mathrm{d}\tau = \int_{-\infty}^{\infty} h(t-\tau)x(\tau)\,\mathrm{d}\tau \qquad (A.19)$$

and the second form follows from the commutativity of the convolution.

We may characterize LTI systems as follows.

- *Causal System*:
  if $h(t) = 0, t < 0$, the system is causal.
- *Dispersive System*:
  if $h(t) \neq c\delta(t - t_0)$ for some $c \in \mathbb{C}, t_0 \in \mathbb{R}$, the system is dispersive and may cause linear (dispersive) distortions of a signal.

Signal-to-Distortion Ratio (SDR)

Given any system S with input $x(t)$ and output $y(t)$, the distortion error is given by $\epsilon(t) \stackrel{\text{def}}{=} y(t) - c\,x(t - t_0)$, where c is a scaling factor and t_0 a time shift. The power of the distortion is defined as the minimum of the average power S_ϵ of $\epsilon(t)$ over the parameters c and t_0 (best-matching via scaling and time shift):

$$D \stackrel{\text{def}}{=} \min_{c,\,t_0} S_\epsilon , \qquad (A.20)$$

with $c_{\min}$ and $t_{0,\min}$ being the arguments leading to this minimum. Thus, $c_{\min}\,x(t - t_{0,\min})$ represents the undistorted part in $y(t)$ and the *signal-to-distortion ratio* is given by

$$\mathrm{SDR} \stackrel{\text{def}}{=} \frac{|c_{\min}|^2 S_x}{D} . \qquad (A.21)$$

Transfer Function of LTI Systems

Instead of characterizing systems in the time domain by their impulse response, they can be characterized in the frequency domain using the *transfer function*. The transfer function of an LTI system is given by

$$H(f) \stackrel{\text{def}}{=} \mathscr{F}\{h(t)\} = \int_{-\infty}^{\infty} h(t)\,\mathrm{e}^{-\mathrm{j}2\pi f t}\,\mathrm{d}t . \qquad (A.22)$$

The input–output relation of the system in the frequency domain reads

$$\begin{array}{c} y(t) = x(t) * h(t) \\ \updownarrow \\ Y(f) = X(f) \cdot H(f) , \end{array} \qquad (A.23)$$

that is, a convolution in the time domain corresponds to a multiplication in the frequency domain.

Conversely, a multiplication in the time domain corresponds to a convolution in the frequency domain:

$$y(t) = x(t) \cdot z(t)$$
$$\circ\!\!-\!\!\bullet \qquad (A.24)$$
$$Y(f) = X(f) * Z(f) \,.$$

Hilbert Transform

The *Hilbert transform* of a signal $x(t)$ is given by

$$\mathcal{H}\{x(t)\} \stackrel{\text{def}}{=} x(t) * \frac{1}{\pi t} = \frac{1}{\pi} \int_{-\infty}^{\infty} \frac{x(\tau)}{t - \tau} \, d\tau \,. \qquad (A.25)$$

A.1.3 Discrete-Time Signals and Systems

Signal and Spectrum

For *discrete-time signals* (sequences), the discrete-time variable is k. The signal can be real- or complex-valued:

$$x[k] \in \mathbb{C}, \quad k \in \mathbb{Z} \,. \qquad (A.26)$$

As usual, square brackets "[·]" are used to characterize discrete-time signals, in contrast to round brackets "(·)" for continuous-time signals and curly brackets "{·}" for sets.

The corresponding *z-transform* is denoted by the corresponding uppercase letter and is defined as

$$X(z) \stackrel{\text{def}}{=} \mathcal{Z}\{x[k]\} = \sum_{k=-\infty}^{\infty} x[k] \, z^{-k} \,, \qquad (A.27)$$

or the corresponding *discrete-time Fourier transform* ($\mathcal{F}_*$ transform) or *spectrum* (assuming a sampling period T) as

$$X\left(e^{j2\pi fT}\right) = \mathcal{F}_*\{x[k]\} \stackrel{\text{def}}{=} \mathcal{Z}\{x[k]\}\Big|_{z=e^{j2\pi fT}} = \sum_{k=-\infty}^{\infty} x[k] \, e^{-j2\pi fTk} \,. \qquad (A.28)$$

The correspondence between the time-domain signal $x[k]$ and its z-transform $X(z)$ is written in short as

$$\text{time domain} \qquad x[k] \circ\!\!-\!\!\bullet X(z) \qquad \text{frequency domain.}$$

System

Let

$$\delta[k] \stackrel{\text{def}}{=} \begin{cases} 1, & k = 0 \\ 0, & k \in \mathbb{Z} \setminus \{0\} \end{cases} \qquad (A.29)$$

be the discrete-time unit pulse (or discrete-time Delta function). For a discrete-time LTI system with *impulse response*, that is, the output signal corresponding to $x[k] = \delta[k]$:

$$h[k] \stackrel{\text{def}}{=} \mathcal{S}\{\delta[k]\}, \tag{A.30}$$

the input–output relation is given by

$$y[k] = x[k] * h[k] = h[k] * x[k] \tag{A.31}$$

where "$*$" denotes convolution:

$$y[\kappa] = \sum_{\kappa=-\infty}^{\infty} h[\kappa]x[k-\kappa] = \sum_{\kappa=-\infty}^{\infty} h[k-\kappa]x[\kappa]. \tag{A.32}$$

The *transfer function* of a discrete-time LTI system is given by

$$H(z) \stackrel{\text{def}}{=} \mathcal{Z}\{h[k]\} = \sum_{k=-\infty}^{\infty} h[k] z^{-k}. \tag{A.33}$$

Sampling

Let a continuous-time signal $x(t)$ with spectrum $X(f)$ be given. By *sampling* (sampling period T), a discrete-time signal is generated, that is

$$x[k] = x(t = kT). \tag{A.34}$$

Then, the spectrum of the sequence is given by

$$X\left(e^{j2\pi fT}\right) = \frac{1}{T} \sum_{n=-\infty}^{\infty} X(f - n/T), \tag{A.35}$$

that is, sampling generates replicas at all integer multiples of the sampling frequency $1/T$.

A.2 Some Basic Correspondences and Properties

Table A.1 gives some basic correspondences of the Fourier transform and Table A.2 gives some basic correspondences of the z-transform.

Given the Fourier pair $x(t) \circ\!\!-\!\!\bullet X(f)$, the following correspondences hold:

$$x^*(t) \circ\!\!-\!\!\bullet X^*(-f),$$

$$x^*(-t) \circ\!\!-\!\!\bullet X^*(f),$$

$$x^*(T_d - t) \circ\!\!-\!\!\bullet X^*(f) e^{-j2\pi fT_d}.$$

Table A.1 Basic correspondences of the Fourier transform.

$x(t)$	○—●	$X(f)$
1		$\delta(f)$
$\delta(t)$		1
$e^{j2\pi f_0 t}$		$\delta(f - f_0)$
$\sin(2\pi f_0 t)$		$\frac{1}{2j}(\delta(f - f_0) - \delta(f + f_0))$
$\cos(2\pi f_0 t)$		$\frac{1}{2}(\delta(f - f_0) + \delta(f + f_0))$
$\mathrm{rect}(t/T)$		$\lvert T \rvert\, \mathrm{si}(\pi f T)$
$\mathrm{si}(\pi t/T)$		$\lvert T \rvert\, \mathrm{rect}(fT)$
$\varepsilon(t)$		$\frac{1}{2}\delta(f) + \frac{1}{j2\pi f}$
$\mathrm{sgn}(t)$		$\frac{1}{j\pi f}$
$\frac{1}{t}$		$-j\pi\, \mathrm{sgn}(f)$

Table A.2 Basic correspondences of the z-transform.

$x[k]$	○—●	$X(z)$
$\delta[k]$		$1, \quad z \in \mathbb{C}$
$\varepsilon[k]$		$\dfrac{z}{z-1}, \quad \lvert z \rvert > 1$
$z_\infty^k\, \varepsilon[k]$		$\dfrac{z}{z-z_\infty}, \quad \lvert z \rvert > \lvert z_\infty \rvert$

Given the z-transform pair $x[k]$ ○—● $X(z)$, the following correspondences hold:

$$x^*[k] \;\circ\!\!-\!\!\bullet\; X^*(z^*) ,$$

$$x^*[-k] \;\circ\!\!-\!\!\bullet\; X^*(1/z^*) ,$$

$$x^*[K_\mathrm{d} - k] \;\circ\!\!-\!\!\bullet\; X^*(1/z^*)\, z^{-K_\mathrm{d}} .$$

A.3 Causality, Minimum-Phase Systems, and Stability

In some derivations, we require (basic) properties of causal signals and causal, minimum-phase systems. For completeness, the necessary relations are subsequently collected.

A.3.1 Causal Signals

By analogy to *causal linear time-invariant (LTI) systems*, where the impulse response is zero on the negative part of the time axis, a signal $h(t)$ is said to be *causal* if

$$h(t) = 0, \qquad t < 0. \tag{A.36}$$

Let the corresponding *Fourier transform* be denoted as

$$H(f) = \mathscr{F}\{h(t)\} = \int_0^\infty h(t)\, e^{-j2\pi ft}\, dt\ . \tag{A.37}$$

If we decompose the spectrum according to $H(f) = H_I(f) + jH_Q(f)$ into in-phase (real) and quadrature (imaginary) parts,[3] the following relations hold:

$$H_I(f) = H_I(\infty) + \mathscr{H}\{H_Q(f)\}\ , \tag{A.38}$$
$$H_Q(f) = \qquad -\mathscr{H}\{H_I(f)\}\ , \tag{A.39}$$

where $\mathscr{H}\{x(t)\}$ is again the Hilbert transform of $x(t)$ as defined in (A.25).

Box A.1 Spectra of Causal Signals

The relations (A.38), (A.39) can be shown as follows. Let the step function be denoted by $\varepsilon(t)$. For causal signals it holds that (as long as $x(t)$ has no Dirac delta function $\delta(t)$ at $t = 0$)

$$x(t) = x(t) \cdot \varepsilon(t)\ . \tag{A.40}$$

Applying the Fourier transform on both sides of the equation leads to

$$X(f) = X(f) * \left(\frac{1}{2}\delta(f) + \frac{1}{j2\pi f}\right)$$
$$= \frac{1}{2}X(f) + X(f) * \frac{1}{j2\pi f}\ ,$$

which gives

$$X(f) = X(f) * \frac{1}{j\pi f}\ . \tag{A.41}$$

For the real (in-phase) part $X_I(f) \stackrel{\text{def}}{=} \text{Re}\{X(f)\}$ and the imaginary (quadrature) part $X_Q(f) \stackrel{\text{def}}{=} \text{Im}\{X(f)\}$ of the spectrum, we consequently have

$$X_I(f) + jX_Q(f) = (X_I(f) + jX_Q(f)) * \frac{1}{j\pi f}$$
$$= X_Q(f) * \frac{1}{\pi f} - jX_I(f) * \frac{1}{\pi f}\ . \tag{A.42}$$

A comparison of the real and imaginary parts on the left- and right-hand sides of this equation immediately reveals (A.38) and (A.39).

Hence, the real and imaginary parts of the spectrum of a causal signal are coupled via the Hilbert transform. If the signal has a Dirac component at $t = 0$, which gives a constant offset $H(\infty)$ in the spectrum, this constant has to be taken into account in the real part.

---

[3] As for complex-valued signals, in a slight abuse of the original meaning, we use "in-phase" and "quadrature" for the real and imaginary parts of the respective spectra, too.

The Hilbert transform of a real-valued time-domain signal $x(t)$ can be written in a different, equivalent form, when taking into account that for real-valued signals the component $X_I(f)$ is an even function and $X_Q(f)$ an odd function in f. We have

$$X_I(f) = \frac{1}{\pi} \int_{-\infty}^{\infty} \frac{X_Q(\nu)}{f-\nu} d\nu = \frac{2}{\pi} \int_0^{\infty} \frac{1}{2}\left(\frac{X_Q(\nu)}{f-\nu} + \frac{X_Q(-\nu)}{f+\nu}\right) d\nu$$

$$= \frac{2}{\pi} \int_0^{\infty} \frac{\nu X_Q(\nu)}{f^2 - \nu^2} d\nu, \qquad (A.43)$$

$$X_Q(f) = -\frac{1}{\pi} \int_{-\infty}^{\infty} \frac{X_I(\nu)}{f-\nu} d\nu = -\frac{2}{\pi} \int_0^{\infty} \frac{1}{2}\left(\frac{X_I(\nu)}{f-\nu} + \frac{X_I(-\nu)}{f+\nu}\right) d\nu$$

$$= -\frac{2f}{\pi} \int_0^{\infty} \frac{X_I(\nu)}{f^2 - \nu^2} d\nu. \qquad (A.44)$$

A.3.2 Minimum-Phase Systems

Continuous-Time Systems

Let a *causal, stable continuous-time LTI system* with impulse response $h(t)$ be given. We denote the corresponding *Laplace transform*, the *transfer function*, by

$$\mathsf{H}(s) = \mathscr{L}\{h(t)\} = \int_0^{\infty} h(t) e^{-st} dt. \qquad (A.45)$$

Note that we distinguish clearly between *Fourier* and *Laplace transforms*. For the former, roman-italic letters are used, whereas for the latter, a sans-serif font is employed. Assuming existence, the well-known relation

$$\mathsf{H}(s = j2\pi f) = H(f) \qquad (A.46)$$

holds.

Moreover, we assume that $\mathsf{H}(s) = \mathsf{N}(s)/\mathsf{D}(s)$ is a rational function with numerator polynomial $\mathsf{N}(s)$ and denominator polynomial $\mathsf{D}(s)$. Owing to the assumed stability, all poles $s_{\infty,j}$ (zeros of $\mathsf{D}(s)$) are located in the open left half s-plane, that is, $\mathrm{Re}\{s_{\infty,j}\} < 0, \forall j$.

For *minimum-phase systems* $\mathsf{H}_M(s)$, the zeros $s_{0,i}$ (zeros of $\mathsf{N}(s)$) are also only located in the open left half s-plane, that is, $\mathrm{Re}\{s_{0,i}\} < 0, \forall i$.

As in the case of causality, we denote the impulse response of a *minimum-phase system* as a *minimum-phase signal*.

Consider the following correspondence of the Fourier transform:

$$\varphi(t) \stackrel{\text{def}}{=} h(t) * h^*(-t) \quad \circ\!\!-\!\!\bullet \quad H(f) \cdot H^*(f) \stackrel{\text{def}}{=} |H(f)|^2 \stackrel{\text{def}}{=} \Phi(f). \qquad (A.47)$$

Since $\mathscr{L}\{h^*(-t)\} = H^*(-s^*)$, the above relation is written in the Laplace domain as

$$h(t) * h^*(-t) \quad \circ\!\!-\!\!\bullet \quad H(s) \cdot H^*(-s^*) \stackrel{\text{def}}{=} \Phi(s) \,. \tag{A.48}$$

Since replacing s by $-s^*$ corresponds to mirroring at the imaginary axis,

$$\Phi(s) = H(s) \cdot H^*(-s^*) = \frac{N(s) \cdot N^*(-s^*)}{D(s) \cdot D^*(-s^*)} \tag{A.49}$$

has the same poles and zeros as $H(s)$ and all poles and zeros mirrored at the imaginary axis.

Let $\bar{H}(s)$ be the transfer function of a system where, compared to $H(s)$, some (say, l) zeros are mirrored at the imaginary axis, that is, ($G(s)$ contains the remaining zeros and all poles):

$$H(s) = G(s) \left(s - s_{0,i_1} \right) \cdots \left(s - s_{0,i_l} \right) , \tag{A.50}$$

$$\bar{H}(s) = G(s) \left(s + s^*_{0,i_1} \right) \cdots \left(s + s^*_{0,i_l} \right) . \tag{A.51}$$

Then, we have

$$H(s) \cdot H^*(-s^*) = \Phi(s) = \bar{H}(s) \cdot \bar{H}^*(-s^*) , \tag{A.52}$$

or

$$|H(f)|^2 = \Phi(\mathrm{j} 2\pi f) = |\bar{H}(f)|^2 , \tag{A.53}$$

which means that mirroring zeros[4] does not affect $\Phi(s)$ and, thus, $\Phi(f) = \Phi(\mathrm{j} 2\pi f)$. Given these two systems, $H(s)$ and $\bar{H}(s)$, the same (squared) magnitude spectrum is present. Consequently, given $\Phi(s)$ (or a magnitude spectrum), the transfer function $H(s)$ (and, thus, $H(f)$) can only be recovered if we agree on the position of the zeros – in particular, if we demand minimum-phase systems $H_\mathrm{M}(s)$, where all zeros are located in the left half plane.

Subsequently, we need the notion of the *cepstrum*. The cepstrum corresponding to $H(s)$ is given by

$$c(t) \stackrel{\text{def}}{=} \mathscr{F}^{-1}\{\log(H(f))\}, \tag{A.54}$$

or

$$c(t) \quad \circ\!\!-\!\!\bullet \quad \log(H(f)) \,. \tag{A.55}$$

Subsequently, three important properties of minimum-phase systems

$$h_\mathrm{M}(t) \quad \circ\!\!-\!\!\bullet \quad H_\mathrm{M}(f) \tag{A.56}$$

are derived.

- *Causality.* The cepstrum of a minimum-phase system is causal, that is

$$c(t) = \mathscr{F}^{-1}\{\log(H_\mathrm{M}(f))\} = 0 \,, \quad t < 0 \,. \tag{A.57}$$

---

[4] Poles are not allowed to be mirrored since we want to have stable systems.

- *Relation between Magnitude and Phase.* For minimum-phase systems, the magnitude response $|H_M(f)|$ and phase response $\phi_M(f)$, that is, $H_M(f) = |H_M(f)|e^{j\phi_M(f)}$, are related by

$$\phi_M(f) = -\mathcal{H}\{\log(|H_M(f)|)\} \ . \quad (A.58)$$

- *Energy Concentration.* For a minimum-phase system, the energy is concentrated at earlier time positions than for any other system $h(t) \circ\!\!-\!\!\bullet H(f)$ with the same magnitude spectrum ($|H(f)| = |H_M(f)|$), that is

$$\int_0^t |h_M(\tau)|^2 \, d\tau \geq \int_0^t |h(\tau)|^2 \, d\tau \ . \quad (A.59)$$

Derivations

Causal Cepstrum Let the rational transfer function of a causal, stable, minimum-phase system be given as (c is a scaling constant)

$$H(s) = c \cdot \frac{\prod_i \left(1 - s_{0,i} s^{-1}\right)}{\prod_j \left(1 - s_{\infty,j} s^{-1}\right)}, \quad \text{with} \quad \begin{cases} \mathrm{Re}\{s_{0,i}\} < 0, & \text{min. phase} \\ \mathrm{Re}\{s_{\infty,j}\} < 0, & \text{stable} \end{cases} . \quad (A.60)$$

The following identity can be proven by taking the following relation into account:

$$\log(1 - x/s) = \int_0^\infty \frac{1 - e^{xt}}{t} e^{-st} \, dt \ . \quad (A.61)$$

Taking the log of the transfer function gives

$$\begin{aligned}
\log(H(s)) &= \log\left(c \cdot \frac{\prod_i \left(1 - s_{0,i} s^{-1}\right)}{\prod_j \left(1 - s_{\infty,j} s^{-1}\right)}\right) \\
&= \log(c) + \sum_i \log\left(1 - s_{0,i} s^{-1}\right) - \sum_j \log\left(1 - s_{\infty,j} s^{-1}\right) \\
&= \log(c) + \sum_i \int_0^\infty \frac{1 - e^{s_{0,i} t}}{t} e^{-st} \, dt - \sum_j \int_0^\infty \frac{1 - e^{s_{\infty,j} t}}{t} e^{-st} \, dt \\
&= \log(c) + \int_0^\infty \left(\sum_i \frac{1 - e^{s_{0,i} t}}{t} - \sum_j \frac{1 - e^{s_{\infty,j} t}}{t}\right) e^{-st} \, dt \\
&= \int_0^\infty \left(\log(c)\delta(t) + \sum_i \frac{1 - e^{s_{0,i} t}}{t} - \sum_j \frac{1 - e^{s_{\infty,j} t}}{t}\right) e^{-st} \, dt \ . \quad (A.62)
\end{aligned}$$

Since the integral starts at $t = 0$ and the integrand is the time function corresponding to $\log(H(s))$, this shows that the cepstrum corresponding to a minimum-phase system is causal.

Relation between Magnitude and Phase Let the transfer function of a minimum-phase system be $H_M(f) = |H_M(f)|e^{j\phi_M(f)}$. Then

$$C(f) \stackrel{\text{def}}{=} \log(H_M(f)) = \log\left(|H_M(f)|e^{j\phi_M(f)}\right)$$
$$= \log\left(|H_M(f)|\right) + j\phi_M(f) \tag{A.63}$$

corresponds to a causal system. According to (A.38) and (A.39), the real and imaginary parts are related for causal signals, thus, *log-amplitude* and *phase* are related for minimum-phase signals.

If $H_M(f)$ has no spectral zeros (including $f = \infty$), we have for the phase

$$\phi_M(f) = -\mathcal{H}\{\log(|H_M(f)|)\} = -\tfrac{1}{2}\mathcal{H}\{\log(|H_M(f)|^2)\}, \tag{A.64}$$

and for the negative attenuation $A(f) \stackrel{\text{def}}{=} \log(|H_M(f)|)$:

$$A(f) = A(\infty) + \mathcal{H}\{\phi_M(f)\}. \tag{A.65}$$

If $H_M(f)$ has spectral zeros (i.e., $H_M(f_0) = 0$ for some f_0), (A.43) and (A.44) may be used, leading to the following forms:

$$\phi_M(f) = -\frac{f}{\pi}\int\frac{\log(|H_M(\nu)|^2)}{\nu^2 - f^2}\,d\nu, \tag{A.66}$$

$$A(f) = A(0) + \frac{f^2}{\pi}\int\frac{\phi_M(\nu)}{\nu(\nu^2 - f^2)}\,d\nu. \tag{A.67}$$

Energy Concentration Let $H_M(s)$ be a causal, minimum-phase system and $H(s)$ a causal system with the same transfer function $|H(f)| = |H_M(f)|$. The corresponding impulse responses are denoted by $h_M(t)$ and $h(t)$, respectively. We define the energy up to time instant t by

$$w_M(t) = \int_0^t |h_M(\tau)|^2\,d\tau \quad \text{and} \quad w(t) = \int_0^t |h(\tau)|^2\,d\tau. \tag{A.68}$$

Owing to Parseval's theorem, the following holds:

$$w_M(\infty) = \int_0^\infty |h_M(\tau)|^2\,d\tau = \int_{-\infty}^\infty |H_M(f)|^2\,df$$
$$= \int_{-\infty}^\infty |H(f)|^2\,df = \int_0^\infty |h(\tau)|^2\,d\tau = w(\infty), \tag{A.69}$$

that is, both impulse responses have the same total energy.

We now study $\Delta w(t) \stackrel{\text{def}}{=} w_M(t) - w(t)$. To that end, we assume that $H(s)$ has a single zero at the right half of the s-plane. The general case is obtained by repeatedly applying the subsequent arguments. Then, with $\text{Re}\{s_{0,x}\} < 0$, the transfer functions can be written as

A.3 Causality, Minimum-Phase Systems, and Stability

$$H_M(s) = G(s)\left(s - s_{0,x}\right), \tag{A.70}$$
$$H(s) = G(s)\left(s + s_{0,x}^*\right). \tag{A.71}$$

Moreover, assume that $g(0^+) = \lim_{s\to\infty} s \cdot G(s) = 0$ with $g(t) \circ\!\!-\!\!\bullet\, G(s)$. Owing to causality we also have $g(0^-) = 0$. Then, the following correspondences hold $\left(g'(t) = \frac{d}{dt}g(t)\right)$:

$$H_M(s) = s\,G(s) - s_{0,x}G(s) \quad\bullet\!\!-\!\!\circ\quad h_M(t) = g'(t) - s_{0,x}\,g(t), \tag{A.72}$$
$$H(s) = s\,G(s) + s_{0,x}^*G(s) \quad\bullet\!\!-\!\!\circ\quad h(t) = g'(t) + s_{0,x}^*\,g(t). \tag{A.73}$$

This gives

$$\Delta w(t) = w_M(t) - w(t)$$
$$= \int_0^t |h_M(\tau)|^2\, d\tau - \int_0^t |h(\tau)|^2\, d\tau$$
$$= \int_0^t \left|g'(\tau) - s_{0,x}\,g(\tau)\right|^2 d\tau - \int_0^t \left|g'(\tau) + s_{0,x}^*\,g(\tau)\right|^2 d\tau$$
$$= \int_0^t \Big(g'(\tau)g'^*(\tau) - g'(\tau)s_{0,x}^*\,g^*(\tau) - s_{0,x}\,g(\tau)g'^*(\tau) + s_{0,x}\,g(\tau)s_{0,x}^*\,g^*(\tau)$$
$$\quad -g'(\tau)g'^*(\tau) - g'(\tau)s_{0,x}\,g^*(\tau) - s_{0,x}^*\,g(\tau)g'^*(\tau) - s_{0,x}^*\,g(\tau)s_{0,x}\,g^*(\tau)\Big)\,d\tau$$
$$= -\int_0^t \Big(g(\tau)\,g'^*(\tau)(s_{0,x} + s_{0,x}^*) + g'(\tau)\,g^*(\tau)(s_{0,x} + s_{0,x}^*)\Big)\,d\tau$$
$$= -2\mathrm{Re}\{s_{0,x}\}\int_0^t \Big(g(\tau)\,g'^*(\tau) + g'(\tau)\,g^*(\tau)\Big)\,d\tau$$

with $\dfrac{d}{dt}|g(t)|^2 = \dfrac{d}{dt}\Big(g(t)\cdot g^*(t)\Big) = g'(t)\,g^*(t) + g(t)\,g'^*(t)$

$$= -2\mathrm{Re}\{s_{0,x}\}\Big[|g(\tau)|^2\Big]_0^t$$

and using $g(0) = 0$

$$= -2\mathrm{Re}\{s_{0,x}\}|g(t)|^2. \tag{A.74}$$

Since $\mathrm{Re}\{s_{0,x}\} < 0$, the difference energy is non-negative:

$$\Delta w(t) = -2\mathrm{Re}\{s_{0,x}\}|g(t)|^2 \geq 0, \quad \forall t, \tag{A.75}$$

and we have

$$\int_0^t |h_M(\tau)|^2\, d\tau \geq \int_0^t |h(\tau)|^2\, d\tau, \quad \forall t, \tag{A.76}$$

which means that the (same total) energy is concentrated at earlier time positions.

Discrete-Time Systems

The same statements as above also apply to *causal, stable, discrete-time LTI systems* with impulse response $h[k]$. Here, the z-transform is given by

$$H(z) = \mathcal{Z}\{h[k]\} = \sum_{k=0}^{\infty} h[k]\, z^{-k} \,. \tag{A.77}$$

Moreover, we assume that $H(z) = N(z)/D(z)$ is a rational function with numerator polynomial $N(z)$ and denominator polynomial $D(z)$. Owing to the assumed stability, all poles $z_{\infty,j}$ (zeros of $D(z)$) are located within the unit circle in the z-plane, that is, $|z_{\infty,j}| < 1$, $\forall j$.

For *minimum-phase systems* $H_M(z)$, the zeros $z_{0,i}$ (zeros of $N(z)$) are also only located within the unit circle in the z-plane, that is, $|z_{0,i}| < 1$, $\forall i$.

Consider the following correspondence of the discrete-time Fourier transform:

$$\varphi[k] \stackrel{\text{def}}{=} h[k] * h^*[-k] \quad \circ\!\!-\!\!\bullet \quad H\left(e^{j2\pi fT}\right) \cdot H^*\left(e^{j2\pi fT}\right) = \left|H\left(e^{j2\pi fT}\right)\right|^2 \stackrel{\text{def}}{=} \Phi\left(e^{j2\pi fT}\right).$$

Since $\mathcal{Z}\{h^*[-k]\} = H^*\left(z^{-*}\right)$ (note that $z^{-*} \stackrel{\text{def}}{=} \left(z^{-1}\right)^* = (z^*)^{-1}$), the above relation is written more generally as

$$h[k] * h^*[-k] \quad \circ\!\!-\!\!\bullet \quad H(z) \cdot H^*(z^{-*}) \stackrel{\text{def}}{=} \Phi(z) \,. \tag{A.78}$$

Since replacing z by z^{-*} corresponds to mirroring at the unit circle,

$$\Phi(z) = H(z) \cdot H^*(z^{-*}) = \frac{N(z) \cdot N^*(z^{-*})}{D(z) \cdot D^*(z^{-*})} \tag{A.79}$$

has the same poles and zeros as $H(z)$ and all poles and zeros mirrored at the unit circle. As in the continuous-time case, given $\Phi(z)$ (or a magnitude spectrum), the transfer function $H(z)$ can only be recovered if we agree on the position of the zeros, in particular, if we demand a minimum-phase system $H_M(z)$.

Causal Cepstrum Let the rational transfer function of a causal, stable, minimum-phase system be given as (c is a scaling constant)

$$H(z) = c \cdot \frac{\prod_i \left(1 - z_{0,i} z^{-1}\right)}{\prod_j \left(1 - z_{\infty,j} z^{-1}\right)}, \quad \text{with} \quad \begin{cases} |z_{0,i}| < 1, & \text{min. phase} \\ |z_{\infty,j}| < 1, & \text{stable} \end{cases}. \tag{A.80}$$

The following identity holds

$$\log\left(1 - xz^{-1}\right) = -\sum_{k=1}^{\infty} \frac{x^k}{k} z^{-k}, \quad |x| < 1 \,. \tag{A.81}$$

A.3 Causality, Minimum-Phase Systems, and Stability

Taking the logarithm of the transfer function gives

$$\log(H(z)) = \log\left(c \cdot \frac{\prod_i \left(1 - z_{0,i} z^{-1}\right)}{\prod_j \left(1 - z_{\infty,j} z^{-1}\right)}\right)$$

$$= \log(c) + \sum_i \log\left(1 - z_{0,i} z^{-1}\right) - \sum_j \log\left(1 - z_{\infty,j} z^{-1}\right)$$

$$= \log(c) + \sum_i \sum_{k=1}^{\infty} \frac{z_{0,i}^k}{k} z^{-k} - \sum_j \sum_{k=1}^{\infty} \frac{z_{\infty,j}^k}{k} z^{-k}$$

$$= \log(c) z^{-0} + \sum_{k=1}^{\infty} \left(\sum_i \frac{z_{0,i}^k}{k} - \sum_j \frac{z_{\infty,j}^k}{k}\right) z^{-k} . \quad (A.82)$$

Since the sum starts at $k = 1$ and the summand is the time function corresponding to $\log(H(z))$, this shows that the cepstrum corresponding to a minimum-phase system is causal.

Spectral Factorization The task of spectral factorization is, given $\Phi(z)$, to find a causal, minimum-phase system $B(z) = H_M(z)$ such that (A.78) is fulfilled. Taking the logarithm we have

$$\log(\Phi(z)) = \log(B(z)) + \log(B^*(z^{-*})) , \quad (A.83)$$

with

$$\log(\Phi(z)) = \sum_{k=-\infty}^{\infty} \hat{\varphi}_{xx}[k] z^{-k} . \quad (A.84)$$

Since $B(z)$ is minimum phase, we can unambiguously divide the terms according to

$$\log(B(z)) = \log\left(b[0] + b[1] z^{-1} + b[2] z^{-2} + \cdots\right)$$

$$= \frac{1}{2}\hat{\varphi}_{xx}[0] + \hat{\varphi}_{xx}[1] z^{-1} + \hat{\varphi}_{xx}[2] z^{-2} + \cdots , \quad (A.85)$$

$$\log(B^*(z^{-*})) = \log\left(b^*[0] + b^*[1] z^{+1} + b^*[2] z^{+2} + \cdots\right)$$

$$= \frac{1}{2}\hat{\varphi}_{xx}[0] + \hat{\varphi}_{xx}[-1] z^{+1} + \hat{\varphi}_{xx}[-2] z^{+2} + \cdots . \quad (A.86)$$

For $z \to \infty$, we have

$$\log(B(z))\Big|_{z \to \infty} = \log(b[0]) = \frac{1}{2}\hat{\varphi}[0] , \quad \text{or} \quad b[0] = \exp\left\{\frac{1}{2}\hat{\varphi}[0]\right\} . \quad (A.87)$$

Finally, taking into account that

$$\hat{\varphi}[0] = T \int_{-\frac{1}{2T}}^{\frac{1}{2T}} \log\left(\Phi_{xx}\left(e^{j2\pi fT}\right)\right) df , \quad (A.88)$$

we arrive at

$$b[0] = \exp\left\{\frac{1}{2} T \int_{-\frac{1}{2T}}^{\frac{1}{2T}} \log\left(\Phi\left(e^{j2\pi fT}\right)\right) df\right\} . \quad (A.89)$$

Practical Algorithm If $\Phi\left(e^{j2\pi fT}\right)$ is of finite order O, that is, $\varphi[k] = 0$, $|k| > O$, an easy-to-implement algorithm for performing the spectral factorization is available. Given $\varphi[k]$, $k = -O, \ldots, O$, the matrix

$$\Phi = [\varphi_{i,j}]_{\substack{i=1,\ldots,n \\ j=1,\ldots,n}} \stackrel{\text{def}}{=} \begin{bmatrix} \varphi[0] & \varphi[1] & \cdots & \varphi[n-1] \\ \varphi[-1] & \varphi[0] & & \vdots \\ \vdots & & \ddots & \\ \varphi[-n+1] & \cdots & & \varphi[0] \end{bmatrix} \tag{A.90}$$

with $n \geq O + 1$ is formed. Note that Φ has *Toeplitz structure*, is *Hermitian* (i.e., $\Phi^H = \Phi$), and is positive definite. Hence, a *Cholesky factorization* can be performed; the matrix is decomposed either according to

$$\Phi \stackrel{\text{def}}{=} BB^H, \tag{A.91}$$

where B is *upper triangular*, or according to

$$\Phi \stackrel{\text{def}}{=} GG^H, \tag{A.92}$$

where G is *lower triangular*. Note that B and G are related by $B = JG^*J$, where J is the anti-diagonal identity matrix (reversal matrix, backward identity), in other words the one matrix is obtained from the other by complex conjugation and reversing the order of the rows and of the columns.

Let the matrix B in detail be

$$B = \begin{bmatrix} b_{1,1} & b_{1,2} & & b_{1,n} \\ 0 & b_{2,2} & & \\ \vdots & & \ddots & b_{n-1,n} \\ 0 & & & b_{n,n} \end{bmatrix}. \tag{A.93}$$

The product then elaborates as

$$\Phi = BB^H = \begin{bmatrix} b_{1,1} & b_{1,2} & \cdots & b_{1,n} \\ 0 & b_{2,2} & & \vdots \\ \vdots & & \ddots & b_{n-1,n} \\ 0 & \cdots & 0 & b_{n,n} \end{bmatrix} \cdot \begin{bmatrix} b_{1,1}^* & 0 & \cdots & 0 \\ b_{1,2}^* & b_{2,2}^* & & \vdots \\ \vdots & & \ddots & 0 \\ b_{1,n}^* & \cdots & b_{n-1,n}^* & b_{n,n}^* \end{bmatrix}, \tag{A.94}$$

thus

$$\begin{aligned} \varphi_{1,1} &= |b_{1,1}|^2 + |b_{1,2}|^2 + \cdots + |b_{1,n}|^2, \\ \varphi_{1,2} &= b_{1,2}b_{2,2}^* + b_{1,3}b_{2,3}^* + \cdots + b_{1,n}b_{2,n}^*, \\ \varphi_{1,3} &= b_{1,3}b_{3,3}^* + b_{1,4}b_{3,4}^* + \cdots + b_{1,n}b_{3,n}^*, \\ &\vdots \\ \varphi_{1,n} &= b_{1,n}b_{n,n}^*. \end{aligned} \tag{A.95}$$

Comparing the Cholesky factorization to

$$\begin{aligned}\Phi(z) &= B(z) \cdot B^*(z^{-*}) \\ &= \left(b[0] + b[1]\,z^{-1} + b[2]\,z^{-2} + \cdots\right) \cdot \left(b^*[0] + b^*[1]\,z^1 + b^*[2]\,z^2 + \cdots\right) \\ &= \cdots + (b[0]b^*[1] + b[1]b^*[2] + \cdots)\,z^1 \\ &\quad + (|b[0]|^2 + |b[1]|^2 + |b[2]|^2 + \cdots) \\ &\quad + (b[1]b^*[0] + b[2]b^*[1] + \cdots)\,z^{-1} \\ &\quad + (b[2]b^*[0] + b[3]b^*[1] + \cdots)\,z^{-2} \\ &\quad + \ldots \\ &= \cdots + \varphi[-1]\,z^1 + \varphi[0] + \varphi[1]\,z^{-1} + \varphi[2]\,z^{-2} + \cdots \end{aligned} \qquad (A.96)$$

reveals that if the dimension of Φ is at least $O+1$, the first row of B in the Cholesky factorization will be equal to the coefficients of the polynomial $B(z)$ (or the last row of G contains the coefficients of the polynomial in reversed order), that is

$$b_{1,1} = b[0]\,, \quad b_{1,2} = b[1]\,, \quad b_{1,3} = b[2]\,, \quad \ldots \qquad (A.97)$$

In other words, the Cholesky factorization of the Toeplitz matrix Φ is asymptotically (dimension large enough) equivalent to the spectral factorization of $\Phi(z)$.

The Cholesky factorization is usually stated for the lower-triangular matrix G. The *Cholesky–Banachiewicz algorithm* (which coincides with the factorization approach given by F. L. Bauer) performs the decomposition by going row-wise (loop m) through the matrix. The following steps have to be performed:

$$g_{1,1} = \sqrt{\varphi_{1,1}}\,,$$

$m = 2, 3, \ldots, n$

$$g_{m,i} = \left(\varphi_{1,m-i+1} - \sum_{k=1}^{i-1} g_{m,k}\,g_{i,k}^*\right)/g_{i,i}^*\,, \qquad i = 1, \ldots, m-1\,, \qquad (A.98)$$

$$g_{m,m} = \sqrt{\varphi_{1,1} - \sum_{k=1}^{m-1} |g_{m,k}|^2}\,.$$

Finally, we note that direct calculations show

$$\varphi_{i,i} = \sum_{l=i}^{n} |b_{i,l}|^2\,. \qquad (A.99)$$

Since Φ has Toeplitz structure, $\varphi_{i,i} = \varphi_{1,1}$, $\forall i$, and all terms in the sum are non-negative, we have

$$|b_{1,1}|^2 \le |b_{2,2}|^2 \le \cdots \le |b_{n,n}|^2\,. \qquad (A.100)$$

Bound on the Performance As shown in Section 8.2, the normalized minimum squared Euclidean distance of MLSE is proportional to the (restricted) shortest vector in the lattice spanned by the matrix B. Since B is upper triangular, a

Gram–Schmidt orthogonal basis $f_1, \ldots, f_n$ (where we use row vectors) can be given as

$$\underbrace{\begin{bmatrix} b_{1,1} & & b_{1,n} \\ \vdots & \ddots & \\ 0 & \cdots & b_{n,n} \end{bmatrix}}_{B} = \underbrace{\begin{bmatrix} 1 & & m_{1,n} \\ & \ddots & \\ 0 & & 1 \end{bmatrix}}_{M} \underbrace{\begin{bmatrix} f_{1,1} & & 0 \\ & \ddots & \\ 0 & & f_{n,n} \end{bmatrix}}_{F} = M \underbrace{\begin{bmatrix} f_1 \\ \vdots \\ f_n \end{bmatrix}}_{F} . \quad (A.101)$$

Direct inspection shows that $f_{i,i} = b_{i,i}$ and $\|f_i\|^2 = |f_{i,i}|^2$, $i = 1, \ldots, n$, hold.

Using $z = [z_1, \ldots, z_n] \in \mathbb{Z}^n$ (for ASK signaling; for QAM signaling $z \in (\mathbb{Z} + j\mathbb{Z})^n$ has to be used), any lattice point can be written as $\lambda = zB$, which gives for the squared norm

$$\begin{aligned} \|\lambda\|^2 &= \|zB\|^2 = \|zMF\|^2 \\ &= |z_1|^2 \|f_1\|^2 + \cdots + |m_{1,n}z_1 + \cdots + z_n|^2 \|f_n\|^2 \\ &= |z_1|^2 |f_{1,1}|^2 + \cdots + |m_{1,n}z_1 + \cdots + z_n|^2 |f_{n,n}|^2 . \end{aligned} \quad (A.102)$$

Considering (A.100), and letting only element z_1 in z be nonzero, we arrive at

$$\begin{aligned} \|\lambda\|^2 &\geq |z_1|^2 |f_{1,1}|^2 + \cdots + |m_{1,n}z_1|^2 |f_{n,n}|^2 \\ &\geq |z_1|^2 |f_{1,1}|^2 , \end{aligned} \quad (A.103)$$

since all terms are positive. Finally, using $|z_j|^2 \geq 1$ and $f_{1,1} = b_{1,1}$ gives

$$\min_{\lambda = zB, \ z \in \mathbb{Z}^n \setminus \{0\}} \|\lambda\|^2 \geq |b_{1,1}|^2 . \quad (A.104)$$

Using this result, the normalized minimum squared Euclidean distance of MLSE can be lower-bounded by (cf. (8.39) for the definition of the set $\mathcal{D}$ of difference vectors of ASK; for QAM signaling a respective definition is easily possible)

$$\begin{aligned} d_{\min}^2 &= \frac{E_g}{2 E_b} \cdot \min_{\forall \Delta_a \in \mathcal{D}} \|\Delta_a B\|^2 \geq \frac{E_g}{2 E_b} \cdot \min_{\lambda = zB, \ z \in \mathbb{Z}^n \setminus \{0\}} \|2zB\|^2 \\ &\geq \frac{E_g}{2 E_b} \cdot 4 |b_{1,1}|^2 = \frac{E_g}{E_b} 2 |b_{1,1}|^2 . \end{aligned} \quad (A.105)$$

Implementation An algorithm for determination of the shortest vector and, thus, the minimum distance can be derived from the *sphere decoder*, which has already been modified for finding the shortest vector. Only the additional restriction to a finite range of the components (and the first element being positive) has to be included. The entire algorithm is given in Table A.3. Here, $H = G^{-1}$ (lower triangular) and M are the input parameters and DeltaMin (the difference sequence Δ_a leading to the minimum) and the minimum minlength are the output parameters.

Table A.3 MATLAB® implementation of the algorithm to determine the restricted shortest vector.

```
function [DeltaMin, minlength] = restrictedshortestvect(H, M)
%
n = size(H,1);
minlength = 1e100;
k = n;
l = zeros(n,1);
e = zeros(n,n);
a = zeros(1,n);
y = 0;
step = zeros(n,1); step(k) = 1;
while 1,
  newlength = l(k) + y^2;
  if ( (newlength < minlength) & (a(k) > -M) & (a(k) < M) ),
    if k ~= 1,
      for i=1:k-1, e(k-1,i) = e(k,i)-y*H(k,i); end;
      k = k-1;
      l(k) = newlength;
      a(k) = round(e(k,k));
      if k==1,
        if a(k) > M-1, a(k) = M-1; end;
        if a(k) < 1,   a(k) = 1;   end;
      else,
        if a(k) > M-1, a(k) = M-1; end;
        if a(k) < 1-M, a(k) = 1-M; end;
      end;
      y = (e(k,k)-a(k))/H(k,k);
      step(k) = signn(y);
    else,
      if newlength ~= 0,
        DeltaMin = a;
        minlength = newlength;
        k = k+1;
      end;
      a(k) = a(k)+step(k);
      y = (e(k,k)-a(k))/H(k,k);
      step(k) = - step(k)-signn(step(k));
      if (a(k) < -M | a(k) > M),
        a(k) = a(k) + step(k);
        step(k) = - step(k)-signn(step(k));
      end;
    end;
  else,
    if k == n,
      return;
    else,
      k = k+1;
      a(k) = a(k)+step(k);
      y = (e(k,k)-a(k))/H(k,k);
      step(k) = - step(k)-signn(step(k));
      if (a(k) < -M | a(k) > M),
        a(k) = a(k) + step(k);
        step(k) = - step(k)-signn(step(k));
      end;
    end;
  end;
end;

function y = signn(x)
%
if x > 0, y = 1; else, y = -1; end;
```

A.3.3 Stable Systems

Let the systems $H_1(s)$ and $H_2(s)$ be given and assume that $H_1(s)H_2(s)$ is stable, that is, it has no poles with $\text{Re}\{s_0\} \geq 0$. Given these systems, the new system

$$H(s) = \frac{H_1(s)}{1 + H_1(s)H_2(s)} \tag{A.106}$$

is created. According to the *Nyquist stability criterion* of control theory, $H(s)$ is stable if and only if the trajectory of $H_1(j2\pi f)H_2(j2\pi f)$, $-\infty < f < \infty$, in the complex plane does not encircle the point -1.

For $H_1(s) = 1$ this reduces to the fact that $H(s) = \frac{1}{1+H_2(s)}$ is stable if and only if the trajectory of $H_2(j2\pi f)$, $-\infty < f < \infty$, does not encircle the point -1. If $H(s)$ is stable, its poles (which do not have to be known explicitly) have a negative real part. However, the poles of $H(s)$ are the zeros of $1 + H_2(s)$. Thus, $1 + H_2(s)$ is *minimum-phase* if the trajectory of $H_2(j2\pi f)$, $-\infty < f < \infty$, does not encircle the point -1.

This gives that $A + G(s) = A\left(1 + \frac{1}{A}G(s)\right)$ is minimum-phase if the trajectory of $\frac{1}{A}G(j2\pi f)$, $-\infty < f < \infty$, does not encircle the point -1, which is identical to the demand that the trajectory of $G(j2\pi f)$ does not encircle the point $-A$, or that the trajectory of $A + G(j2\pi f)$ does not encircle the origin.

Hence, given a *causal* and *stable system* $G(s)$, by choosing A sufficiently large, the system $A + G(s)$ can be made *minimum-phase*. Then, knowing $|A + G(f)|^2$ is sufficient, since (cf. (A.64))

$$\arg(A + G(f)) = -\frac{1}{2}\mathcal{H}\{\log(|A + G(f)|^2)\}. \tag{A.107}$$

A.3.4 Application of the Time–Frequency Duality

Owing to the duality between the time and the frequency domain in the Fourier transform, that is, for $x(t) \multimap X(f)$ we have the correspondence $X(t) \multimap x(-f)$, the above statement can be rephrased as

> Given a time function $g(t)$ that corresponds to a causal and stable spectrum $G(f)$, by choosing A sufficiently large, $A + g(t)$ can be made minimum-phase. Then, knowing $|A + g(t)|^2$ is sufficient, since $\arg(A + g(t)) = \frac{1}{2}\mathcal{H}\{\log(|A + g(t)|^2)\}$.

This fact is also known in physics as *Kramers–Kronig relation* in electron spectroscopy using X-rays. In mathematics, these relations are known as the *Sokhotski–Plemelj theorem* or, simply, *Hilbert transform*.

A.4 Stochastic Processes

Besides (deterministic) signals and LTI systems, in communications we rely heavily on the concept of *stochastic processes*. Thus, we recapitulate the basic concepts of random processes in this section.

Definition A.1 *Stochastic Process (Random Process)*

Let H = $\{\eta_1, \eta_2, \ldots\}$ be the set of (disjoint) elementary events η of a random experiment.

A *stochastic process* assigns a *signal*

$$x(t) = x(\eta, t)$$

to each elementary event η (see Figure A.2). In other words, a time-domain signal is randomly drawn from a set of signals, or the outcome of the random experiment is a function $x(\eta, t)$ of time.

The same holds for discrete-time random sequences.

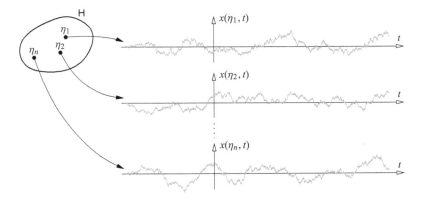

Figure A.2 Concept of a stochastic process.

Note that, when considering *joint stochastic processes*, two (or more) signals, for example, $x(\eta,t)$, $y(\eta,t)$, ..., are assigned to an elementary event η (i.e., are drawn simultaneously).

In this sense, when dealing with a *complex-valued stochastic process*

$$z(\eta, t) = x(\eta, t) + jy(\eta, t), \tag{A.108}$$

real and imaginary parts are drawn jointly.

Please strictly distinguish the following two items:

- x(η,t) (in sans-serif font) denotes the process (i.e., the *ensemble of realizations*);
- $x(\eta, t)$ (in regular font) denotes a particular realization or *sample function*.

Subsequently, if it is not required to emphasize the dependency on the elementary event (i.e., on the randomness), we drop the argument "η" and simply write x(t) and $x(t)$ instead of x(η,t) and $x(\eta, t)$.

A.4.1 Parameters of Stochastic Processes

Despite the fact that the actual realization is drawn randomly, there are some important parameters characterizing the behavior (or "the rules of the game").

Figure A.3
Visualization of a stochastic process. Three sample functions are shown. When fixing the time instant, a random variable $x = x(t_1)$ is obtained.

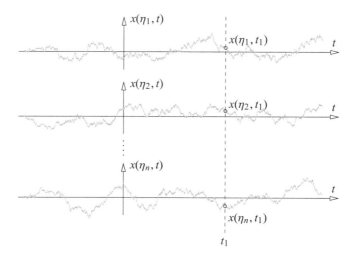

Probability Density Function (pdf)

We first consider a fixed time instant $t_1 \in \mathbb{R}$. In this case, the process $x(t)$ reduces to a *random variable* $x = x(t_1)$ (see Figure A.3).

A *real-valued* random variable (and thus the process at time instant t_1) can be characterized by the *probability density function (pdf)*

$$f_x(x,t_1) \stackrel{\text{def}}{=} \frac{\mathrm{d}}{\mathrm{d}x} \Pr\{x(t_1) \le x\} = \frac{\mathrm{d}}{\mathrm{d}x} F_x(x,t_1) , \qquad (A.109)$$

where $\Pr\{\cdot\}$ denotes the probability and $F_x(x,t_1)$ is the *cumulative distribution function (cdf)*.

Joint real-valued processes $x(t)$ and $y(t)$ considered at time instants t_1 and t_2, respectively, are characterized by the *joint pdf*

$$f_{xy}(x,y,t_1,t_2) \stackrel{\text{def}}{=} \frac{\partial^2}{\partial x \partial y} \Pr\{x(t_1) \le x \wedge y(t_2) \le y\} . \qquad (A.110)$$

A *complex-valued* process $z(t) = z_\mathrm{R}(t) + \mathrm{j} z_\mathrm{I}(t)$ with real-valued components $z_\mathrm{R}(t)$ and $z_\mathrm{I}(t)$ is characterized by the *joint pdf*

$$f_{z_\mathrm{R} z_\mathrm{I}}(z_\mathrm{R}, z_\mathrm{I}, t) \stackrel{\text{def}}{=} \frac{\partial^2}{\partial z_\mathrm{R} \partial z_\mathrm{I}} \Pr\{z_\mathrm{R}(t) \le z_\mathrm{R} \wedge z_\mathrm{I}(t) \le z_\mathrm{I}\} . \qquad (A.111)$$

This pdf is often *formally* written as

$$f_z(z,t) \qquad (A.112)$$

but has to be understood in the sense of (A.111).

Mean, Variance, and Power

The pdf gives the distribution of the signal amplitudes. In some cases, an average characterization is sufficient. Such a single quantifying number is the *mean* or *expected value* of the process at a given time instant t_1. We have

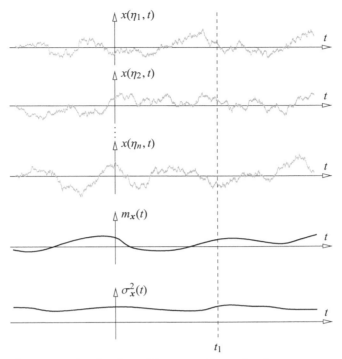

Figure A.4 Visualization of the mean of a stochastic process. The expectation is carried out over the sample functions.

$$m_x(t_1) \stackrel{\text{def}}{=} E\{x(t_1)\} = \int_{-\infty}^{\infty} x\, f_x(x,t_1)\, dx\ . \tag{A.113}$$

Here, $E\{\cdot\}$ denotes the *expectation*.

Note that the expectation is carried out over the sample functions, that is, across the process (vertically in Figure A.4).

Besides the mean, which is the first-order moment, second-order moments are of interest. The *variance* of the process at time instant t_1 is defined as (see also Figure A.4)

$$\sigma_x^2(t_1) \stackrel{\text{def}}{=} E\{|x(t_1) - m_x(t_1)|^2\} = \int_{-\infty}^{\infty} |x - m_x(t_1)|^2 f_x(x,t_1)\, dx\ . \tag{A.114}$$

The (normalized) power of the process at time instant t_1 is given by

$$S_x(t_1) = E\{|x(t_1)|^2\}\ , \tag{A.115}$$

which can be written as

$$S_x(t_1) = \sigma_x^2(t_1) + |m_x(t_1)|^2\ , \tag{A.116}$$

that is, the power is given as the sum of the variance (AC power) and the squared mean (DC power).

Autocorrelation Function (acf)

The above quantities characterize the process at a particular time instant. In order to characterize the dependencies of different time instants (i.e., the evolution over time), the *autocorrelation function (acf)* is of interest. The acf of $x(t)$ between the time instants t_1 and t_2 is defined as

$$\phi_{xx}(t_1, t_2) \stackrel{\text{def}}{=} E\{x(t_1) \cdot x^*(t_2)\} . \tag{A.117}$$

For $t_1 = t_2$, the acf reduces to the power, that is

$$S_x(t_1) = \phi_{xx}(t_1, t_1) . \tag{A.118}$$

Similarly, the *crosscorrelation function (ccf)* of joint processes $x(t)$ and $y(t)$ at time instants t_1 and t_2, respectively, can be defined as

$$\phi_{xy}(t_1, t_2) \stackrel{\text{def}}{=} E\{x(t_1) \cdot y^*(t_2)\} . \tag{A.119}$$

Please clearly distinguish between the *autocorrelation* of a deterministic, energy-limited signal and the *autocorrelation function* of a (wide-sense) stationary stochastic process.

A.4.2 Wide-Sense Stationarity

The above-defined quantities are valid for any stochastic process. However, in practice, we are often interested in stochastic processes following stricter rules, in particular processes that are stationary (i.e., where the characteristic quantities do not vary over time).

A stochastic process is *(wide-sense) stationary*, if:

- the mean is constant over time

$$m_x(t_1) = m_x = \text{const.} \qquad \forall t_1 \in \mathbb{R}; \tag{A.120}$$

- acf and ccf depend only on the *time difference* $\tau \stackrel{\text{def}}{=} t_1 - t_2$

$$\phi_{xx}(t_1, t_2) = \phi_{xx}(t_2 + \tau, t_2) \qquad \forall t_2, \tau \in \mathbb{R} . \tag{A.121}$$

There also exists a definition of stationarity in the strict sense; we only require the above-mentioned weaker form (wide-sense) of stationarity.

The autocorrelation function and crosscorrelation function of wide-sense stationary processes are then written as

$$\phi_{xx}(\tau) \stackrel{\text{def}}{=} E\{x(t + \tau) \cdot x^*(t)\} , \tag{A.122}$$

$$\phi_{xy}(\tau) \stackrel{\text{def}}{=} E\{x(t + \tau) \cdot y^*(t)\} . \tag{A.123}$$

Remember that $\phi_{xx}(\tau) = \phi_{xx}^*(-\tau)$ and $\phi_{xy}(\tau) = \phi_{yx}^*(-\tau)$ and that power also does not vary over time as $S_x = \phi_{xx}(\tau = 0)$.

Instead of characterizing a (wide-sense) stationary process by its autocorrelation function, the *Fourier transform* of the acf $\phi_{xx}(\tau)$ (which is a function of a time (difference) parameter) can be specified, called the *power spectral density (psd)*

$$\Phi_{xx}(f) \stackrel{\text{def}}{=} \mathcal{F}\{\phi_{xx}(\tau)\} = \int_{-\infty}^{\infty} \phi_{xx}(\tau)\,e^{-j2\pi f\tau}\,d\tau\,. \tag{A.124}$$

From the properties of signals and their spectra, we have

$$\Phi_{xx}(f) \in \mathbb{R}\,, \qquad \Phi_{xx}(f) \geq 0\,. \tag{A.125}$$

By integrating over the power spectral density (power per frequency), the power of the process is obtained as

$$S_x = \phi_{xx}(\tau = 0) = \int_{-\infty}^{\infty} \Phi_{xx}(f)\,df\,. \tag{A.126}$$

For joint processes, the Fourier transform of the ccf is known as *cross power spectral density*, defined as

$$\Phi_{xy}(f) \stackrel{\text{def}}{=} \mathcal{F}\{\phi_{xy}(\tau)\}\,. \tag{A.127}$$

A.4.3 Gaussian Processes

Owing to the central limit theorem, the sum of independent processes tends to be Gaussian. Hence, Gaussian processes are very good and widely used models for noise and often even for signals.

Consider a real-valued stationary stochastic process $x(t)$ with mean m_x and acf $\phi_{xx}(\tau)$ and N arbitrary time instants $t_1, \ldots, t_N$. We define $\mathbf{x} \stackrel{\text{def}}{=} [x(t_1), \ldots, x(t_N)]$. A *real-valued* stochastic process is a *Gaussian process* if, for any $\mathbf{x}$, the joint pdf is given by

$$f_{\mathbf{x}}(\mathbf{x}) = \frac{1}{\sqrt{(2\pi)^N |\mathbf{M}|}} \cdot \exp\left(-\frac{1}{2}(\mathbf{x} - \mathbf{m_x})\mathbf{M}^{-1}(\mathbf{x} - \mathbf{m_x})^{\mathsf{T}}\right), \tag{A.128}$$

where the *covariance matrix* is defined as

$$\mathbf{M} = [\mu_{ik}]\,, \qquad \mu_{ik} = \phi_{xx}(t_i - t_k) - m_x^2\,, \qquad i,k = 1,\ldots,N\,. \tag{A.129}$$

Hence, Gaussian processes are fully characterized by the mean (first-order moment) and the acf (second-order moment).

Example A.1 Gaussian Processes

For $N = 1$, the general form (A.128) specializes to a *real-valued* Gaussian random variable n with pdf

$$f_n(n) = \frac{1}{\sqrt{2\pi\sigma_n^2}} \cdot \exp\left(-\frac{(n - m_n)^2}{2\sigma_n^2}\right) \tag{A.130}$$

with mean m_n and variance σ_n^2.

For $N = 2$, zero mean $m_x = 0$, and covariance matrix $M = \sigma_x^2 \begin{bmatrix} 1 & \rho \\ \rho & 1 \end{bmatrix}$, with correlation coefficient $\rho \stackrel{\text{def}}{=} \phi_{xx}(t_1 - t_2)/\sigma_x^2$, we have

$$f_x(x_1, x_2) = \frac{1}{2\pi\sigma_x^2\sqrt{1-\rho^2}} \cdot \exp\left(-\frac{x_1^2 - 2\rho x_1 x_2 + x_2^2}{2\sigma_x^2(1-\rho^2)}\right). \tag{A.131}$$

This pdf is visualized in Figure A.5 for different correlation coefficients ρ.

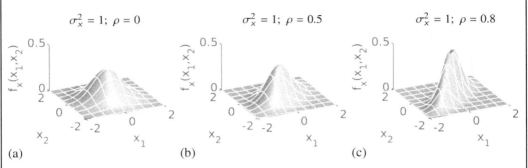

Figure A.5 Two-dimensional Gaussian pdf with different correlation coefficients ρ: (a) $\rho = 0$; (b) $\rho = 0.5$; (c) $\rho = 0.8$.

Complex-valued Gaussian noise is the most important noise model in the ECB domain. Complex Gaussian noise can be seen as two-dimensional real-valued Gaussian noise.

Assume independent real (n_I) and imaginary (n_Q) parts, both Gaussian with the same variance $\sigma_{n_I}^2 = \sigma_{n_Q}^2 = \sigma_n^2/2$ and the means m_{n_I} and m_{n_Q}, respectively. The *complex-valued* Gaussian random variable $n = n_I + j\,n_Q$ then has the (two-dimensional) joint pdf

$$\begin{aligned}f_{n_I n_Q}(n_I, n_Q) &= \frac{1}{\sqrt{2\pi\sigma_{n_I}^2}} \exp\left(-\frac{(n_I - m_{n_I})^2}{2\sigma_{n_I}^2}\right) \cdot \frac{1}{\sqrt{2\pi\sigma_{n_Q}^2}} \exp\left(-\frac{(n_Q - m_{n_Q})^2}{2\sigma_{n_Q}^2}\right) \\ &= \frac{1}{2\pi\sigma_{n_I}^2} \exp\left(-\frac{(n_I - m_{n_I})^2 + (n_Q - m_{n_Q})^2}{2\sigma_{n_I}^2}\right). \end{aligned} \tag{A.132}$$

This pdf can *formally* be written as

$$f_n(n) = \frac{1}{\pi\sigma_n^2} \exp\left(-\frac{|n - m_n|^2}{\sigma_n^2}\right), \tag{A.133}$$

with $m_n \stackrel{\text{def}}{=} m_{n_I} + j\,m_{n_Q}$.

Figure A.5 visualizes the pdf of complex-valued Gaussian noise for $m_n = 2 + 1j$ and $\sigma_n^2 = 1$.

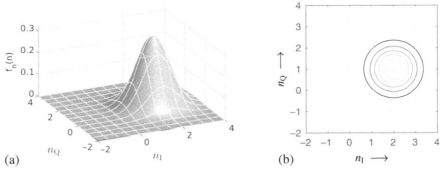

(a)   (b)

Figure A.6 Visualization of the pdf of complex-valued Gaussian noise with $m_n = 2 + 1\mathrm{j}$ and $\sigma_n^2 = 1$. (a) Three-dimensional view. (b) Contour plot (isolines, i.e., curves of constant value, are shown).

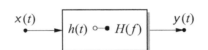

Figure A.7 Filtering of a stochastic process.

A.4.4 LTI Systems and Stochastic Processes

Let an LTI system with impulse response $h(t)$ or transfer function $H(f) = \mathscr{F}\{h(t)\}$ be given. If a (wide-sense stationary) stochastic process $x(t)$ is the input to the system, the output $y(t)$ is also a (wide-sense stationary) stochastic process; each sample function of $x(t)$ corresponds to a sample function of $y(t)$. The situation is depicted in Figure A.7.

Then, the following relations hold:

$$\phi_{xy}(\tau) = \phi_{xx}(\tau) * h(-\tau) \quad \circ\!\!-\!\!\bullet \quad \Phi_{xy}(f) = \Phi_{xx}(f) \cdot H^*(f), \quad (A.134)$$

$$\phi_{yx}(\tau) = \phi_{xx}(\tau) * h(\tau) \quad \circ\!\!-\!\!\bullet \quad \Phi_{yx}(f) = \Phi_{xx}(f) \cdot H(f), \quad (A.135)$$

$$\phi_{yy}(\tau) = \phi_{xx}(\tau) * \varphi_{hh}(\tau) \quad \circ\!\!-\!\!\bullet \quad \Phi_{yy}(f) = \Phi_{xx}(f) \cdot |H(f)|^2, \quad (A.136)$$

where $\varphi_{hh}(\tau) = h(\tau) * h^*(-\tau)$ is the *autocorrelation* of the impulse response $h(t)$.

A.4.5 Sampling and Interpolation of Stochastic Processes

In digital communications (as can be seen in Chapter 3), the transition between continuous-time and discrete-time signals (and vice versa) is of great importance. Hence, the connection between continuous-time and discrete-time processes is required.

Let $x(t)$ be a (wide-sense stationary) *continuous-time* process with acf $\phi_{xx}(\tau) = \mathrm{E}\{x(t + \tau) \cdot x^*(t)\}$ and psd $\Phi_{xx}(f) = \mathscr{F}\{\phi_{xx}(\tau)\}$.

When sampling $x(t)$ with the sampling interval T, a *discrete-time* process

$$x[k] \stackrel{\text{def}}{=} x(t = kT) \quad (A.137)$$

is obtained. It is straightforward to show that its autocorrelation sequence and power spectral density (the discrete-time Fourier transform (cf. (A.28)) of the acf) are given by

$$\phi_{xx}[\kappa] = \phi_{xx}(\tau = \kappa T),\tag{A.138}$$

$$\Phi_{xx}\left(e^{j2\pi fT}\right) = \frac{1}{T} \sum_{\mu} \Phi_{xx}(f - \mu/T).\tag{A.139}$$

Let $x[k]$ be a (wide-sense stationary) *discrete-time* process with acf $\phi_{xx}[\kappa] = \mathrm{E}\{x[k+\kappa] \cdot x^*[k]\}$ and psd $\Phi_{xx}(e^{j2\pi fT}) = \mathscr{F}_*\{\phi_{xx}[\kappa]\} = \mathcal{Z}\{\phi_{xx}[\kappa]\}\big|_{z=e^{j2\pi fT}}$. When interpolating $x[k]$ with the interpolation (reconstruction) function $g(t)$, we obtain the *continuous-time* process

$$x(t) \stackrel{\text{def}}{=} \sum_{k} x[k] g(t - kT).\tag{A.140}$$

Note that, in general, this process has a T-periodic structure and, thus, is a cyclo-stationary process.

A.4.6 Cyclo-stationary Stochastic Processes

As can be seen in Chapter 3, in digital communications the statistic properties of the signals typically vary periodically over time (symbol period T). In particular, the sample functions of pulse-amplitude modulated (PAM) signals (see Figure A.8) have a structure where the timing (symbol-by-symbol structure) is "visible." By drawing all sample functions on top of each other, typically a T-periodic structure will be visible, which is called the *eye pattern*.

Such signals are not drawn from a stationary process but from a *cyclo-stationary stochastic process* – the characteristic quantities (mean, power, acf) are not constant over time but fluctuate periodically (with period T) over time. Hence, we can write

$$m_x(t) = \mathrm{E}\{x(t)\} = \mathrm{E}\{x(t + kT)\}$$
$$= m_x(t - kT), \qquad \forall k \in \mathbb{Z},\tag{A.141}$$

$$\phi_{xx}(t + \tau, t) = \mathrm{E}\{x(t+\tau)x^*(t)\} = \mathrm{E}\{x(t+kT+\tau)x^*(t+kT)\}$$
$$= \phi_{xx}(t + kT + \tau, t + kT).\tag{A.142}$$

For cyclo-stationary processes, quantities averaged over one period are of interest. In particular, the *average acf* is defined as[5]

$$\bar{\phi}_{xx}(\tau) \stackrel{\text{def}}{=} \frac{1}{T} \int_0^T \phi_{xx}(t + \tau, t) \, dt.\tag{A.143}$$

Such an averaging can be interpreted as taking the union of the cyclo-stationary process with all its time-shifted versions, where the shifts are uniformly drawn from the interval $[0, T]$. Such a procedure is called *phase randomization*.

---

[5] Averaging can be done over any interval of width T. The interval $[-T/2, T/2]$ is also commonly used.

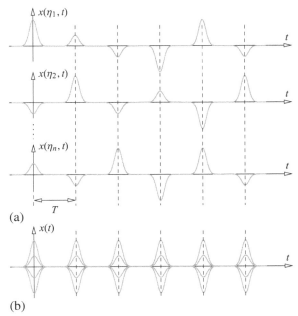

Figure A.8 (a) Exemplary sample functions of a pulse-amplitude modulation scheme. The information is represented in the amplitude of the pulse that is sent every T. The corresponding process is cyclo-stationary. (b) Corresponding eye pattern.

Having the average acf, the *average psd* is given by ($\mathscr{F}_\tau\{\cdot\}$ denotes the transformation w.r.t. τ)

$$\bar{\Phi}_{xx}(f) \overset{\text{def}}{=} \mathscr{F}_\tau\{\bar{\phi}_{xx}(\tau)\} = \frac{1}{T} \int_0^T \mathscr{F}_\tau\{\phi_{xx}(t+\tau,t)\}\, dt \,. \qquad (A.144)$$

When, for example, measuring the psd of PAM transmit signals, the average psd will be obtained.

A.5 Prediction

In this appendix, we briefly consider the task of *prediction*, that is, the problem of estimating a sample at the current time step based on known samples from previous time steps.

To that end, we assume that a discrete-time wide-sense stationary stochastic process $x[k]$ with the *autocorrelation sequence*

$$\phi_{xx}[\kappa] = \mathrm{E}\{x[k+\kappa]\,x^*[k]\} \qquad (A.145)$$

or *power spectral density* $\Phi_{xx}\left(\mathrm{e}^{\mathrm{j}2\pi fT}\right) = \mathscr{F}_*\{\phi_{xx}[\kappa]\}$ is given. The task of linear prediction is to find a *linear combination* of the past p samples (prediction order p) that "best" (in the mean-square sense) matches the actual sample $x[k]$. Using

$$\boldsymbol{p} \stackrel{\text{def}}{=} \begin{bmatrix} p^*[1] \\ p^*[2] \\ \vdots \\ p^*[\rho] \end{bmatrix}, \quad \boldsymbol{x}[k] \stackrel{\text{def}}{=} \begin{bmatrix} x[k-1] \\ x[k-2] \\ \vdots \\ x[k-\rho] \end{bmatrix}, \tag{A.146}$$

this linear combination can be written as

$$\tilde{x}[k] \stackrel{\text{def}}{=} \sum_{\kappa=1}^{\rho} p[\kappa]\, x[k-\kappa] = \boldsymbol{p}^{\mathsf{H}} \boldsymbol{x}[k]. \tag{A.147}$$

Hence, coefficients $p[1], p[2], \ldots, p[\rho]$ are sought, such that the *prediction error*

$$\epsilon[k] \stackrel{\text{def}}{=} x[k] - \tilde{x}[k] \tag{A.148}$$

has minimum variance, that is

$$\sigma_\epsilon^2 = \mathrm{E}\{|\epsilon[k]|^2\} = \mathrm{E}\{|x[k] - \boldsymbol{p}^{\mathsf{H}} \boldsymbol{x}[k]|^2\} \to \min. \tag{A.149}$$

Optimum Predictor

The variance of the prediction error can be written as

$$\begin{aligned}
\sigma_\epsilon^2 &= \mathrm{E}\{|x[k] - \boldsymbol{p}^{\mathsf{H}} \boldsymbol{x}[k]|^2\} \\
&= \mathrm{E}\{(x[k] - \boldsymbol{p}^{\mathsf{H}} \boldsymbol{x}[k])(x^*[k] - \boldsymbol{x}^{\mathsf{H}}[k]\boldsymbol{p})\} \\
&= \mathrm{E}\{|x[k]|^2\} - \boldsymbol{p}^{\mathsf{H}} \mathrm{E}\{\boldsymbol{x}[k]\, x^*[k]\} - \mathrm{E}\{x[k]\, \boldsymbol{x}^{\mathsf{H}}[k]\}\boldsymbol{p} + \boldsymbol{p}^{\mathsf{H}} \mathrm{E}\{\boldsymbol{x}[k]\,\boldsymbol{x}^{\mathsf{H}}[k]\}\boldsymbol{p} \\
&= \sigma_x^2 - \boldsymbol{p}^{\mathsf{H}} \boldsymbol{\phi}_{xx} - \boldsymbol{\phi}_{xx}^{\mathsf{H}} \boldsymbol{p} + \boldsymbol{p}^{\mathsf{H}} \boldsymbol{\Phi}_{xx} \boldsymbol{p},
\end{aligned} \tag{A.150}$$

with the obvious definitions of the vector $\boldsymbol{\phi}_{xx}$ and the correlation matrix $\boldsymbol{\Phi}_{xx}$. Completing the square, we obtain

$$\begin{aligned}
\sigma_\epsilon^2 &= \sigma_x^2 - \boldsymbol{p}^{\mathsf{H}} \boldsymbol{\Phi}_{xx} \boldsymbol{\Phi}_{xx}^{-1} \boldsymbol{\phi}_{xx} - \boldsymbol{\phi}_{xx}^{\mathsf{H}} \boldsymbol{\Phi}_{xx}^{-1} \boldsymbol{\Phi}_{xx} \boldsymbol{p} \\
&\quad + \boldsymbol{p}^{\mathsf{H}} \boldsymbol{\Phi}_{xx} \boldsymbol{\Phi}_{xx}^{-1} \boldsymbol{\Phi}_{xx} \boldsymbol{p} + \boldsymbol{\phi}_{xx}^{\mathsf{H}} \boldsymbol{\Phi}_{xx}^{-1} \boldsymbol{\phi}_{xx} - \boldsymbol{\phi}_{xx}^{\mathsf{H}} \boldsymbol{\Phi}_{xx}^{-1} \boldsymbol{\phi}_{xx} \\
&= \sigma_x^2 - \boldsymbol{\phi}_{xx}^{\mathsf{H}} \boldsymbol{\Phi}_{xx}^{-1} \boldsymbol{\phi}_{xx} + (\boldsymbol{p}^{\mathsf{H}} \boldsymbol{\Phi}_{xx} - \boldsymbol{\phi}_{xx}^{\mathsf{H}}) \boldsymbol{\Phi}_{xx}^{-1} (\boldsymbol{\Phi}_{xx} \boldsymbol{p} - \boldsymbol{\phi}_{xx}).
\end{aligned} \tag{A.151}$$

The correlation matrix $\boldsymbol{\Phi}_{xx}$ of a regular wide-sense stationary process $x[k]$ is positive (semi)definite; the inverse $\boldsymbol{\Phi}_{xx}^{-1}$ of a positive (semi)definite matrix is also positive (semi)definite. Hence, the "quadratic term" (last term in (A.151)) is always greater than or equal to zero. The minimum is attained if this term is zero, which obtained for

$$\boldsymbol{\Phi}_{xx} \boldsymbol{p} - \boldsymbol{\phi}_{xx} \stackrel{!}{=} 0. \tag{A.152}$$

This set of equations is called the *Yule–Walker equations*. In detail, we have

$$\begin{bmatrix} \phi_{xx}[0] & \phi_{xx}[1] & \cdots & \phi_{xx}[\rho-1] \\ \phi_{xx}[-1] & \ddots & & \vdots \\ \vdots & & \ddots & \phi_{xx}[1] \\ \phi_{xx}[-\rho+1] & \cdots & \phi_{xx}[-1] & \phi_{xx}[0] \end{bmatrix} \cdot \begin{bmatrix} p[1] \\ p[2] \\ \vdots \\ p[\rho] \end{bmatrix} = \begin{bmatrix} \phi_{xx}[1] \\ \phi_{xx}[2] \\ \vdots \\ \phi_{xx}[\rho] \end{bmatrix} . \quad (A.153)$$

The optimum solution for the predictor coefficients is thus given by

$$\boldsymbol{p}_{\text{opt}} = \boldsymbol{\Phi}_{xx}^{-1} \boldsymbol{\phi}_{xx} . \quad (A.154)$$

Looking at (A.151), the variance of the prediction error using the optimum prediction coefficients then amounts to

$$\begin{aligned} \sigma_{\epsilon,\text{opt}}^2 &= \sigma_x^2 - \boldsymbol{\phi}_{xx}^H \boldsymbol{\Phi}_{xx}^{-1} \boldsymbol{\phi}_{xx} \\ &= \sigma_x^2 - \boldsymbol{\phi}_{xx}^H \boldsymbol{p}_{\text{opt}} \\ &= \underbrace{\phi_{xx}[0]}_{\text{variance before prediction}} - \underbrace{\sum_{\kappa=1}^{\rho} p^*[\kappa] \phi_{xx}[\kappa]}_{\text{variance reduction using optimal prediction}} . \end{aligned} \quad (A.155)$$

Discussion

The calculation of the predicted sample $\tilde{x}[k]$ and that of the prediction error $\epsilon[k]$ can be interpreted as linear filtering, as depicted in Figure A.9. As can be seen from (A.147), which it a convolution sum, the predicted sample $\tilde{x}[k]$ is the output of the strictly causal *prediction filter*

$$P(z) \stackrel{\text{def}}{=} \sum_{k=1}^{\rho} p[k] z^{-k} , \quad (A.156)$$

where the sequence $\langle x[k] \rangle$ is input. The prediction error $\epsilon[k]$ is the output of the *prediction error filter*

$$E(z) \stackrel{\text{def}}{=} 1 - P(z) = 1 - \sum_{k=1}^{P} p[k] z^{-k} . \quad (A.157)$$

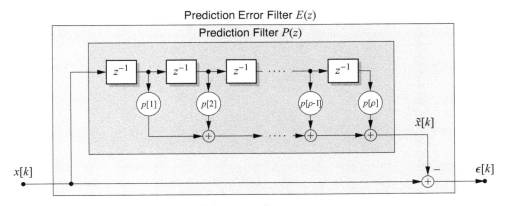

Figure A.9 Prediction filter and prediction error filter.

It can be shown that increasing the prediction order ρ, the prediction error $\epsilon[k]$ approaches a white stochastic process. As long as $\langle\epsilon[k]\rangle$ is not a white sequence, not all correlations in $\langle x[k]\rangle$ have been exploited and an increase of ρ will give a lower variance σ_ϵ^2. Hence, for $\rho \to \infty$, the prediction error filter is a *whitening filter* and we have

$$\left|E\left(e^{j2\pi fT}\right)\right|^2 \Phi_{xx}\left(e^{j2\pi fT}\right) = \sigma_\epsilon^2 . \tag{A.158}$$

Moreover, it can be shown that $E(z)$ is minimum-phase when the coefficients are adjusted via the Yule–Walker equations.

APPENDIX B
Short Introduction to Channel Coding

Channel coding is an advanced technology in digital communications that is based on sophisticated mathematical theories. An elaborate treatment of this topic is far beyond the scope of this introductory book on digital communications – already the essentials of channel coding fill textbooks. However, in order to understand the operation of today's communication systems, some basic knowledge of channel coding is indispensable.

In this appendix, a brief introduction to *block coding* is first provided. We outline the basics of linear block codes, especially of *low-density parity-check codes (LDPC codes)* and *polar codes*. An introduction to the mathematical background of iterative *soft-decision decoding* of linear codes is given. Moreover, we look a little deeper into *trellis coding*, because this is an inherent part of many modulation schemes and dispersively distorted signals. Finally, a specific version of mapping, that is, the transition of code symbols to signal numbers, namely *modulus conversion*, is described.

B.1 Binary Block Codes

B.1.1 Basic Definitions

Without loss of generality, the exposition is restricted to *binary* codes, as non-binary codes may be generated via a subsequent mapping. In order to design good codes and to efficiently implement encoding and decoding for long codes, suitable algebraic operations are required. To that end, the binary symbols are identified with the elements of the *binary finite field* $\mathbb{F}_2$. Quantities from a finite field are typeset in Fraktur font (cf. also Chapter 1). The addition $\mathfrak{c}_1 \oplus \mathfrak{c}_2$ of two binary symbols $\mathfrak{c}_1$ and $\mathfrak{c}_2$ is given by the XOR operation and the multiplication $\mathfrak{c}_1 \odot \mathfrak{c}_2 = \mathfrak{c}_1 \mathfrak{c}_2$ by the AND operation, in detail

$$\mathbb{F}_2 = \{0, 1\} \quad \text{with} \quad \begin{array}{ll} 0 \oplus 0 = 0 & 0 \odot 0 = 0 \\ 0 \oplus 1 = 1 & 0 \odot 1 = 0 \\ 1 \oplus 0 = 1 & 1 \odot 0 = 0 \\ 1 \oplus 1 = 0 & 1 \odot 1 = 1 \end{array}. \quad \text{(B.1)}$$

Note that 1 is its own inverse element with respect to addition, that is, in $\mathbb{F}_2$, subtraction is the same as addition.

Encoding and Code

Block encoding is defined as the bijective mapping of words of length K of binary source or information symbols $\mathfrak{q}_k$ (bits) onto *codewords*, words of length N ($K \le N$)

of binary code symbols c_n, that is

$$\mathbf{q} = [\,q_0\ q_1\ q_2\ \ldots\ q_{K-1}\,] \mapsto \mathbf{c} = [\,c_0\ c_1\ c_2\ \ldots\ c_{N-1}\,]\,. \tag{B.2}$$

For $K < N$, *redundancy* is introduced – not all words of length N with code symbols from $\mathbb{F}_2$ are used for transmission. Thereby, error detection and/or error correction is enabled.

The (block) *code* $\mathbf{C}$ is the set of codewords $\mathbf{c}$ that can be generated by the encoding process (and which are then used for transmission), that is

$$\mathbf{C} = \{\,\mathbf{c} \mid \mathbf{q} \mapsto \mathbf{c},\ \forall \mathbf{q}\,\}\,. \tag{B.3}$$

Decoding

Decoding is the estimation of the binary sourceword $\hat{\mathbf{q}}$ and/or the binary codeword $\hat{\mathbf{c}}$ from the block of decision variables $\mathbf{d} = [\,d_0\ d_1\ d_2\ \ldots\ d_{N-1}\,]$ of length N associated with the code symbols at the receiver side. An estimation of the entire codeword is performed, rather than individual symbol-by-symbol detection, followed by inverting the encoding.

In block coding, encoding and decoding, respectively, are independently done for each block, that is, block by block separately. There is no memory between the blocks.

Parameters

The most important parameters of a channel code are

- codelength $\qquad\qquad\qquad\qquad\qquad\qquad N$
- message length; code dimension[1] $\qquad\quad K$
- code rate $\qquad\qquad\qquad\qquad\qquad\qquad R_c = \frac{1}{N}\log_2\left(2^K\right) = \frac{K}{N}\ \left[\frac{\text{bit}}{\text{code symbol}}\right].$
  (average information per code symbol)

Example B.1 Simple Block Codes

The simplest example of a code is the binary $(N, K) = (N, 1)$ *repetition code*. Here, $K = 1$, thus $R_c = 1/N$. The encoding is visualized in Figure B.1. The source symbol q_0 is simply repeated N times.

Figure B.1 Encoding of the repetition code.

The next simple example is a binary $(N, K) = (N, N - 1)$ *single parity-check code*. Here, $K = N - 1$, thus $R_c = (N - 1)/N$. The encoding is visualized in Figure B.2.

---

[1] The denomination *code dimension* for the amount of information represented by one codeword (i.e., the number of *information bits*) originates from the theory of *linear codes*, where the code constitutes a K-dimensional linear subspace of the vector space of all possible words of length N over $\mathbb{F}_2$.

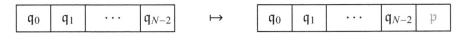

Figure B.2 Encoding of a single parity-check code.

A *parity bit* $\mathfrak{p}$ is appended, such that the number of symbols "1" is even. This calculation of the parity bit can be expressed as $\mathfrak{p} = \mathfrak{q}_0 \oplus \mathfrak{q}_1 \oplus \ldots \oplus \mathfrak{q}_{n-2}$. Using this code, a single bit error can be *detected*.

A further, well-known example of a binary code is the (7,4) *Hamming code*, where $K = 4$ and $N = 7$, thus $R_c = 4/7$. The encoding is visualized in Figure B.3.

| $\mathfrak{q}_0$ | $\mathfrak{q}_1$ | $\mathfrak{q}_2$ | $\mathfrak{q}_3$ | $\mapsto$ | $\mathfrak{q}_0$ | $\mathfrak{q}_1$ | $\mathfrak{q}_2$ | $\mathfrak{q}_3$ | $\mathfrak{p}_1$ | $\mathfrak{p}_2$ | $\mathfrak{p}_3$ |

Figure B.3 Encoding of the (7,4) Hamming code.

Here, three parity bits are appended to a block of four source bits, calculated as follows:

$$\mathfrak{p}_1 = \mathfrak{q}_1 \oplus \mathfrak{q}_2 \oplus \mathfrak{q}_3$$
$$\mathfrak{p}_2 = \mathfrak{q}_0 \oplus \mathfrak{q}_2 \oplus \mathfrak{q}_3$$
$$\mathfrak{p}_3 = \mathfrak{q}_0 \oplus \mathfrak{q}_1 \oplus \mathfrak{q}_3 \,.$$

Note that each of these equations demand even parity within the set of involved symbols. For example, the first equation can also be written as $\mathfrak{p}_1 \oplus \mathfrak{q}_1 \oplus \mathfrak{q}_2 \oplus \mathfrak{q}_3 = 0$, that is, the $\mathbb{F}_2$-sum over the four symbols has to be zero, which is achieved if the number of "1"s is even within the set $\{\mathfrak{p}_1, \mathfrak{q}_1, \mathfrak{q}_2, \mathfrak{q}_3\}$.

Using this code, a single bit error can be *corrected*.

Example B.2 Systematic Binary Linear Block Encoding for Error Control

The encoding is called *systematic* if the source bits are contained unaltered within the codewords and $N - K$ binary *parity-check symbols* $\mathfrak{p}_i$ are appended, that is

$$\mathbf{q} = [\mathfrak{q}_0 \; \mathfrak{q}_1 \; \ldots \; \mathfrak{q}_{K-1}] \mapsto \mathbf{c} = [\mathfrak{q}_0 \; \mathfrak{q}_1 \; \ldots \; \mathfrak{q}_{K-1} \; \mathfrak{p}_0 \; \ldots \; \mathfrak{p}_{N-K-1}] \,.$$

These check symbols are generated by $N - K$ *parity-check equations* in which different subsets of information symbols are involved. For *linear codes*, the parity of the checks always has to be even. Here, the receiver-side extraction of the sourceword from an estimated codeword is obvious.

Example B.3 Redundancy of Binary Block Codes

For (binary) block codes, only 2^K out of 2^N possible words are usable codewords (i.e., elements of the code). Thus, for each codeword, there are 2^{N-K} words that do not belong to the code. The probability that a noisy channel produces an erroneous valid codeword vanishes exponentially with increasing codeword length N (cf. Tables B.1 and B.2) (roughly approximated using the rule of thumb $2^n \approx 10^{0.3n}$).

Table B.1 Number of words, codewords, and unused words per codeword. $R_c = 1/2$.

N	2^N	2^K	2^{N-K}
100	10^{30}	10^{15}	10^{15}
1000	10^{300}	10^{150}	10^{150}
10 000	10^{3000}	10^{1500}	10^{1500}

Table B.2 Number of words, codewords, and unused words per codeword. $R_c = 3/4$.

N	2^N	2^K	2^{N-K}
100	10^{30}	$10^{22.5}$	$10^{7.5}$
1000	10^{300}	10^{225}	10^{75}
10 000	10^{3000}	10^{2250}	10^{750}

The numbers of codewords and non-valid words per one codeword soon grow far beyond what one may call an "astronomic" number, even in everyday technical applications.

B.1.2 Linear Binary Block Codes

For efficient operations, the sourceword $\mathbf{q}$ and the codeword $\mathbf{c}$ are identified with row vectors[2] in vector spaces over $\mathbb{F}_2$:

$$\mathbf{q} \in \mathbb{F}_2^K, \qquad \mathbf{c} \in \mathbb{F}_2^N. \tag{B.4}$$

Sums of vectors, scalar products, and vector/matrix products are given in the usual way, but all operations are carried out in the arithmetics of $\mathbb{F}_2$.

A code $\mathbf{C}$ is *linear* if the superposition principle holds, which includes the fact that a sum of codewords is also a valid codeword:

$$\begin{matrix}\mathbf{q}_1 \mapsto \mathbf{c}_1 \\ \mathbf{q}_2 \mapsto \mathbf{c}_2\end{matrix} \quad \Rightarrow \quad \mathbf{q}_1 \oplus \mathbf{q}_2 \mapsto \mathbf{c}_1 \oplus \mathbf{c}_2. \tag{B.5}$$

As the sum of a binary codeword with itself yields the all-zero word, a linear code $\mathbf{C}$ forms the mathematical structure of a *group* with respect to vector addition. Therefore, the nomination *group code* is also used. A linear code $\mathbf{C}$ is a linear subspace $\mathbb{F}_2^K$ of dimension K of the vector space $\mathbb{F}_2^N$, which also explains the term "dimension" for the length of the sourceword.

---

[2] In coding theory, row vectors are preferably used as they intuitively correspond better to words than column vectors.

Generator Matrix

Any set of K linearly independent codewords $\mathbf{c}_k$, $k = 0,\ldots,K-1$, is a valid *basis* of this subspace. K codewords that form a basis are arranged into a $K \times N$ matrix $\mathfrak{G}$, referred to as a *generator matrix* of $\mathbf{C}$:

$$\mathfrak{G} = \begin{bmatrix} \mathbf{c}_0 \\ \vdots \\ \mathbf{c}_{K-1} \end{bmatrix}. \tag{B.6}$$

Encoding of a sourceword $\mathbf{q}$ is done by

$$\mathbf{c} = \mathbf{q}\,\mathfrak{G}. \tag{B.7}$$

The code $\mathbf{C}$ is thus the row space of the generator matrix

$$\mathbf{C} = \{\mathbf{c} = \mathbf{q}\,\mathfrak{G} \mid \mathbf{q} \in \mathbb{F}_2^K\}. \tag{B.8}$$

Note that there are as many different generator matrices (i.e., encoding rules) for a given linear code $\mathbf{C}$ as there are sets of K linearly independent codewords times $K!$ the number of possible permutations. Specifically, elementary operations (additions of two rows, permutations of rows) on the rows of $\mathfrak{G}$ result in another valid generator matrix.

A generator matrix for *systematic encoding*, where the sourceword is contained unaltered (in the first K positions) in the codeword, has the form

$$\mathfrak{G} = [\,\mathfrak{I}_K \;\; \mathfrak{P}_{K \times (N-K)}\,]. \tag{B.9}$$

Here, $\mathfrak{I}_K$ is the identity matrix of dimension K; the parity symbols are generated by the $K \times (N-K)$ matrix $\mathfrak{P}$.

For any linear code, a systematic encoding exists. However, a permutation of the code symbols may be necessary if the source symbols should appear at the first positions. A code with a permuted order of the symbols is equivalent (but not identical) to the original code and has identical error control capabilities.

Example B.4 Generator Matrices of Simple Block Codes

The generator matrix of the $(N, K) = (N, 1)$ repetition code is given by

$$\mathfrak{G} = [\,1\;1\;\cdots\;1\,], \tag{B.10}$$

that of the $(N, K) = (N, N-1)$ single parity-check code by

$$\mathfrak{G} = \begin{bmatrix} \mathfrak{I}_{N-1} & \begin{matrix} 1 \\ \vdots \\ 1 \end{matrix} \end{bmatrix}, \tag{B.11}$$

and that of the $(7, 4)$ Hamming code by

$$\mathfrak{G} = \begin{bmatrix} 1 & 0 & 0 & 0 & 0 & 1 & 1 \\ 0 & 1 & 0 & 0 & 1 & 0 & 1 \\ 0 & 0 & 1 & 0 & 1 & 1 & 0 \\ 0 & 0 & 0 & 1 & 1 & 1 & 1 \end{bmatrix}, \tag{B.12}$$

which follows directly from the equations for calculating the parity symbols.

Parity-Check Matrix

The linear subspace $\mathbf{C}$ of dimension K of the space $\mathbb{F}_2^N$ can also be defined by means of its *kernel* or *nullspace* $\mathbf{C}^\perp$ of dimension $N - K$, that is, the set of all vectors $\mathfrak{h}$ that are orthogonal to all codewords (Hesse normal form)

$$\mathfrak{c} \odot \mathfrak{h}^\mathsf{T} = \mathfrak{o}, \quad \forall \mathfrak{c} \in \mathbf{C}, \quad \mathfrak{h} \in \mathbf{C}^\perp. \tag{B.13}$$

Any set of $N - K$ linearly independent row vectors $\mathfrak{h}_k$, $k = 0,\ldots,N - K - 1$, forms a basis of the nullspace of $\mathfrak{G}$. The arrangement of such a basis into a $(N - K) \times N$ matrix

$$\mathfrak{H} = \begin{bmatrix} \mathfrak{h}_0 \\ \mathfrak{h}_1 \\ \vdots \\ \mathfrak{h}_{N-K-1} \end{bmatrix} \tag{B.14}$$

is called the *parity-check matrix* for the code $\mathbf{C}$. According to (B.13), each row of $\mathfrak{H}$ represents a parity-check equation (parity even) on a subset of code symbols, or compactly

$$\mathfrak{c}\,\mathfrak{H}^\mathsf{T} \stackrel{!}{=} \mathfrak{o}, \quad \forall \mathfrak{c} \in \mathbf{C}. \tag{B.15}$$

As, according to (B.7), the codewords are given by $\mathfrak{c} = \mathfrak{q}\mathfrak{G}$, (B.14) can also be written as $\mathfrak{q}\mathfrak{G}\mathfrak{H}^\mathsf{T} = \mathfrak{o}, \forall \mathfrak{q}$, which is only possible if

$$\mathfrak{G}\mathfrak{H}^\mathsf{T} = \mathfrak{o}_{K \times (N-K)}. \tag{B.16}$$

Note that the parity-check matrix is not unique. Any set of $N - K$ linearly independent check equations (i.e., vectors $\mathfrak{h}_i$ that span the nullspace $C^\perp$) are suitable. If the generator matrix $\mathfrak{G}$ of a linear code $\mathbf{C}$ is given in systematic form (B.9), a parity-check matrix can immediately be found by

$$\mathfrak{H} = [\,\mathfrak{P}^\mathsf{T} \ \mathfrak{I}_{N-K}\,], \tag{B.17}$$

since

$$\mathfrak{G}\mathfrak{H}^\mathsf{T} = [\,\mathfrak{I}\ \mathfrak{P}\,][\,\mathfrak{P}^\mathsf{T}\ \mathfrak{I}\,]^\mathsf{T} = [\,\mathfrak{I}\ \mathfrak{P}\,]\begin{bmatrix}\mathfrak{P}\\\mathfrak{I}\end{bmatrix}^\mathsf{T} = \mathfrak{P} \oplus \mathfrak{P} = \mathfrak{o}. \tag{B.18}$$

Example B.5 Parity-Check Matrices of Simple Block Codes

The parity-check matrix of the $(N,K) = (N,1)$ repetition code is given by

$$\mathfrak{H} = \begin{bmatrix} \mathfrak{I}_{N-1} & \begin{matrix}1\\\vdots\\1\end{matrix} \end{bmatrix}, \tag{B.19}$$

that of the $(N,K) = (N, N-1)$ single parity-check code by

$$\mathfrak{H} = [\,1\ 1\ \cdots\ 1\,], \tag{B.20}$$

and that of the $(7,4)$ Hamming code by

$$\mathfrak{H} = \begin{bmatrix} 0 & 1 & 1 & 1 & 1 & 0 & 0 \\ 1 & 0 & 1 & 1 & 0 & 1 & 0 \\ 1 & 1 & 0 & 1 & 0 & 0 & 1 \end{bmatrix}. \tag{B.21}$$

When a parity-check matrix $\mathfrak{H}$ of a code $\mathbf{C}$ is used as generator matrix, the so-called *dual code* $C^\perp$ is generated. The parity-check matrix of the dual code $C^\perp$ is the generator matrix of the original code $\mathbf{C}$. Many properties of a code can be derived from properties of the dual code (e.g., the minimum Hamming distance or the weight distribution).

Note that the repetition code and the single parity-check code are dual to each other.

Minimum Hamming Distance and Weight Distribution

The number of symbols in which two codewords of a code $\mathbf{C}$ differ is called the *Hamming distance* between these codewords. The number of symbols "1" in a codeword is denoted as the *weight* of this codeword. The *weight distribution* A_j, $j = 0, \ldots, N$, of a code gives the number of codewords with weight j. Obviously, $A_0 = 1$ (a single all-zero codeword) and $\sum_{j=0}^{N} A_j = 2^K$ (total number of codewords).

As a linear code forms a group, the number of codewords with Hamming distance j from any codeword to all other codewords is equal to the weight distribution A_j.

An important parameter of a code is its *minimum Hamming distance* $\delta_{\min}$ between any two codewords. For linear codes, the minimum Hamming distance is identical to the minimum (nonzero) weight over the codewords, that is, $\delta_{\min}$ is the smallest $j > 0$ for which $A_j > 0$.

There exist very sophisticated theories to construct codes with a high minimum Hamming distance for a given length and rate. Among these are the Reed–Solomon (RS) codes and the Bose–Chaudhuri–Hocquenghem (BCH) codes. The required advanced algebra goes far beyond the scope of this book. However, these algebraic approaches do not lead to channel coding schemes that asymptotically approach limits from information theory. It turned out that a high minimum Hamming distance is not the decisive parameter for performance.

The breakthrough in channel coding towards theoretical limits was the introduction of randomly constructed schemes like *Turbo codes* (convolutional codes concatenated by a quasi-random interleaver) and *low-density parity-check (LDPC) codes*. For LDPC codes, a sparse parity-check matrix with given optimized degree distributions may be generated randomly. Moreover, efficient soft-decision decoding is possible, see the next section.

Decoding Principles for Linear Binary Block Codes

Two basic types of decoding philosophies exist. On the one hand, *hard-decision decoding* is widely studied. There, (preliminary) decisions $\mathfrak{r}$ on code symbols are generated and fed to the decoder. On the other hand, in *soft-decision decoding*, no decisions are taken (and thereby no errors are generated at this stage) but the actual real-valued decision symbols d are passed to the decoder. The distance of d from the decision boundaries provides some reliability information about the corresponding code symbol; a soft-decision decoder utilizes this reliability information.

Using a soft-decision decoder, a further aspect comes into play. Often, it is desirable to not only produce an estimate on the codeword or estimates on the code symbols, but also to deliver information on how confident the decoder is about the estimate, that is, to produce *reliability information* on the code symbols. Such a procedure is called *soft-input/soft-output decoding*.

B.1.3 Hard-Decision Decoding of Linear Binary Block Codes

When taking decisions and applying the inverse of the mapping of binary code symbols to signal points, the receive word of finite-field symbols can be written as

$$\mathfrak{r} = \mathfrak{c} \oplus \mathfrak{e}, \qquad (B.22)$$

where $\mathfrak{c} \in \mathbf{C}$ is the transmitted codeword and $\mathfrak{e} \in \mathbb{F}_2^N$ is the superimposed error word.

Error Detection

In this case, it is detected whether an error occurred (i.e., whether $\mathfrak{r}$ is a valid codeword or not). Owing to the linearity of the code, all error patterns that are not valid codewords, $\mathfrak{e} \notin \mathbf{C}$, can be detected.

For checking $\mathfrak{r}$ to be a codeword, the *syndrome* is formed:

$$\mathfrak{s} \stackrel{\text{def}}{=} \mathfrak{r}\mathfrak{H}^\mathsf{T} = (\mathfrak{c} \oplus \mathfrak{e})\mathfrak{H}^\mathsf{T} = \mathfrak{c}\mathfrak{H}^\mathsf{T} \oplus \mathfrak{e}\mathfrak{H}^\mathsf{T} = \mathfrak{e}\mathfrak{H}^\mathsf{T}. \qquad (B.23)$$

If and only if $\mathfrak{s} \stackrel{!}{=} \mathfrak{o}_{1\times(N-K)}$, $\mathfrak{r}$ is a valid codeword and will be accepted by the receiver. Note that the syndrome only depends on the error pattern. As $\mathfrak{e} \in \mathbb{F}_2^N$ but $\mathfrak{s} \in \mathbb{F}_2^{N-K}$, the error pattern uniquely determines the syndrome but not the other way round.

By a code with minimum distance $\delta_{\min}$, it is guaranteed that up to

$$\delta_{\min} - 1 \qquad (B.24)$$

symbol errors per codeword can be detected.

Forward Error Correction

In forward error correction, from the receive word $\mathfrak{r}$ a decision $\hat{\mathfrak{c}}$ on the codeword is generated. A simple strategy is to calculate the syndrome $\mathfrak{s}$ and choose the minimum-weight error pattern $\mathfrak{e}$ for which $\mathfrak{s} = \mathfrak{e}\mathfrak{H}^\mathsf{T}$ holds. This is justified since on channels with a symbol error probability smaller than 1/2, a small number of errors is more likely than a large number of errors. Having this estimate $\hat{\mathfrak{e}}$ on the error word, the codeword estimate is calculated as

$$\hat{\mathfrak{c}} = \mathfrak{r} \oplus \hat{\mathfrak{e}}. \qquad (B.25)$$

If $N - K$ is small, a lookup table for $\mathfrak{s} \to \mathfrak{e}$ may be used; otherwise, algebraic methods should be used to produce $\hat{\mathfrak{e}}$ and thus $\hat{\mathfrak{c}}$.

Sophisticated theories exist to construct good error-correcting codes and corresponding hard-decision decoders. The reader is referred to the rich literature on this topic.

By a code with minimum distance $\delta_{\min}$ it is guaranteed that up to

$$\left\lfloor \frac{\delta_{\min}-1}{2} \right\rfloor \tag{B.26}$$

symbol errors per codeword can be corrected.

B.1.4 Soft-Decision Decoding of Simple Linear Binary Block Codes

In practice, soft-decision decoding should be preferred, since it provides more reliable decoding results and thus a better power efficiency compared to hard-decision decoding. Efficient soft-decision decoding of algebraic codes like RS and BCH codes is still an issue. However, for the well-performing code classes of low-density parity-check codes and polar codes, efficient decoders exist. Thereby, the decoding can be broken down into the basic principles of decoding single parity-check codes and repetition codes, respectively.

We assume a memoryless channel, that is, the code symbols are transmitted independently; the detection variable d_j only depends on the code symbol $\mathfrak{c}_j$ and the randomness of the channel at this position.

In soft-decision decoding, in a first step, reliabilities, specifically the *a-posteriori probabilities* for the code symbols given the respective observation

$$p_j \stackrel{\text{def}}{=} \Pr\{\mathfrak{c}_j = \mathrm{o} \mid d_j\}, \tag{B.27}$$

are calculated for each code symbol $\mathfrak{c}_j$ from the corresponding (analog) detection variable d_j.

Before we turn to the specific cases, a general principle utilized in channel decoding is explained. In Figure B.4, the received word $[d_0, d_1, \ldots, d_{N-1}]$ corresponding to a codeword $\mathfrak{c}$ is visualized schematically.

When decoding, the following fact has to be considered. Using a code, each code symbol $\mathfrak{c}_j$ is represented twice.

- First, directly by itself. The probability (B.27) for $\mathfrak{c}_j$ taking on a specific value (here always o) deduced from observation d_j is called *intrinsic probability*.

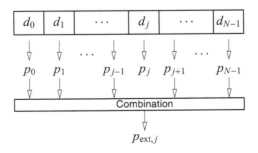

Figure B.4 Intrinsic and extrinsic probabilities.

- Second, by code constraints. Combining the intrinsic probabilities p_i, $i = 0, \ldots, j-1, j+1, \ldots, N-1$, of all other symbols, thereby taking the code constraints into account, the *extrinsic probability* of the code symbol c_j can be calculated.

As the channel is assumed to be memoryless, both assertions on the symbol c_j, intrinsic and extrinsic, are independent.

Note that it is easy to show that for linear block codes the possible code symbols "o" and "1" have equal *a-priori probability* at all positions within the codeword, that is

$$p_{\text{apri},j} \stackrel{\text{def}}{=} \Pr\{c_j = \text{o}\} = \tfrac{1}{2}, \qquad \forall j. \tag{B.28}$$

Expecting binary bipolar signaling (equal a-priori probabilities) with a mapping of $c = \text{o}$ to $a = +1$ and $c = 1$ to $a = -1$, and a transmission over an AWGN channel ($d = a + n$) with noise variance $\sigma_{n_\text{I}}^2$ per real dimension, the intrinsic probability (B.27) is given by

$$\begin{aligned}
\Pr\{c = \text{o} \mid d\} &= \frac{f_d(d \mid c = \text{o}) \Pr\{c = \text{o}\}}{f_d(d \mid c = \text{o}) \Pr\{c = \text{o}\} + f_d(d \mid c = 1) \Pr\{c = 1\}} \\
&= \frac{\frac{1}{\sqrt{2\pi\sigma_{n_\text{I}}^2}} e^{-(d-1)^2/(2\sigma_{n_\text{I}}^2)} \tfrac{1}{2}}{\frac{1}{\sqrt{2\pi\sigma_{n_\text{I}}^2}} e^{-(d-1)^2/(2\sigma_{n_\text{I}}^2)} \tfrac{1}{2} + \frac{1}{\sqrt{2\pi\sigma_{n_\text{I}}^2}} e^{-(d+1)^2/(2\sigma_{n_\text{I}}^2)} \tfrac{1}{2}} \\
&= \frac{e^{-(d-1)^2/(2\sigma_{n_\text{I}}^2)}}{e^{-(d-1)^2/(2\sigma_{n_\text{I}}^2)} + e^{-(d+1)^2/(2\sigma_{n_\text{I}}^2)}} \\
&= \frac{1}{1 + e^{(-(d+1)^2+(d-1)^2)/(2\sigma_{n_\text{I}}^2)}} \\
&= \frac{1}{1 + e^{-2d/\sigma_{n_\text{I}}^2}}.
\end{aligned} \tag{B.29}$$

Soft-Decision Decoding of the Repetition Code

We consider the *repetition code* of length N and having observations $d_0, \ldots, d_{N-1}$. Without loss of generality, we consider the code symbol c_0; the intrinsic probability is given by p_0. The extrinsic probability is calculated from the other $J = N - 1$ probabilities p_i. We have

$$p_{\text{ext,rep}} = \Pr\{c_0 = \text{o} \mid d_1, \ldots, d_J\}$$

Bayes' rule

$$= \frac{f_d(d_1, \ldots, d_J \mid c_0 = \text{o}) \Pr\{c_0 = \text{o}\}}{f_d(d_1, \ldots, d_J)}$$

B.1 Binary Block Codes

equal a-priori probabilities and law of total probability

$$= \frac{f_d(d_1, \ldots, d_J \mid c_0 = 0) \frac{1}{2}}{f_d(d_1, \ldots, d_J \mid c_0 = 0) \frac{1}{2} + f_d(d_1, \ldots, d_J \mid c_0 = 1) \frac{1}{2}}$$

observations are independent

$$= \frac{\prod_{j=1}^{J} f_y(d_j \mid c = 0)}{\prod_{j=1}^{J} f_y(d_j \mid c = 0) + \prod_{j=1}^{J} f_y(d_j \mid c = 1)}$$

Bayes' rule $f_d(d_j \mid c_j) = \frac{\Pr\{c_j \mid d_j\} f_d(d_j)}{\Pr\{c_j\}} = 2 \Pr\{c_j \mid d_j\} f_d(d_j)$

$$= \frac{\prod_{j=1}^{J} 2 \Pr\{c_j = 0 \mid d_j\} f_d(d_j)}{\prod_{j=1}^{J} 2 \Pr\{c_j = 0 \mid d_j\} f_d(d_j) + \prod_{j=1}^{J} 2 \Pr\{c_j = 1 \mid d_j\} f_d(d_j)}$$

simplify the fraction

$$= \frac{\prod_{j=1}^{J} \Pr\{c_j = 0 \mid d_j\}}{\prod_{j=1}^{J} \Pr\{c_j = 0 \mid d_j\} + \prod_{j=1}^{J} \Pr\{c_j = 1 \mid d_j\}}$$

using the definition of the a-posteriori probabilities and $\Pr\{c_j = 1 \mid y_j\} = 1 - \Pr\{c_j = 0 \mid y_j\}$

$$= \frac{\prod_{j=1}^{J} p_j}{\prod_{j=1}^{J} p_j + \prod_{j=1}^{J} (1 - p_j)} \ . \tag{B.30}$$

Note that for the repetition code, this formula with index j running from 1 to $J = N - 1$ gives the extrinsic probability for c_0. Having intrinsic and extrinsic probabilities (two independent observations on the same symbol, which correspond to a repetition code of length 2), the same formula gives the optimum combination of these two probabilities to the final decoding probability for c_0. Finally, it is easy to see that combining these two steps into a single calculation, the probability for $c_0 = 0$ considering all observations d_j and the code constraint is given by (B.30) with index running from 0 to J.

Soft-Decision Decoding of Single Parity-Check Codes

First, we consider the single parity-check code of length $N = 3$ and, without loss of generality, address code symbol c_0; the intrinsic probability is given by p_0. The extrinsic probability is calculated from the probabilities p_1 and p_2. We have

$$p_{\text{ext,pcc}} = \Pr\{c_1 \oplus c_2 = 0 \mid d_1, d_2\}$$

two combinations are possible

$$= \Pr\{c_1 = 0 \wedge c_2 = 0 \mid d_1, d_2\} + \Pr\{c_1 = 1 \wedge c_2 = 1 \mid d_1, d_2\}$$

observations are independent and using the definition of the a-posteriori probabilities

$$= p_1 p_2 + (1-p_1)(1-p_2) \tag{B.31}$$

or with basic rearrangements

$$= \tfrac{1}{2} + \tfrac{1}{2}(2p_1 - 1)(2p_2 - 1) . \tag{B.32}$$

The situation for other codelengths is shown by induction. As *base case* we presume

$$p_{\text{ext,pcc},J} = \Pr\{c_1 \oplus \cdots \oplus c_J = 0 \mid d_1, \ldots, d_J\} = \tfrac{1}{2} + \tfrac{1}{2}\prod_{j=1}^{J}(2p_j - 1) . \tag{B.33}$$

Direct inspection shows that this formula holds for $J = 1$ ($p_{\text{ext,pcc},1} = p_1$) and $J = 2$ (cf. (B.32)). The *induction step* shows

$$p_{\text{ext,pcc},J+1} = \Pr\{c_1 \oplus \cdots \oplus c_J \oplus c_{J+1} = 0 \mid d_1, \ldots, d_J, d_{J+1}\}$$

two combinations are possible and observations are independent

$$= \Pr\{c_1 \oplus \cdots \oplus c_J = 0 \mid d_1, \ldots, d_J\} \cdot \Pr\{c_{J+1} = 0 \mid d_{J+1}\}$$
$$+ \Pr\{c_1 \oplus \cdots \oplus c_J = 1 \mid d_1, \ldots, d_J\} \cdot \Pr\{c_{J+1} = 1 \mid d_{J+1}\}$$

using the definition of the probabilities

$$= p_{\text{ext,pcc},J}\, p_{J+1} + (1 - p_{\text{ext,pcc},J})(1 - p_{J+1})$$

or with basic rearrangements

$$= \tfrac{1}{2} + \tfrac{1}{2}(2p_{\text{ext,pcc},J} - 1)(2p_{J+1} - 1) = \tfrac{1}{2} + \tfrac{1}{2}\prod_{i=1}^{J+1}(2p_i - 1). \tag{B.34}$$

We note that (B.33) can also be written as

$$2p_{\text{ext,pcc},J} - 1 = \prod_{j=1}^{J}(2p_j - 1) . \tag{B.35}$$

In summary, for $J = N - 1$, (B.33) gives the extrinsic probability for code symbol c_0; the intrinsic probability is given by p_0. The probability for $c_0 = 0$ considering all observations d_j and the code constraint is given by combining these two probabilities via (B.30), leading to

$$\Pr\{c_0 = 0 \mid d_0, \ldots, d_{N-1}\} = \frac{p_0\, p_{\text{ext,pcc},N-1}}{p_0\, p_{\text{ext,pcc},N-1} + (1 - p_0)(1 - p_{\text{ext,pcc},N-1})} \tag{B.36}$$

or with some rearrangements

$$= \frac{p_0 \left(\tfrac{1}{2} + \tfrac{1}{2}\prod_{j=1}^{N-1}(2p_j - 1)\right)}{\tfrac{1}{2} + \tfrac{1}{2}\prod_{j=0}^{N-1}(2p_j - 1)} . \tag{B.37}$$

Finally, we consider a widely used approximation for the calculation of the extrinsic probability of parity-check equations. To that end we start with (B.35) and write it as

$$2p_{\text{ext,pcc},J} - 1 = \prod_{j=1}^{J}(2p_j - 1) = \prod_{j=1}^{J}\text{sgn}(2p_j - 1)\prod_{j=1}^{J}|2p_j - 1|. \quad \text{(B.38)}$$

For reliable symbols p_j is either close to 0 or 1 and, thus, $|2p_j - 1|$ is close to 1 in both cases. The product of magnitudes is dominated by unreliable symbols where p_j is close to $\frac{1}{2}$ or $|2p_j - 1|$ is close to zero. This leads to the approximation

$$2p_{\text{ext,pcc},J} - 1 \approx \prod_{j=1}^{J}\text{sgn}(2p_j - 1)\,\min_j |2p_j - 1|. \quad \text{(B.39)}$$

B.1.5 Tanner Graph and Basics of Iterative Decoding

Tanner Graph

The check properties of a parity-check matrix $\mathfrak{H}$ of a linear code **C** can be illustrated graphically. To that end, one side (circular nodes) shows N *variable nodes* c_i that represent the code symbols/columns of $\mathfrak{H}$; the other side (square nodes) shows $N-K$ *check nodes* $\mathfrak{p}_i$ that represent the check equations/rows of $\mathfrak{H}$. Since such a graph has two different types of nodes, it is called a *bipartite graph*.

Variable nodes c_i that are involved in a check equation (a "1" in $\mathfrak{H}$) are linked by edges to the corresponding check node $\mathfrak{p}_i$. Such a graph is called the *Tanner graph* for the parity-check matrix $\mathfrak{H}$. As the parity-check matrix of a code is not unique, the Tanner graph of a code is not unique; there exist many different Tanner graphs for a given linear code **C**.

Example B.6 Tanner Graph

In Figure B.5, the Tanner graph for the parity-check matrix (B.21) of the (7,4) Hamming code is depicted.

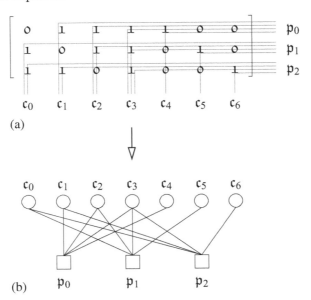

Figure B.5 (a) Given parity check matrix for the (7, 4) Hamming code and (b) its Tanner graph.

The number of edges that merge into a node is called its *degree (from a node perspective)*. A parity-check matrix $\mathfrak{H}$ or a Tanner graph, respectively, is denoted to be *regular* if all variable nodes have the same degree N_v and all check nodes have the same degree N_c. Irregular parity-check matrices are characterized by *degree distributions* $V_i, i = 1, \ldots, N - K$, for variable nodes and $C_i, i = 1, \ldots, N$, for check nodes denoting the relative frequency of nodes with degree i.

Example B.7 Node Degrees

The parity-check matrix (B.21) or the Tanner graph of Figure B.5, respectively, for the (7,4) Hamming code is irregular with respect to the variable nodes with degree distribution

$$V_1 = 3/7, \quad V_2 = 3/7, \quad V_3 = 1/7, \tag{B.40}$$

whereas all check nodes have degree 4, that is, $C_4 = 1$.

Often, an explicit parity-check matrix is not specified but only codelength, rate, and degree distributions are. These parameters specify an *ensemble* of parity-check equations and thus an ensemble of codes. A performance analysis is then done for these ensembles and not for a specific code realization.

For a *regular* parity-check matrix, code rate and node degrees are coupled via the number of entries 1 in the parity-check matrix $\mathfrak{H}$. Each row represents a check equation with N_c entries 1. Thus, the matrix $\mathfrak{H}$ contains $(N - K) \cdot N_c$ entries 1. Each column of $\mathfrak{H}$ represents a code symbol that has N_v connections to check equations. Thus, the number of "1"s in $\mathfrak{H}$ can also be written as $N \cdot N_v$. Hence, the code rate can be expressed as

$$(N - K) \cdot N_c \stackrel{!}{=} N \cdot N_v \quad \Rightarrow \quad R_c = \frac{K}{N} = 1 - \frac{N_v}{N_c}. \tag{B.41}$$

For irregular codes, the same formula holds for the average variable and check node degrees

$$R_c = 1 - \frac{\bar{N}_v}{\bar{N}_c}, \quad \text{with} \quad \bar{N}_v \stackrel{\text{def}}{=} \sum_{i=1}^{N-K} i V_i, \quad \bar{N}_c \stackrel{\text{def}}{=} \sum_{i=1}^{N} i C_i. \tag{B.42}$$

Basics of Iterative Decoding

Using the pictorial tool Tanner graph and knowledge of how the repetition code and the single parity-check code are soft-input/soft-output decoded, general linear block codes can be decoded. To that end, we presume that the Tanner graph or the parity-check matrix $\mathfrak{H}$ is given.

After having received the block of detection variables d_i corresponding to a codeword, for initialization, the intrinsic probabilities $p_i, i = 0, \ldots, N-1$, are calculated (cf. (B.29) for bipolar 2-ary transmission). The probabilities (beliefs) at the check nodes p_j are set to 1/2.

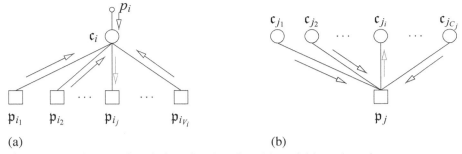

(a)   (b)

Figure B.6 Basic operations in iterative decoding. (a) Variable-node update; (b) check-node update.

For iterative decoding, two basic operations are performed (see Figure B.6).

- *Variable-Node Update*

  A single variable node i (the ith column in the parity-check matrix; see Figure B.6(a)) is considered. V_i check nodes and the node representing the intrinsic probability from the channel (depicted at the top of Figure B.6(a)) are connected to the variable node. All respective edges represent the same data. Hence, a repetition code is present. Knowing the intrinsic probability p_i and the probabilities the connected check nodes hold about the variable, extrinsic probabilities for each check node can be calculated. To that end, (B.30) is applied, where the index runs over all connected check nodes except the considered one ($j = i_l$, $l \in \{1,\ldots,V_i\} \setminus \{j\}$). This is done for all check nodes.

- *Check-Node Update*

  A single check node j (the jth row in the parity-check matrix; see Figure B.6(b)) is considered. C_j variable nodes are connected to the check node. The $\mathbb{F}_2$-sum over the variables corresponding to the respective edges has to be zero. Hence, a single parity-check code is present. Knowing the probabilities the connected variable nodes hold about the variable, extrinsic probabilities for each variable node can be calculated. To that end, (B.33) is applied, where the index runs over all connected variable nodes except the considered one ($i = j_l, l \in \{1,\ldots,C_j\}\setminus\{i\}$). This is done for all variable nodes.

These two update operations are carried out according to some scheduling (often all variable nodes are updated in sequence and then all check nodes are updated in sequence) and iterated several times. This process is continued until either hard decisions on the code symbols represent a valid codeword, that is, all parity-check equations are satisfied (syndrome zero), or a maximum number of iterations is reached. Usually, for moderate-to-high signal-to-noise ratios, only 10 to 40 iterations are necessary for successful decoding. The final probabilities on the code symbols are given by combining all probabilities via (B.30) obtained from the connected check nodes and the respective intrinsic probability.

This process of distributing and consolidating the different beliefs about each symbol is called *belief propagation*.

In the derivation of the combination equations (B.30) and (B.33), statistical independence of the observations has been assumed. If (as is usually the case) the Tanner graph contains cycles, feedback cannot be avoided and the independence of the observations is lost. If the parity-check matrix is sparse (low-density parity-check code), that is, the Tanner graph contains only a few edges per node, short cycles can be avoided and a performance loss due to a violation of the assumptions can be avoided to a large extent. The length of the shortest cycle in the Tanner graph is called the *girth*.

If it were possible to avoid any information feedback by cycles, this iterative process would achieve the performance of optimum maximum-likelihood symbol-by-symbol decoding after complete belief propagation throughout the Tanner graph.

B.1.6 Log-Likelihood Ratio and Practical Considerations

As discussed above, update equations (B.30) and (B.33) or (B.35), respectively, have to be applied alternately. These equations require a large number of multiplications and divisions. Since the logarithm transforms multiplications into additions, logarithmic representations of the probabilities are favorable for implementations.

Log-Likelihood Ratio

The expression of a probability p by its *log-likelihood ratio (LLR)* is defined as

$$L \stackrel{\text{def}}{=} \log\left(\frac{p}{1-p}\right). \tag{B.43}$$

The LLR provides an expansion of the probability interval $[0, 1]$ to the entire real number line; in particular, probabilities close to 0 or 1 are expanded to large negative or positive numbers, respectively. This is beneficial for numerical stability.

Note that the probability corresponding to a log-likelihood ratio, the inverse function to (B.43), is

$$p = \frac{1}{1 + e^{-L}}. \tag{B.44}$$

A further, favorable representation of a probability p is its corresponding *bipolar likelihood variable*

$$b \stackrel{\text{def}}{=} 2p - 1, \tag{B.45}$$

which maps the probability interval $[0, 1]$ to the interval $[-1, 1]$. Considering (B.35), the calculation of the extrinsic probability of parity-check equations is given by multiplying such quantities. The inverse mapping is here

$$p = \frac{1}{2} + \frac{1}{2} b. \tag{B.46}$$

Using the mathematical functions

$$\tanh(x) = \frac{e^x - e^{-x}}{e^x + e^{-x}}, \qquad \operatorname{arctanh}(x) = \frac{1}{2} \log\left(\frac{1+x}{1-x}\right), \tag{B.47}$$

one can write

$$b = 2p - 1 = \frac{2}{1+e^{-L}} - 1 = \frac{1-e^{-L}}{1+e^{-L}} = \frac{e^{L/2} - e^{-L/2}}{e^{L/2} + e^{-L/2}} = \tanh(L/2) \quad \text{(B.48)}$$

or

$$L = 2\operatorname{arctanh}(b) = 2\operatorname{arctanh}(2p-1) . \quad \text{(B.49)}$$

Intrinsic Probability

For binary, bipolar ASK with amplitudes ± 1 over the AWGN channel, the LLR of the intrinsic probability (B.29) is given by

$$L_j \stackrel{\text{def}}{=} \log\left(\frac{p_j}{1-p_j}\right) = \log\left(\frac{\Pr\{c_j = 0 \mid d_j\}}{1 - \Pr\{c_j = 0 \mid d_j\}}\right) = \log\left(\frac{\frac{1}{1+e^{-2d/\sigma_{n_I}^2}}}{1 - \frac{1}{1+e^{-2d/\sigma_{n_I}^2}}}\right)$$

$$= \log\left(\frac{1}{e^{-2d_j/\sigma_{n_I}^2}}\right) = \frac{2}{\sigma_{n_I}^2} d_j . \quad \text{(B.50)}$$

The LLR is simply proportional to the decision variable d_j.

Repetition Code

The combination (B.30) of the individual observations in case of a repetition code can be written in LLRs as

$$L_{\text{ext,rep}} = \log\left(\frac{p_{\text{ext,rep}}}{1 - p_{\text{ext,rep}}}\right) \stackrel{\text{(B.30)}}{=} \log\left(\frac{\frac{\prod_{j=1}^J p_j}{\prod_{j=1}^J p_j + \prod_{j=1}^J (1-p_j)}}{1 - \frac{\prod_{j=1}^J p_j}{\prod_{j=1}^J p_j + \prod_{j=1}^J (1-p_j)}}\right)$$

$$= \log\left(\frac{\prod_{j=1}^J p_j}{\prod_{j=1}^J (1-p_j)}\right) = \log\left(\prod_{j=1}^J \frac{p_j}{1-p_j}\right)$$

$$= \sum_{j=1}^J \log\left(\frac{p_j}{1-p_j}\right) \stackrel{\text{(B.43)}}{=} \sum_{j=1}^J L_j , \quad \text{(B.51)}$$

that is, a simple summation of the LLRs has to be done.

This transform of the update equation for variable nodes, that is, the combination of independent probabilities for the same binary variable, into an addition is the main benefit of using LLRs.

Single Parity-Check Code

The combination (B.33) of the individual observations in case of a single parity-check code can be written in LLRs as

$$L_{\text{ext,pcc}} \stackrel{(B.49)}{=} 2\operatorname{arctanh}(2p_{\text{ext,pcc}} - 1) \stackrel{(B.33)}{=} 2\operatorname{arctanh}\left(\prod_{j=1}^{J} \underbrace{(2p_j - 1)}_{b_j}\right)$$

$$\stackrel{(B.48)}{=} 2\operatorname{arctanh}\left(\prod_{j=1}^{J} \tanh(L_j/2)\right). \quad (B.52)$$

Taking into account that $\tanh(\cdot)$ and $\operatorname{arctanh}(\cdot)$ are strictly increasing, point-symmetric functions, we first obtain $\operatorname{sgn}(2p_j - 1) = \operatorname{sgn}(2\operatorname{arctanh}(2p_j - 1)) = \operatorname{sgn}(L_j)$. Approximation (B.39) then leads to

$$L_{\text{ext,pcc}} \stackrel{(B.39)}{\approx} 2\operatorname{arctanh}\left(\prod_{j=1}^{J} \operatorname{sgn}(2p_j - 1) \min_{j} |2p_j - 1|\right)$$

$$= \prod_{j=1}^{J} \operatorname{sgn}(2p_j - 1)\, 2\operatorname{arctanh}\left(\min_{j} |2p_j - 1|\right)$$

$$= \prod_{j=1}^{J} \operatorname{sgn}(L_j) \min_{j} |L_j|. \quad (B.53)$$

Practical Considerations

If LLRs are employed for representing probabilities, the nonlinear functions $\tanh(\cdot)$ and its inverse $\operatorname{arctanh}(\cdot)$ have to be implemented in a numerically efficient way or by means of approximations. In the majority of cases, an implementation by means of lookup tables will provide sufficient accuracy.

From (B.52), it can be seen that multiplications still have to be performed for the check-node updates. These are avoided by using the logarithm of the bipolar likelihood variables, separated into magnitude m and sign s:

$$m \stackrel{\text{def}}{=} \log(|b|), \qquad s = \operatorname{sgn}(b) \in \{-1, 1\}. \quad (B.54)$$

Here, the update for a single parity-check code (check nodes) reads

$$m_{\text{ext,pcc}} = \sum_{j=1}^{J} m_j, \qquad s_{\text{ext,pcc}} = \prod_{j=1}^{J} s_j. \quad (B.55)$$

(The multiplication of ± 1 is trivial.)

If neither an approximation (B.53) nor (nontrivial) multiplications are accepted, the pair m, s can be used for a representation of probabilities. Here, the two nonlinear functions $\log(|\tanh(\cdot)|)$ and $\operatorname{arctanh}(\exp(\cdot))$ have to be properly approximated or implemented by means of lookup tables.

B.1.7 Remarks on LDPC Codes and Belief Propagation

Some remarks on LDPC codes and belief propagation may be made.

- For a sparse matrix only a few check (variable) nodes are connected to a variable (check) node. Hence, the number N_v of probabilities/log-likelihood ratios that have to be combined per node is typically small.
- Although the total implementation effort for iterative decoding may be high, the number of calculations per decoded code symbol is independent of the length of the code. Thus, iterative decoding can be applied even to very long codes.
- Employing well-designed degree distributions for variable and check nodes in the Tanner graph, local centers of variable nodes with a high degree together with check equations with a low degree may arise at which a successful updating of reliability information is possible, even at very low signal-to-noise ratios. The iterative process may then recover the entire codeword out of such centers of convergence. Therefore, very power-efficient coding schemes, which closely achieve limits from information theory, are based on irregular codes.
- Variable nodes of degree 1 should be avoided, because for such symbols the reliabilities will never be improved after the first iteration. Therefore, the construction of a parity-check matrix obtained directly from a systematic encoding (cf. (B.17)) is not suitable for iterative decoding.
- As for a long linear binary code a huge number of different parity-check matrices exist, parallel iterative decoding processes may be implemented based on different parity-check matrices. This is called multiple-bases belief propagation. As there is some probability that special channel noise situations that cannot be handled by one decoder can be resolved by another one, additional gains in power efficiency are possible by multiple-bases belief propagation.
- LDPC codes may be constructed pseudo-randomly while satisfying given degree distributions for variable and check nodes. Powerful design methods are available.
- Regular LDPC codes together with iterative belief-propagation decoding were first proposed by Robert Gallager in 1962 but were later forgotten. In 1996, David McKay recovered LDPC codes in the context of the upcoming interest in Turbo codes. Since that time, LDPC codes together with iterative decoding have found a lot of applications. Optimum designs for LDPC codes were developed by Richardson and Urbanke.

B.1.8 Polar Codes

Basic Construction and Generator Matrix

Besides LDPC codes, *polar codes* are very attractive in practice because they permit simple decoding and show good performance. In contrast to LDPC codes, polar codes are defined by their generator matrix.

Figure B.7 Basic building block of polar codes.

The basic building block is the 2×2 Hadamard-type matrix

$$\mathfrak{G}_2 = \begin{bmatrix} 1 & 0 \\ 1 & 1 \end{bmatrix}. \quad (B.56)$$

Figure B.7 shows the corresponding encoding structure.

A generator matrix of dimension $N = 2^\mu$ is generated as the μ-fold *Kronecker product* of $\mathfrak{G}_2$, that is

$$\mathfrak{G}_N^{\text{ext}} = \underbrace{\mathfrak{G}_2 \otimes \mathfrak{G}_2 \otimes \cdots \otimes \mathfrak{G}_2}_{\mu \text{ factors}}, \quad (B.57)$$

where $\otimes$ denotes the Kronecker product (here in the arithmetic of the binary finite field). In the spirit of the radix-2 *fast Fourier transform (FFT)*, the successive generation of larger matrices from matrices of half the dimension gives rise to a very simple and efficient structure for encoding.

Note that $\mathfrak{G}_N^{\text{ext}}$ is an $N \times N$ square matrix; as $K = N$ holds, no redundancy is present. To obtain an error-correcting code, some *rows* of this extended matrix have to be deleted. Equivalently, and more common when dealing with polar codes, the positions in the extended sourceword $\mathbf{u}$ are set to 0 – in the language of polar codes, the positions are *frozen*.[3] As any number of positions can be *frozen*, any rate from $1/N$ to $(N-1)/N$ can be adjusted in a very flexible way.

Example B.8 Polar Code – Matrix and Structure

For $N = 8 = 2^3 = 2^\mu$, the generator matrix reads

$$\mathfrak{G}_8^{\text{ext}} = \begin{bmatrix} 1 & 0 & 0 & 0 & 0 & 0 & 0 & 0 \\ 1 & 1 & 0 & 0 & 0 & 0 & 0 & 0 \\ 1 & 0 & 1 & 0 & 0 & 0 & 0 & 0 \\ 1 & 1 & 1 & 1 & 0 & 0 & 0 & 0 \\ 1 & 0 & 0 & 0 & 1 & 0 & 0 & 0 \\ 1 & 1 & 0 & 0 & 1 & 1 & 0 & 0 \\ 1 & 0 & 1 & 0 & 1 & 0 & 1 & 0 \\ 1 & 1 & 1 & 1 & 1 & 1 & 1 & 1 \end{bmatrix} \quad (B.58)$$

and a visualization of the encoding is depicted in Figure B.8. Here, the structure corresponding to the procedure of *decimation in frequency* is shown.

---

[3] In principle, the positions may also be set to 1 or some to 0 and some to 1. If the positions are set to 0, a linear code is obtained. Otherwise, a *coset* of a linear code is present.

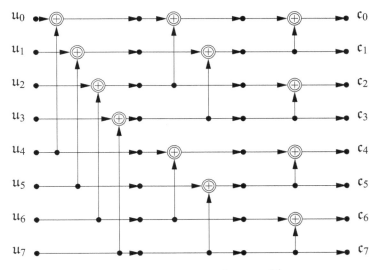

Figure B.8 Encoding structure corresponding to $\mathfrak{G}_8^{\text{ext}}$.

To obtain a rate-1/2 code ($K = 4$), the positions 0, 1, 2, and 4 may be frozen. Then, the extended sourceword **u** is related to the sourceword **q** by

$$[u_0, u_1, u_2, u_3, u_4, u_5, u_6, u_7] = [0, 0, 0, q_0, 0, q_1, q_2, q_3]. \qquad (B.59)$$

The question of which positions should be frozen can be answered when the basic operation principle has been clarified. To that end, we return to the 2×2 case. From (B.56) and Figure B.7, we have

$$[c_0, c_1] = [u_0 \oplus u_1, u_1]. \qquad (B.60)$$

Further, let us assume that the code symbols c_j are transmitted using binary bipolar ASK ("o" is mapped to +1 and "1" to −1) via two independent time steps. The situation in the two-dimensional signal space is shown in Figure B.9(a). With respect to the code symbols c_j, a *Gray mapping* is present. When drawing the same diagram with respect to the source symbols u_i (see Figure B.9(b)) a mapping by so-called *set partitioning* shows up, where a hierarchy of reliability for the binary label elements is present.

In a successive decoding procedure, first u_0 may be decoded ignoring the other symbols. This situation is depicted in Figure B.9(c). Both possible values of the symbol u_0 are represented twice in the signal space. Nevertheless, a decision on u_0 can be taken. Since each point has two nearest neighbors, the reliability is reduced compared to individual binary bipolar transmission where only a single nearest neighbor is active.

However, knowing u_0 (we assume that no error occurred), the symbol u_1 profits. This situation is depicted in Figure B.9(d), assuming $u_0 = 0$. The Euclidean distance is increased by a factor of $\sqrt{2}$ compared to individual binary bipolar transmission.

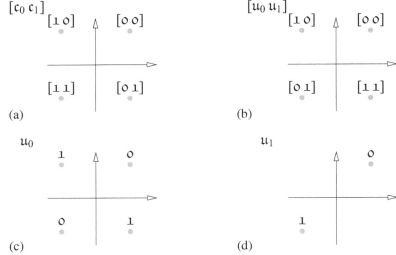

Figure B.9 Situation for a polar code with $N = 2$ in the two-dimensional signal space.

This fact is the fundamental principle of polar codes: starting from two identical channels (same reliability), one channel with poorer and one with better quality are generated. This procedure is repeated due to the iterated construction of the generator matrix. It has been shown in the literature that asymptotically ($N \to \infty$) either very good (error-free) or completely useless channels are generated. The fraction of these two types of channels is given exactly by the capacity of the underlying channel.

Hence, the design of polar codes amounts to identifying the positions of the good and bad channels. The positions in **u** corresponding to the bad channels are set to zero (frozen), and those corresponding to the good channels are filled with the source symbols q_i. Meanwhile, efficient methods for identifying the set of bad channels are available.

Decoding of Polar Codes

The successive principle discussed above can be generalized for soft-decision decoding of polar codes of arbitrary length N. Before we turn to this task, we first have another look at the basic 2×2 building block. To that end, the situations of encoding and decoding are visualized in Figure B.10.

The encoding is depicted in part (a). In part (b), the decoding of u_0 is given; the decision on u_1 is postponed for the moment. The structure is obtained from

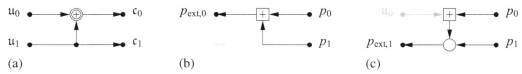

Figure B.10 Polar code with $N = 2$. (a) Encoding; (b) decoding of u_0; (c) decoding of u_1 when u_0 is known.

the encoding structure by reversing all signal paths. As is always the case, from the channel output *a-posteriori probabilities* p_j for the code symbols are calculated (see (B.27)). Remember that $c_0 = u_0 \oplus u_1$, that is, a *parity check* is present. From p_0 (which characterizes $u_0 \oplus u_1$) and p_1 (which characterizes $u_1 = c_1$), an extrinsic probability on u_0 can be calculated as derived in Section B.1.4. Repeating (B.32), the probability on u_0 is calculated according to

$$p_{\text{ext},0} = \tfrac{1}{2} + \tfrac{1}{2}(2p_0 - 1)(2p_1 - 1), \tag{B.61}$$

or from (B.52) when using LLRs

$$L_{\text{ext},0} = 2\operatorname{arctanh}(\tanh(L_0/2)\tanh(L_1/2)), \tag{B.62}$$

when L_0 and L_1 are the LLRs corresponding to p_0 and p_1, respectively.

Having $p_{\text{ext},0}$ or $L_{\text{ext},0}$, a decision on u_0 can be taken. This decision is then taken as certain, which means that further on $p_{\text{ext},0}$ is considered to be either 1 (if $u_0 = 0$) or 0 (if $u_0 = 1$) (or $L_{\text{ext},0}$ is either $+\infty$ or $-\infty$). If position 0 is frozen, the above calculations do not have to be carried out, since u_0 is known in advance.

Using the (perfect) knowledge on u_0 ($p_{\text{ext},0}$ is 0 or 1) and considering that $u_1 = c_0 \oplus u_0$, an extrinsic probability on u_1 can be calculated (top operation in Figure B.10(c)). To that end, again (B.61) or (B.62) is used. However, these equations simplify to

$$p_{0|u_0} = \begin{cases} p_0, & u_0 = 0 \\ 1 - p_0, & u_0 = 1 \end{cases}, \tag{B.63}$$

or when using LLRs

$$L_{0|u_0} = \begin{cases} L_0, & u_0 = 0 \\ -L_0, & u_0 = 1 \end{cases}. \tag{B.64}$$

Then, two (independent) observations on u_1 are available, that is, a *repetition code* is present. Consequently, a final extrinsic probability on u_1 can be calculated as derived in Section B.1.4. Repeating (B.30), the probability on u_1 is calculated as

$$p_{\text{ext},1} = \frac{p_{0|u_0} p_1}{p_{0|u_0} p_1 + (1 - p_{0|u_0})(1 - p_1)} = \begin{cases} \frac{p_0 p_1}{p_0 p_1 + (1-p_0)(1-p_1)}, & u_0 = 0 \\ \frac{(1-p_0) p_1}{(1-p_0) p_1 + p_0(1-p_1)}, & u_0 = 1 \end{cases}, \tag{B.65}$$

or from (B.51) when using LLRs as

$$L_{\text{ext},1} = L_{0|u_0} + L_1 = \begin{cases} L_0 + L_1, & u_0 = 0 \\ -L_0 + L_1, & u_0 = 1 \end{cases}. \tag{B.66}$$

Having $p_{\text{ext},1}$ or $L_{\text{ext},1}$, a decision on u_1 can be taken.

Since the encoding of polar codes is successively composed of the 2×2 building blocks (cf. Figure B.8 for $N = 8 = 2^3$ with three stages), decoding is done by repeatedly applying the above derived calculations. Owing to the structure, first the "upper" binary symbol (u_0) has to be processed. When a decision on this symbol is available (or when it is a frozen symbol), the "lower" binary symbol (u_1) can be processed. Going through the encoding structure, it can be deduced that the symbols of the extended sourceword **u** should be processed in *bit-reversed order*. This means that the indices $j = 0, 1, \ldots, N - 1$ are expressed as binary numbers (base 2) – these 0/1-words of length μ are reversed and converted again to decimal numbers. The obtained sequence gives the decoding order.

In summary, as for LDPC codes, the decoding of polar codes can be broken down into the decoding of single parity-check and repetition codes and only two basic combination equations are required.

Example B.9 Decoding of Polar Codes

We consider the decoding for $N = 8 = 2^3 = 2^\mu$ and $K = 4$, that is, a rate-1/2 code. The frozen positions have the indices $i \in \{0, 1, 2, 4\}$. The respective encoding structure has already been depicted in Figure B.8.

Via bit-reversal the decoding order is obtained as $i = 0, 4, 2, 6, 1, 5, 3, 7$. As the positions $i \in \{0, 4, 2\}$ are frozen, the first symbol to be decoded is u_6. Reversing all signal paths (except that corresponding to known symbols) in the encoding structure of Figure B.8, the respective decoding structure is obtained (cf. Figure B.11).

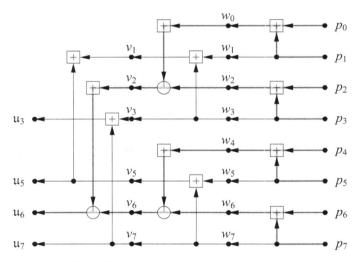

Figure B.11 Decoding structure for $N = 8$ and the frozen positions $i \in \{0, 1, 2, 4\}$. Situation when decoding u_6.

The frozen (and thus known) quantities are shown in light gray and, as they are elements from $\mathbb{F}_2$, are typeset in Fraktur font. The flow of information (either probabilities or LLRs; real-valued quantities, thus typeset in italic) from right to left is shown in shades of dark gray. Here, first the intermediate quantities w_0, w_2, w_4, and w_6 are calculated from p_j using (B.61) or (B.62). Since u_0 and u_4 are known to be $\mathfrak{o}$, $\mathfrak{v}_0 = \mathfrak{o}$ is known. Using (B.63) or (B.64), v_2 can be calculated from w_0 and w_2. The arguments hold for $\mathfrak{v}_4 = \mathfrak{o}$; here v_6 can be calculated. Finally, since $\mathfrak{v}_2 = \mathfrak{o}$, from v_2 and v_6 the likelihood on u_6 can be calculated and a decision can be made.

This decision is propagated back (from left to right); see the light gray parts of Figure B.12. Now, the intermediate quantities w_1, w_3, w_5, and w_7 can be calculated using (B.63) or (B.64), since w_0, w_2, w_4, and w_6 are known. Then, via (B.61) or (B.62), v_1 and v_5 are calculated, from which, finally, as u_1 is known, the likelihood on u_4 can be calculated and a decision can be made.

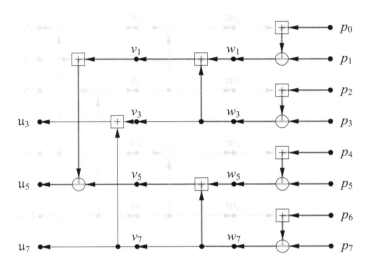

Figure B.12 Decoding structure for $N = 8$ and the frozen positions $i \in \{0, 1, 2, 4\}$. u_6 has already been decoded. Situation when decoding u_5.

Again, the decision is propagated back (from left to right); see the light gray parts of Figure B.13. The intermediate quantities v_3 and v_7 can be calculated and from that the likelihood on u_3.

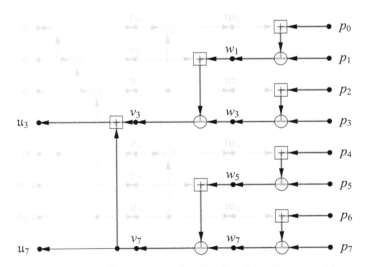

Figure B.13 Decoding structure for $N = 8$ and the frozen positions $i \in \{0, 1, 2, 4\}$. u_6 and u_5 have already been decoded. Situation when decoding u_3.

Finally (not shown), since u_3 is now known, using v_3 and v_7 the likelihood on u_7 can be calculated and the last decision can be taken.

B.2 Trellis Codes
B.2.1 Basic Definitions

In *trellis coding*, a sequence of M_y-ary output symbols $\mathfrak{y}[\ell]$ is created from a sequence of M_x-ary input symbols $\mathfrak{x}[\ell]$ in a recursive manner employing a *Mealy machine*. Previous input symbols $\mathfrak{x}[\ell - \lambda]$ influence the coding process for the actual symbol $\mathfrak{x}[\ell]$. For simplicity, but without loss of generality, we restrict our considerations to

$$M_x = 2^K, \quad M_y = 2^N, \quad R_c = \frac{K}{N},$$

that is, the encoder input symbols are blocks of K information bits and the encoder output symbols are blocks of N binary code symbols. Usually, K and N are chosen to be very small, for example, $K = 1$ and $N = 2$. The Mealy machine possesses Z different internal states $S[\ell] \in \{0, 1, \ldots, Z-1\}, \ell \in \mathbb{Z}$.

The recursive encoding process of a Mealy machine is depicted in Figure B.14.

The encoder is fully specified by its *output function*

$$\mathfrak{y}[\ell] = \mathrm{fct}_o(\mathfrak{x}[\ell], S[\ell]) \tag{B.67}$$

and its *state-transition function*

$$S[\ell + 1] = \mathrm{fct}_s(\mathfrak{x}[\ell], S[\ell]) . \tag{B.68}$$

These functions may be implemented by means of lookup tables without any further restrictions.

The functioning of a trellis encoder can be visualized by a *state-transition diagram* where the transitions are labeled by the input symbols causing a state transition and by the related output symbols (see the example in Figure B.16). The state-transition diagram represents the entire set of arguments and results for the state-transition function and the output function of the Mealy machine.

The state transitions are called *branches* and are labeled by the pair of input and output symbols as in Figure B.15.

The sequence of states forms a *Markov chain*. Evolving the state-transition diagram with respect to the discrete time index ℓ, the *trellis diagram* of a trellis encoder results. This diagram, looking like a trellis used for the growing of plants (e.g.,

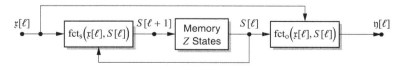

Figure B.14 Trellis encoder as Mealy machine.

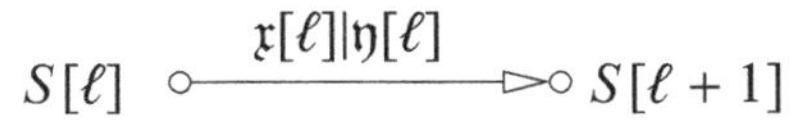

$$S[\ell] \circ \xrightarrow{\mathfrak{x}[\ell]\|\mathfrak{y}[\ell]} \circ S[\ell+1]$$

Figure B.15 Branch with starting and ending state and branch label.

roses) in front of a wall, gave this encoding principle its name. It is preferably used for the decoding of trellis codes at the receiver side.

The message to be encoded (i.e., the sequence of source symbols) is represented by a consecutive sequence of branches called a *path* through the trellis. The fact that not all paths are possible expresses the redundancy of the code. The sequence of output symbols along a path is a *codeword* of infinite length, at least in theory. In practice, the trellis encoding process is usually terminated and restarted at some synchronization instant, such that in fact a kind of block encoding results even for trellis encoding.

Example B.10 State-Transition Diagram and Trellis Diagram

An example of a state-transition diagram is shown in Figure B.16.

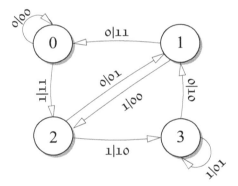

Figure B.16 Example of a state-transition diagram of a trellis encoder with four states, binary input ($K = 1$) and 4-ary output, represented by pairs of binary symbols ($N = 2$).

The corresponding trellis diagram is shown in Figure B.17. A particular path is highlighted in light gray.

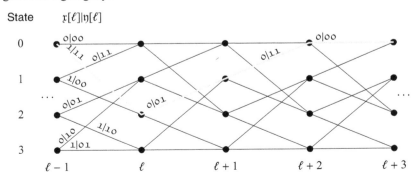

Figure B.17 Trellis diagram corresponding to the above state-transition diagram.

The task of the *trellis decoder* at the receiver side is to find the path through the trellis with highest a-posteriori probability using detection variables that are calculated from the observed channel output. A highly efficient implementation of a trellis decoder, the *Viterbi algorithm (VA)*, will be addressed below.

Subsequently, two widespread types of trellis encoders are briefly introduced.

B.2.2 Convolutional Codes

The encoder of a *convolutional code* is a *linear, dispersive* $K \times N$ multiple-input/multiple-output system for binary sequences. The output sequences are generated from convolutions of the input sequences with the impulse responses like in usual linear time-invariant systems, but the arithmetic is carried out over the binary field $\mathbb{F}_2$. Thus, convolutional codes are linear codes and a particular subclass of trellis codes.

The individual linear systems performing the convolutions may have a *non-recursive* (finite impulse response, FIR) or a *recursive* (infinite impulse response, IIR) structure, as is well known from digital signal processing. The states of the binary delay elements in the system implementation all together form the state of the Mealy machine.

Example B.11 Convolutional Encoder with Non-recursive Structure

Figure B.18 sketches the block diagram of a non-recursive convolutional encoder with rate $R_c = 1/2$ and four states.

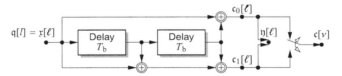

Figure B.18 Non-recursive encoder for a binary convolutional code.

Here, $K = 1$ (hence, the discrete time index ℓ of the encoder coincides with the discrete time index l of the source symbols) and $N = 2$ (combined in $\eta[\ell]$). In addition, the serialized version of the output, $c[\nu]$, is shown too. Here, the discrete time (index ν) runs twice as fast. Note that the encoding is non-systematic as the binary information symbols are not contained unaltered in the code sequence.

All possible combinations for the content of the shift-register elements form the memory states:

State	Memory
0	00
1	01
2	10
3	11

The state diagram and the trellis diagram for the code generated with this encoder have already been given in Figures B.16 and B.17.

The encoding of an example source sequence is given in Figure B.19 (at the start, the memory element contains the values oo). If the light gray source symbol in the fourth time interval is changed, due to the FIR structure, the produced output only differs in three time steps (also in light gray).

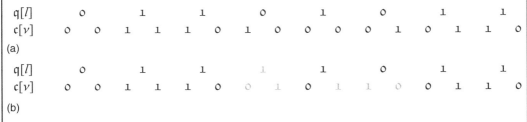

(a)

(b)

Figure B.19 Example of encoding a source sequence using the above convolutional encoder. (a) Example sequence; (b) effect of changing a single source bit (light gray).

Example B.12 Convolutional Encoder with Recursive Structure

Figure B.20 sketches the block diagram of an encoder for a binary convolutional code, that is, $M_x = 2$. Here, $N = 2$, $K = 1$, thus $R_c = 1/2$. It is a discrete-time, recursive, linear time-invariant system (infinite impulse response system, IIR), working over the finite field $\mathbb{F}_2$. As in Example B.11, the output symbols $\eta[l]$ from an M_y-ary alphabet are equivalently serialized into the binary sequence $c[v]$.

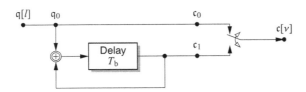

Figure B.20 Encoder for a binary trellis code.

Owing to the *recursive* structure, the mapping $q_0 \mapsto [c_0, c_1]$, that is, of the source bit ($K = 1$) onto the pair ($N = 2$) of coded bits, depends on the *entire past* of the source sequence. A change of a single source bit will change the entire output sequence from the modified position on.

The encoding of an example source sequence is given in Figure B.21 (at the start, the memory element contains the value o). Looking at the encoder, for even time indices v, the serialized stream of coded symbols is equal to the source bits (the upper direct branch in the encoder) and for odd time indices, it is equal to the state of the delay (memory) element. If the light gray source symbol in the fourth time interval is changed, the subsequent output (here all symbols at even positions) changes (also in light gray).

```
q[l]        0       1       1       0       1       0       1       1
c[ν]    0   0   1   0   1   1   0   0   1   0   0   1   1   1   1   0
(a)
q[l]        0       1       1       1       1       0       1       1
c[ν]    0   0   1   0   1   1   1   0   1   1   0   0   1   0   1   1
(b)
```

Figure B.21 Example of encoding a source sequence using the above trellis encoder. (a) Example sequence; (b) effect of changing a single source bit (light gray).

The encoding is systematic as the binary information symbols are evidently contained within the sequence of code symbols. Remember that the encoder shown in Figure B.20 had appeared already several times in Chapters 4, 6, and 7 for differential encoding or for establishing signal continuity.

Example B.13 Trellis Representation of Block Codes

Block codes may be interpreted as terminated trellis codes: for these, trellis diagrams over N steps can be derived. This point of view is favorable for the efficient soft-decision decoding using trellis-based algorithms like the Viterbi algorithm.

A straightforward method to define such a trellis for binary linear codes with systematic encoding is to take all 2^{N-K} different tuples of parity symbols as trellis states. Starting from the all-zero codeword, the trellis is evolved by filling successively the positions of information bits and combining this with trellis branches to the corresponding current status of the parity sequence. When, after K steps, all combinations of information bits are mapped to a trellis path, the states remain fixed until step N.

Example B.14 Signal with Intersymbol Interference (ISI)

In many situations, the required temporal orthogonality of signal elements is not present, either not implemented at the transmitter or lost due to signal distortions in the channel. In such situations, intersymbol interference (ISI) has to be processed at the receiver. Any sort of intersymbol interference, either linear or nonlinear, can be modeled by a trellis encoder with a shift register memory (FIR structure) combined with a set of signal elements representing all possible signal trajectories within the modulation interval T.

When ISI is spread over L modulation intervals and M-ary modulation is used, up to M^L signal elements exist that may be represented in a suitably defined signal space, see Chapter 6.

For PAM and linear signal distortion, a signal space with one dimension per modulation step exists (cf. Chapter 8). Its basis function is the impulse response of the whitened-matched filter, matched to the end-to-end pulse that violates the temporal orthogonality condition. An optimum receiver for ISI implies a trellis decoder for an encoder with FIR structure.

B.2.3 The Viterbi Algorithm for Trellis Decoding

The *Viterbi algorithm* (named after Andrew J. Viterbi) is a "shortest-path" algorithm, which means it finds the "best" path through a trellis. To each branch in the trellis a *cost* or *metric* is assigned – the cost of a path through the trellis is the *sum* of the *branch costs* along the way. Note that the cost can be any real number (positive or negative). The algorithm can either find the path with the smallest or the largest accumulated cost. Without loss of generality, the algorithm is subsequently explained for searching for the path with *minimum accumulated cost*.

Basic Operation

In order to visualize the operation of the algorithm, a two-state trellis with states denoted as A and B is considered. The associated branch costs (for simplicity, all positive) are depicted in Figure B.22. The generalization to any number of states is straightforward.

The Viterbi algorithm works successively over the time steps k. Assume that we are at time step $k + 1$. The accumulated metric of state $S \in \{A, B\}$ at time index k is denoted as $\Lambda^{(S)}[k]$ and the branch metric for the branch connecting state S at time index k and state T at time index $k + 1$ is denoted as $\lambda^{(S) \to (T)}[k]$.

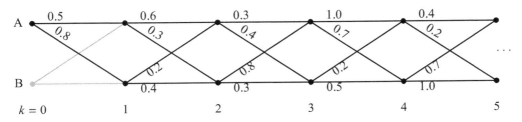

Figure B.22 Two-state trellis and associated branch costs.

For each state, the following operations are carried out.

- *Add.* The accumulated metric of each path merging into the considered state is calculated. This metric is simply the accumulated metric of the state from which the branch emerges, plus the branch metric of the branch. For example, for state A we have $\Lambda^{(A)}[k] + \lambda^{(A)\to(A)}[k]$ and $\Lambda^{(B)}[k] + \lambda^{(B)\to(A)}[k]$.
- *Compare.* The accumulated metrics of the *incoming* paths are compared.
- *Select.* The path with the smallest accumulated metric is selected. For example, for state A we have

$$\Lambda^{(A)}[k+1] = \min\{\Lambda^{(A)}[k] + \lambda^{(A)\to(A)}[k],\ \Lambda^{(B)}[k] + \lambda^{(B)\to(A)}[k]\}.$$

This gives the *survivor* for the state (if the path with the largest metric has to be searched, at this step the path with the largest accumulated metric is selected instead). This selection is justified as a path that is inferior to a competing path at the merge can never become the better one later on.

Example B.15 Viterbi Algorithm

As an example, we perform the Viterbi algorithm for the above example trellis. Assume that the path has to start in the upper state A. The progress of the algorithm is depicted in Figure B.23.

At time index $k = 1$, there is only one branch merging into state A and B, respectively. Since the accumulated metric of state A at time $k = 0$ is assumed to be $\Lambda^{(A)}[0] = 0$, the paths (hence, states) have the accumulated metrics $\Lambda^{(A)}[1] = 0.5$ and $\Lambda^{(B)}[1] = 0.8$ (highlighted in dark gray).

At any time index k, state A can be reached from state A via the branch A $\to$ A or from state B via the branch B $\to$ A. A similar situation is present for state B. Hence, at time index $k = 2$, the two paths with the accumulated metrics $\Lambda^{(A)}[1] + \lambda^{(A)\to(A)}[1] = 0.5 + 0.6 = 1.1$ and $\Lambda^{(B)}[1] + \lambda^{(B)\to(A)}[1] = 0.8 + 0.2 = 1.0$ compete. Since $1.0 < 1.1$, the path coming from state B wins. The survivor is highlighted. For state B, the two incoming paths have the accumulated metrics $\Lambda^{(A)}[1] + \lambda^{(A)\to(B)}[1] = 0.5 + 0.3 = 0.8$ and $\Lambda^{(B)}[1] + \lambda^{(B)\to(B)}[1] = 0.8 + 0.4 = 1.2$. Here, the path from state A wins.

The procedure is repeated over the time steps. The survivor is always highlighted. At $k = 5$, the decoding result up to $k = 4$ is available because all (here two) surviving paths have an identical history, that is, the decision on these symbols will never change again. It is the path through the sequence of states A, A, B, B, A.

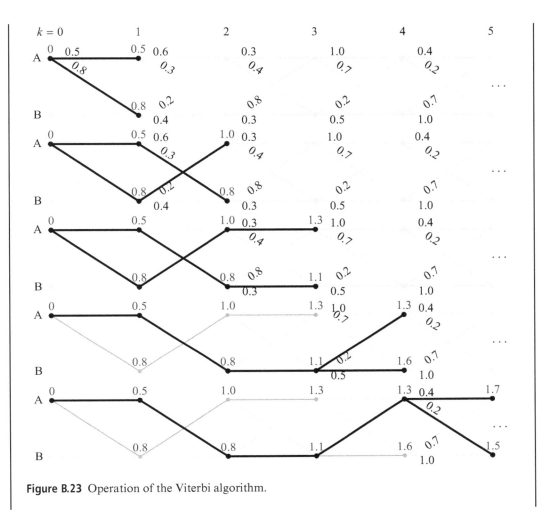

Figure B.23 Operation of the Viterbi algorithm.

This effect holds in general: at time step k, the decoding result is only unique up to some earlier time step with high probability. In other words: the decoding needs some looking ahead.

In practice, tests for unique path histories are usually avoided. Instead, final decisions are delivered from an arbitrary path and going a sufficient number of steps backward. Thus, one output symbol per time step is generated in a continuous flow. The sufficient number of steps looking back is minimized by taking the path with the currently minimum accumulated metric. A rule of thumb is to look $5 \log_2(\text{\#trellis states})$ time steps back in this path. Additionally, all accumulated metrics are usually normalized by subtraction of a constant from all values in the current step in order to avoid metric overflow in the long term. This constant may be the minimum of all actual accumulated metrics.

Finally, please be aware that the Viterbi algorithm works in the same way if the trellis has a larger number of states. Typically, not all branches from any previous to any subsequent state exist in this case (see the examples in Chapters 7 and 8). Of course, the add/compare step is only done for existing branches.

Metric for Trellis Decoding in Digital Communications

For ML soft-decision decoding of a trellis code in digital communications, the cost metric for a trellis branch should be proportional to the negative logarithm of the probability of the observed detection variable given the signal element that is associated with this branch. The random process of generating the detection variables has to be statistically independent over the modulation steps, such that these probabilities can be multiplied, that is, their logarithms can be added up along the path, as additivity of branch metrics is essential for the Viterbi algorithm. For the AWGN channel, the temporal orthogonality of signal elements and their corresponding matched filters guarantees this property. As derived in Section 6.2, the squared Euclidean distance between detection variable and signal element is an appropriate branch metric for the Viterbi algorithm, as well as the (negative) correlations with the signal elements. For hard-decision decoding, the number of symbols in which the received sequence differs from the encoder output symbols associated with a branch is an appropriate metric.

If the trellis branches are not a-priori equiprobable and, therefore, MAP decoding is desired instead of ML decoding, the suitably scaled negative logarithm of the a-priori probabilities of the branches has to be added to the ML branch metrics (cf. (3.85)).

Sequence vs. Symbol Detection

The Viterbi algorithm determines the entire path through a trellis that has maximum probability. Thus, either *maximum-likelihood sequence detection (MLSD)* or *maximum-a-posteriori sequence detection (MAPSD)* is efficiently implemented. Note that neither probabilities for individual symbols are calculated, nor optimum individual decisions on symbols are performed. Therefore, the Viterbi algorithm will usually not deliver individual symbol decisions at lowest possible error probabilities. This task, that is, *ML or MAP symbol-by-symbol trellis decoding (MLSSD or MAPSSD)*, is performed in an optimum way by the closely related algorithm of *Bahl–Cook–Jelinek–Raviv (BCJR)*.

B.3 General Version of Mapping

As discussed in Chapter 6, in the general version of digital communications, the binary source sequence is first channel-encoded and the code symbols are mapped to signal numbers. These signal numbers are then fed to the modulator.

Often, the channel code is binary and the set of signal elements has cardinality M equal to a power of two. Here, simple tuples of $\log_2(M)$ coded bits establish the signal numbers. In some cases, M is not a power of two; then, blocks of U coded binary symbols are mapped to blocks of V signal numbers (M-ary elements), where

$$2^U \leq M^V \tag{B.69}$$

has to hold (cf. Section 6.1.3). This can be done algorithmically via so-called *modulus conversion*. Thereby, lookup tables (which may be applicable for small U) are

Table B.3 Modulus conversion in pseudo-code notation.
Left: mapping. Right: inverse mapping.

Input $[c_{U-1}, \ldots, c_0]$	Input $[m_{V-1}, \ldots, m_0]$
Let $I = 0$	Let $I = 0$
for $i = U-1, U-2, \ldots, 0$	for $i = V-1, V-2, \ldots, 0$
$\quad I = I \cdot 2 + c_i$	$\quad I = I \cdot M + m_i$
endfor	endfor
for $i = 0, 1, \ldots, V-1$	for $i = 0, 1, \ldots, U-1$
$\quad m_i = I \bmod M$	$\quad c_i = I \bmod 2$
$\quad I = (I - m_i)/M$	$\quad I = (I - c_i)/2$
endfor	endfor
Output $[m_{V-1}, \ldots, m_0]$	Output $[c_{U-1}, \ldots, c_0]$

avoided. As the name suggests, the main principle is the conversion of a binary number into a representation with base M.

First, the binary code symbols c_i are here interpreted as real-valued integers from the set $\{0, 1\}$ and the tuple of U binary code symbols $c_0, c_1, \ldots, c_{U-1}$ is interpreted as an integer in binary notation

$$I = c_{U-1} \cdot 2^{U-1} + \cdots + c_2 \cdot 2^2 + c_1 \cdot 2 + c_0, \quad c_j \in \{0, 1\}$$
$$= [\, c_{U-1}, \ldots, c_2, c_1, c_0\,]_2 \, . \tag{B.70}$$

The task of the mapping is to find the V-tuple (block) of signal numbers in base-M representation of this integer, that is

$$I = [\, m_{V-1}, \ldots, m_2, m_1, m_0\,]_M, \quad m_i \in \{0, 1, \ldots, M-1\} \, . \tag{B.71}$$

The mapping $([c_{U-1}, \ldots, c_0]_2 \to [m_{V-1}, \ldots, m_0]_M)$ and the inverse mapping $([m_{V-1}, \ldots, m_0]_M \to [c_{U-1}, \ldots, c_0]_2)$ can be done as given in Table B.3, which implements standard number conversion.

Example B.16 Modulus Conversion

Let $M = 3$ (ternary modulation). For $U = 3$ and $V = 2$ ($2^3 < 3^2$), the modulus conversion gives the mapping displayed in Table B.4. As can be seen, the surplus combination $[m_1, m_0] = [2, 2]$ is not used.

Table B.4 Mapping from binary 3-tuples to ternary 2-tuples.

$[c_2, c_1, c_0]$	$[m_1, m_0]$
0 0 0	0 0
0 0 1	0 1
0 1 0	0 2
0 1 1	1 0
1 0 0	1 1
1 0 1	1 2
1 1 0	2 0
1 1 1	2 1

APPENDIX C

Supplements to OFDM

This appendix collects supplements to OFDM. First, the average power spectral density of OFDM is derived. Then, a continuous-time variant of frequency multiplexing is considered and its average psd is derived. Both variants are compared. Finally, some remarks on unique-word OFDM are made.

C.1 Preliminaries

We assume an orthogonal frequency-division multiplexing (OFDM) scheme with D carriers. The symbol duration without guard interval is denoted as $T_u = DT$. Thereby, T is the symbol duration of the "chips" or transmit pulse shape $g_T(t) \circ\!\!-\!\!\bullet G_T(f)$. A guard interval of length $T_g = D_g T$ is appended. Hence, the total length of an OFDM symbol is $T_s = T_u + T_g$. With $D_s = D + D_g$, the block lengths are related by $T_s = D_s T = DT + D_g T$. Figure C.1 visualizes the definitions of the block lengths.

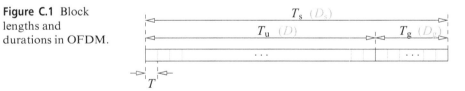

Figure C.1 Block lengths and durations in OFDM.

The data symbols to be communicated are denoted by $a_j[k_s]$, where $k_s \in \mathbb{Z}$ is the symbol index and $j = 0, 1, \ldots, D-1$ the carrier index. We assume i.i.d. (w.r.t. k_s and j) zero-mean data symbols $a_j[k_s]$ with variance $\sigma_{a,j}^2 = \mathrm{E}\{|a_j[k_s]|^2\}$, that is

$$\mathrm{E}\left\{a_{j,1}[k_{s_1}] \cdot a_{j_2}^*[k_{s_2}]\right\} = \begin{cases} \sigma_{a,j}^2, & j_1 = j_2 = j, \ k_{s_1} = k_{s_2} \\ 0, & \text{else} \end{cases}. \tag{C.1}$$

All signals and systems are given in the ECB domain.
We use the definitions

$$\mathrm{si}(x) \stackrel{\text{def}}{=} \frac{\sin(x)}{x} \tag{C.2}$$

and

$$\text{rectc}_{T_u}(t) \stackrel{\text{def}}{=} \begin{cases} 1, & 0 \le t < T_u \\ 0, & \text{else} \end{cases}, \tag{C.3}$$

$$\text{rectc}_D[k] \stackrel{\text{def}}{=} \begin{cases} 1, & 0 \le k < D, \\ 0, & \text{else} \end{cases}, \tag{C.4}$$

for the *causal* continuous-time and discrete-time rectangular pulse, respectively.

C.2 Discrete-Time OFDM

Figure C.2 shows the discrete-time variant of OFDM (DT-OFDM) typically used in practice. We always identify "OFDM" with this scheme.

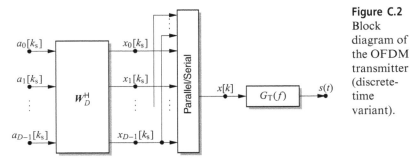

Figure C.2 Block diagram of the OFDM transmitter (discrete-time variant).

Given the data symbols, with the inverse DFT (W_D denotes the normalized DFT matrix) the channel symbols are calculated. After insertion of the cyclic prefix (guard interval), the discrete-time transmit sequence is given by

$$x[k] = x[D_s k_s + k_c]$$

$$= \sum_{k_s=-\infty}^{\infty} \text{rectc}_{D_s}[k - D_s k_s] \frac{1}{\sqrt{D}} \sum_{j=0}^{D-1} a_j[k_s] \cdot e^{+j\frac{2\pi}{D} j k_c}, \tag{C.5}$$

where we have written $k = D_s k_s + k_c$, with $k_s \in \mathbb{Z}$ being the discrete-time block index and $k_c = 0, \ldots, D_s - 1$ counts the symbols within one block.

After pulse shaping (transmit filtering) the continuous-time transmit signal, which is a PAM signal, reads

$$s(t) = \sum_{k=-\infty}^{\infty} x[k] g_T(t - kT). \tag{C.6}$$

As detailed in Chapter 9, OFDM in its discrete-time implementation is a PAM scheme (symbol interval T) with blockwise preprocessing. Hence, the average power spectral density is given by (cf. (3.10))

$$\bar{\Phi}_{ss}(f) = \bar{\Phi}_{xx}\left(e^{j2\pi fT}\right) \cdot \frac{|G_T(f)|^2}{T}, \tag{C.7}$$

where $\bar{\Phi}_{xx}(e^{j2\pi fT})$ is the (average) psd of the discrete-time transmit sequence $x[k]$ and $G_T(f)$ is the transfer function of the pulse-shaping filter.

The autocorrelation sequence of the transmit sequence $x[k]$ calculates as

$$\begin{aligned}
\tilde{\phi}_{xx}[k_1, k_2] &= \tilde{\phi}_{xx}[D_s k_{s_1} + k_{c_1}, D_s k_{s_2} + k_{c_2}] \\
&= \mathrm{E}\left\{x[D_s k_{s_1} + k_{c_1}] \cdot x^*[D_s k_{s_2} + k_{c_2}]\right\} \\
&= \frac{1}{D}\mathrm{E}\left\{\sum_{k_{s_1}=-\infty}^{\infty} \mathrm{rectc}_{D_s}[k - k_{s_1} D_s] \sum_{j_1=0}^{D-1} a_{j_1}[k_{s_1}] \cdot e^{+j\frac{2\pi}{D}j_1 k_{c_1}} \right. \\
&\qquad \left. \cdot \sum_{k_{s_2}=-\infty}^{\infty} \mathrm{rectc}_{D_s}[k - k_{s_2} D_s] \sum_{j_2=0}^{D-1} a_{j_2}^*[k_{s_2}] \cdot e^{-j\frac{2\pi}{D}j_2 k_{c_2}}\right\} \\
&= \frac{1}{D}\sum_{k_{s_1}=-\infty}^{\infty}\sum_{k_{s_2}=-\infty}^{\infty}\sum_{j_1=0}^{D-1}\sum_{j_2=0}^{D-1} \mathrm{E}\left\{a_{j_1}[k_{s_1}] \cdot a_{j_2}^*[k_{s_2}]\right\} \\
&\qquad \cdot \mathrm{rectc}_{D_s}[k - k_{s_1} D_s]\, \mathrm{rectc}_{D_s}[k - k_{s_2} D_s]\, e^{+j\frac{2\pi}{D}(j_1 k_{c_1} - j_2 k_{c_2})} \\
&\overset{(C.1)}{=} \frac{1}{D}\sum_{k_s=-\infty}^{\infty}\sum_{j=0}^{D-1} \sigma_{a,j}^2 \cdot \mathrm{rectc}_{D_s}[k - k_s D_s]\, \mathrm{rectc}_{D_s}[k - k_s D_s]\, e^{+j\frac{2\pi}{D}j(k_{c_1} - k_{c_2})}.
\end{aligned}$$
(C.8)

The autocorrelation sequence depends on $k_{s_1} - k_{s_2}$ and on $k_{c_1} - k_{c_2}$. Since it does not depend on $(D_s k_{s_1} + k_{c_1}) - (D_s k_{s_2} + k_{c_2})$ only, the sequence $x[k]$ is *non-stationary*. However, because for all $k' \in \mathbb{Z}$

$$\tilde{\phi}_{xx}[D_s k_{s_1} + k_{c_1} + k' D_s, D_s k_{s_2} + k_{c_2} + k' D_s] = \tilde{\phi}_{xx}[D_s k_{s_1} + k_{c_1}, D_s k_{s_2} + k_{c_2}],$$
(C.9)

the transmit sequence is *cyclo-stationary*.

The average autocorrelation sequence is obtained by averaging $\tilde{\phi}_{xx}[k_1, k_2]$ over one period of length D_s. Defining $k_{c_2} = k$, $k_{c_1} - k_{c_2} = \kappa$, the average autocorrelation sequence is given by

$$\bar{\phi}_{xx}[\kappa] = \frac{1}{D_s}\sum_{k=0}^{D_s-1} \tilde{\phi}_{xx}[k + \kappa, k].$$
(C.10)

Using

$$S = \frac{1}{D_s}\sum_{k=0}^{D_s-1} \mathrm{rectc}_{D_s}[k - k_{s_1} D_s]\, \mathrm{rectc}_{D_s}[k - k_{s_2} D_s]$$

$$= \begin{cases} \frac{D_s - |\kappa|}{D_s}, & |\kappa| < D_s,\ k_s = 0 \\ 0, & \text{else} \end{cases},$$
(C.11)

the average autocorrelation sequence can be written as

$$\bar{\phi}_{xx}[\kappa] = \frac{1}{D} \frac{D_s - |\kappa|}{D_s} \sum_{j=0}^{D-1} \sigma_{a,j}^2 \, e^{+j\frac{2\pi}{D} j\kappa}, \qquad |\kappa| < D_s. \qquad (C.12)$$

The average psd of the sequence $x[k]$ is then given as

$$\begin{aligned}
\Phi_{xx}\left(e^{j2\pi fT}\right) &= \sum_{\kappa=-\infty}^{+\infty} \bar{\phi}_{xx}[\kappa] \cdot e^{-j2\pi fT\kappa} \\
&= \sum_{\kappa=-D_s}^{D_s} \frac{D_s - |\kappa|}{D_s} \cdot \frac{1}{D} \sum_{j=0}^{D-1} \sigma_{a,j}^2 \cdot e^{+j\frac{2\pi}{D} j\kappa} \cdot e^{-j2\pi fT\kappa} \\
&= \frac{1}{D} \sum_{j=0}^{D-1} \sigma_{a,j}^2 \cdot \sum_{\kappa=-D_s}^{D_s} \frac{D_s - |\kappa|}{D_s} e^{-j2\pi(fT - j/D)\kappa}. \qquad (C.13)
\end{aligned}$$

Using a result derived in Section C.5, we have

$$\begin{aligned}
\sum_{\kappa=-D_s}^{D_s} \frac{D_s - |\kappa|}{D_s} \cdot e^{-j2\pi(fT - j/D)\kappa} &= D_s \sum_{\mu=-\infty}^{+\infty} \left(\frac{\sin(\pi(fT - j/D - \mu)D_s)}{\pi(fT - j/D - \mu)D_s}\right)^2 \\
&= \frac{1}{D_s} \frac{\sin^2(\pi(fT - j/D)D_s)}{\sin^2(\pi(fT - j/D))}. \qquad (C.14)
\end{aligned}$$

Putting all this together, the average psd of the OFDM transmit signal calculates as

$$\bar{\Phi}_{ss}(f) = \frac{|G_T(f)|^2}{T} \cdot \frac{D_s}{D} \sum_{j=0}^{D-1} \sigma_{a,j}^2 \sum_{\mu=-\infty}^{+\infty} \mathrm{si}^2\left(\pi(fT_s - jT_s/T_u - \mu D_s)\right) \qquad (C.15)$$

$$= \frac{|G_T(f)|^2}{T} \cdot \frac{1}{D_s D} \sum_{j=0}^{D-1} \sigma_{a,j}^2 \frac{\sin^2\left(\pi(fT_s - jT_s/T_u)\right)}{\sin^2\left(\pi(fT_s - jT_s/T_u)/D_s\right)}. \qquad (C.16)$$

C.3 Continuous-Time Variant

Figure C.3 shows a variant of multi-carrier modulation, which can frequently be found in the literature. It is a system-theoretical view employing continuous-time

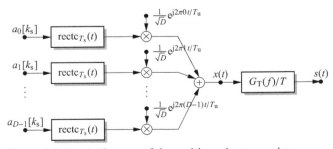

Figure C.3 Block diagram of the multi-carrier transmitter (continuous-time variant).

exponentials (possibly followed by filtering with $G_T(f)/T$). We denote this variant as *continuous-time OFDM* (CT-OFDM).

Here, the baseband transmit signal is given by

$$s(t) = \frac{1}{T} g_T(t) * x(t)$$
$$= \frac{1}{T} g_T(t) * \frac{1}{\sqrt{D}} \sum_{k_s=-\infty}^{\infty} \sum_{j=0}^{D-1} a_j[k_s] \, \text{rectc}_{T_s}(t - k_s T_s) \, e^{j2\pi j t / T_u} , \quad \text{(C.17)}$$

with the obvious definition of the signal $x(t)$.

The autocorrelation function of the transmit signal $x(t)$ calculates as
$$\tilde{\phi}_{xx}(t_1, t_2) = \mathrm{E}\{x(t_1) \cdot x^*(t_2)\}$$
$$= \frac{1}{D} \mathrm{E} \left\{ \sum_{k_{s_1}=-\infty}^{\infty} \sum_{j_1=0}^{D-1} a_{j_1}[k_{s_1}] \, \text{rectc}_{T_s}(t_1 - k_{s_1} T_s) \, e^{j2\pi j_1 t_1 / T_u} \right.$$
$$\left. \cdot \sum_{k_{s_2}=-\infty}^{\infty} \sum_{j_2=0}^{D-1} a^*_{j_2}[k_{s_2}] \, \text{rectc}_{T_s}(t_2 - k_{s_2} T_s) \, e^{-j2\pi j_2 t_2 / T_u} \right\}$$
$$= \frac{1}{D} \sum_{k_{s_1}=-\infty}^{\infty} \sum_{k_{s_2}=-\infty}^{\infty} \sum_{j_1=0}^{D-1} \sum_{j_2=0}^{D-1} \mathrm{E}\{a_{j_1}[k_{s_1}] \cdot a^*_{j_2}[k_{s_2}]\}$$
$$\cdot \text{rectc}_{T_s}(t_1 - k_{s_1} T_s) \cdot \text{rectc}_{T_s}(t_2 - k_{s_2} T_s) \, e^{j2\pi (j_1 t_1 - j_2 t_2)/T_u}$$
$$\stackrel{(C.1)}{=} \frac{1}{D} \sum_{k_s=-\infty}^{\infty} \sum_{j=0}^{D-1} \sigma^2_{a,j} \cdot \text{rectc}_{T_s}(t_1 - k_s T_s) \cdot \text{rectc}_{T_s}(t_2 - k_s T_s) \, e^{j2\pi j (t_1 - t_2)/T_u} .$$
(C.18)

As the autocorrelation function does not only depend on $t_1 - t_2$, the transmit signal (due to the block structure) is *non-stationary*.

However, because for all $k'_s \in \mathbb{Z}$

$$\tilde{\phi}_{xx}(t_1 + k'_s T_s, t_2 + k'_s T_s) = \tilde{\phi}_{xx}(t_1, t_2) , \quad \text{(C.19)}$$

the transmit signal is *cyclo-stationary* with period T_s.

Defining $t_2 = t$, $t_1 - t_2 = \tau$, the average autocorrelation function is given by

$$\bar{\phi}_{xx}(\tau) = \frac{1}{T_s} \int_0^{T_s} \tilde{\phi}_{xx}(t + \tau, t) \, dt . \quad \text{(C.20)}$$

Using

$$I = \frac{1}{T_s} \int_0^{T_s} \text{rectc}_{T_s}(t + \tau - k_s T_s) \, \text{rectc}_{T_s}(t - k_s T_s) \, dt$$
$$= \begin{cases} \frac{T_s - |\tau|}{T_s}, & |\tau| < T_s, \; k_s = 0 \\ 0, & \text{else} \end{cases} , \quad \text{(C.21)}$$

the average autocorrelation function can be written as

$$\bar{\phi}_{xx}(\tau) = \frac{1}{D} \frac{T_s - |\tau|}{T_s} \sum_{j=0}^{D-1} \sigma_{a,j}^2 \, e^{j2\pi j \tau/T_u} \,. \tag{C.22}$$

The average psd of the signal $x(t)$ is then given as

$$\bar{\Phi}_{xx}(f) = \int_{-\infty}^{\infty} \bar{\phi}_{xx}(\tau) \, e^{-j2\pi f \tau} \, d\tau$$

$$= \int_{-T_s}^{T_s} \frac{T_s - |\tau|}{T_s} \frac{1}{D} \sum_{j=0}^{D-1} \sigma_{a,j}^2 \, e^{j2\pi j \tau/T_u} e^{-j2\pi f \tau} \, d\tau$$

$$= \frac{1}{D} \sum_{j=0}^{D-1} \sigma_{a,j}^2 \cdot \int_{-T_s}^{T_s} \frac{T_s - |\tau|}{T_s} \, e^{-j2\pi(f - j/T_u)\tau} \, d\tau \,. \tag{C.23}$$

Using a result derived in Section C.5, we have

$$\int_{-T_s}^{T_s} \frac{T_s - |\tau|}{T_s} \, e^{-j2\pi(f - j/T_u)\tau} \, d\tau = T_s \, \text{si}^2 \left(\pi(f T_s - j D_s/D)\right) \,. \tag{C.24}$$

Putting all this together, the average psd of the CT-OFDM transmit signal calculates as

$$\bar{\Phi}_{ss}(f) = \frac{|G_T(f)|^2 T_s}{T^2} \frac{D-1}{D} \sum_{j=0}^{D-1} \sigma_{a,j}^2 \, \text{si}^2 \left(\pi(f T_s - j D_s/D)\right) \tag{C.25}$$

$$= \frac{|G_T(f)|^2}{T} \frac{D_s}{D} \sum_{j=0}^{D-1} \sigma_{a,j}^2 \, \frac{\sin^2(\pi(f T_s - j T_s/T_u))}{(\pi(f T_s - j T_s/T_u))^2} \,. \tag{C.26}$$

CAP-Like Variant

For the continuous-time variant, the transmit signal can also be given by

$$x(t) = \frac{1}{\sqrt{D}} \sum_{k_s=-\infty}^{\infty} \sum_{j=0}^{D-1} a_j[k_s] \, \text{rectc}_{T_s}(t - k_s T_s) \, e^{j2\pi j(t - k_s T_s)/T_u} \,, \tag{C.27}$$

where in the spirit of carrierless AM/PM (CAP, cf. Section 4.2.1) the rect pulse used for pulse shaping is modulated instead of pulse shaping followed by modulation.

Here, the autocorrelation function calculates as

$$\tilde{\phi}_{xx}(t_1,t_2) = \mathrm{E}\{x(t_1) \cdot x^*(t_2)\}$$

$$= \frac{1}{D} \sum_{k_{s_1}=-\infty}^{\infty} \sum_{k_{s_2}=-\infty}^{\infty} \sum_{j_1=0}^{D-1} \sum_{j_2=0}^{D-1} \mathrm{E}\{a_{j_1}[k_{s_1}] \cdot a_{j_2}^*[k_{s_2}]\}$$
$$\cdot \mathrm{rectc}_{T_s}(t_1 - k_{s_1}T_s) \, \mathrm{rectc}_{T_s}(t_2 - k_{s_2}T_s)$$
$$\cdot e^{j2\pi j_1(t_1-k_{s_1}T_s)/T_u} e^{-j2\pi j_2(t_2-k_{s_2}T_s)/T_u}$$

$$\stackrel{(C.1)}{=} \frac{1}{D} \sum_{k_s=-\infty}^{\infty} \sum_{j=0}^{D-1} \sigma_{a,j}^2 \cdot \mathrm{rectc}_{T_s}(t_1 - k_sT_s) \, \mathrm{rectc}_{T_s}(t_2 - k_sT_s) \cdot e^{j2\pi j(t_1-t_2)/T_u} , \qquad (C.28)$$

which is the same expression as above. Hence, the CAP philosophy results in the same average power spectral density as the conventional continuous-time OFDM modulation.

C.4 Comparison and Discussion

The two variants (discrete-time and continuous-time) do not have the same average power spectral density. In the continuous-time version the term

$$\mathrm{si}^2\left(\pi(fT_s - jT_s/T_u)\right) = \frac{\sin^2\left(\pi(fT_s - jT_s/T_u)\right)}{\left(\pi(fT_s - jT_s/T_u)\right)^2} \qquad (C.29)$$

is present, whereas in the discrete-time version

$$\sum_{\mu=-\infty}^{+\infty} \mathrm{si}^2\left(\pi(fT_s - jD_s/D - \mu D_s)\right) = \frac{1}{D_s^2} \frac{\sin^2\left(\pi(fT_s - jT_s/T_u)\right)}{\sin^2\left(\pi(fT - jT/T_u)\right)} \qquad (C.30)$$

appears. This is due to the fact that discrete-time sequences have frequency-periodic spectra. The main period is cut out by the pulse-shaping filter.

Even if $G_T(f)$ is a perfect low-pass filter with bandwidth $1/T$, both variants do not coincide! In the continuous-time variant, D sinc-functions are superimposed – in the conventionally used discrete-time variant, periodic replica a of each carrier are present. If carriers are set to zero ($\sigma_{a,j}^2 = 0$), these replicas may have a small impact on the band used.

This effect is visualized in Figure C.4. Out of the $D = 8$ carriers (their contributions are shown in light gray), two are set to zero; no guard interval is introduced. A pulse-shaping filter with a broad roll-off is used. The differences at the band edges due to the spectral replicas in the discrete-time case are clearly visible.

Moreover, in both variants, when a guard interval is introduced, the carriers are not orthogonal to each other; the frequency spacing is $\frac{D_s}{D}\frac{1}{T_s} = \frac{1}{DT} = \frac{1}{T_u}$, whereas the width of the main lobe is $\frac{1}{T_s} < \frac{1}{T_u}$. If $T_g = T$, that is, $T_s = 2T_u$, spectral zeros between adjacent carriers would appear.

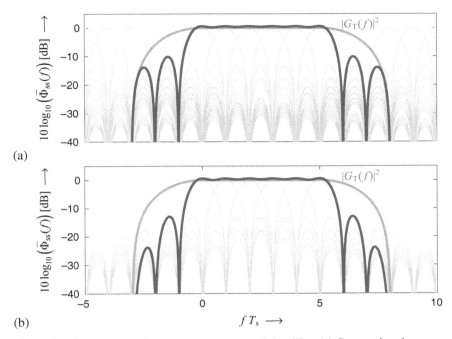

Figure C.4 Comparison of average power spectral densities. (a) Conventional discrete-time case; (b) theoretical continuous-time version.

C.5 Derivations

From basic correspondences of the Fourier transform, we have

$$c(t) = \operatorname{rect}(t) \circ\!\!\!-\!\!\bullet C(f) = \frac{\sin(\pi f)}{\pi f} = \operatorname{si}(\pi f), \tag{C.31}$$

$$d(t) = \operatorname{tri}(t) \circ\!\!\!-\!\!\bullet D(f) = \left(\frac{\sin(\pi f)}{\pi f}\right)^2 = \operatorname{si}^2(\pi f), \tag{C.32}$$

where $\operatorname{tri}(t) \stackrel{\text{def}}{=} \operatorname{rect}(t) * \operatorname{rect}(t) = 1 - |t|, |t| \le 1$.

C.5.1 Continuous Time

Using the above relation and the correspondence for time scaling (T is here any scaling factor), the Fourier transform of the signal

$$d(t/T) = \begin{cases} \frac{T-|t|}{T}, & |t| < T \\ 0, & \text{else} \end{cases} \tag{C.33}$$

is this given by

$$\int_{-T}^{T} \frac{T - |t|}{T} \, e^{-j2\pi f t} \, dt = T \left(\frac{\sin(\pi f T)}{\pi f T}\right)^2. \tag{C.34}$$

C.5.2 Discrete Time

We are now interested in the discrete-time Fourier transform of the sequence

$$d[k] = \begin{cases} \frac{K-|k|}{K}, & |k| < K \\ 0, & \text{else} \end{cases}, \qquad (C.35)$$

which is defined by

$$D\left(e^{j2\pi fT}\right) \stackrel{\text{def}}{=} \sum_{k=-\infty}^{\infty} d[k] \, e^{-j2\pi fTk} = \sum_{k=-K}^{K} \frac{K-|k|}{K} e^{-j2\pi fTk} . \qquad (C.36)$$

On the one hand, an analytic expression of this so-called *Fejér kernel* can be obtained by considering

$$F_K(x) = \sum_{k=-K}^{K} \frac{K-|k|}{K} e^{jxk} = \frac{1}{K} \sum_{l=0}^{K-1} \sum_{k=-l}^{l} e^{jxk} . \qquad (C.37)$$

When writing $\omega \stackrel{\text{def}}{=} e^{jx}$, the inner sum (which is called the *Dirichlet kernel*) is given by

$$\sum_{k=-l}^{l} \omega^k = \omega^{-l} \sum_{k=0}^{2l} \omega^k = \omega^{-l} \frac{1-\omega^{2l+1}}{1-\omega} = \frac{\omega^{-l}-\omega^{l+1}}{1-\omega} . \qquad (C.38)$$

This leads to

$$\begin{aligned} F_K(x) &= \frac{1}{K} \sum_{l=0}^{K-1} \frac{\omega^{-l}-\omega^{l+1}}{1-\omega} = \frac{1}{K(1-\omega)} \left(\sum_{l=0}^{K-1} \omega^{-l} - \sum_{l=0}^{K-1} \omega^{l+1} \right) \\ &= \frac{1}{K(1-\omega)} \left(\frac{1-\omega^{-K}}{1-\omega^{-1}} - \omega \frac{1-\omega^{K}}{1-\omega^{1}} \right) = \frac{1}{K(1-\omega)} \omega \left(\frac{\omega^{-K}-1}{1-\omega} - \frac{1-\omega^{K}}{1-\omega} \right) \\ &= \frac{1}{K(1-\omega)^2 \omega^{-1}} \left(\omega^{-K} - 2 + \omega^{K} \right) = \frac{1}{K(\omega^{1/2}-\omega^{-1/2})^2} \left(\omega^{K/2} - \omega^{-K/2} \right)^2 \\ &= \frac{1}{K} \frac{\left(e^{jxK/2}-e^{-jxK/2}\right)^2}{(e^{jx/2}-e^{-jx/2})^2} = \frac{1}{K} \frac{\sin^2(xK/2)}{\sin^2(x/2)} , \end{aligned} \qquad (C.39)$$

and thus

$$D\left(e^{j2\pi fT}\right) = F_K(x = -2\pi fT) = \frac{1}{K} \frac{\sin^2(\pi fTK)}{\sin^2(\pi fT)} . \qquad (C.40)$$

On the other hand, sampling in the time domain (sampling interval $T = 1/K$) corresponds to a periodic continuation in the frequency domain, that is, for $d(t) \circ\!\!-\!\!\bullet D(f)$:

$$d(k/K) \circ\!\!-\!\!\bullet K \sum_{\mu} D(f - \mu K) . \qquad (C.41)$$

This immediately gives

$$d[k] = d(k/K) \circ\!\!-\!\!\bullet K \sum_{\mu=-\infty}^{\infty} \left(\frac{\sin(\pi(fT-\mu)K)}{\pi(fT-\mu)K} \right)^2 . \qquad (C.42)$$

Finally, comparing (C.36), (C.40), and (C.42):

$$\sum_{k=-K}^{K} \frac{K-|k|}{K} e^{-j2\pi fTk} = \frac{1}{K} \frac{\sin^2(\pi fTK)}{\sin^2(\pi fT)} = K \sum_{\mu=-\infty}^{\infty} \left(\frac{\sin(\pi(fT-\mu)K)}{\pi(fT-\mu)K} \right)^2 . \quad \text{(C.43)}$$

C.6 Noise Variance in OFDM

We are interested in the noise variance in the D parallel subchannels in OFDM. The discrete-time version is considered.

Let the autocorrelation sequence of the additive noise $n[k]$ at the output of the receive filter (cf. Figure 9.6) in the general case be given by

$$\phi_{nn}[\kappa] = \mathrm{E}\{ n[k+\kappa]n^*[k] \} . \quad \text{(C.44)}$$

Blocks are processed at the receiver (see Figure 9.8). The noise vector $\boldsymbol{n}_\mathrm{c}[k_\mathrm{s}]$ in (9.15) contains D subsequent samples of the noise. After applying the DFT, the noise vector of interest

$$\boldsymbol{n}[k_\mathrm{s}] = [n_0[k_\mathrm{s}], \ldots, n_{D-1}[k_\mathrm{s}]] \stackrel{\text{def}}{=} \boldsymbol{n}_\mathrm{c}[k_\mathrm{s}] \boldsymbol{W}_D \quad \text{(C.45)}$$

is obtained. The elements $n_j[k_\mathrm{s}]$, $j = 0,\ldots,D-1$, of this vector are the noise samples present in the parallel, independent channels.

White noise. First we consider the special case of (zero-mean) white noise $n[k]$ with variance σ_n^2, which means that $\phi[\kappa] = \phi_{nn}[\kappa] = \sigma_n^2 \delta[\kappa]$. Here, the elements of the noise vector $\boldsymbol{n}_\mathrm{c}[k_\mathrm{s}]$ are uncorrelated (and due to the Gaussian assumption statistically independent) and the blocks (block index k_s) are independent.

The correlation matrix of $\boldsymbol{n}_\mathrm{c}[k_\mathrm{s}]$ is thus given by

$$\boldsymbol{\Phi}_{n_\mathrm{c} n_\mathrm{c}}[\kappa_\mathrm{s}] \stackrel{\text{def}}{=} \mathrm{E}\{ \boldsymbol{n}_\mathrm{c}^\mathsf{T}[k_\mathrm{s}+\kappa_\mathrm{s}] \cdot \boldsymbol{n}_\mathrm{c}^*[k_\mathrm{s}] \} = \sigma_n^2 \boldsymbol{I} \, \delta[\kappa_\mathrm{s}] . \quad \text{(C.46)}$$

The correlation matrix of the transformed noise $\boldsymbol{n}[k_\mathrm{s}]$ reads

$$\begin{aligned}
\boldsymbol{\Phi}_{nn}[\kappa_\mathrm{s}] &\stackrel{\text{def}}{=} \mathrm{E}\{ \boldsymbol{n}^\mathsf{T}[k_\mathrm{s}+\kappa_\mathrm{s}] \cdot \boldsymbol{n}^*[k_\mathrm{s}] \} \\
&= \mathrm{E}\{ \boldsymbol{W}_D^\mathsf{T} \boldsymbol{n}_\mathrm{c}^\mathsf{T}[k_\mathrm{s}+\kappa_\mathrm{s}] \cdot \boldsymbol{n}_\mathrm{c}^*[k_\mathrm{s}] \boldsymbol{W}_D^* \} \\
&= \boldsymbol{W}_D^\mathsf{T} \mathrm{E}\{ \boldsymbol{n}_\mathrm{c}^\mathsf{T}[k_\mathrm{s}+\kappa_\mathrm{s}] \cdot \boldsymbol{n}_\mathrm{c}^*[k_\mathrm{s}] \} \boldsymbol{W}_D^* \\
&= \boldsymbol{W}_D^\mathsf{T} (\sigma_n^2 \boldsymbol{I} \, \delta[\kappa_\mathrm{s}]) \boldsymbol{W}_D^* \\
&= \sigma_n^2 \boldsymbol{I} \, \delta[\kappa_\mathrm{s}] \boldsymbol{W}_D^\mathsf{T} \boldsymbol{W}_D^*
\end{aligned}$$

and using that $\boldsymbol{W}_D$ is a unitary matrix, that is, $\boldsymbol{W}_D^\mathsf{T} \boldsymbol{W}_D^* = (\boldsymbol{W}_D^\mathsf{H} \boldsymbol{W}_D)^* = \boldsymbol{I}$, we arrive at

$$\boldsymbol{\Phi}_{nn}[\kappa_\mathrm{s}] = \sigma_n^2 \boldsymbol{I} \, \delta[\kappa_\mathrm{s}] . \quad \text{(C.47)}$$

Hence, the noise samples $n_j[k_\mathrm{s}]$, $j = 0,\ldots,D-1$, are white with respect to the subcarrier index j and the block index k_s.

General case. We now turn to the general case of non-white noise. If the correlations extend more than the length of the guard interval, the noise samples in subsequent noise vectors $\boldsymbol{n}_c[k_s]$ will be correlated. After transformation with the DFT matrix, the noise samples $n_j[k_s]$ will be correlated over the subcarrier index j and the block index k_s.

Usually, correlations in the discrete-time noise are not too strong and we can assume that samples further apart than the length of the guard interval are uncorrelated, that is, $\phi_{nn}[\kappa] = 0$, $|\kappa| > D_g$. Then, as the blocks are separated by the guard interval, the noise vectors are uncorrelated with respect to k_s. However, correlations within the block are present. The correlation matrix has Toeplitz structure and is now given by

$$\boldsymbol{\Phi}_{n_c n_c}[k_s] = \begin{bmatrix} \phi_{nn}[0] & \phi_{nn}[-1] & \cdots & \phi_{nn}[-D+1] \\ \phi_{nn}[1] & \ddots & & \vdots \\ \vdots & & \ddots & \vdots \\ \phi_{nn}[D-1] & \cdots & \cdots & \phi_{nn}[0] \end{bmatrix} \delta[k_s]. \quad (C.48)$$

After transformation to the frequency domain according to (C.45), we have

$$\boldsymbol{\Phi}_{nn}[k_s] = \boldsymbol{W}_D^\mathsf{T} \boldsymbol{\Phi}_{n_c n_c}[k_s] \boldsymbol{W}_D^*$$

$$= \boldsymbol{W}_D^\mathsf{T} \left[\frac{1}{\sqrt{D}} \sum_{k=0}^{D-1} \phi_{nn}[l-k] e^{+j\frac{2\pi}{D} mk} \phi_{nn}[l-k] \right]_{\substack{l=0,\ldots,D-1 \\ m=0,\ldots,D-1}} \delta[k_s]$$

$$= \left[\frac{1}{\sqrt{D}} \sum_{l=0}^{D-1} \frac{1}{\sqrt{D}} \sum_{k=0}^{D-1} \phi_{nn}[l-k] e^{+j\frac{2\pi}{D} mk} \phi_{nn}[l-k] e^{-j\frac{2\pi}{D} il} \right]_{\substack{i=0,\ldots,D-1 \\ m=0,\ldots,D-1}} \delta[k_s]$$

$$= \frac{1}{D} \left[\sum_{l=0}^{D-1} \sum_{k=0}^{D-1} \phi_{nn}[l-k] e^{-j\frac{2\pi}{D}(il-mk)} \right]_{\substack{i=0,\ldots,D-1 \\ m=0,\ldots,D-1}} \delta[k_s]. \quad (C.49)$$

The off-diagonal elements of this correlation matrix are in general nonzero, thus the noise samples are correlated. Only if D is very large compared to the extent of the correlations of the initial noise sequence do the noise samples after transformation tend to be uncorrelated (the DFT matrix is only approximately the modal matrix of a Toeplitz matrix).

We are particularly interested in the noise powers in the individual channels. These are given by the main-diagonal elements ($i = m = j$) of the correlation matrix in (C.49), that is,

$$\sigma_{n,j}^2 = \frac{1}{D} \sum_{l=0}^{D-1} \sum_{k=0}^{D-1} \phi_{nn}[l-k] e^{-j\frac{2\pi}{D} j(l-k)} \quad (C.50)$$

or with a change of variables $\kappa = l - k$ and $\lambda = l + k$

$$= \frac{1}{D} \sum_{\kappa=-D+1}^{D-1} \sum_{\lambda=|\kappa|:2:(2D-2)-|\kappa|} \phi_{nn}[\kappa] e^{-j\frac{2\pi}{D} j\kappa}, \quad j = 0,\ldots,D-1, \quad (C.51)$$

where $\lambda = |\kappa| : 2 : (2D-2) - |\kappa|$ denotes that λ runs from $|\kappa|$ in steps of two to $(2D-2)-|\kappa|$. The summand is independent of the summation index λ; the sum over λ simply gives the number of elements over which λ runs, which is $D - |\kappa|$. Thus, we arrive at

$$\sigma_{n,j}^2 = \sum_{\kappa=-D+1}^{D-1} \left(1 - \frac{|\kappa|}{D}\right) \phi_{nn}[\kappa] \, e^{-j\frac{2\pi}{D}\kappa j} \,, \qquad j = 0, \ldots, D-1 \,. \tag{C.52}$$

Note that the term $1 - \frac{|\kappa|}{D}$ is a triangular window. If the autocorrelation sequence $\phi_{nn}[\kappa]$ decreases fast in comparison to the triangular window, we have approximately for the noise power in the jth subchannel

$$\sigma_{n,j}^2 \approx \sum_{\kappa=-D+1}^{D-1} \phi_{nn}[\kappa] \, e^{-j\frac{2\pi}{D}\kappa j}$$

and since $\phi_{nn}[\kappa] = 0$, $|\kappa| \geq D_{\mathrm{g}}$, we can extend the sum and obtain

$$= \sum_{\kappa=-\infty}^{+\infty} \phi_{nn}[\kappa] \, e^{-j\frac{2\pi}{D}\kappa j} \;=\; \Phi_{nn}\left(e^{j2\pi j/D}\right) \,, \qquad j = 0, \ldots, D-1 \,. \tag{C.53}$$

The noise powers in the parallel channels are thus given by samples of the noise power spectral density of the initial discrete-time noise $n[k]$.

We finally note that the effect of weighting the autocorrelation sequence with a triangular window is inherent to the rectangular windowing of the receive sequence and similarly appears when performing spectral estimation via (averaged) periodograms.

C.7 Remarks on Unique-Word OFDM

In *unique-word OFDM* (UW-OFDM), the guard interval of D_{g} samples is not filled with a (partial) cyclic repetition of the block to be transmitted, but by a fixed, that is, *unique word* (UW). Thus, contrary to a cyclic prefix, this signal is not data-dependent and well known at the receiver. Therefore, the UW can be exploited for further purposes beyond the conversion of the linear convolution into a cyclic convolution (e.g., for frame synchronization and/or estimation of the impulse response of the present dispersive channel). Moreover, the UW can even be efficiently exploited for an improved data reliability as the insertion of the UW operates in an analogous way to algebraic channel encoding.

The block structure in UW-OFDM is depicted in Figure C.5. As the preceding block also ends with the UW, the interference in the current block by the causal, dispersive channel impulse response is identical to that caused by the end of this block, that is, a cyclic convolution results in the same way (but for blocks of $D + D_{\mathrm{g}}$ samples) as using a cyclic prefix.

In contrast to conventional OFDM, the guard interval of D_{g} samples has to be included in the block that is processed at the transmitter and the receiver side by the IDFT and the DFT, respectively, of length $D + D_{\mathrm{g}}$. Conceptually, first a block

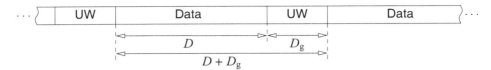

Figure C.5 Structure of data blocks in the time domain for UW-OFDM.

of D_g zeros is generated in the vector x_{zero} of time-domain transmit symbols – this poses constraints on the frequency-domain data vector a. Thereafter, the zero block is replaced by a UW with desired properties. Therefore, the main task of the transmitter-side processing in UW-OFDM is to construct a suitable frequency-domain vector a, such that $x_{\text{zero}} = W_{D+D_g}^H a$ contains the desired block of zeros at the last D_g positions (W_{D+D_g} is the DFT matrix of indicated dimension). For this purpose, $D_r \geq D_g$ redundant subcarriers at positions $k_{r_i}, i \in \{1, \ldots, D_r\}$, have to be reserved in the frequency-domain vector a of dimension $D + D_g$.

Moreover, as usual in OFDM schemes, D_{zero} subcarriers at the band edge (w.l.o.g. the upper band edge) are set to zero. In summary, $D_{\text{data}} = D + D_g - D_r - D_{\text{zero}}$ symbols can be used for data transmission; the symbols are collected in the data vector a_{data}. Figure C.6 shows the sequence of processing in the UW-OFDM transmitter.

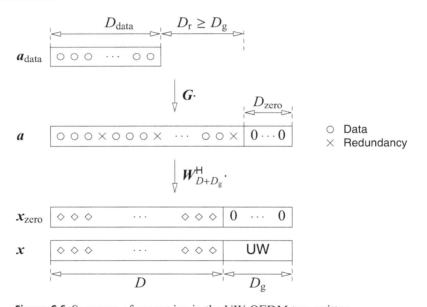

Figure C.6 Sequence of processing in the UW-OFDM transmitter.

As is well known from linear coding theory, the vector a (codeword) is calculated via a $(D + D_g) \times D_{\text{data}}$ generator matrix G from the vector a_{data} that contains the D_{data} information-carrying amplitude coefficients to be transmitted by the actual block

$$a = G a_{\text{data}} .$$

In case of *systematic encoding* (i.e., when the amplitude coefficients of the data vector a_{data} appear explicitly within the vector a), the generator matrix has the form

$$G = \begin{bmatrix} P \\ 0 \end{bmatrix} \cdot \begin{bmatrix} I \\ T \end{bmatrix}.$$

Here, I is the identity matrix of dimension D_{data} and T a $D_r \times D_{\text{data}}$ matrix generating the D_r amplitude coefficients for the redundant subcarriers. P is a permutation matrix of dimension $D_{\text{data}} + D_r$ in order to favorably distribute the redundant subcarriers within the frequency-domain vector and 0 is the $D_{\text{zero}} \times (D_{\text{data}} + D_r)$ all-zero matrix generating the null subcarriers.

The positions k_{r_i} of the D_r redundant subcarriers turn out to be crucial with regard to the energy in these subcarriers for the production of the all-zero block within the vector x_{zero} after the IDFT in the time domain. Optimization procedures prove that a spacing of the redundant subcarriers as uniformly as possible along the frequency axis is favorite (see Figure C.6).

If more redundant subcarriers are reserved than essentially necessary (i.e., $D_r > D_g$), an under-determined set of linear equations for generating D_g consecutive zeros is present. Among the set of solutions, that with the minimum energy of the redundant subcarriers should be used.

An IDFT of length $D + D_g$ is applied to vector a generating the vector x_{zero}. The unique word is placed into the block of D_g consecutive zeros within x_{zero}. The resulting time-domain vector x is transmitted by PAM as usual for OFDM.

At the receiver side, the guard interval is not discarded; a DFT of length $D + D_g$ is applied to the entire corresponding received vector.

The introduction of a block of D_g consecutive zeros in one domain with respect to a DFT corresponds to the construction of a *Reed–Solomon (RS) code* in the other domain. Here, the code is not defined over a finite field, but over the complex numbers. More specifically, a coset of a complex-number RS code results after the insertion of the chosen UW within the guard space, but the error control capabilities remain the same as for the linear code. In contrast to RS codes over finite fields, it is not favorable to implement an algorithm for algebraic error correction for complex-number RS codewords in the frequency domain because of an extreme sensitivity with respect to the calculation of error values. Instead, the error control properties of the RS code are preferably exploited by means of Wiener filtering. Owing to the interdependences between the subcarrier amplitude coefficients, values in deep attenuation notches can very efficiently be recovered by such a filtering. The losses in rate and energy due to the redundant subcarriers are more than compensated by means of this coding and filtering. Together with typical channel coding schemes for OFDM, a highly efficient, serially concatenated coding scheme results. For more details we refer to the literature on UW-OFDM.

APPENDIX D

Time-Invariant and Time-Varying Channel Models

In this appendix, we briefly discuss the characterization of communication channels by means of channel models. On the one hand, time-invariant channels are considered, which are typically present in cable (wireline) transmission scenarios. The transfer functions of typical examples are given. On the other hand, time-variant channel models are studied, which are inherent in mobile wireless situations. Here, the stochastic modeling of the impulse response is addressed.

D.1 Linear Time-Invariant Channel Models

As discussed in Chapter 1, the channel contains all parts of the communication system that have to be accepted as they are, that is, cannot be modified. The end-to-end model incorporates the actual transmission media, amplifiers at transmitter and receiver, and in case of carrier-modulated transmission I/Q mixers at both ends of the transmission link. The inclusion of up- and downconversion gives rise to a channel model in the equivalent complex baseband (ECB) domain.

A common assumption that is essential to make things manageable is that all parts of the link – hardware at the transmitter and the receiver but also potential nonlinear media such as optical fibers or magnetic storage media – are operated in their linear range. Thus, it is expected that clipping and saturation effects and other nonlinear distortions are avoided.

A further assumption, which leads to an easy-to-handle description, is that the channel is *time-invariant*, that is, it does not change its properties over time, at least for a considered transmission burst.

Under these premises, a frequently used, generic channel model is given by a *linear time-invariant (LTI)* system with additive white Gaussian noise. A Gaussian random process for noise results from the central limit theorem of probability theory. Moreover, it is proven in information theory that white Gaussian noise describes a worst-case scenario. This model (ECB domain) is depicted in Figure D.1. Thereby, all linear, dispersive effects of the channel are summarized in the *impulse response* $h_C(t)$ or its Fourier transform $H_C(f) = \mathcal{F}\{h_C(t)\}$, the *transfer function*. All disturbances (noise and interferences) are summarized in the *noise* $n(t)$, which is effective at the output of the LTI system. Without loss of generality, we may assume that the noise is white, that is, the noise power spectral density is constant with value N_0 in the ECB domain (see Chapter 2). In case of non-white

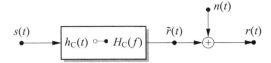

Figure D.1 Block diagram of the equivalent (linearized) channel model (ECB domain).

noise, a *whitening filter* may guarantee whiteness associated with a change of the impulse response/transfer function.

For a given transmit signal $s(t)$, due to the properties of LTI systems, the receive signal is then given by

$$r(t) = s(t) * h_C(t) + n(t) \;. \tag{D.1}$$

When the channel has no dispersive characteristics and introduces no intersymbol interference, the impulse response is simply $h_C(t) = \delta(t)$ (a transmission delay is ignored). Then, the simple *AWGN channel model* with

$$r(t) = s(t) + n(t) \tag{D.2}$$

is present.

D.2 Models for Cable Transmission

Transmission over wirelines is widely used in practice. An example is the so-called *digital subscriber line (DSL) technology*, where cables (symmetric wire pairs), initially installed for analog voice communication ("plain old telephone"), are used to connect customers to the internet ("last-mile" link). For the same purpose, the *data over cable service interface specification (DOCSIS)* uses the TV cable network, where typically coaxial cables are installed.

D.2.1 Coaxial Cables

A lateral cut through a coaxial cable is depicted in Figure D.2.

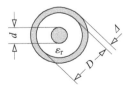

Figure D.2 Lateral cut through a coaxial cable.

For cable lengths $\ell > 100$ m and frequencies $f \geq 1$ MHz, the transfer function is well approximated by

$$H_C(f) \approx 10^{-a_{\text{dB}}\ell \sqrt{f/f_{\text{ref}}}/20} \cdot e^{-j2\pi f \ell/c_r} \tag{D.3}$$

ℓ: cable length in (km)
a_{dB}:   attenuation per kilometer (in dB) at frequency f_{ref}
f_{ref}:   normalization frequency
c_r: propagation speed $c_r = c_0/\sqrt{\varepsilon_r \mu_r}$ (speed of light $c_0 = 3 \times 10^8$ m/s, dielectric constant $\varepsilon_r > 1$, permeability $\mu_r \approx 1$).

Note that the attenuation according to $\sqrt{f}$ (i.e., $-20 \log_{10}(|H_C(f)|) = \text{const.} \cdot \sqrt{f}$) is caused by the skin effect (alternating current is concentrated near the surface of a conductor).

Typical types are the standard coaxial cable and the mini coaxial cable. Table D.1 collects the parameters for both types for $f_{\text{ref}} = 1$ MHz.

Table D.1 Parameters for standard coaxial cable and mini coaxial cable.

	D (mm)	d (mm)	$a_{\text{dB}}\left(\frac{\text{dB}}{\text{km}}\right)$
Standard coaxial cable	9.5	2.6	2.46
Mini coaxial cable	4.4	1.2	5.31

Example D.1 Attenuation of Coaxial Cables

We assume quaternary (4-ary) baseband transmission of 500 Mbit/s over standard coaxial cable of length 2 km. Since two bits are transmitted per symbol, the symbol frequency (baud rate) is $\frac{1}{T} = \frac{500}{2}$ MHz $= 250$ MHz. Ignoring the roll-off, the transmission band extends from $f = 0$ to $f_N = \frac{1}{2T} = 125$ MHz, the so-called Nyquist frequency.

Over the band, the signal attenuation increases from

$$-20 \log_{10} |H_C(0)| \approx 0 \text{ dB}$$

to

$$-20 \log_{10} |H_C(f_N)| = a_{\text{dB}} \ell \sqrt{f_N/f_{\text{ref}}} = 55 \text{ dB} .$$

Hence, a significant variation of the attenuation over the transmission band is present; the spectral components are attenuated differently (non-flat transfer function), leading to (severe) linear distortions.

D.2.2 Symmetric Wire Pairs

In the (classical) telephone network and in local area networks, twisted-pair lines are used. A lateral cut through a (twisted) pair is depicted in Figure D.3. For better symmetry of the conductors, that is, for less electromagnetic coupling of the signals

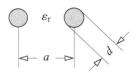

Figure D.3 Lateral cut through a (twisted) pair.

in multi-pair cables, the wire pairs are twisted. In some countries (e.g., Germany), the lines are twisted in quads ("star quad"). Twisted wire pairs are usually bundled in cables containing up to some thousand pairs and are installed for telephone links (nowadays DSL links) within subscriber line areas.

The diameter d of the conductor is typically 0.35, 0.4, or 0.6 mm. For cable lengths $\ell > 100$ m and frequencies 200 kHz $\leq f \leq$ 10 MHz, the transfer function is well approximated by

$$|H_C(f)| = 10^{-a_{\mathrm{dB}}(f)\cdot\ell/20}, \tag{D.4}$$

with

$$a_{\mathrm{dB}}(f) = a_0 + a_1 \left(\frac{f}{f_{\mathrm{ref}}}\right)^{b_1} \quad \left[\frac{\mathrm{dB}}{\mathrm{km}}\right]. \tag{D.5}$$

Table D.2 collects the parameters for typical cables and $f_{\mathrm{ref}} = 1$ MHz.

Table D.2 Parameters for twisted-pair cables.

	$a_0 \left(\frac{\mathrm{dB}}{\mathrm{km}}\right)$	$a_1 \left(\frac{\mathrm{dB}}{\mathrm{km}}\right)$	b_1
0.4 mm cable	4.5	19	0.7
0.6 mm cable	2.2	13	0.65

In a multi-pair cable, near-end and far-end crosstalk are the main sources of disturbance.

Compared to coaxial cables, twisted-pair lines have a much higher attenuation. Given a standard coaxial cable of length ℓ_{coax}, a twisted-pair cable of length $\ell_{\mathrm{twistpair}}$ (0.4 mm diameter) has the same attenuation at frequency f if

$$2.46 \, (f/f_{\mathrm{ref}})^{0.5} \, \ell_{\mathrm{coax}} \stackrel{!}{=} (4.5 + 19 \, (f/f_{\mathrm{ref}})^{0.7}) \, \ell_{\mathrm{twistpair}}.$$

Let $\mathcal{B} = [0, f_N]$ be the transmission band of interest. Then, both cable types show a rather similar average attenuation within $\mathcal{B}$ if

$$J = \frac{1}{f_N} \int_0^{f_N} \left| 2.46 \, (f/f_{\mathrm{ref}})^{0.5} - (4.5 + 19 \, (f/f_{\mathrm{ref}})^{0.7}) \frac{\ell_{\mathrm{twistpair}}}{\ell_{\mathrm{coax}}} \right|^2 \mathrm{d}f$$

is minimized over the ratio of the cable lengths. For $f_N = 8.499$ MHz (which corresponds to the VDSL-2 standard G.993.2, profile 8d), we have

$$\ell_{\mathrm{twistpair}} \approx 0.086 \cdot \ell_{\mathrm{coax}}. \tag{D.6}$$

Example D.2 Transfer Function of Cables

The magnitudes of the transfer functions (in dB) of coaxial and symmetric cables (both of length 2 km) are compared in Figure D.4.

It can be seen that symmetric cables (twisted pairs) have a much higher attenuation (negative of the channel gain in dB) compared to coaxial cables. The smaller the diameter of the conductor (the "cheaper" the cable), the higher the attenuation and the increase in attenuation over the frequency.

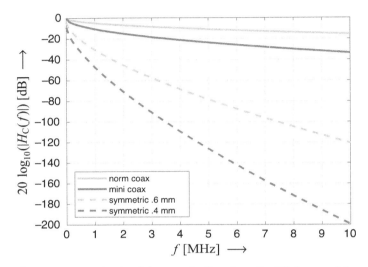

Figure D.4 Magnitude of the transfer functions (in dB) for coaxial and symmetric cables both of length 2 km.

In Figure D.5 the attenuation is compared for corresponding cable lengths.

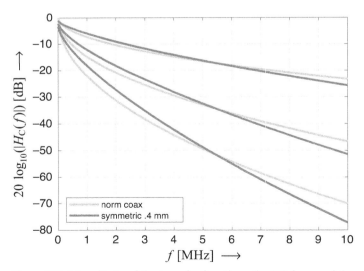

Figure D.5 Magnitude of the transfer functions (in dB) for coaxial and symmetric cables.

The length of the coaxial cable is chosen to be $\ell_{\text{coax}} = 3, 6,$ and 9 km (top to bottom). The corresponding lengths of the twisted-pair cable are then $\ell_{\text{twistpair}} = 258, 516,$ and 774 m.

D.3 Radio Transmission

When dealing with radio-frequency transmission at a carrier frequency f_c, the usual assumption is that only the *far-field model* is considered. This means that the distance d between transmitter and receiver is significantly larger than the wavelength λ (speed of light $c_0 = 3 \times 10^8$ m/s):

$$d \gg \lambda = \frac{c_0}{f_c}, \tag{D.7}$$

such that a *planar wave* can be expected. Moreover, it is common to model that the power is radiated *isotropically* (uniformly in each direction). If the antenna has an antenna pattern (directive antenna), an *equivalent isotropic radiator* with an equivalent power (effective isotropic radiated power, EIRP) is specified. Thus, seen from the location of the receiver, for analysis purposes, the transmitter is modeled as an omni-directional radiator with the equivalent signal power EIRP.

Two basic models for radio transmission are widely used.

- *Pure Propagation Loss*
  The transmitted signal is attenuated (only a small fraction of the radiated power is captured by the receive antenna) but not distorted. Terrestrial links typically have to bridge attenuations up to 160 dB, whereas in satellite communications this even goes up far beyond 200 dB.
- *Multipath Propagation*
  At the receive antenna, several delayed and scaled versions of the transmit (TX) signal may be superimposed, that is, multiple propagation paths exist from the transmitter to the receiver (RX). This leads to linear distortions.

The situation in a multipath-propagation scenario is depicted in Figure D.6. A direct path between transmitter and receiver may be present. Additionally, reflected paths may contribute. Thereby, the wave may simply be reflected like a mirror, or it can be scattered (deviate in a random fashion from the straight reflection). Finally, the waves may be diffracted, that is, bent around corners or edges.

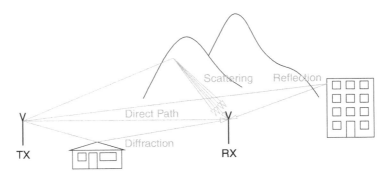

Figure D.6 Multipath-propagation scenario.

For the moment we assume that the scenario is static, that is, no movement of any object is present. Then the transmission characteristics from the transmitter to the receiver are time-invariant and the channel can be modeled with the impulse response of an LTI system.

The channel impulse response is given by the superposition (addition) of the contributions of the different paths; each path is individually attenuated and (due to the different lengths of the paths) delayed. In the ECB domain, the impulse response can be written as

$$h_C(t) = \sum_{i=1}^{L} c_i\, \delta(t - T_i), \qquad c_i \in \mathbb{C} \tag{D.8}$$

c_i : complex gain factor of the ith propagation path
T_i : delay of the ith propagation path.

An exemplary impulse response is depicted in Figure D.7.

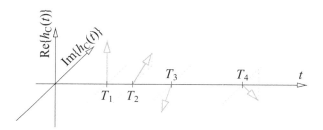

Figure D.7 Exemplary impulse response (ECB).

Instead of specifying the impulse response, equivalently, the transfer function of the channel can be given. Using basic relations of the Fourier transform (in particular, linearity and $\delta(t - T_i) \circ\!\!-\!\!\bullet\, e^{-j2\pi f T_i}$), we obtain from (D.8)

$$H_C(f) = \mathscr{F}\{h_C(t)\} = \mathscr{F}\left\{\sum_{i=1}^{L} c_i\, \delta(t - T_i)\right\} = \sum_{i=1}^{L} c_i\, e^{-j2\pi f T_i}. \tag{D.9}$$

A sketch of the magnitude of a channel transfer function is shown in Figure D.8.

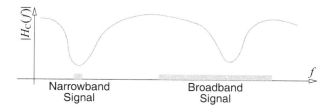

Figure D.8 Exemplary channel transfer function.

Looking at this figure, the effects (linear distortions) caused by multipath propagation can be characterized as follows.

- *Narrowband Signals/Narrowband Channels*
  If the bandwidth of the signal is very small in comparison to typical fluctuations of the channel transfer function, extremely varying attenuation (including deep

fading) may occur. Thus, there is usually a probability that the signal is completely lost. However, within the signal bandwidth, the channel transfer function is (almost) flat and no linear distortions will be present.

When transmitter, receiver, and/or the reflecting objects are moving, the channel transfer function and thereby the location of the deep fades on the frequency axis will change over time. Hence, a *frequency-non-selective* or *flat-fading* channel is present.

- *Broadband Signals/Broadband Channels*
  If the bandwidth of the signal is large in comparison to typical fluctuations of the channel transfer function, the transfer function will exhibit significant fluctuations within the frequency band in use. This may cause severe linear distortions, which requires the application of sophisticated equalization techniques. However, a complete deep attenuation of the entire signal (i.e., its complete loss) will be very unlikely for broadband signals.

  Hence, a *frequency-selective* or *multipath-fading* channel is present.

D.4 Linear Time-Varying Systems

Up to now we have expected that the channel is time-invariant – this property is dropped subsequently. However, linear behavior is still assumed. For the mathematical characterization of the channel, in particular the input/output relation, we have to resort to the theory of *linear time-varying (LTV) systems*. In the following, a brief description is given.

A system $\mathcal{S}$ that transforms an input signal into an output signal is *linear* if for any signals $x_i(t)$ and scalars $c_i \in \mathbb{C}$, $i = 1, 2$, the following relation holds:

$$\mathcal{S}\{c_1 x_1(t) + c_2 x_2(t)\} = c_1 \mathcal{S}\{x_1(t)\} + c_2 \mathcal{S}\{x_2(t)\}. \tag{D.10}$$

A system is *time-varying* if, for $y(t) = \mathcal{S}\{x(t)\}$:

$$y(t - t_0) \neq \mathcal{S}\{x(t - t_0)\}, \tag{D.11}$$

for all $t_0 \in \mathbb{R}$.

D.4.1 Time-Domain Characterization of LTV Systems

A (deterministic) LTV system is only fully characterized if the reactions of the system to Dirac impulses at all times $t = \tau_e$, $\forall \tau_e \in \mathbb{R}$, are known. The *time-variant impulse response* – which depends on the two parameters (observation) time t and excitation time τ_e – is defined as

$$h_0(\tau_e, t) \stackrel{\text{def}}{=} \mathcal{S}\{\delta(t - \tau_e)\}. \tag{D.12}$$

Figure D.9 gives a visualization of the excitation at $t = \tau_e$ and the reactions (impulse responses) assuming a causal system (the reaction starts at the time of excitation).

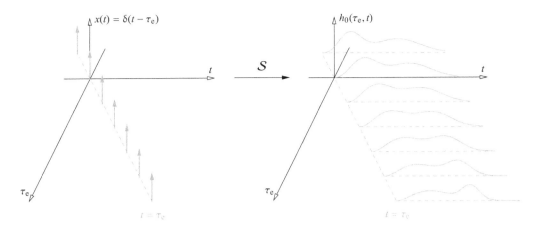

Figure D.9 Time-variant impulse response (shown only for some exemplary excitation times).

Owing to the sifting property of the Dirac function, we can write

$$x(t) = \int_{-\infty}^{\infty} x(\tau_e)\,\delta(t - \tau_e)\,d\tau_e\,. \tag{D.13}$$

Taking the assumed linearity of the system into account, we have

$$y(t) = \mathcal{S}\{x(t)\} = \mathcal{S}\left\{\int_{-\infty}^{\infty} x(\tau_e)\,\delta(t - \tau_e)\,d\tau_e\right\} = \int_{-\infty}^{\infty} x(\tau_e)\,\mathcal{S}\{\delta(t - \tau_e)\}\,d\tau_e\,, \tag{D.14}$$

and the *input/output relation* of an LTV system is given by

$$y(t) = \int_{-\infty}^{\infty} x(\tau_e)\,h_0(\tau_e,t)\,d\tau_e\,, \tag{D.15}$$

or substituting $\tau_e = t - \tau$:

$$y(t) = \int_{-\infty}^{\infty} x(t - \tau)\,h_0(t - \tau,t)\,d\tau\,. \tag{D.16}$$

It is common to define a transformed version of $h_0(\tau_e,t)$, specifically

$$h(\tau,t) \stackrel{\text{def}}{=} h_0(t - \tau,t)\,, \tag{D.17}$$

which leads to the following form of the input/output relation:

$$y(t) = \int_{-\infty}^{\infty} x(t - \tau)\,h(\tau,t)\,d\tau\,. \tag{D.18}$$

Thereby, $h(\tau,t)$ is the *time-variant weight function* of the LTV system. Note that $h(\tau,t)$ is *not* an impulse response (although it is usually called such in the literature). In no case is $h(\tau,t)$ the reaction of the time-variant system to an impulse!

D.4 Linear Time-Varying Systems

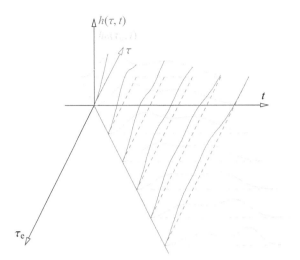

Figure D.10
Time-variant weight function and connection to the time-variant impulse response.

Figure D.10 visualizes the time-variant weight function. For a fixed time position t, the τ-axis runs in the negative direction of τ_e, with $\tau = 0$ for $t = \tau_e$ (the dark gray diagonal axis).

Both descriptions of the LTV system, the impulse response $h_0(\tau_e, t)$ and the weight function $h(\tau, t)$, are equivalent, but they give rise to different interpretations or models of the system. Thinking of the integrals to be discretized, that is, $\int f(\tau) \, d\tau \approx \sum_l f(l\Delta\tau) \cdot \Delta\tau$, and the impulse response of finite duration, that is, $h_0(\tau_e, t) = 0, t \neq [\tau_e, \tau_e + T_h], \forall \tau_e$, where T_h is the maximum *delay spread*, the structures depicted in Figure D.11 correspond to the input/output relations (D.15) and (D.18), respectively.

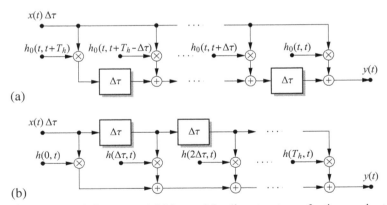

Figure D.11 (a) Output and (b) input delay line structure of a time-variant linear system (discretized depiction).

The time-variant weight function $h(\tau, t)$ corresponds directly to a structure in which the system could be implemented. It is the well-known (from discrete-time FIR systems) *input delay line structure*. However, in contrast to time-*in*variant linear systems, the coefficients are not constant with respect to t, that is, only dependent on the *delay time* τ, but change over time. Conversely, the description via the time-variant impulse response $h_0(\tau_e, t)$ corresponds to an *output delay line structure*. It is usually more convenient to work with the input delay line structure, since the history of the input signal is contained in the delay line of which a linear combination is built. Hence, we prefer the weight function $h(\tau, t)$.

D.4.2 Frequency-Domain Characterization of LTV Systems

As for LIT systems, in some situations it is more convenient to resort to a *frequency-domain* description of the LTV system. Here, we restrict ourselves to the time-variant weight function. As $h(\tau, t)$ is a function of two parameters, three different transforms exist – with respect to to τ, t, or both variables. The diagram in Figure D.12 visualizes the respective functions.

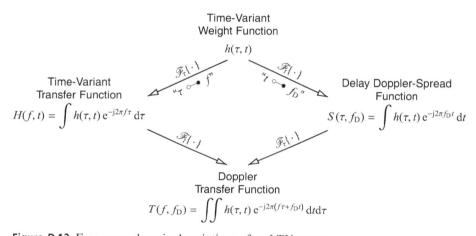

Figure D.12 Frequency-domain descriptions of an LTV system.

Time-Variant Transfer Function

Often, one is interested in the *time-variant transfer function*, which is obtained by taking the Fourier transform of $h(\tau, t)$ with respect to the parameter τ, the *delay in the input delay line* ("$\tau \circ\!\!-\!\!\bullet\, f$")

$$H(f, t) \stackrel{\text{def}}{=} \mathscr{F}_\tau\{h(\tau, t)\} = \int_{-\infty}^{\infty} h(\tau, t)\, e^{-j 2\pi f \tau}\, d\tau\,. \tag{D.19}$$

The name is due to the fact that for a mono-frequent input signal, that is, $x(t) = e^{j 2\pi f_0 t}$, taking (D.18) into account, the output is given by

$$y(t) = \int_{-\infty}^{\infty} e^{j 2\pi f_0 (t-\tau)} h(\tau, t)\, d\tau = e^{j 2\pi f_0 t} \int_{-\infty}^{\infty} e^{-j 2\pi f_0 \tau} h(\tau, t)\, d\tau = e^{j 2\pi f_0 t} \cdot H(f_0, t)\,.$$

$$\tag{D.20}$$

D.4 Linear Time-Varying Systems

An arbitrary input signal $x(t)$ with Fourier transform $X(f) = \mathscr{F}\{x(t)\}$ is the weighted "sum" of mono-frequent components $e^{j2\pi ft}$

$$x(t) = \int_{-\infty}^{\infty} X(f) \, e^{j2\pi ft} \, df \tag{D.21}$$

and each component contributes $e^{j2\pi ft} H(f,t)$ to the output signal. Owing to the linearity of the system, the output signal is given by

$$y(t) = \int_{-\infty}^{\infty} X(f) H(f,t) \, e^{j2\pi ft} \, df \, . \tag{D.22}$$

Its spectrum can thus be written as

$$\begin{aligned} Y(f) &= \int_{-\infty}^{\infty} y(t) \, e^{-j2\pi ft} \, dt \\ &= \int_{-\infty}^{\infty} \int_{-\infty}^{\infty} X(\nu) \, H(\nu,t) \, e^{-j2\pi(f-\nu)t} \, d\nu \, dt \, . \end{aligned} \tag{D.23}$$

Delay Doppler-Spread Function

The *delay Doppler-spread function* is obtained by taking the Fourier transform of $h(\tau,t)$ with respect to the parameter t, the *observation time* ("$t \multimap f_D$")

$$S(\tau, f_D) \stackrel{\text{def}}{=} \mathscr{F}\{h(\tau,t)\} = \int_{-\infty}^{\infty} h(\tau,t) \, e^{-j2\pi f_D t} \, dt \, . \tag{D.24}$$

The spectrum of the output signal $y(t)$ can be written as

$$\begin{aligned} Y(f) = \mathscr{F}\{y(t)\} &= \mathscr{F}\left\{ \int_{-\infty}^{+\infty} x(t-\tau) \, h(\tau,t) \, d\tau \right\} \\ &= \int_{-\infty}^{+\infty} \mathscr{F}\{x(t-\tau)\} * \mathscr{F}\{h(\tau,t)\} \, d\tau \\ &= \int_{-\infty}^{+\infty} X(f) \, e^{-j2\pi f\tau} * S(\tau,f) \, d\tau \\ &= \int_{-\infty}^{+\infty} \int_{-\infty}^{+\infty} X(f-\nu) S(\tau,\nu) \, e^{-j2\pi(f-\nu)\tau} \, d\nu \, d\tau \, . \end{aligned} \tag{D.25}$$

The name of the function is due to the fact that for a mono-frequent input signal, that is, $X(f) = \delta(f - f_0)$, the output spectrum is given by

$$Y(f) = \int_{-\infty}^{+\infty} \int_{-\infty}^{+\infty} \delta(f - f_0 - \nu) S(\tau, \nu) \, e^{-j2\pi(f-\nu)\tau} \, d\nu \, d\tau$$

$$= \int_{-\infty}^{+\infty} S(\tau, f - f_0) \, e^{-j2\pi f_0 \tau} \, d\tau \, . \tag{D.26}$$

For each delay τ, the Dirac spectrum of the input signal is broadened to $S(\tau, f - f_0)$.

D.5 Stochastic, Time-Varying Channel Models

The above exposition gives the description of a *deterministic* linear time-varying system. To describe channels, however, channel parameters such as impulse response or weight function are random quantities and need to be characterized adequately. To that end we distinguish between two basic situations, a *flat-fading channel* and a *frequency-selective channel*.

D.5.1 Flat-Fading Channel

If the channel introduces no dispersion, the weight function reduces to $h(\tau, t) \stackrel{\text{def}}{=} h(t) \, \delta(t - \tau)$. The ECB model of a flat-fading (frequency non-selective) channel is depicted in Figure D.13. Besides the unavoidable additive noise $n(t)$, the multi-

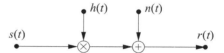

Figure D.13 Flat-fading channel.

plicative channel gain $h(t) \in \mathbb{C}$ is present. Both quantities are drawn from random processes and have to be characterized adequately. As usual, we assume the noise to be white (constant psd) and Gaussian (specific pdf).

Distribution of the Fading Gain

Owing to different physical phenomena, the characterization of the distribution of the gain factor $h(t)$ is divided into slow- and fast-fading effects. In the following, we describe both effects separately. The total channel gain is the product of both parts.

Slow Fading Due to Shadowing Slow variations are typically introduced by *shading* effects. Owing to its nature, shadowing is non-stationary. The common model to characterize shadowing is to assume the (time-variant) gain $h(t) = h_\text{s}$ is real-valued, positive, and *Gaussian distributed* in dB. That is, $h_\text{s,dB} \stackrel{\text{def}}{=} -20 \log_{10}(h_\text{s})$ has the pdf

$$f_{h_\text{s,dB}}(h_\text{s,dB}) = \frac{1}{\sqrt{2\pi\sigma^2_{h_\text{s,dB}}}} \, e^{-(h_\text{s,dB} - m_{h_\text{s,dB}})^2 / (2\sigma^2_{h_\text{s,dB}})} \, , \tag{D.27}$$

where mean $m_{h_{s,dB}}$ and variance $\sigma^2_{h_{s,dB}}$ characterize the random variable completely. Note that the (instantaneous) power $h_s^2 = 10^{-h_{s,dB}/10}$ is then *log-normal distributed*.

Fast Fading Due to Scattering Presuming that a large number of scatterers are present and thus a large number of individually fading effects contribute to the entire (except shadowing) complex-valued fading gain h, a usual assumption following the central limit theorem is that h is *Gaussian distributed*:

$$f_h(h) = \frac{1}{\pi \sigma_h^2} e^{-|h-m_h|^2/\sigma_h^2},$$

with mean m_h and variance σ_h^2.

If no line-of-sight component is active, we have $m_h = 0$. Then, the channel gain h is zero-mean complex Gaussian and $|h|$ is Rayleigh distributed – a *flat Rayleigh-fading channel* is present.

If a line-of-sight component is active ($m_h \neq 0$) the magnitude of the channel gain $|h|$ is Rician distributed – a *flat Rician-fading channel* is present. It is characterized by the *Rician factor* $K \stackrel{\text{def}}{=} m_h^2/\sigma_h^2$. Note that for $K \to \infty$ an AWGN channel results.

Autocorrelation Function/Power Spectral Density

To fully specify the fast-fading process $h(t)$, the temporal variations have to be known. We ignore the slow fading and assume that the fast fading is stationary. The random process from which $h(t)$ is drawn is then fully characterized by the autocorrelation function

$$\phi_{hh}(\tau) = \mathrm{E}\{h(t+\tau)\,h^*(t)\} \tag{D.28}$$

or by the power spectral density

$$\Phi_{hh}(f_\mathrm{D}) = \mathscr{F}_\tau\{\phi_{hh}(\tau)\}. \tag{D.29}$$

We now consider an important special case of the psd when a single path with a random *Doppler shift* is present. Whenever transmitter and receiver are moving relative to each other, a Doppler shift occurs. The situation of a moving object (maybe a car) and an incident planar wave is shown in Figure D.14.

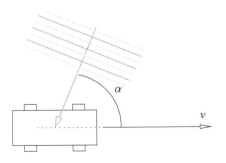

Figure D.14 Moving object and incident planar wave.

From basic geometry, the Doppler shift f_D, that is, the change in frequency compared to the carrier frequency f_c of the transmitted signal, calculates as

$$f_D = \frac{v}{c_0} f_c \cos(\alpha) \stackrel{\text{def}}{=} f_{D,\max} \cos(\alpha) \qquad (D.30)$$

v: velocity
α: angle of incidence
c_0: speed of light ($c_0 = 3 \times 10^8$ m/s)
f_c: carrier frequency
$f_{D,\max} = \frac{v}{c_0} f_c$: maximum Doppler shift.

When the Doppler shift is the main effect, the fading process $h(t)$ can be approximated by

$$h(t) \approx h\, e^{j2\pi f_D(t) t}, \qquad (D.31)$$

where h is the zero-mean fading gain (variance σ_h^2) and $f_D(t)$ the (stationary) process characterizing the Doppler shift. Since different physical processes are responsible, we may assume that h and $f_D(t)$ are statistically independent.

Under these assumptions, the psd of the random process $h(t)$ is given by

$$\Phi_{hh}(f_D) = \mathscr{F}_\tau\{\mathrm{E}\{h\, e^{j2\pi f_D(t)(t+\tau)}\, h^*\, e^{-j2\pi f_D(t) t}\}\}$$

$\mathscr{F}\{\cdot\}$ and $\mathrm{E}\{\cdot\}$ are linear operator

$$= \mathrm{E}\{|h|^2\, \mathscr{F}_\tau\{e^{j2\pi f_D(t)\tau}\}\}$$

$\mathscr{F}_\tau\{e^{j2\pi f_D(t)\tau}\} = \delta(f_D - f_D(t))$

$$= \mathrm{E}\{|h|^2 \delta(f_D - f_D(t))\}$$

independence

$$= \mathrm{E}\{|h|^2\} \cdot \mathrm{E}\{\delta(f_D - f_D(t))\}$$
$$= \sigma_h^2 \cdot \mathrm{E}\{\delta(f_D - f_D(t))\}. \qquad (D.32)$$

Knowing the pdf $f_{f_D}(f_D)$ of the Doppler process, the last expectation can be written as

$$\mathrm{E}\{\delta(f_D - f_D(t))\} = \int \delta(f_D - x)\, f_{f_D}(x)\, \mathrm{d}x = f_{f_D}(f_D). \qquad (D.33)$$

In summary, the psd of the fading process is given by

$$\Phi_{hh}(f_D) = \sigma_h^2 \cdot f_{f_D}(f_D). \qquad (D.34)$$

In order to obtain an analytic expression for the pdf $f_{f_D}(f_D)$ of the Doppler shift, it is common to assume that the angle of incidence α is uniformly distributed, that is

$$f_\alpha(\alpha) = \frac{1}{2\pi}, \qquad \alpha \in [-\pi, \pi). \qquad (D.35)$$

Via (D.30), α is transformed into f_D.

The pdf $f_y(y)$ of a random variable y obtained from the random variable x with pdf $f_x(x)$ via the function $y = g(x)$ is given by

$$f_y(y) = \sum_{x \in \mathcal{X}} \frac{f_x(x)}{|g'(x)|}, \quad \text{with} \quad \mathcal{X} = \{x \mid y = g(x)\}. \quad \text{(D.36)}$$

For $y = g(x) = a\cos(x)$ we have $g'(x) = -a\sin(x) = -a\sqrt{1-\cos^2(x)} = -\sqrt{a^2 - y^2}$ and for x uniform over the interval $[-\pi, \pi)$ the equation $y = g(x)$ has two solutions ($\mathcal{X}$ has cardinality two). This gives

$$f_y(y) = \frac{1}{\sqrt{a^2 - y^2}} \left(\frac{1}{2\pi} + \frac{1}{2\pi} \right) = \frac{1}{\pi\sqrt{a^2 - y^2}}, \quad \text{(D.37)}$$

and thus for the Doppler pdf ($a = f_{D,\max}$, $y = f_D$)

$$f_{f_D}(f_D) = \frac{1}{\pi\sqrt{f_{D,\max}^2 - f_D^2}} = \frac{1}{\pi f_{D,\max} \sqrt{1 - (f_D/f_{D,\max})^2}}. \quad \text{(D.38)}$$

Finally, the psd of the fading process is given by

$$\Phi_{hh}(f_D) = \sigma_h^2 \cdot \frac{1}{\pi f_{D,\max} \sqrt{1 - (f_D/f_{D,\max})^2}}, \quad \text{(D.39)}$$

or its autocorrelation function (inverse Fourier transform of the psd) reads

$$\phi_{hh}(\tau) = \sigma_h^2 \cdot J_0(\tau \pi f_{D,\max}/2), \quad \text{(D.40)}$$

where $J_0(x)$ is the zeroth-order Bessel function of the first kind. Both quantities are depicted in Figure D.15.

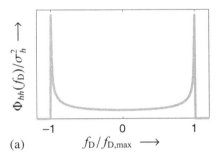
(a)

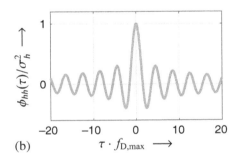
(b)

Figure D.15 (a) Power spectral density and (b) autocorrelation function of the fading process in case of Doppler shift.

In the literature, (D.39) is often denoted as *Jakes' spectrum*. However, note that it is the *power spectral density* of the fading process that is proportional (cf. (D.34)) to the *probability density function* $f_{f_D}(f_D)$ of the Doppler shift. Since it is a pdf of a random variable that is a frequency, it is a quantity over the frequency axis.

D.5.2 Frequency-Selective Channel

In the above case of a flat-fading channel, a single fading coefficient is present and its stochastic nature has to be characterized. In case of frequency-selective channels, the statistical properties of the weight function $h(\tau,t)$ have to be specified. The starting point is the input delay-line model shown in Figure D.11.

According to the central limit theorem, it is reasonable to assume that the pdf of $h(\tau,t)$ is Gaussian for all parameters τ and t. Thus, for a complete characterization, the autocorrelation function or equivalently the power spectral density has to be given. As $h(\tau,t)$ is a two-variable function, acf and psd would depend on four parameters. However, some common assumptions lead to a feasible model.

First, it is expected that the weight function is a stationary random process with respect to the *observation time t*. This reduces the number of required parameters to three. The acf would then be given by

$$\phi_{hh}(\tau, \Delta\tau, \Delta t) = \mathrm{E}\{h(\tau + \Delta\tau, t + \Delta t) \cdot h^*(\tau,t)\} . \tag{D.41}$$

Second, it is assumed that the weights $h(\tau,t)$ at different *delay times* τ are caused by different, individually moving scatterers and thus are uncorrelated. This gives

$$\phi_{hh}(\tau, \Delta\tau, \Delta t) = \phi_{hh}(\tau, \Delta t)\,\delta(\Delta\tau) , \tag{D.42}$$

that is, this assumption reduces the number of required parameters to two. These two steps result in the so-called *wide-sense stationarity (WSS)* and *uncorrelated scatterer (US)* model, in short *WSSUS model*, which is widely used when modeling the channel in mobile communications.

For further description, this two-parameter acf is transformed into the frequency domain with respect to the observation time difference Δt. This gives the *scattering function* ("$\Delta t \multimap f_\mathrm{D}$")

$$\Phi_{hh}(\tau, f_\mathrm{D}) \stackrel{\mathrm{def}}{=} \mathscr{F}_{\Delta t}\{\phi_{hh}(\tau, \Delta t)\} = \int_{-\infty}^{\infty} \phi_{hh}(\tau, \Delta t)\, \mathrm{e}^{-\mathrm{j}2\pi f_\mathrm{D} \Delta t}\, \mathrm{d}\Delta t . \tag{D.43}$$

Finally, it is assumed that this two-parameter scattering function can be separated into the product of two one-parameter functions in the form

$$\Phi_{hh}(\tau, f_\mathrm{D}) \stackrel{!}{=} \phi_{hh}(\tau) \cdot \Phi_{hh}(f_\mathrm{D}) , \tag{D.44}$$

with the *average power distribution over the delay time*, called the *power delay profile (pdp)*

$$\phi_{hh}(\tau) \stackrel{\mathrm{def}}{=} \phi_{hh}(\tau,0) = \mathrm{E}\{|h(\tau,t)|^2\} . \tag{D.45}$$

Note that due to the properties of the Fourier transform, the pdp can also be written as

$$\phi_{hh}(\tau) = \int_{-\infty}^{\infty} \Phi_{hh}(\tau, f_\mathrm{D})\, \mathrm{d}f_\mathrm{D} , \tag{D.46}$$

and with the average *Doppler power spectral density*

$$\Phi_{hh}(f_\mathrm{D}) \stackrel{\mathrm{def}}{=} \frac{1}{c_\mathrm{D}} \int_{-\infty}^{\infty} \Phi_{hh}(\tau, f_\mathrm{D})\, \mathrm{d}\tau , \tag{D.47}$$

where $c_\mathrm{D} = \int_{-\infty}^{\infty} \int_{-\infty}^{\infty} \Phi_{hh}(\tau, f_\mathrm{D})\, \mathrm{d}\tau \mathrm{d}f_\mathrm{D} = \int_{-\infty}^{\infty} \phi_{hh}(\tau)\, \mathrm{d}\tau$ normalizes the integral over the Doppler psd to one.

D.5 Stochastic, Time-Varying Channel Models

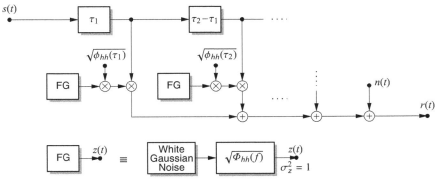

Figure D.16 Input-delay-line structure for a time-varying multipath channel obeying the WSSUS model.

This decomposition gives rise to a convenient and practical model of a time-varying multipath channel under the WSSUS assumption. It is characterized by the

- power delay profile $\phi_{hh}(\tau)$ and the
- average Doppler power spectral density $\Phi_{hh}(f_D)$.

Figure D.16 shows the input-delay-line structure that realizes the channel for a discrete version of the power delay profile ($\phi_{hh}(\tau)$ is only nonzero for $\tau \in \{\tau_1, \tau_2, \ldots\}$). The time variation of the fading gains is generated according to psd $\Phi_{hh}(f_D)$ by filtering white Gaussian noise (fading generator FG). The random signals with variance one (due to the above normalization) are then scaled by $\sqrt{\phi_{hh}(\tau)}$ to adjust the respective power.

Example D.3 Power Delay Profiles

Figure D.17 sketches the power delay profiles according to COST 207 (Coopération européenne dans le domaine de la recherche scientifique et technique). It shows continuous-time models for *digital land mobile radio communications* for a typical bad urban area (a) and for a typical hilly terrain (b).

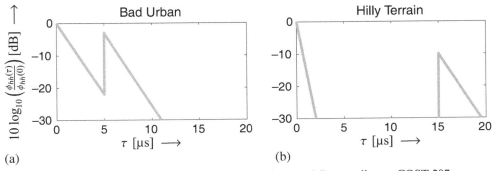

Figure D.17 Power delay profiles (continuous-time model) according to COST 207.

Figure D.18 shows the scattering function according to COST 207 for a "typical urban" scenario (discrete-time model) on a linear scale. The separation between the dimensions, power delay profile and Doppler power spectral density (in the form of the Jakes spectrum), respectively, is clearly visible.

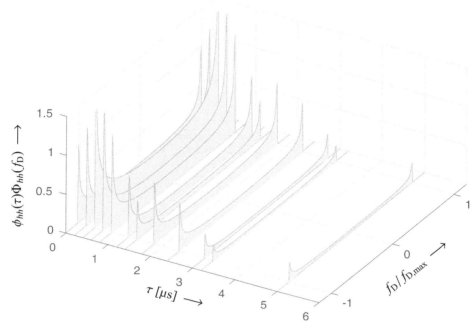

Figure D.18 Scattering function (discrete-time model) according to COST 207.

APPENDIX E
Implementation of PAM Transmission in Software

This appendix presents a MATLAB® experiment that will extend our knowledge of PAM schemes. A communication system using digital PAM will be implemented in MATLAB®. This enables us to modify the various components in a very flexible way. The numerical simulation technique that is being used in this project is standard nowadays in the analysis (e.g., error ratio simulations) and design of digital communication schemes.

Starting from a given MATLAB® simulation script file, you will implement the required function blocks step by step and, in the end, be able to simulate the bit error ratio (BER) performance of the scheme over a given channel.

We assume that the reader has some basic knowledge of MATLAB® and the programming of m-functions.

E.1 Preliminaries and Simulation Environment

E.1.1 Transmission System

A block diagram of the entire transmission system to be considered is depicted in Figure E.1.

As usual, the entire scheme is basically divided into three main blocks: *transmitter* (top), *channel*, and *receiver* (bottom). Additionally, a *source*, producing a random bit sequence, and the corresponding *sink* are present. Via a comparison of the transmitted and received binary sequences, q[l] and q̂[l], respectively, the error ratio can be estimated. Note that the different clocking rates in the PAM scheme are indicated as well.

E.1.2 MATLAB® Simulation Environment

In this project we write MATLAB® functions to realize a numerical simulation of a PAM transmission scheme. To that end, MATLAB® script files and MATLAB® functions are available on the companion website.

We will successively complete the respective functions to finally obtain a complete PAM simulation environment. Table E.1 gives an overview of the required m-files and whether they are given or have to be written.

The main file of the MATLAB® simulation environment is `pam_simul.m`. It implements the system one-to-one in software. First, the relevant parameters are defined.

E Implementation of PAM Transmission in Software

Table E.1 m-Files required for this project.

m-File	Status
channel_baseband.m	to be written in Problem E.11
channel_discrete.m	to be written in Problem E.14
channel_rf.m	to be written in Problem E.8
g_rect.m	to be written in Problem E.4
g_sqrtcos.m	given
pam_map.m	given
pam_mf.m	given
pam_pulseshape.m	given
pam_quantdemap.m	given
pam_simul.m	fragment given; to be completed
rf_demodulate.m	to be written in Problem E.10
rf_modulate.m	to be written in Problem E.6

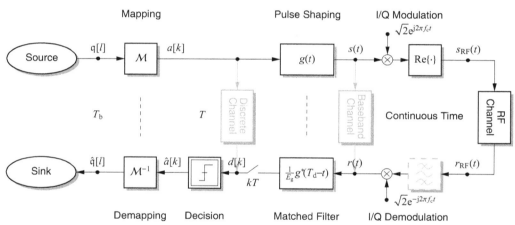

Figure E.1 Block diagram of the digital PAM transmission scheme. The gray blocks are subsequently explained in the text.

The PAM parameters "constellation" Const and "basic pulse shape" g are specified in the struct PAMparam. Basically, all function blocks shown in Figure E.1 are present. Note that the signals are denoted in MATLAB® the same way as in the block diagram. Moreover, the bypassing of the radio-frequency (RF) channel and the continuous-time part (see gray blocks in Figure E.1), respectively, is also included.

Note that all signals have to be normalized (i.e., unitless) in MATLAB®. We normalize the continuous time t to the symbol duration T and all signals to the voltage 1 V. Moreover, without loss of generality, we set the pulse energy to $E_g/T = 1$ V^2.

Problem

E.1 Preparation
(a) Download the files provided on the website.
(b) Start the graphical user interface of MATLAB®, create a subfolder, store all m-files into this working directory, and change into this directory.
(c) Familiarize yourself with the given MATLAB® script file and MATLAB® functions.

E.2 PAM Transmitter

The transmitter of a conventional PAM scheme mainly consists of the following three blocks: *mapping*, *pulse shaping*, and *modulation*.

E.2.1 Mapping

Fundamentals and Theory

The task of *bit mapping* (denoted by $\mathcal{M}$ in Figure E.1) is to map binary m-tuples to (complex) amplitude coefficients $a[k]$, drawn from an M-ary *constellation* $\mathcal{A}$. In order to address M points, the binary labels are of length $m = \log_2(M)$.

There are different ways[1] to assign the binary labels to the signal points. This assignment is denoted as *labeling*. Constellation and labeling together form the mapping.

As examples, in Figure E.2, two different labelings for 8ASK and 16QAM, respectively, are displayed. On the one hand, a *natural labeling* is shown; the signal points are simply labeled in the natural sequence of binary counting. Using *Gray labeling*, the labels are arranged in such a way that the labels of neighboring signal points differ in a single binary symbol.

Problem

E.2 Mapping
(a) What is the main advantage of Gray labeling compared to other labeling strategies?
(b) The following bit sequence is given:

$$101100010011$$

Map this sequence onto a sequence of complex PAM constellation points employing bipolar 8ASK and 16QAM using natural labeling and Gray labeling from Figure E.2, respectively.

---
[1] Ignoring symmetry, there are $M!$ different ways to label a constellation.

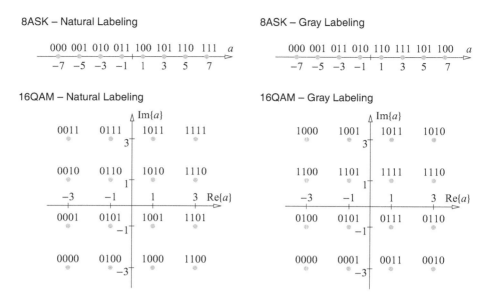

Figure E.2 Natural and Gray labelings for bipolar 8ASK and 16QAM, respectively.

Write down the results in the following table. How long are the sequences of signal points for these examples?

	Natural Labeling	Gray Labeling
8ASK		
16QAM		

Implementation and Simulation

In the simulations, first the required PAM parameters have to be defined.

A generic mapping routine is given by the function pam_map.m, which is called as follows:

```
>> a = pam_map( q, PAMparam )
```

It maps a sequence (vector) q of binary symbols to the sequence (vector) a of amplitude coefficients. The field Const in the struct PAMparam has to represent the constellation/mapping as a vector of length M. The binary label $[q_{m-1}, \ldots, q_0]$ is interpreted as an integer, which simply addresses[2] the respective vector element in the constellation. For example, for Gray labeled 16QAM:

$$\mathcal{M}([0\ 0\ 0\ 0]) \mapsto \text{Const}(0+1) = -3 - 3j$$
$$\mathcal{M}([0\ 1\ 1\ 0]) \mapsto \text{Const}(6+1) = 3 - j$$

---

[2] Since MATLAB® addresses vectors starting from one, the indices always have to be incremented by one.

Problem

E.3 16QAM Mapping
(a) Define the mapping vectors `Const` for (cf. Figure E.2)
- 8ASK with natural labeling
- 8ASK with Gray labeling
- 16QAM with natural labeling
- 16QAM with Gray labeling.
(b) Test the mapping function with your constellations for a random bit sequence `q`.

Plot the obtained sequence `a` of amplitude coefficients as points in the complex plane (e.g., `plot( ..., 'o' )`).

E.2.2 Pulse Shaping

Fundamentals and Theory

Via *pulse shaping* the discrete-time sequence $a[k]$ (symbol interval T) is transformed into a continuous-time transmit signal $s(t)$, given in the *equivalent complex baseband*. In PAM, this transformation is realized by *filtering* the sequence $a[k]$ with the *basic pulse shape* $g(t)$. As is well known, for transmission over the AWGN channel, the pulse shape $g(t)$ should have square-root Nyquist $(\sqrt{\text{Nyquist}})$ characteristics. The pulse energy calculates as

$$E_g = \int_{-\infty}^{\infty} |g(t)|^2 \, dt \,. \tag{E.1}$$

The transmit signal (ECB or RF domain) is always a *continuous-time* (i.e., analog) signal. In numerical simulations, continuous-time signals cannot be represented. MATLAB® works entirely on vectors and matrices – it is very well suited for discrete-time signals but not directly for continuous ones. However, obeying Shannon's sampling theorem, any bandlimited continuous-time signal can be represented without error by samples taken at a sufficiently high sampling frequency.

Hence, instead of generating/working on/storing the signal $s(t)$ or $s_{\text{RF}}(t)$, we work on the sampled versions

$$s(\kappa T_A), \quad s_{\text{RF}}(\kappa T_A), \quad \kappa \in \mathbb{Z},$$

where T_A is the sampling interval and κ the respective discrete time index. It is advantageous to link the PAM symbol interval T to the sampling interval T_A and define the *oversampling factor*

$$\text{OSF} = T/T_A \,. \tag{E.2}$$

Note that for baseband signals this factor is typically chosen on the order of 4 to 8; for carrier-modulated signals it has to be adjusted to four or eight times the product of carrier frequency and symbol duration.

Implementation and Simulation

The MATLAB® function `pam_pulseshape.m` is predefined and available. Given the sequence `a` of amplitude coefficients it returns the (sampled, oversampling factor `OSF`, time vector `t`) continuous-time transmit signal `s` in the ECB domain, that is

```
>> [s, t] = pam_pulseshape( a, OSF, PAMparam );
```

This function needs to know the basic pulse shape $g(t)$ to be used – it is specified via the string `g` in the `PAMparam` struct. If, for example, `g = 'whatpulse'` is assigned, the function `pam_pulseshape` expects that an m-file `g_whatpulse.m` is available, which, given a time vector `t` (normalized to T), returns the pulse at the samples specified in the time vector.

The function `g_sqrtcos.m` for the cosine-roll-off $\sqrt{\text{Nyquist}}$ pulse (roll-off factor α)

$$g(t) = \sqrt{\frac{E_g}{T}} \frac{4\alpha t \cos(\pi(1+\alpha)t/T) + T\sin(\pi(1-\alpha)t/T)}{\pi t (1 - (4\alpha t/T)^2)} \tag{E.3}$$

is already available.

Problem

E.4 Pulse Shaping
(a) Write a function `g_rect.m` for the rectangular pulse in the time domain (hard keying)

$$g(t) = \sqrt{\frac{E_g}{T}} \, \text{rect}(t/T) \, .$$

The function header is `function g = g_rect( t )`, where `t` is the normalized (to T) time variable.

Test the routines by plotting the pulse shape.

(b) Test the concatenation of mapping and pulse shaping for a random bit sequence (`OSF = 64`).

Visualize the continuous-time transmit signal $s(t)$ for the different constellations and for the different pulse shapes (`N_bits = 120`).

E.2.3 Modulation to Radio Frequency

Fundamentals and Theory

In the final step, the baseband signal $s(t)$ is modulated to radio frequency via

$$s_{\text{RF}}(t) = \sqrt{2} \, \text{Re}\{s(t) e^{j2\pi f_c t}\} \, . \tag{E.4}$$

The spectrum of the real-valued RF transmit signal $s_{\text{RF}}(t)$ is concentrated around the *carrier frequency* f_c.

Problem

E.5 Modulation
(a) Explain why it is reasonable to include the factor $\sqrt{2}$ in the modulation equation (E.4).
(b) Give an equation for the radio-frequency signal $s_{\mathrm{RF}}(t)$ where the real-valued quadrature components of the baseband signal, $s_I(t)$ and $s_Q(t)$, appear explicitly.

Implementation and Simulation

A function that performs the modulation shall be written next.

Problem

E.6 Modulation
(a) Create a MATLAB® function for the modulation of the baseband signal s to the radio-frequency signal s_rf. It should be called as follows:

    >> s_rf = rf_modulate(s, t, fc)

where t is the time vector (returned by pam_pulseshape) and fc the carrier frequency normalized to T.

(b) Test your function using the random ECB signal $s(t)$ from above and visualize the continuous-time radio-frequency transmit signal $s_{\mathrm{RF}}(t)$ for the different mappings and for the different pulse shapes (OSF = 64, fc = 8).

Compare (visually) $|s(t)|$ and $s_{\mathrm{RF}}(t)$ (N_bits = 120). How are these two signals related?

E.3 Channel

Fundamentals and Theory

In this project, we consider the *additive white Gaussian noise (AWGN)* channel model, see Figure E.3, where the received signal in the radio-frequency domain is given as

$$r_{\mathrm{RF}}(t) = s_{\mathrm{RF}}(t) + n_{\mathrm{RF}}(t) \tag{E.5}$$

Figure E.3 AWGN channel model.

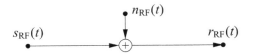

or in the *equivalent complex baseband (ECB)* as (cf. "Baseband Channel" in Figure E.1)

$$r(t) = s(t) + n(t) \,. \tag{E.6}$$

The transmitted signal is hence "only" corrupted by superimposed Gaussian noise that has a constant power spectral density. As usual in digital communications, the two-sided noise power spectral density is denoted as $N_0/2$, hence the autocorrelation of the noise in the radio-frequency domain, $n_{\mathrm{RF}}(t)$, is given by $\phi_{n_{\mathrm{RF}} n_{\mathrm{RF}}}(\tau) = N_0/2 \cdot \delta(\tau)$.

Problem

E.7 Power Spectral Density and Noise Power
(a) Let the basic pulse shape $g(t)$ with energy E_g, the (zero-mean) constellation $\mathcal{A} = \{a_1, \ldots, a_M\}$ with variance $\sigma_a^2 = \mathrm{E}\{|a_i|^2\}$, and the desired signal-to-noise ratio $(E_b/N_0)_{[\mathrm{dB}]} = 10 \log_{10}(E_b/N_0)$ be specified. How is the (two-sided) noise power spectral density $N_0/2$ of the white noise calculated from these given parameters?
(b) In the simulations, we represent all continuous-time signals by sampled versions with sampling interval T_A. For that, conceptually, the continuous-time signals have to be passed through an ideal low-pass anti-aliasing filter with (two-sided) bandwidth $1/T_A$.

How large is the noise power N (to be adjusted in the simulations) after this filtering? State an expression using the quantities given in part (a), T, and OSF.

Implementation and Simulation

A function that simulates the RF channel shall be written next.

Problem

E.8 AWGN RF Channel
(a) Write a MATLAB® function channel_rf.m that should be called as follows:

    >> r_rf = channel_rf(s_rf, N)

Thereby, N is the power (cf. Problem E.7, part (b)) of the zero-mean white Gaussian noise to be used in the simulation via oversampling.
(b) Test your function by visually comparing the signals s_rf and r_rf. To that end, plot both signals in the same figure.

E.4 Receiver

The receiver of a digital PAM scheme has already been depicted in Figure E.1. Basically, the function blocks one-to-one correspond to the respective blocks at the transmitter. Step by step we will implement the function blocks corresponding to the respective blocks at the transmitter.

E.4.1 Demodulation to Baseband

Fundamentals and Theory

First, the radio-frequency signal is – in practice after some bandpass filtering for preselection (gray bandpass in Figure E.1) – demodulated to the baseband. All subsequent operations are carried out using complex baseband signals. Mathematically, we calculate

$$r(t) = r_{\text{RF}}(t) \cdot \sqrt{2} e^{-j2\pi f_c t} , \qquad (E.7)$$

which is a valid baseband signal if the bandwidth of the RF signal meets the condition $B_{\text{RF}} < 2f_c$, and some form of low-pass filtering, suppressing the mixing products at twice the carrier frequency, follows.

Problem

E.9 Demodulated Noise
Let real-valued white Gaussian noise $n_{\text{RF}}(t)$ be given. Describe the statistical properties of the corresponding ECB signal $n(t)$.

Implementation and Simulation

A function that implements the demodulation shall be written next, and the baseband channel model is studied.

Problems

E.10 Demodulation
(a) Create a MATLAB® function

```
>> r = rf_demodulate( r_rf, t, fc )
```

for the demodulation of a radio-frequency signal to the complex baseband.
(b) Test the demodulation with the already available portion of your PAM transmission scheme.

E.11 Baseband Channel Model

(a) As shown in Figure E.1 (cf. (E.5) and (E.6)), modulation, RF channel, and demodulation can be replaced by an equivalent complex-valued baseband channel.

Create a MATLAB® function that implements this ECB AWGN channel

```
>> r = channel_baseband( s, N )
```

Here, N is again the power (cf. Problem E.7, part (b)) of the zero-mean white Gaussian noise to be used in the simulation via oversampling. How does the power for the equivalent baseband noise relate to N?

(b) Let $N = 0$ (i.e., $E_b/N_0 \to \infty$) for the moment. Compare (visually) the ECB signal $r(t)$ when simulating the transmission over the RF channel and when simulating the baseband channel only (use the same transmit signal $s(t)$ in both cases, i.e., simulate both transmission variants in parallel).

How can the difference be explained? How can it be removed?

E.4.2 Matched Filter and Sampling

Fundamentals and Theory

The counterpart to the pulse shaping at the transmitter is the *matched filter* at the receiver. For a basic pulse shape $g(t)$, the matched filter reads

$$h_{(\text{MF})}(t) = \frac{1}{E_g} g^*(T_d - t) , \qquad (E.8)$$

where E_g is the pulse energy. Note that the *detection delay* T_d is introduced in order to enable a causal implementation of the filter.

After the matched filter, samples spaced by T are taken, that is, one detection variable $d[k]$ per modulation interval is generated.

Problem

E.12 Matched Filter

Consider transmission over the AWGN channel (two-sided noise power spectral density $N_0/2$) employing a $\sqrt{\text{Nyquist}}$ pulse shape $g(t)$.

(a) Give the end-to-end discrete-time model from $a[k]$ at the transmitter to $d[k]$ at the receiver (cf. "Discrete Channel" in Figure E.1).

(b) Calculate the power of the noise sequence $n[k]$ present in the decision variable $d[k]$. How does this power relate to the quantity "E_b/N_0"?
Hint: Express this relation first with E_b/N_0 on a linear scale, then convert E_b/N_0 into decibels.

Implementation and Simulation

The matched filter shall be included in the simulations now and shall be tested. Additionally, the discrete-time end-to-end channel model is studied.

Problems

E.13 Matched Filter and Sampling
A MATLAB® function for matched filtering is already available and has the calling syntax

```
>> [d, dc, td] = pam_mf( r, t, OSF, PAMparam )
```

It outputs the discrete-time detection sequence d and, optionally, the continuous-time detection signal $d(t)$ together with the corresponding time vector (in dc and td, respectively).

(a) Include this function in your simulation file and visualize the continuous-time detection signal for a randomly generated PAM transmit signal (N_bits = 120). Compare the continuous-time detection signal with the discrete-time detection sequence (real parts only). What do you observe? Now vary the noise power.

(b) Let $N = 0$ (i.e., $E_b/N_0 \to \infty$) for the moment. Compare (visually) the detection signal $d(t)$ when simulating the transmission over the RF channel and when simulating the baseband channel only (use the same transmit signal $s(t)$ in both cases). What do you expect?

E.14 Discrete End-to-End Channel Model
Remember, for $\sqrt{\text{Nyquist}}$ basic pulses $g(t)$, the end-to-end channel including pulse shaping, matched filtering, and sampling is given by the discrete-time AWGN channel model.

(a) Create a MATLAB® function

```
>> d = channel_discrete( a, sigmansq )
```

that implements this discrete-time AWGN channel.

(b) How large is the variance σ_n^2 (sigmansq in MATLAB®) of the complex noise sequence when given E_b/N_0 (linear or in dB), the pulse shape $g(t)$, and the signal constellation $\mathcal{A}$?

E.4.3 Threshold Decision and Demapping

Fundamentals and Theory

In the final step, decisions on the transmitted data symbols $a[k]$ are made. Assuming uncoded transmission, these estimates are generated using a *threshold device*. Finally, given the estimates $\hat{a}[k]$ on the signal points, via the inverse mapping $\mathcal{M}^{-1}$, estimates $\hat{q}[l]$ on the binary symbols (bits) are delivered.

Problem

E.15 Decision
(a) We want to design decision regions for maximum-likelihood detection for a signal constellation $\mathcal{A}$ with equiprobable signal points. Where are the decision boundaries (in the complex plane) located?

(b) Let the binary sequence

$$0\ 0\ 0\ 1\ 1\ 1\ 1\ 0\ 1\ 0\ 1\ 1$$

be transmitted using 16QAM with natural labeling, and assume that the corresponding sequence of detection variables reads

$$-3.2 + 0.1\mathrm{j} \quad 2.2 + 1.8\mathrm{j} \quad 1.4 + 3.8\mathrm{j}\ .$$

Perform the following steps.
- Give the estimated 16QAM symbols $\hat{a}[k]$ produced by the threshold device.
- Demap the estimated QAM symbols to obtain the estimated sequence $\hat{q}[l]$.
- Calculate the bit error ratio.

(c) Given the transmitted bit sequence q[l] and the estimated one $\hat{q}$[l], how can the bit error ratio be calculated in MATLAB®? Try to solve this using only a single line of MATLAB® code and without a `for` (or any other) loop.

(d) How many bits (as a rule of thumb) should be transmitted in order to be able to reliably estimate the bit error ratio down to values around 10^{-6} with sufficient accuracy?

Implementation and Simulation

The decision device and demapping shall now be included in the simulation environment.

Problem

E.16 Decision Device and Demapping
A MATLAB® function

```
>> [ahat, qhat] = pam_quantdemap( d, PAMparam )
```

taking decisions and performing the inverse mapping is already available. It contains a generic threshold decision and a generic demapping, which works for all constellations specified in `PAMparam.Const`.

(a) Include this step in your PAM simulation file and test your entire PAM transmission scheme.

In the simulation file `pam_simul`, both the transmitted bit sequence `q` and the estimated one `qhat` are available. The same holds for `a` and `ahat`.

(b) Complete the MATLAB® simulation with statements calculating the *bit error ratio (BER)* and the *symbol error ratio (SER)* from these vectors.

Test this counting for the detection sequence given in Problem E.15, part (b).

E.4.4 BER Simulations

We are now ready to perform numerical simulations. Use your PAM simulation file `pam_simul` to simulate the bit and symbol error ratios.

Problems

E.17 Error Ratio Simulations

(a) First, we use 16QAM with Gray labeling, a cosine-roll-off $\sqrt{\text{Nyquist}}$ pulse shape, and we set $E_b/N_0 \cong 10$ dB. Simulate (`N_bits = 1e6` or larger)
   - the entire transmission system including the RF channel
   - transmission via the continuous-time ECB channel
   - transmission via the discrete-time channel.

   What error ratios do you measure? What do you expect from the theory?

(b) What is the main advantage of a pure T-spaced (discrete-time, one sample per T) simulation of PAM schemes?

(c) Perform a T-spaced simulation ($E_b/N_0 \cong 10$ dB) of
   - 16QAM with Gray labeling, cosine-roll-off $\sqrt{\text{Nyquist}}$ pulse
   - 16QAM with natural labeling, cosine-roll-off $\sqrt{\text{Nyquist}}$ pulse
   - 16QAM with Gray labeling, rectangular pulse
   - 16QAM with natural labeling, rectangular pulse.

   Compare the respective bit and symbol error ratios. What do you observe? How can this be explained? What is the advantage of Gray labeling?

(d) Compare the results with the theoretical calculations.
   Hint: The error function erfc(x) is available in MATLAB® (`erfc`). The relation to the complementary Gaussian integral function is given by $Q(x) = \frac{1}{2}\text{erfc}\left(x/\sqrt{2}\right)$.

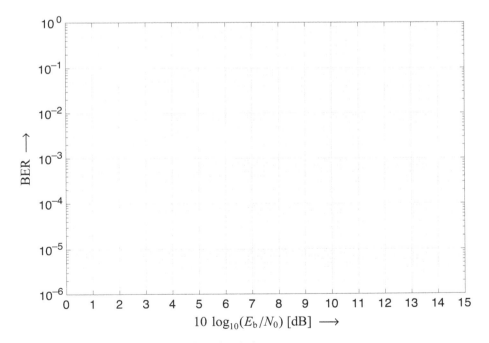

Figure E.4 Diagram for the numerical simulations.

E.18 Error Ratio Curves over the SNR

(a) Perform a T-spaced bit error ratio simulation for 16QAM with Gray labeling and cosine-roll-off $\sqrt{\text{Nyquist}}$ pulse for various values of E_b/N_0 in the range from 0 to 15 dB.

Draw the results on the diagram of Figure E.4.

(b) Down to which error ratio do you expect reliable results when, for example, simulating 10^6 bits?

Bibliography

Textbooks

Anderson, J. B. *Digital Transmission Engineering*, 2nd ed. Wiley-Interscience, Hoboken, NJ, 2005.

Anderson, J. B., Johannesson, R. *Understanding Information Transmission*. Wiley-IEEE Press, Piscataway, NJ, 2005.

Anderson, J. B., Aulin, T., Sundberg, C.-E. *Digital Phase Modulation*. Plenum Press, New York, 1986.

Benvenuto, N., Cherubini, G. *Algorithms for Communications Systems and their Application*. Wiley, New York, 2002.

Barry, J. R., Lee, E. A., Messerschmitt, D. G. *Digital Communication*, 3rd ed. Kluwer Academic Press, Boston, MA, 2003.

Benedetto, S., Biglieri, E., Catellani, V. *Digital Transmission Theory*. Prentice-Hall, Englewood Cliffs, NJ, 1987.

Benedetto, S., Biglieri, E. *Principles of Digital Transmission*. Kluwer Academic Press, Norwell, MA, 1999.

Bhattacharya, A. *Digital Communication*. McGraw Hill, Uttar Pradesh, India, 2017.

Biglieri, E. *Coding for Wireless Channels*. Springer, New York, 2005.

Biglieri, E., Divsalar, D., McLane, P., Simon, M. K. *Introduction to Trellis-Coded Modulation with Applications*. Macmillan, New York, 1991.

Blahut, R. E. *Digital Transmission of Information*. Addison-Wesley, Reading, MA, 1990.

Blahut, R. E. *Modem Theory: An Introduction to Telecommunications*. Cambridge University Press, Cambridge, 2009.

Bossert, M. *Channel Coding for Telecommunications*. Wiley, New York, 1999.

Chung, K. L. *Markov Chains with Stationary Probabilities*. Springer, Berlin, 1967.

Clark, C. G., Cain, J. B. *Error-Correction Coding for Digital Communications*. Springer, New York, 1981.

Conway, J. H., Sloane, N. J. A. *Sphere Packings, Lattices and Groups*, 3rd ed. Springer, New York, 1999.

Cover, T. M., Thomas, J. A. *Elements of Information Theory*, 2nd ed. Wiley, New York, 2006.

Ferrari, G., Colavolpe, G., Raheli, R. *Detection Algorithms for Wireless Communications*. Wiley, Chichester, 2004.

Fischer, R. F. H. *Precoding and Signal Shaping for Digital Transmission*. Wiley, New York, 2002.

Franks, L. E. *Signal Theory*. Prentice-Hall, Englewood Cliffs, NJ, 1969.

Gallager, R. G. *Information Theory and Reliable Communication*. Wiley, New York, 1968.

Gallager, R. G. *Principles of Digital Communications*. Cambridge University Press, New York, 2008.

Glover, I. A., Grant, P. M. *Digital Communications*, 3rd ed. Prentice-Hall, Harlow, 2010.

Goldsmith, A. *Wireless Communications*. Cambridge University Press, Cambridge, 2005.

Haykin, S. *Digital Communication Systems*. Wiley, New York, 2013.

Haykin, S., Moher, M. *Communication Systems*, 5th ed. Wiley, New York, 2009.

Heegard, C., Wicker, S. B. *Turbo Coding*. Springer, New York, 1998.

Höher, P. A. *Grundlagen der digitalen Informationsübertragung*, 2nd ed. B. G. Springer-Vieweg, Wiesbaden, 2013 (in German).

Honig, M. L., Messerschmitt, D. G. *Adaptive Filters: Structures, Algorithms, and Applications*, 3rd ed. Kluwer Academic Press, Boston, MA, 1985.

Horn, R. A., Johnson, C. R. *Matrix Analysis*, 2nd ed. Cambridge University Press, New York, 2013.

Huber, J. *Trelliscodierung*. Springer, Berlin, 1992 (in German).

Jayant, N. S., Noll, P. *Digital Coding of Waveforms*. Prentice-Hall, Englewood Cliffs, NJ, 1984.

Johannesson, R., Zigangirov, K. Sh. *Fundamentals of Convolutional Coding*, 2nd ed. Wiley-IEEE Press, Piscataway, NJ, 2015.

Kammeyer, K. D. *Nachrichtenübertragung*, 5th ed. B. G. Teubner, Stuttgart, 2010 (in German).

Kühn, V. *Wireless Conmmunications over MIMO Channels: Applications to CDMA and Multiple Antenna Systems*. Wiley, New York, 2006.

Lapidoth, A. *A Foundation in Digital Communications*. Cambridge University Press, Cambridge, 2009.

Lin, S., Costello, D. J. *Error Contol Coding: Fundamentals and Applications*, 2nd ed. Pearson/Prentice-Hall, Upper Saddle River, NJ, 2004.

Madhow, U. *Fundamentals of Digital Communication*. Cambridge University Press, Cambridge, 2008.

MacKay, D. J. C. *Information Theory, Inference, and Learning Algorithms*. Cambridge University Press, New York, 2003.

Mengali, U., D'Andrea, A. N. *Synchronization Techniques for Digital Receivers*. Plenum Press, New York, 1997.

Meyr, H., Moeneclaey, M., Fechtel, S. A. *Digital Communication Receivers: Synchronization*. Wiley, New York, 1998.

Molisch, A. F. *Wireless Communications*. Wiley-IEEE Press, Piscataway, NJ, 2006.

Neubauer, A., Freudenberger, J., Kühn, V. *Coding Theory: Algorithms, Architectures, and Applications*. Wiley, Chichester, 2007.

Papoulis, A. *Signal Analysis*. McGraw-Hill, New York, 1977.

Papoulis, A., Pillai, S. U. *Probability, Random Variables, and Stochastic Processes*, 4th ed. McGraw-Hill, New York, 2002.

Proakis, J. G., Salehi, M. *Digital Communications*, 5th ed. McGraw-Hill, New York, 2008.

Proakis, J. G., Salehi, M. *Communication Systems Engineering*, 2nd ed. Prentice-Hall, Upper Saddle River, NJ, 2002.

Rice, M. *Digital Communications: A Discrete-Time Approach*. Pearson/Prentice-Hall, Upper Saddle River, NJ, 2009.

Richardson, T., Urbanke, R. *Modern Coding Theory*. Cambridge University Press, Cambridge, 2008.

Rimoldi, B. *Principles of Digital Communication: A Top-Down Approach*. Cambridge University Press, Cambridge, 2016.

Ryan, W. E., Lin, S. *Channel Codes: Classical und Modern*. Cambridge University Press, Cambridge, 2009.

Shannon, C. E., Weaver, W. *The Mathematical Theory of Communications*, 16th ed. University of Illinois Press, Champaign, IL, 1971.

Schlegel, C., Perez, L. *Trellis and Turbo Coding: Iterative and Graph-Based Error Control Coding*, 2nd ed. IEEE & Wiley, Hoboken, NJ, 2015.

Simon, M. K., Alouini, M. S. *Digital Communications over Fading Channels*, 2nd ed. Wiley-Interscience, Hoboken, NJ, 2005.

Sklar, B., Harris, F. *Digital Communications, Fundamentals and Applications*. Pearson/Addison-Wesley, Boston, MA, 2021.

Trees, H. L. van *Detection, Estimation, and Modulation Theory – Part III: Radar-Sonar Signal Processing and Gaussian Signals in Noise*. Wiley, New York, 2001.

Verdú, S. *Multiuser Detection*. Cambridge University Press, New York, 1998.

Vucetic, B., Yuan, J. *Turbo Codes: Principles and Applications.* Kluwer Academic Press, Boston, MA, 2000.

Wozencraft, J. M., Jacobs, I. M. *Principles of Communication Engineering.* Wiley, New York, 1965.

Ziemer, R. E., Peterson, R. L. *Introduction to Digital Communication*, 2nd ed. Prentice-Hall, Upper Saddle River, NJ, 2001.

Ziemer, R. E., Tranter, W. H. *Principles of Communications: Systems, Modulation, and Noise*, 7th ed. Wiley, New York, 2014.

Selected Scientific Papers

Abbasfar, A., Divsalar, D., Yao, K. Accumulate–repeat–accumulate codes. *IEEE Transactions on Communications*, vol. 55, pp. 692–702, 2007.

Al-Dhahir, N., Cioffi, J. M. MMSE decision-feedback equalizer: Finite-length results. *IEEE Transactions on Information Theory*, vol. 41, pp. 961–975, 1995.

Armstrong, J. Peak-to-average power reduction for OFDM by repeated clipping and frequency domain filtering. *Electronics Letters*, vol. 38, pp. 246–247, 2002.

Anderson, J. B., Mohan, S. Sequential coding algorithms: A survey and cost analysis. *IEEE Transactions on Communications*, vol. 32, pp. 169–176, 1984.

Arıkan, E. Channel polarization: A method for constructing capacity-achieving codes for symmetric binary-input memoryless channels. *IEEE Transactions on Information Theory*, vol. 55, no. 7, pp. 3051–3073, 2009.

Arıkan, E., Teletar, E. On the rate of channel polarization. In *Proceedings of 2009 IEEE International Symposium on Information Theory*, pp. 1493–1495, 2009.

Ashikhmin, A., Kramer, G., ten Brink, S. Extrinsic information transfer functions: Model and erasure channel properties. *IEEE Transactions on Information Theory*, vol. 50, pp. 2657–2673, 2004.

Bahl, L. R., Cocke, J., Jelinek, F., Raviv, J. Optimal decoding of linear codes for minimizing symbol error rate. *IEEE Transactions on Information Theory*, vol. 20, pp. 284–287, 1974.

Bäuml, R. W., Fischer, R. F. H., Huber, J. B. Reducing the peak-to-average power ratio of multicarrier modulation by selected mapping. *Electronics Letters*, vol. 32, pp. 2056–2057, 1996.

Belfiore, C. A., Park, J. H. Decision feedback equalization. *Proceedings of the IEEE*, vol. 67, pp. 1143–1156, 1979.

Bello, P. A. Characterization of randomly time-variant channels. *IEEE Transactions on Communication Systems*, vol. 11, pp. 360–393, 1963.

Benedetto, S., Montorsi, G. Unveiling turbo-codes: Some results on parallel concatenated codes. *IEEE Transactions on Information Theory*, vol. 42, pp. 409–428, 1996.

Benedetto, S., Divsalar, D., Montorsi, G., Pollara, F. Serial concatenation of interleaved codes: Performance analysis, design, and iterative decoding. *IEEE Transactions on Information Theory*, vol. 44, pp. 909–926, 1998.

Barry, J. R., Lee, E. A., Messerschmitt, D. G. Capacity penalty due to ideal zero-forcing decision-feedback equalization. *IEEE Transactions on Information Theory*, vol. 42, pp. 1062–1071, 1996.

Berrou, C., Glavieux, A. Near optimum error correcting coding and decoding: Turbo-codes. *IEEE Transactions on Communications*, vol. 44, pp. 1261–1271, 1996.

Berrou, C., Glavieux, A., Thitimajshima, P. Near Shannon limit optimum error-correcting coding and decoding: Turbo-codes. In *Proceedings of the IEEE International Conference on Communications (ICC)*, pp. 1064–1070, 1993.

Betts, W., Calderbank, A. R., Laroia, R. Performance of nonuniform constellations on the Gaussian channel. *IEEE Transactions on Information Theory*, vol. 40, pp. 1633–1638, 1994.

Bingham, J. A. C. Multicarrier modulation for data transmission: An idea whose time has come. *IEEE Communications Magazine*, vol. 28, pp. 5–14, 1990.

Bose, R. C., Ray-Chauduri, D. K. On a class of error correcting binary group codes. *Information and Control*, vol. 8, pp. 68–79, 1960.

Breiling, M., Müller-Weinfurtner, S. H., Huber, J. B. SLM peak-power reduction without explicit side information. *IEEE Communications Letters*, vol. 5, pp. 239–241, 2001.

Calderbank, A. R., Ozarow, L. H. Nonequiprobable signaling on the Gaussian channel. *IEEE Transactions on Information Theory*, vol. 36, pp. 726–740, 1990.

Cain, J. B., Clark, G. C., Geist, J. M. Punctured convolutional codes of rate (n-1)/n and simplified maximum-likelihood decoding. *IEEE Transactions on Information Theory*, vol. 25, pp. 97–100, 1979.

Caire, G., Taricco, G., Biglieri, E. Bit-interleaved coded modulation. *IEEE Transactions on Information Theory*, vol. 44, no. 3, pp. 927–946, 1998.

Chevillat, P. R., Eleftheriou, E. Decoding of trellis-encoded signals in the presence of intersymbol interference and noise. *IEEE Transactions on Communications*, vol. 37, no. 7, pp. 669–676, 1989.

Chow, P. S., Cioffi, J. M., Bingham, J. A. C. A practical discrete multitone transceiver loading algorithm for data transmission over spectrally shaped channels. *IEEE Transactions on Communications*, vol. 43, pp. 773–775, 1995.

Chung, S. Y., Forney, G. D., Richardson, T. J., Urbanke, R. L. On the design of low-density check-codes within 0.0045 dB from Shannon limit. *IEEE Communications Letter*, vol. 5, pp. 58–60, 2001.

Cioffi, J. M., Dudevoir, G. P., Eyuboğlu, M. V., Forney, G. D. MMSE decision-feedback equalizers and coding – Part I: Equalization results, Part II: Coding results. *IEEE Transactions on Communications*, vol. 43, pp. 2582–2604, 1995.

Costa, M. H. M. Writing on dirty paper. *IEEE Transactions on Information Theory*, vol. 29, no. 5, pp. 439–441, 1983.

Davey, M. C., MacKay, D. Low density parity check codes over GF(q). *Proceedings of the IEEE Communication Letters*, vol. 2, pp. 165–167, 1998.

deBuda, R. Coherent demodulation of frequency-shift keying with low deviation. *IEEE Transactions on Communications*, vol. 20, pp. 429–435, 1972.

Divsalar, D., Simon, M. K. Multiple-symbol differential detection of MPSK. *IEEE Transactions on Communications*, vol. 38, pp. 300–308, 1990.

Divsalar, D., Jin, H., McEliece, R. J. Coding theorems for "Turbo-like" codes. *Proceedings of the 36th Allerton Conference on Communication, Control and Computing*, pp. 201–210, 1998.

Divsalar, D., Jones, C., Dolinar, S., Thorpe, J. Protograph based LDPC codes with minimum distance linearly growing with block size. In *Proceedings of IEEE GLOBECOM'05*, pp. 1152–1156, 2005.

Duel-Hallen, A., Heegard, C. Delayed decision-feedback sequence estimation. *IEEE Transactions on Communications*, vol. 37, pp. 428–436, 1989.

Van Eetvelt, P., Wade, G., Tomlinson, M. Peak to average power reduction for OFDM schemes by selective mapping. *Electronics Letters*, vol. 32, pp. 1963–1964, 1996.

Erez, U., Zamir, R. Achieving $\frac{1}{2}\log(1+\text{SNR})$ on the AWGN channel with lattice encoding and decoding. *IEEE Transactions on Information Theory*, vol. 50, no. 10, pp. 2293–2314, 2004.

Eyuboğlu, M. V., Qureshi, S. U. H. Reduced-state sequence estimation with set partitioning and decision feedback. *IEEE Transactions on Communications*, vol. 36, pp. 13–20, 1988.

Eyuboğlu, M. V. Detection of coded modulation signals on linear, severely distorted channels using decision-feedback noise prediction with interleaving. *IEEE Transactions on Communications*, vol. 36, pp. 401–409, 1988.

Eyuboğlu, M. V., Qureshi, S. U. H. Reduced-state sequence estimation for coded modulation on intersymbol interference channels. *IEEE Journal on Selected Areas in Communications*, JSAC-7, pp. 989–995, 1989.

Eyuboğlu, M. V., Forney, G. D. Trellis precoding: Combined coding, precoding and shaping for intersymbol interference channels. *IEEE Transactions on Information Theory*, vol. 38, pp. 301–314, 1992.

Fano, R. M. A heuristic discussion of probabilistic coding. *IEEE Transactions on Information Theory*, vol. 9, pp. 64–74, 1963.

Fischer, R. F. H., Huber, J. B., Komp, G. Coordinated digital transmission: Theory and examples. *Archiv für Elektronik und Übertragungstechnik (International Journal of Electronics and Communications)*, vol. 48, pp. 289–300, 1994.

Fischer, R. F. H., Gerstacker, W., Huber, J. B. Dynamics limited precoding, shaping, and blind equalization for fast digital transmission over twisted pair lines. *IEEE Journal on Selected Areas in Communications*, JSAC-13, pp. 1622–1633, 1995.

Fischer, R. F. H., Huber, J. B. A new loading algorithm for discrete multitone transmission. In *Proceedings of the IEEE Global Telecommunications Conference 1996*, pp. 724–728, 1996.

Fischer, R. F. H. and Huber, J. B. Comparison of precoding schemes for digital subscriber lines. *IEEE Transactions on Communications*, vol. 45, pp. 334–343, 1997.

Fischer, R. F. H. Calculation of shell frequency distributions obtained with shell-mapping schemes. *IEEE Transactions on Information Theory*, vol. 45, pp. 1631–1639, 1999.

Fischer, R. F. H., Windpassinger, C., Lampe, A., Huber, J. B. Space-time transmission using Tomlinson–Harashima precoding. In *Proceedings of the 4th International ITG Conference on Source and Channel Coding*, Berlin, 2002.

Fischer, R. F. H. The modulo-lattice channel: The key feature in precoding Schemes. *Archiv für Elektronik und Übertragungstechnik (International Journal of Electronics and Communications)*, vol. 59, no. 4, pp. 244–253, 2005.

Fischer, R. F. H., Siegl, C. Lattice-reduction-aided equalization for transmission over intersymbol-interference channels. *IET Electronics Letters*, vol. 41, pp. 969–970, 2005.

Fischer, R. F. H. From Gram–Schmidt orthogonalization via sorting and quantization to lattice reduction. In *Joint Workshop on Coding and Communications (JWCC)*, Santo Stefano Belbo, Italy, October 2010.

Fischer, R. F. H. Efficient lattice-reduction-aided MMSE decision-feedback equalization. In *International Conference on Acoustics, Speech and Signal Processing*, Prag, Czech Republic, May 2011.

Fischer, R. F. H., Windpassinger, C., Stierstorfer, C., Siegl, C., Schenk, A., Abay, Ü. Lattice-reduction-aided MMSE equalization and the successive estimation of correlated data. *Archiv für Elektronik und Übertragungstechnik (International Journal of Electronics and Communications)*, vol. 65, no. 8, pp. 688–693, 2011.

Fischer, R. F. H., Cyran, M., Stern, S. Factorization approaches in lattice-reduction-aided and integer-forcing equalization. In *2016 International Zurich Seminar on Communications*, Zurich, Switzerland, March 2016.

Fischer, R. F. H., Huber, J. B., Stern, S., Guter, P. Multilevel codes in lattice-reduction-aided equalization. In *2018 International Zurich Seminar on Communications*, Zurich, Switzerland, February 2018.

Fischer, R. F. H., Stern, S., Huber, J. B. Lattice-reduction-aided and integer-forcing equalization – structures, criteria, factorization, and coding. *Foundations and Trends in Communication and Information Theory*, vol. 16, no. 1–2, pp. 1–155, 2019.

Forney, G. D. Convolutional codes I: Algebraic structure. *IEEE Transactions on Information Theory*, vol. 16, pp. 720–738, 1970.

Forney, G. D. Maximum likelihood sequence estimation of digital sequences in the presence of intersymbol interference. *IEEE Transactions on Information Theory*, vol. 18, pp. 363–378, 1972.

Forney, G. D. The Viterbi algorithm. *Proceedings of the IEEE*, vol. 61, pp. 268–278, 1973.

Forney, G. D. Structural analysis of convolutional codes via dual codes. *IEEE Transactions on Information Theory*, vol. 19, pp. 512–518, 1973.

Forney, G. D. Convolutional codes II: Maximum likelihood decoding. *Information and Control*, vol. 25, pp. 222–266, 1974.

Forney, G. D. Convolutional codes III: Sequential decoding. *Information and Control*, vol. 25, pp. 267–297, 1974.

Forney, G. D. Multidimensional constellations – Part II: Vornonoi constellations. *IEEE Journal on Selected Areas in Communications*, JSAC-7, pp. 941–958, 1989.

Forney, G. D., Gallager, R. G., Lang, G. R., Longstaff, F. M., Qureshi, S. U. H. Efficient modulation for band-limited channels. *IEEE Journal on Selected Areas in Communications*, JSAC-2, pp. 632–647, 1984.

Forney, G. D. Coset codes – Part I: Introduction and geometrical classification. *IEEE Transactions on Information Theory*, vol. 34, pp. 1123–1151, 1988.

Forney, G. D. Coset codes – Part II: Binary lattices and related codes. *IEEE Transactions on Information Theory*, vol. 34, pp. 1152–1187, 1988.

Forney, G. D., Wei, L.-F. Multidimensional constellations – Part I: Introduction, figures of merit, and generalized cross constellations. *IEEE Journal on Selected Areas in Communications*, JSAC-7, pp. 877–892, 1989.

Forney, G. D. Geometrically uniform codes. *IEEE Transactions on Information Theory*, vol. 37, pp. 1241–1260, 1991.

Forney, G. D. Trellis shaping. *IEEE Transactions on Information Theory*, vol. 38, pp. 281–300, 1992.

Forney, G. D., Eyuboğlu, M. V. Combined equalization and coding using precoding. *IEEE Communications Magazine*, vol. 29, pp. 25–34, 1991.

Forney, G. D., Ungerböck, G. Modulation and coding for linear Gaussian channels. *IEEE Transactions on Information Theory*, vol. 44, no. 6, pp. 2384–2415, 1998.

Forney, G. D., Trott, M. D., Chung, S.-Y. Sphere-bound-achieving coset codes and multilevel coset codes. *IEEE Transactions on Information Theory*, vol. 46, no. 3, pp. 820–850, 2000.

Forney, G. D. On the role of MMSE estimation in approaching the information-theoretic limits of linear Gaussian channels: Shannon meets Wiener. In *Proceedings of the 41st Allerton Conference on Communication, Control and Computing*, pp. 430–439, 2003.

Franks, L. E. Carrier and bit synchronization in data communication – a tutorial review. *IEEE Transactions on Communications*, vol. 28, no. 8, pp. 1107–1121, 1980.

Gallager, R. G. Low-density parity-check codes. *IRE Transactions on Information Theory*, vol. 8, pp. 21–28, 1962.

Gallager, R. G. *Low-Density Parity-Check Codes*. PhD thesis, Cambridge, MA, 1963.

Gerstacker, W. An alternative approach to minimum mean-squared error DFE with finite length constraints. *Archiv für Elektronik und Übertragungstechnik (International Journal of Electronics and Communications)*, vol. 50, no. 1, pp. 27–31, 1996.

Graf, K. P., Huber, J. B. Design and performance of an all-digital adaptive 2.048 Mbit/s data transmission system using noise prediction. In *Proceedings of ISCAS'89*, pp. 1808–1812, 1989.

Gray, R. M., Neuhoff, D. L. Quantization. *IEEE Transactions on Information Theory*, vol. 44, no. 6, pp. 2325–2383, 1998.

Gray, R. M. Toeplitz and circulant matrices: A review. *Foundations and Trends in Communications and Information Theory*, vol. 2, no. 3, pp. 155–239, 2006.

Grettenberg, T. Representation theorem for complex normal processes (corresp.). *IEEE Transactions on Information Theory*, vol. 11, no. 2, pp. 305–306, 1965.

Han, S. H., Lee, J. H. An overview of peak-to-average power ratio reduction techniques for multicarrier transmission. *IEEE Wireless Communications*, pp. 56–65, 2005.

Hagenauer, J. Rate-compatible punctured convolutional codes (RCPC codes) and their applications. *IEEE Transactions on Communications*, vol. 36, pp. 389–400, 1988.

Hagenauer, J., Höher, P. A Viterbi-alogrithm with soft outputs and its applications. In *Proceedings of IEEE GLOBECOM'89*, pp. 47.1.1–47.1.7, 1989.

Hagenauer, J., Offer, E., Papke, L. Iterative decoding of binary block and convolutional codes. *IEEE Transactions on Information Theory*, vol. 42, pp. 429–445, 1996.

Hamming, R. W. Error detecting and correcting codes. *Bell System Technical Journal*, vol. 29, pp. 147–160, 1950.

Harashima, H., Miyakawa, H. Matched-transmission technique for channels with intersymbol interference. *IEEE Transactions on Communications*, vol. 20, pp. 774–780, 1972.

Hehn, T., Huber, J. B. LDPC codes and convolutional codes with equal structural delay: A comparison. *IEEE Transactions on Communications*, vol. 57, no. 6, pp. 1683–1692, 2009.

Henkel, W., Wagner, B. Trellis shaping for reducing the peak-to-average ratio of multitone signals. In *IEEE International Symposium on Information Theory*, Ulm, Germany, June/July, 1997.

Hofbauer, C. *Design and Analysis of Unique Word OFDM*. PhD thesis, University of Klagenfurt, Austria, 2016.

Huber, J., Liu, W. An alternative approach to reduced-complexity CPM-receivers. *IEEE Journal on Selected Areas in Communications*, vol. 7, no. 9, pp. 1437–1449, 1989.

Huber, J. B., Rüppel, A. K. Zuverlässigkeitsschätzung für die Ausgangssysmbole von Trellis-Decoder. *Archiv für Elektronik und Übertragungstechnik (International Journal of Electronics and Communications)*, vol. 44, pp. 8–21, 1990 (in German).

Huber, J., Liu, W. Data-aided synchronization of coherent CPM-receivers. *IEEE Transactions on Communications*, vol. 40, pp. 178–189, 1992.

Huemer, M., Onic, A., Hofbauer, C. Classical and Bayesian linear data estimators for unique word OFDM. *IEEE Transactions on Signal Processing*, vol. 59, pp. 6073–6085, 2011.

Huemer, M., Hofbauer, C., Onic, A., Huber, J. B. On the exploitation of redundant energy in UW-OFDM: LMMSE versus sphere detection. *IEEE Signal Processing Letters*, vol. 19, pp. 340–343, 2012.

Huemer, M., Hofbauer, C., Huber, J. B. Non-systematic complex number RS coded OFDM by unique word prefix. *IEEE Transactions on Signal Processing*, vol. 60, pp. 285–299, 2012.

Huemer, M., Hofbauer, C., Onic, A., Huber, J. B. Design and analysis of UW-OFDM signals. *Archiv für Elektronik und Übertragungstechnik (International Journal of Electronics and Communications)*, vol. 68, no. 10, pp. 958–968, 2014.

Huettinger, S., Huber, J. B. Analysis and design of power efficient coding schemes with parallel concatenated convolutional codes. *IEEE Transactions on Communications*, vol. 54, pp. 1251–1258, 2006.

Hughes-Hartogs, D. Ensemble modem structure for imperfect transmission media. U.S. Patents Nos. 4 679 227, July 1987; 4 731 816, March 1988; and 4 833 796, May 1989.

Imai, H., Hirakawa, S. A new multilevel coding method using error correcting codes. *IEEE Transactions on Information Theory*, vol. 23, pp. 371–377, 1977.

Jelinek, F. Fast sequential decoding using a stack. *IBM Journal of Research and Development*, vol. 13, pp. 675–685, 1969.

Khandani, A. K., Kabal, P. Shaping multidimensional signal spaces – Part I: Optimum shaping, shell mapping, Part II: Shell-addressed constellations. *IEEE Transactions on Information Theory*, vol. 39, pp. 1799–1819, 1993.

Korada, S. B., Sasoglu, E., Urbanke, R. L. Polar codes: Characterization of exponent, bounds, and constructions. *IEEE Transactions on Information Theory*, vol. 56, no. 12, pp. 6253–6264, 2010.

Koetter, R., Singer, C. A., Tüchler, M. Turbo equalization. *IEEE Signal Processing Magazine*, vol. 21, pp. 67–80, 2004.

Krongold, B. S., Jones, D. L. PAR reduction in OFDM via active constellation expansion. *IEEE Transactions on Broadcasting*, vol. 49, pp. 258–268, 2003.

Kschischang, F. R., Pasupathy, S. Optimal nonuniform signaling for Gaussian channels. *IEEE Transactions on Information Theory*, vol. 39, pp. 913–929, 1993.

Kschischang, F. R., Frey, B. J., Loeliger, H. A. Factor graphs and sum–product algorithm. *IEEE Transactions on Information Theory*, vol. 47, pp. 498–519, 2001.

Kühne, F. Peak envelope modulation PAM-QAM-signals. *Archiv für Elektronik und Übertragungstechnik (International Journal of Electronics and Communications)*, vol. 36, pp. 138–139, 1982.

Land, I., Huettinger, S., Hoeher, P., Huber, J. B. Bounds on information combining. *IEEE Transactions on Information Theory*, vol. 51, pp. 612–619, 2005.

Land, I., Huber, J. B. Information combining. *Foundations and Trends in Communications and Information Theory*, vol. 3, no. 3, pp. 227–330, 2006.

Laroia, R. Coding for intersymbol interference channels – combined coding and precoding. *IEEE Transactions on Information Theory*, vol. 42, pp. 1053–1061, 1996.

Lee, P. J. New short constraint length rate 1/n convolutional codes which minimize the required SNR for given bit error rate. *IEEE Transactions on Communications*, vol. 33, pp. 171–177, 1985.

Lenstra, A. K., Lenstra, H. W., Lovász, L. Factoring polynomials with rational coefficients. *Mathematische Annalen*, vol. 261, no. 4, pp. 515–534, 1982.

Massey, J. L. Coding and modulation in digital communications. *1974 International Zürich Seminar on Digital Communications*, Zürich, Switzerland, March 1974.

Maiya, S. V., Costello, J. D., Fuja, T. E. Low latency coding: Convolutional vs. LDPC codes. *IEEE Transactions on Communications*, vol. 60, pp. 1215–1225, 2012.

Messerschmitt, D. G. A geometrical theory of intersymbol interference. Part I and II. *Bell System Technical Journal*, vol. 52, no. 9, pp. 1483–1539, 1973.

MacKay, D. J. C., Neal, R. M. Near Shannon limit performance of low density parity check codes. *IEE Electronic Letters*, vol. 33, pp. 457–458, 1997.

MacKay, D. J. Good error-correcting codes based on very sparse matrices. *IEEE Transactions on Information Theory*, vol. 45, pp. 399–431, 1999.

Mueller, K. H., Müller, M. Timing recovery in digital synchronous receivers. *IEEE Transactions on Communications*, vol. 24, pp. 516–531, 1976.

Müller, S. H., Huber, J. B. OFDM with reduced peak-to-average power ratio by optimum combination of partial transmit sequences. *Electronics Letters*, vol. 33, pp. 368–369, 1997.

Müller, S. H., Bäuml, R. W., Fischer, R. F. H., Huber, J. B. OFDM with reduced peak-to-average power ratio by multiple signal representation. *Annals of Telecommunications*, vol. 52, no. 1–2, pp. 58–67, 1997.

Murugan, A. D., Gamal, H. El, Damen, M. O., Caire, G. A unified framework for tree search decoding: Rediscovering the sequential decoder. *IEEE Transactions on Information*

Theory, vol. 53, no. 3, pp. 933–953, 2006.

Nyquist, H. Certain factors affecting telegraph speed. *Bell System Technical Journal*, vol. 3, pp. 324–346, 1924.

Nyquist, H. Certain topics in telegraph transmission theory. *Transactions of the American Institute of Electrical Engineering*, vol. 47, pp. 617–644, 1928.

Paaske, E. Short binary convolutional codes with maximal free distance for rates 2/3 and 3/4. *IEEE Transactions on Information Theory*, vol. 20, pp. 683–689, 1974.

Polyanskiy, Y., Poor, V., Verdú, S. Channel coding rate in the finite blocklength regime. *IEEE Transactions on Information Theory*, vol. 56, no. 5, pp. 2307–2359, 2010.

Prabhu, V. K., Rowe, H. E. Spectra of digital phase modulation by matrix methods. *Bell System Technical Journal*, vol. 53, pp. 899–935, 1974.

Rachinger, C., Huber, J. B., Müller, R. R. Comparison of convolutional and block codes for low structural delay. *IEEE Transactions on Communications*, vol. 63, no. 12, pp. 4629–4638, 2015.

Rachinger, C., Huber, J. B., Gerstacker, W. Comparison of OFDM and single-carrier transmission for ultra-low delay applications. *2018 IEEE GLOBECOM Workshops*, Abu Dhabi, 2018.

Rapp, Ch. Coded and uncoded M-PSK and CPM signals on nonlinear bandpass channels. *Proceedings of IEEE GLOBECOM'90*, pp. 501.5.1–501.1.5, 1990.

Reed, I. S., Solomon, G. Polynomial codes over certain fields. *Journal of Society for Industrial and Applied Mathematics*, vol. 8, pp. 300–304, 1960.

Rice, S. O. Envelopes of narrow-band signals. *Proceedings of the IEEE*, vol. 70, no. 7, pp. 692–699, 1982.

Richardson, T. J., Shokrollahi, A., Urbanke, R. L. Design of capacity-approaching irregular low-density parity-check codes. *IEEE Transactions on Information Theory*, vol. 47, pp. 619–637, 2001.

Richardson, T. J., Urbanke, R. L. The capacity of low-density parity-check codes under message passing. *IEEE Transactions on Information Theory*, vol. 47, pp. 599–618, 2001.

Rimoldi, B. A decomposition approach to CPM. *IEEE Transactions on Information Theory*, vol. 34, pp. 260–270, 1988.

Rimoldi, B., Urbanke, R. L. A rate-splitting approach to the Gaussian multiple-access channel. *IEEE Transactions on Information Theory*, vol. 42, pp. 364–375, 1996.

Sason, I., Shamai, S. Performance analysis of linear codes under maximum-likelihood decoding: A tutorial. *Foundations and Trends in Communications and Information Theory*, 3(1/2), pp. 1–225, 2006.

Sato, Y. A method of self-recovering equalization for multilevel amplitude-modulation systems. *IEEE Transactions on Communications*, vol. 23, pp. 679–682, 1975.

Seidl, M., Schenk, A. M., Stierstorfer, C., Huber, J. B. Polar-coded modulation. *IEEE Transactions on Communications*, vol. 61, pp. 4108–4119, 2013.

Shannon, C. E. A mathematical theory of communication. *Bell System Technical Journal*, pt. I, vol. 27, no. 3, pp. 379–423, 1948; pt. II, vol. 27, no. 4, pp. 623–656, 1948.

Shannon, C. E. Communications in the presence of noise. *Proceedings of the Institute of Radio Engineers*, 37, pp. 10–21, 1949.

Simon, M. K., Divsalar, D. On the implementation and performance of angle and double differential detection schemes. *IEEE Transactions on Communications*, vol. 40, pp. 278–291, 1992.

Steendam, H. Theoretical performance evaluation and optimization of UW-OFDM. *IEEE Transactions on Communications*, vol. 64, pp. 1739–1750, 2016.

Stern, S., Fischer, R. F. H. Joint algebraic coded modulation and lattice-reduction-Aided Preequalization. *Electronic Letters*, vol. 52, no. 7, pp. 523–525, 2016.

Stern, S., Fischer, R. F. H. Lattice-reduction-aided precoding for coded modulation over algebraic signal constellations. In *Proceedings of the 20th International ITG Workshop on Smart Antennas*, pp. 356–363, 2016.

Stern, S., Fischer, R. F. H. Optimal factorization in lattice-reduction-aided and integer-forcing linear equalization. In *11th International ITG Conference on Systems, Communications, and Coding (SCC)*, Hamburg, Germany, February 2017.

Stierstorfer, C., Fischer, R. F. H. Lattice-reduction-aided Tomlinson–Harashima precoding for point-to-multipoint transmission. *Archiv für Elektronik und Übertragungstechnik (International Journal of Electronics and Communications)*, vol. 60, pp. 328–330, 2006.

Sutskover, I., Shamai(Shitz), S., Ziv, J. Extremes of information combining. *IEEE Transactions on Information Theory*, vol. 51, pp. 1313–1325, 2005.

Sutskover, I., Shamai(Shitz), S., Ziv, J. Constrained information combining: Theory and applications for LDPC coded systems. *IEEE Transactions on Information Theory*, vol. 53, pp. 1617–1643, 2007.

Tal, I., Vardy, A. How to construct polar codes. *IEEE Transactions on Information Theory*, vol. 59, no. 10, pp. 6562–6582, 2013.

ten Brink, S. Convergence behavior of iteratively decoded parallel concatenated codes. *IEEE Transactions on Communications*, vol. 49, pp. 1727–1737, 2001.

ten Brink, S., Kramer, G., Ashikhmin, A. Design of low-density parity-check codes for modulation and detection. *IEEE Transactions on Communications*, vol. 52, pp. 670–678, 2004.

Tomlinson, M. New automatic equaliser employing modulo arithmetic. *Electronic Letters*, vol. 7, pp. 138–139, 1971.

Ungerböck, G. Adaptive maximum-likelihood receiver for carrier-modulated data-transmission systems. *IEEE Transactions on Communications*, vol. 22, pp. 624–636, 1974.

Ungerböck, G., Csajka, I. On improving data-link performance by increasing channel alphabet and introducing sequence coding. In *IEEE International Symposium on Information Theory*, Ronneby, Sweden, June 1976.

Ungerböck, G. Channel coding with multilevel/phase signals. *IEEE Transactions on Information Theory*, vol. 28, pp. 55–67, 1982.

Ungerböck, G. Trellis-coded modulation with redundant signal sets, Part I: Introduction. *IEEE Communications Magazine*, vol. 25, pp. 5–11, 1987.

Ungerböck, G. Trellis-coded modulation with redundant signal sets, Part II: State of the art. *IEEE Communications Magazine*, vol. 25, pp. 12–21, 1987.

Viterbi, A. J. Error bounds for convolutional codes and an asymptotically optimum decoding algorithm. *IEEE Transactions on Information Theory*, vol. 13, pp. 260–269, 1967.

Viterbi, A. J. Convolutional codes and their performance in conmmunication systems. *IEEE Transactions on Communications*, vol. 19, pp. 751–772, 1971.

Viterbi, A. J. An intuitive justification and a simplified implementation of the MAP decoder for convolutional codes. *IEEE Journal on Selected Areas in Communications*, vol. 16, no. 2, pp. 260–264, 1998.

Wachsmann, U., Fischer, R. F. H., Huber, J. B. Multilevel codes: Theoretical concepts and practical design rules. *IEEE Transactions on Information Theory*, vol. 45, no. 5, pp. 1361–1391, 1999.

Werner, J. J. Tutorial on carrierless AM/PM – Part I: Fundamentals and digital CAP transmitter. *AT&T Bell Laboratories Report*, June 1992.

Werner, J. J. Tutorial on carrierless AM/PM – Part II: Performance of bandwidth-efficient line codes. *AT&T Bell Laboratories Report*, February 1993.

Wesel, R. D., Cioffi, J. M. Achievable rates for Tomlinson–Harashima precoding. *IEEE Transactions on Information Theory*, vol. 44, pp. 824–831, 1998.

Wiberg, N. *Codes and Decoding on General Graphs*. PhD thesis, Linköping University, Sweden, 1996.

Windpassinger, C., Fischer, R. F. H., Huber, J. B. Lattice-reduction-aided broadcast precoding. *IEEE Transactions on Communications*, vol. 52, no. 12, pp. 2057–2060, 2004.

Wolf, J. K. Efficient maximum-likelihood decoding of linear block codes using a trellis. *IEEE Transactions on Information Theory*, vol. 24, pp. 76–80, 1978.

Wunder, G., Fischer, R. F. H., Boche, H., Litsyn, S., No, J.-S. The PAPR problem in OFDM transmission: New directions for a long-lasting problem. *IEEE Signal Processing Magazine*, vol. 30, no. 6, pp. 130–144, 2013.

Youla, D. C. On the factorization of rational matrices. *IEEE Transactions on Information Theory*, vol. 7, pp. 172–189, 1961.

Index

A/D conversion, 146
acf, *see* autocorrelation function
alternate mark inversion code, 229
AMI, *see* alternate mark inversion code
amplitude coefficient, 55, 57, 61, 74, 88, 180, 197, 212, 270
amplitude modulation, 27
amplitude-shift keying, 64
    bipolar, 64
    non-coherent demodulation, 169, 172
    normalized minimum squared Euclidean distance, 100
    power/bandwidth plane, 173
    signal constellation, 64
    symbol error ratio, 94
    unipolar, 64
analytic continuation, 310
analytic signal, 142, 143
arg-function, 386
ASK, *see* amplitude-shift keying
attenuation, 12
audio transmission, 6
autocorrelation, 410
    energy-limited signal, 389, 413
    normalized sampled, 293, 295, 298, 303
    of the basic pulse, 59, 74
autocorrelation function, 44, 414
    average, 414
        see also noise autocorrelation function
autocorrelation sequence, 76, 358, 415
    amplitude coefficient, 59, 62
average power, 388
AWGN, *see* noise, additive white Gaussian noise
AWGN channel, *see* channel, AWGN

back-off, 114, 378
Bahl–Cook–Jelinek–Raviv algorithm, 452

bandpass signals, 26
bandwidth, 2, 3, 18
    $X\%$, 121, 264, 280
bandwidth efficiency, 3, 5, 18, 102, 225, 250
bandwidth expansion, 223
bandwidth-excess factor, *see* roll-off factor
baseband transmission, 7, 28, 56
basic period, 10
basic pulse, 54, 66, 133, 289, 348, 488
basis function, 195, 201
baud rate, *see* rate, symbol
Bauer algorithm, 403
belief, 432
belief propagation, 433
BER, *see* bit error ratio
Bessel function, 164, 235, 258
binary symbol, 5
bit, 5
bit error ratio, 15, 92, 99, 498
block coding, 188, 419

calculus of variations, 331
CAP, *see* carrierless amplitude and phase modulation
carrier, 27
carrier frequency, 26, 154, 248, 273, 347
carrier-modulated transmission, 8, 56, 473
carrierless amplitude and phase modulation, 133, 459
    autocorrelation of pulse, 135
    crosscorrelation of pulses, 135
Carson's rule, 267
Cauchy–Schwarz inequality, 70
cdf, *see* cumulative distribution function
central limit theorem, 378, 411, 481
cepstrum, 396, 401
channel, 2, 11, 487
    AWGN, 15, 469, 493
        discrete-time, 76, 291, 317

    parallel, independent, 214, 358
    baseband, 40
    continuous-time, 317
    dispersion, 301
    dispersive, 287, 288, 319, 344
    flat-fading, 362, 475, 480
    frequency-non-selective, 475
    frequency-selective, 288, 475
    impulse response, 13, 288, 344, 468
    input sequence, 325
    model, 13, 75, 468
        discrete-time, 291
        FIR, 321, 326
        stochastic, 480
        vector, 232
    multipath-fading, 475
    radio-frequency, 40, 488
    time-varying, 218, 468
    transfer function, 13, 288, 344, 468
channel coding, 10, 188, 273, 313, 328, 330, 419
channel encoder, 188, 189
    systematic, 421
channel estimation, 324
channel state information, 288, 324, 363
characteristic attenuation, 300, 309
characteristic curve, 235
chip, 348
Cholesky factorization, 299, 402
Cholesky–Banachiewicz algorithm, 403
clipping, 378
coaxial cable, 300, 469
code, 188, 420
    dual code, 425
    generator matrix, 423
    girth, 434
    Hamming code, 421, 423
    Hamming distance, 425
    Hamming weight, 425
    linear code, 422

Index

low-density parity-check code, 419, 425, 434
  minimum Hamming distance, 425
  parity-check code, 223
  parity-check matrix, 424
  polar code, 437
  repetition code, 223, 420, 423, 428
  single parity-check code, 420, 423, 429
  syndrome, 426
  systematic encoding, 423
  weight distribution, 425
codelength, 420
codeword, 419, 445
coding, 180
  inherent, 250, 252
COFDM, *see* orthogonal frequency-division multiplexing, coded
combined coding and modulation, 223
communications, 1
  bidirectional, 1, 324
  digital, 4
  tele-, 1
  unidirectional, 1
complementary Gaussian integral function, 95
complex number, 386
  absolute value, 386
  argument, 386
  Cartesian form, 386
  conjugate complex, 386
  imaginary part, 386
  imaginary unit, 386
  polar form, 386
  real part, 386
complexity, 18
complexity reduction, 242, 281, 324
compression, *see* source coding
constellation expansion, 383
continuous-phase modulation, 247, 270
  coherent receiver, 278
  continuous-phase encoder, 273
  frequency function, 270
  frequency pulse, 270, 271
    L-RC, 271
    L-REC, 271
  modulator, 276
  normalized minimum squared Euclidean distance, 278
  phase function, 270, 274
  phase pulse, 271

phase state, 275
power/bandwidth plane, 280
transmit signal, 270
conversion
  AM/AM, 113, 273
  AM/PM, 113, 273
convolution, 56, 390, 392
  cyclic, 351
  linear, 351
convolutional code, 446
correlation coefficient, 258, 412
correlation matrix, 416
correlation receiver, 219, 222
correlation vector, 221
covariance matrix, 411
CPFSK, *see* frequency-shift keying, continuous phase
CPM, *see* continuous-phase modulation
crest factor, 84, 115, 377
  basic pulse, 85
  signal constellation, 85
cross power spectral density, 411
crosscorrelation, 213
crosscorrelation function, 410
crosstalk, 13
CSI, *see* channel state information
cumulative distribution function, 408
cyclic-prefix, 351

DAB, *see* digital audio broadcasting
data over cable service interface specification, 7, 469
DCPSK, *see* phase-shift keying, differential coherent
decision
  boundary, 91
  region, 91–93, 497
  symbol-by-symbol, 74, 88, 308, 312, 497
  variable, 67, 93
decision-feedback equalization, 287, 313, 373
  channel capacity, 316
  miniumum mean-squared error, 315
  zero-forcing, 315
decision-feedback sequence detection, 323
decoding, 215, 420
  hard-decision, 425
  soft-decision, 425, 427
  soft-input/soft-output, 426, 432

symbol-by-symbol, 452
deep-space mission, 8
delay, *see* latency
delay Doppler-spread function, 479
delay element, 43
delay spread, 477
demodulation, 495
  coherent, 141, 156
  differential, 160
  non-coherent, 142, 154, 231
    symbol-by-symbol, 232
detection, 88
  differential, 237
  instant, 67, 71
  signal, 71
  symbol, 291
  variable, 211
  vector, 211
DFE, *see* decision-feedback equalization
DFSD, *see* decision-feedback sequence detection
DFT, *see* Fourier transform, discrete
difference
  energy, 202, 298, 299
  sequence, 298
  symbol, 298
differential encoding, 157, 228, 275
digital audio broadcasting, 8, 347
digital subscriber lines, 7, 133, 347, 469
  asymmetric, 347
  very-high-bit-rate, 337, 347, 367, 376
digital video broadcasting, 8, 347
directed radio, 8
Dirichlet kernel, 462
discrete multitone, 345
dispersion, *see* channel, dispersion
distortion, 287
distribution
  exponential, 162
  Rayleigh, 161, 162, 357, 362
  Rice, 161, 162, 257
disturbance, 11
DMT, *see* discrete multitone
DOCSIS, *see* data over cable service interface specification
Doppler effect, 154, 481
DPSK, *see* phase-shift keying, differential
DSL, *see* digital subscriber lines
DVB, *see* digital video broadcasting

ECB, *see* equivalent complex baseband
eigenvalue, 356
eigenvector, 356
encoding, 181, 188, 419
end-to-end impulse response, 306
end-to-end model, 75, 212, 320, 326, 333, 349
  discrete-time, 302, 306, 310, 311, 497
  OFDM, 363
end-to-end pulse, 288, 303
end-to-end transfer function, 300, 306, 371
  discrete-time, 291
energy, 388
  average received, 16
  per bit, 16, 18, 96, 99, 218
    PAM signal, 63
  per modulation step, 16
  per symbol, 16, 18
    PAM signal, 63
energy detection, 141
envelope, 38
  constant, 120
  detection, 236
equalization, *see also* decision-feedback equalization; linear equalization; linear pre-equalization; root–root equalization; Tomlinson–Harashima precoding
equivalent complex baseband
  signal, 26, 30, 33
    Cartesian form, 34
    polar form, 34
  stochastic process, 44, 45
  system, 40, 41
  transformation, 30
  properties, 33
equivalent encoding, 191
equivalent signal elements, 191, 258
error
  carrier frequency, 231
  frequency, 39, 154
  phase, 39, 154, 231, 257
error correction, 420
error detection, 420, 426
error function, 96
  complementary, 96
error propagation, 313, 328
Euclidean distance, 91, 216, 223

minimum squared, 217, 321
normalized minimum squared, 96, 99, 100, 129, 218, 297, 299
  of signal constellation, 96
normalized squared, 120
Euclidean norm, 198, 299
Euler's formula, 28
expectation, 409
extrinsic probability, 428
eye opening, 83
eye pattern, 82, 130, 194, 414

fading coefficient, 219, 480
faster-than-Nyquist signaling, 231
FEC, *see* forward error correction
feedback loop, 328
Fejér kernel, 462
FFT, *see* Fourier transform, fast
finite field, 5, 189, 419
finite-state machine, 261, 275, 296, 321, 322, 444
FIR, *see* system, finite impulse response
folded spectral SNR, 304, 325, 328, 362, 372, 373
forward error correction, 223, 426
Fourier transform, 387
  correspondences, 392
  discrete, 355, 356
    inverse, 356, 455
    matrix, 355, 455
  discrete-time, 400
  fast, 359
  inverse, 387
  time-frequency duality, 406
free-space radiation, 12
frequency modulation, 28, 247
  analog, 267
  binary, 122
frequency-division multiplexing
  continuous-time variant, 454, 458
frequency-shift keying, 123, 247
  M-ary, 267
  binary, 253
  bipolar, 248
  bit error ratio, 258
  coherent demodulation, 256
  continuous phase, 124, 228, 250, 260
    coherent demodulation, 263
    normalized minimum squared Euclidean distance, 262
    correlation coefficient, 254
    discontinuous, 250

discontinuous binary, 253
non-coherent demodulation, 256
normalized minimum squared Euclidean distance, 254, 267
phase jump, 250
phasor, 249
power/bandwidth plane, 264
signal elements, 248
signal space representation, 255
symbol error ratio, 268
transmit signal, 250
trellis description, 260
unipolar, 248
Frobenius norm, 198
FSK, *see* frequency-shift keying
function
  Dirac delta, 387
  signum, 388
  step, 387, 394

Gaussian low-pass filter, 126, 130
Gaussian minimum-shift keying, 125, 271
  bandwidth efficiency, 129
  frequency pulse, 126
  normalized minimum squared Euclidean distance, 129
  power efficiency, 129
  sub-optimal receiver, 130
global system for mobile communications, 8, 125
GMSK, *see* Gaussian minimum-shift keying
Gram–Schmidt procedure, 204, 215, 255
GSM, *see* global system for mobile communications
guard interval, 350

hard keying, 77, 247
Hilbert pair, 135
Hilbert transform, 30, 36, 38, 44, 135, 142, 391, 406
Hölder mean, *see* mean, Hölder

i.i.d., *see* sequence, independent and identically distributed
I/Q modulation, 141
IEEE 802.11, 347
IIR, *see* system, infinite impulse response
impulse response, 40
  causal, 310, 326
  minimum-phase, 326, 349

monic, 311, 326, 349
  time-variant, 475
information theory, 14, 20, 213, 317
inherent coding, 260
instantaneous frequency, 27, 122, 247, 248, 270
instantaneous power, 388
interblock interference, 349
interference, 11
  adjacent channel, 13
  co-channel, 13
  multi-user, 13
  multiple-access, 13
interpolation, 414
intersymbol interference, 73, 273, 287, 306, 321, 349, 448
  channel, 320, 344
  free of, 73, 212, 306, 327
  interblock, 350, 371
  intrablock, 350
interval
  bit, 18
  code symbol, 11
  symbol, 18, 55
intrinsic probability, 427
ISI, *see* intersymbol interference
ISO/OSI model, 16

Jakes' spectrum, 483
Jensen's inequality, 336

Kramers–Kronig receiver, 112, 141, 144
Kramers–Kronig relation, 406

labeling
  Gray, 99, 489
  natural, 226
Lagrange function, 332, 365
Laplace transform, 395
latency, 18, 21
lattice, 299
light wave, 7
line coding, 229
linear equalization, 287, 306, 371, 373
  miniumum mean-squared error, 309
  zero-forcing, 307
linear pre-equalization, 326, 375
loading algorithm, 363
loss, 300
LPE, *see* linear pre-equalization

LTI, *see* system, linear time-invariant

magnetic recording, 8
MAP, *see* maximum-a-posteriori
mapping, 10, 55, 188, 189, 489
mapping efficiency, 190
Marcum-Q-function, 165, 258
Markov chain, 253, 444
master clock, 10
matched filter, 69, 71, 133, 257, 294, 301, 496
  bank of, 211, 215, 222, 278
  bound, 304
  demodulator, 222, 235, 292
  receiver, 306
  w.r.t. the end-to-end basic pulse, 293
matrix
  anti-diagonal identity, 402
  circulant, 353
  Hermitian, 355, 358, 402
  modal, 355, 356
  of eigenvectors, 355
  positive definite, 416
  Toeplitz, 299, 402, 403
  triangular, 402
  unitary, 355
maximum-a-posteriori sequence detection, 452
maximum-a-posteriori strategy, 88
maximum-likelihood decoding, 215, 219
maximum-likelihood detection, 90, 215, 217, 233, 321
maximum-likelihood sequence detection, 216, 222, 264, 293, 296, 319, 322, 452
maximum-likelihood strategy, 89
MCM, *see* multi-carrier modulation
Mealy machine, 444
mean
  arithmetic, 304, 305, 335
  generalized, 335
  geometric, 305, 315, 330, 335, 367, 372
  harmonic, 305, 308, 328, 335
  Hölder, 334, 335, 362, 376
  root mean square, 27, 84, 334, 335
  squared harmonic root, 334, 335
memory, 181
metallic wires, 7
metric, 449

accumulated, 295, 296, 320, 322, 449
branch, 296, 321, 322, 449
Forney, 322
increment, 216, 221, 295
sum, 221
Ungerböck, 322
minimum-shift keying, 119, 228, 255
  bandwidth efficiency, 120
  power efficiency, 120
ML, *see* maximum-likelihood
MLD, *see* maximum-likelihood detection
MLSD, *see* maximum-likelihood sequence detection
modulation, 10, 55, 57, 180, 181, 492
  dimensionality of, 197
modulation index, 124, 248, 254, 267, 271
  weak, 278
modulation interval, 248
modulation step, 10
modulo device, 328
modulo operation, 329
modulus conversion, 419, 452
MSDD, *see* multiple-symbol differential detection
MSK, *see* minimum-shift keying
multi-carrier modulation, 345
multi-carrier transmission, 345, 373
multiple-symbol differential detection, 237
  decision-feedback, 242
  symbol error ratio, 241

nearest neighbor, 94, 98, 268, 297, 365
  average number, 98, 99, 218
node degree, 432
  distribution, 432
node update, 433
noise, 11
  additive white Gaussian, 15, 18, 344
  colored, 14
  discrete-time, 76, 212
  Gaussian, 13
  thermal, 13
  white, 13, 47, 213, 310
noise autocorrelation function, 13, 44–46
noise bandwidth, 14
noise enhancement, 317

noise power spectral density, 13, 46–48, 96, 291, 468
noise prediction, 318, 372
Nyquist
  band, 228
  criterion, 73, 74, 304
    in frequency domain, 76
  frequency, 228, 300, 337
  function, 74, 307
    square-root, 75, 134, 289
  stability criterion, 406

OFDM, *see* orthogonal frequency-division multiplexing
offset, *see* error
offset QAM, 115
on/off keying, 169
  error probability, 171
ONF, *see* optimum Nyquist filter
OOK, *see* on/off keying
optical transmission, 7
optimum Nyquist filter, 307, 327, 333, 371
orthogonal frequency-division multiplexing, 344, 345, 454
  block processing, 344, 348, 349, 353
  coded, 363
  cyclic-prefix, 351
  loaded, 370
  optimally loaded, 367
  transmit signal, 348
  transmitter, 347
  unique-word, 350, 465
orthogonality, 184
  temporal, 184
orthogonality condition, 289
  twofold, 195, 212, 213
orthonormal basis, 195
orthonormal expansion, 197
out-of-band radiation, 378
oversampling, 147, 491

PAM, *see* pulse-amplitude modulation
  optimum receive filter, 66
PAPR, *see* peak-to-average power ratio
Parseval's theorem, 202, 389
partial transmit sequences, 382
PCM, *see* pulse-code modulation
pdf, *see* distribution; probability density function

peak-to-average power ratio, 84, 114, 359, 377
  reduction schemes, 381
periodic replica, 460
phase, 27
phase ambiguities, 156
phase cylinder, 128
phase modulation, 27, 247
phase randomization, 414
phase trajectory, 128, 261, 274
phase-shift keying, 65
  differential, 157, 237
    power/bandwidth plane, 166
    receiver, 160
  differential coherent, 159
    error probability, 159
  normalized minimum squared Euclidean distance, 100
  signal constellation, 65
  symbol error ratio, 97
phasor, 34, 116, 155
phasor demo, 82, 118, 124, 251
photo diode, 141
physical layer, 16
post cursor, 313
power, 18, 63, 388
power amplifier, 85, 113, 280, 378
  ECB model, 113
power efficiency, 3, 5, 16, 18, 102, 223, 224, 250
power spectral density, 325, 378, 411, 494
  continuous FSK, 252
  CPM, 280
  discontinuous FSK, 252
  FSK, 252, 264
  general scheme, 187
  OFDM, 359
  PAM, 58, 60, 290
  uncoded PAM, 62
power/bandwidth plane, 19
pre-emphasis/de-emphasis, 331
pre-equalization, 287, 325
precoding, 328
prediction 415
  error, 318, 416, 417
  error filter, 319
  filter, 318, 417
  gain, 318
  order, 319, 415
  see also noise prediction
preprocessing, 325
  filter, 325, 331
probability

a-posteriori, 88
a-priori, 9, 90
probability density function, 13, 408
process, *see* stochastic process
projection, 201, 210
protocol, 16
psd, *see* power spectral density
PSK, *see* phase-shift keying
PTS, *see* partial transmit sequences
pulse
  $\sqrt{\text{Nyquist}}$, 75
  cosine-roll-off, 79, 492
  mutually orthogonal, 134
  rectangular, 77, 388
  time-limited, 78
  triangular, 388
pulse-amplitude modulation, 54, 487
  bit error ratio, 101
  crest factor, 85
  detection, 88
  geometrical interpretation, 91
  normalized minimum squared Euclidean distance, 100
  optimum receiver, 292
  power/bandwidth plane, 102
  power/rate plane, 102
  signal bandwidth, 81
  transmit signal, 56, 288
  transmitter, 180, 287
  transparent scheme, 228
  variants, 112
pulse-code modulation, 4

Q-function, *see* complementary Gaussian integral function
QAM, *see* quadrature-amplitude modulation
  symbol error ratio, 97
quadratic form, 298
quadrature components, 34, 42, 116
  in-phase component, 34
  PAM signal, 57
  quadrature component, 34
quadrature downconversion, 36
  simplified form, 37
quadrature upconversion, 35
quadrature-amplitude modulation, 65
  normalized minimum squared Euclidean distance, 100
  signal constellation, 65
    cross, 65
    square, 65

random experiment, 407
random process, *see* stochastic
    process
random variable, 408
    expected value, 408
    Gaussian, 412
    moment, 409
    variance, 409
rate, 3, 364
    code, 11, 18, 189, 420
    data, 3, 18
    digital communication scheme,
        10, 18, 190
    modulation, 3, 11, 18, 55
    per discrete time step, 3
    symbol, 18
rate loss, 223, 224
receive filter, 291, 302, 310
receive signal, 11
receiver, 2, 15, 487, 495
    reduced-complexity, 242, 281,
        301, 324
reduced-state sequence detection,
    323
redundancy, 9, 181, 188, 189,
    420
reference frequency, 27, 248, 273
reference symbol, 237, 240
regeneration, 5, 6
reliability information, 425
repeated clipping and filtering, 382
RMS, *see* mean, root mean square
roll-off factor, 79, 87, 105
root–root equalization, 333, 375
RRE, *see* root–root equalization
run-length coding, 229

sample function, 82
sampling, 392, 413
    T spaced, 67, 496
sampling frequency, 392
sampling period, 392
sampling phase, 302
satellite communications, 8
scrambler, 9, 382
selected mapping, 381
sequence, 9
    elements of, 9
    independent and identically
        distributed, 5, 9
sequential decoding, 242
Shannon capacity, 317
Shannon limit, 20
shortest vector problem, 299

side information, 381
signal, 11
    analytic, 29
    autocorrelation, 389
    causal, 393
        spectrum, 394
    continuous-time, 387
    discrete-time, 391
    energy-limited, 388
    power, 2
    power-limited, 388
    relevant part, 213
signal constellation, 55, 63, 488
    cardinality, 55
    cross, 65
    square, 65
signal element, 10, 181, 238, 276
    average, 252
    constraints, 184
    coordinates, 197
    cross energy, 185
    energy, 185, 202
    temporally orthogonal, 193
signal number, 181
signal point, 55
signal representation, 195
signal set, 181
    CAP, 183
    FSK, 183
    PAM, 182
signal shaping, 8
signal space representation, 180
    general version of digital
        modulation, 180
    receive signal, 210
    reduced dimensionality, 214
    transmitter, 199
signal-to-distortion ratio, 390
signal-to-noise ratio, 12, 67, 303
    decision-feedback equalization,
        315
    linear equalization, 308
    linear pre-equalization, 328
    OFDM, 362
    OFDM, loaded, 372
    per bit, 18
    pre-emphasis/de-emphasis, 334
    spectral, 290, 304, 315, 325
    Tomlinson–Harashima
        precoding, 330
single-carrier transmission, 345, 373
single-sideband modulation, 142
    compatible, 143
    with carrier, 144

sink, 487
SLM, *see* selected mapping
SNR, *see* signal-to-noise ratio
source, 9, 487
source coding, 4
spectral efficiency, *see* bandwidth
    efficiency
spectral factorization, 401
spectral side lobes, 273, 382
spectrum, 387
sphere decoding, 242, 404
state, 261, 295, 444
stochastic process, 44, 406
    complex-valued, 407
    cyclic-stationary, 82
    cyclo-stationary, 414, 456
    discrete-time, 413
    ensemble, 407
    Gaussian, 411
    joint, 407
    non-stationary, 458
    regular, 416
    rotationally invariant, 46, 233
    sample function, 407
    stationary, 410
storage, 1, 8
subcarrier, 345, 358
subchannel, 345, 357
sufficient statistics, 294
Sunde's FSK, 250, 266, 279
support, 288
survivor, 450
switched telephone lines, 8
symbol error ratio, 92, 218, 297,
    364, 498
    upper bound, 99
synchronization, 17, 39, 148, 154,
    156
    carrier, 17
    frame, 465
    sequence, 157
    symbol timing, 17, 82, 88
system
    causal, 390, 395, 400, 406, 417
    continuous-time, 389
    discrete-time, 392
    dispersive, 390
    finite impulse response, 446
    impulse response, 389, 392
    infinite impulse response, 446
    linear time-invariant, 13, 389,
        413, 468
    linear time-varying, 468, 475
    minimum-phase, 395, 400, 406

system (cont.)
    cepstrum, 396, 397
        energy concentration, 397, 398
        relation magnitude/phase, 397, 398
    non-causal, 72
    stable, 395, 400, 406
    transfer function, 390, 392

Tanner graph, 431
tapped delay line, 320, 477
theorem of Grettenberg, 46
theorem of irrelevant data, 213
THP, *see* Tomlinson–Harashima precoding
threshold device, 327, 497
Tomlinson–Harashima precoding, 328, 375
tone reservation, 382
trade-off, 224, 270
transfer function, 40, 390, 392, 395
    minimum-phase, 311
    time-variant, 478
transmission scenario
    broadcast, 363
    point-to-point, 363
transmit power, 364
transmit pulse, 287, 349
transmit signal, 10, 11, 56, 182, 250, 348
transmitter, 2, 10, 347, 487
trellis
    branch, 261, 296, 322, 444
    diagram, 261, 296, 322, 444
    encoder, 226, 261, 444
    path, 296, 322, 445
trellis coding, 188, 225, 273, 419, 444
trellis decoding, 278, 446
    reduced-state, 282
trellis-coded modulation, 8, 281
TV cable, 7
twisted-pair line, 337, 368, 470

union bound, 97, 218, 268
unit pulse, 392
up/downconversion, 39
useful signal, 212

vector channel model, 214, 232
vector demodulator, 211, 215, 232, 320
vector receiver, *see* vector demodulator
Viterbi algorithm, 264, 273, 278, 282, 296, 321, 322, 446, 449
voice-band modems, 8
voltage-controlled oscillator, 124

weight function, 476
whitened-matched filter, 310, 312, 326, 333, 349
whitening filter, 14, 288, 418, 469
wireless local area network, 347
wireless transmission, 8
WLAN, *see* wireless local area network
WMF, *see* whitened-matched filter

Yule–Walker equations, 318, 416

z-transform, 302, 391
    correspondences, 392

www.ingramcontent.com/pod-product-compliance
Ingram Content Group UK Ltd.
Pitfield, Milton Keynes, MK11 3LW, UK
UKHW011826050125
452985UK00009B/254